Introduction to Group Theory
with Applications

# MATERIALS SCIENCE AND TECHNOLOGY

**EDITORS**

ALLEN M. ALPER
*GTE Sylvania Inc.*
*Precision Materials Group*
*Chemical & Metallurgical Division*
*Towanda, Pennsylvania*

A. S. NOWICK
*Henry Krumb School of Mines*
*Columbia University*
*New York, New York*

*A. S. Nowick and B. S. Berry,* ANELASTIC RELAXATION IN CRYSTALLINE SOLIDS, 1972

*E. A. Nesbitt and J. H. Wernick,* RARE EARTH PERMANENT MAGNETS, 1973

*W. E. Wallace,* RARE EARTH INTERMETALLICS, 1973

*J. C. Phillips,* BONDS AND BANDS IN SEMICONDUCTORS, 1973

*J. H. Richardson and R. V. Peterson (editors),* SYSTEMATIC MATERIALS ANALYSIS, VOLUMES I, II, AND III, 1974; IV, 1978

*A. J. Freeman and J. B. Darby, Jr. (editors),* THE ACTINIDES: ELECTRONIC STRUCTURE AND RELATED PROPERTIES, VOLUMES I AND II, 1974

*A. S. Nowick and J. J. Burton (editors),* DIFFUSION IN SOLIDS: RECENT DEVELOPMENTS, 1975

*J. W. Matthews (editor),* EPITAXIAL GROWTH, PARTS A AND B, 1975

*J. M. Blakely (editor),* SURFACE PHYSICS OF MATERIALS, VOLUMES I AND II, 1975

*G. A. Chadwick and D. A. Smith (editors),* GRAIN BOUNDARY STRUCTURE AND PROPERTIES, 1975

*John W. Hastie,* HIGH TEMPERATURE VAPORS: SCIENCE AND TECHNOLOGY, 1975

*John K. Tien and George S. Ansell (editors),* ALLOY AND MICROSTRUCTURAL DESIGN, 1976

*M. T. Sprackling,* THE PLASTIC DEFORMATION OF SIMPLE IONIC CRYSTALS, 1976

*James J. Burton and Robert L. Garten (editors),* ADVANCED MATERIALS IN CATALYSIS, 1977

*Gerald Burns,* INTRODUCTION TO GROUP THEORY WITH APPLICATIONS, 1977

*L. H. Schwartz and J. B. Cohen,* DIFFRACTION FROM MATERIALS, 1977

*Zenji Nishiyama,* MARTENSITIC TRANSFORMATION, 1978

*Paul Hagenmuller and W. van Gool (editors),* SOLID ELECTROLYTES: GENERAL PRINCIPLES, CHARACTERIZATION, MATERIALS, APPLICATIONS, 1978

*G. G. Libowitz and M. S. Whittingham,* MATERIALS SCIENCE IN ENERGY TECHNOLOGY, 1978

*Otto Buck, John K. Tien, and Harris L. Marcus (editors),* ELECTRON AND POSITRON SPECTROSCOPIES IN MATERIALS SCIENCE AND ENGINEERING, 1979

in preparation

*Lawrence L. Kazmerski (editor),* PROPERTIES OF POLYCRYSTALLINE AND AMORPHOUS THIN FILMS AND DEVICES

*Manbred von Heimendahl,* ELECTRON MICROSCOPY OF MATERIALS: AN INTRODUCTION

# INTRODUCTION TO GROUP THEORY WITH APPLICATIONS

*Gerald Burns*

IBM THOMAS J. WATSON RESEARCH CENTER
YORKTOWN HEIGHTS, NEW YORK

ACADEMIC PRESS New York San Francisco London 1977
A Subsidiary of Harcourt Brace Jovanovich, Publishers

ACADEMIC PRESS, INC.
111 Fifth Avenue, New York, New York 10003

*United Kingdom Edition published by*
ACADEMIC PRESS, INC. (LONDON) LTD.
24/28 Oval Road, London NW1

LIBRARY OF CONGRESS CATALOG CARD NUMBER: 75-19620

ISBN 0-12-145750-8

PRINTED IN THE UNITED STATES OF AMERICA

To my parents
for their patience and kindness

# CONTENTS

# PREFACE

This book was developed from notes used for a group theory course given at the IBM Research Center at three different times over a period of six years. Its goal is to provide a "feel" for the subject. Examples are given for each new concept. Theorems are motivated; they are not simply stated, proved, and used several chapters later. Rather the importance of a theorem is stressed, and examples of its use given. Long proofs, beautifully presented elsewhere (see Wigner), are not included here. This offers the serious student of group theory the opportunity to become acquainted with other books. This volume follows the philosophy that one is motivated to understand proofs and the beauty of theorems if their usefulness is understood beforehand, and in this pragmatic approach, the book differs from others. As Maimonides ("Guide for the Perplexed") said: "We must first learn the truths by tradition, after this we must be taught how to prove them."

The presentation grows progressively more difficult as the introductory chapters (Chapters 1–6 or 7) unfold. The last one or two sections of each chapter are the more difficult ones; they may be skipped on a first reading. The remaining chapters may be read in any order, although there is clearly a connection between Chapters 8 and 13, 9 and 10, as well as 11 and 12.

This book exhibits a leaning toward point groups and the 32 crystallographic point groups. However, this is only partially done because of my involvement in solid state physics. Rather, it is done because the point group operations are easy to grasp and provide a convenient tool to use as examples in the study of group theory. Space groups appear in Chapter 1 as well as in some of the later chapters.

They almost always occur in the last section of the chapter and may be ignored if there is no interest in that topic. On the other hand, if one is interested in space groups, a modest working knowledge can be developed with very little extra effort.

Although the book has been used as a textbook, it also has been successfully used without the aid of an instructor. There are enough examples to apply one's knowledge immediately. The notes, appendixes, and problems at the end of each chapter broaden the subject matter of the chapter itself. As much as possible I have tried to keep the text on the subject of group theory, without teaching or reviewing large chunks of physics or chemistry. This has the advantage of enabling one to concentrate and use more group theory for a given amount of study.

## ACKNOWLEDGMENTS

It is a pleasure to have this opportunity to thank the many people who helped me to improve the presentation. In particular my thanks go to A. M. Glazer and B. A. Scott who have read extensive parts of the mauscript and have helped the clarity and organization considerably. A number of very useful improvements were suggested by E. Burstein, A. Lurio, V. L. Nelson, A. S. Nowick, N. S. Shiren, and F. Stern, and I would like to thank them. F. H. Dacol took care of the logistics of keeping the notes, problems, captions, figures, and appendixes together with the text as well as sketching most of the figures. This took an enormous job off my shoulders and I am indeed grateful to him. Various stages of the manuscript were typed by J. L. Butcher, C. C. Carrington, D. E. Cleveland, S. A. Haeger, C. Russett, and B. A. Smalley and it is a pleasure to acknowledge them. However, the final typing, which was really typesetting on our experimental printer system, was done by L. J. Callahan and I am indeed grateful for her patience and herculean effort, and she and I are also grateful for the help from the experimental printing group particularly P. Archibald and A. C. Hohl.

## Chapter 1

# SYMMETRY OPERATIONS

*... the recognition that almost all the rules of spectroscopy follow from the symmetry of the problem is the most remarkable result.*

*Wigner, "Group Theory", Author's Preface*

Very little formalism need be learned before the power of group theory can be applied to problems in physics and chemistry and much of this formalism is covered in Chapters 2–4. This chapter, independent of group theory, is presented first for two reasons: (1) It is a self-contained topic enabling a scientist to describe, by means of a symbol, the symmetry operations possessed by a molecule or crystal and conversely, to see the symbol and understand the symmetry operations. (2) The symmetry operations of the molecules or crystals can then be used as examples to understand the ideas in group theory better.

## 1-1 Introduction

Symmetry is possessed by atoms, molecules, and infinite crystals (henceforth called crystals until the boundary conditions are discussed in Chapter 12). That is, we can operate on them with a **symmetry operation (or covering operation)** which interchanges the positions of various atoms but results in the molecule or crystal looking exactly the same as before the symmmetry operation (the molecule or crystal is in an equivalent position) and when the operation is continuously repeated the molecule or crystal continues to be in an equivalent position.

Figure 1-1 shows a planar $C_6H_6$ molecule (benzene). The symmetry operations of this molecule are: rotations about the z-axis by $\pm\pi/3$, $\pm 2\pi/3$, $\pi$, $2\pi$; rotations by $\pi$ about six axes in the plane of the molecule

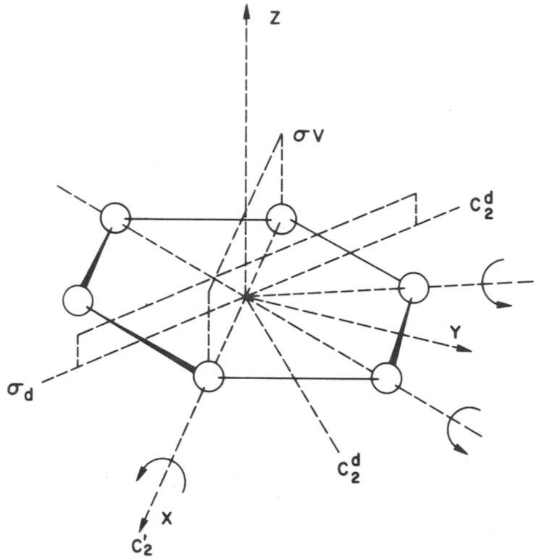


**Fig. 1-1** Planar benzene ($C_6H_6$) showing some of the symmetry operations.

as shown in the figure; reflections across six planes that contain the z-axis, two of which are shown; reflection across the plane of the molecule; inversion through the center of the molecule; and four "improper" rotations which will be discussed later. It will be shown in the next chapter that the set of independent covering operations forms a group in the mathematical sense. It is just these groups which will be studied in most of the remainder of this book. In this chapter we will develop an understanding of the symmetry operations and their notation, and of the classification of molecules and crystals into the point groups (32 in the case of crystals). (Again we emphasize that in the next chapter a group will be defined in the mathematical sense. In this chapter the word is used in a more literal sense.)

## 1-2 Point Symmetry Operations

In this section we discuss symmetry operations for which one point in the body is kept fixed. These are the **point symmetry operations** usually associated with a molecule or crystal. For molecules the number of such operations is finite. At the end of this chapter other types of symmetry operations without any fixed point are discussed as well as several systems with an infinite number of symmetry operations.

### a. Principal axis

For noncubic crystals or for molecules there is usually one axis having higher rotational symmetry than the others (more symmetry operations). For example, an axis with 4-fold symmetry will automatically have 2-fold symmetry and thus have four symmetry operations (rotation by $\pi/2$, $\pi$, $3\pi/2$, $2\pi$). An axis with just 2-fold symmetry has just two symmetry operations (rotation by $\pi$ and $2\pi$). Thus a 4-fold axis is said to have higher symmetry than a 2-fold axis. The axis with the highest symmetry is defined as the **z-axis** or the **principal axis**.

In the case of orthorhombic crystals there are three mutually perpendicular axes of 2-fold symmetry. The choice of z-axis is arbitrary in this case. (Often crystallographers will call the longest cell dimension the x-axis.) Similarly, for other crystals with several principal axes there is a certain arbitrariness but this presents no fundamental problem in the mathematics as will be seen later.

### b. Symmetry operations

The symmetry operations are defined and discussed in Table 1-1. We can summarize them briefly in Schoenflies' notation:

| | | | |
|---|---|---|---|
| E | identity | $\sigma_h$ | reflection in a horizontal plane |
| i | inversion | $\sigma_v$ | reflection in a vertical plane |
| $C_n$ | rotation by $2\pi/n$ (proper rotation) | $\sigma_d$ | reflection in a diagonal plane |
| | | $S_n$ | improper rotation ($S_n = \sigma_h C_n$) |
| $\sigma$ | reflection | | |

Note that all the symmetry operations involve a movement of the body. A **symmetry element** is a geometric entity such as a center of inversion, an axis of rotation, or a plane. Symmetry operations can be carried out with respect to these symmetry elements or combinations of them.

The existence of certain symmetry operations implies the existence of others. For example, if an object has a $C_n$ axis and a $C_2'$ symmetry axis perpendicular to the $C_n$ axis is added, there must be at least $nC_2'$ axes. Similarly, the existence of a $C_n$ axis and one $\sigma_v$ plane implies at least n $\sigma_v$ planes. Also, a $C_n$ axis and $\sigma_h$ imply certain improper rotations.

### c. Examples (see Fig. 1-2)

**Example 1.** $H_2O$. This molecule possesses symmetry elements E and $C_2$ (the 2-fold axis being the principal axis shown). It also possesses

**Table 1-1** Symmetry operations

| | |
|---|---|
| E | **Identity.** The molecule or crystal is not rotated at all or rotated by $2\pi$ about any axis. All objects possess this symmetry operation. Taking **r** as a vector from the origin to any point, E**r** → **r**. |
| i | **Inversion.** The molecule is inverted through some origin, the center of inversion. Benzene (Fig. 1-1) has this symmetry element. (However, your right hand does not. Upon inversion it looks like your left hand, which is clearly distinguishable from what it looked like before the operation.) The operation is $i(x, y, z) \rightarrow (-x, -y, -z)$. |
| $C_n$ | **Rotation**. Sometimes called a proper rotation. A rotation of the molecule by $2\pi/n$ about an axis, in the sense of a right hand screw by **convention**. If the axis is not the principal axis, often, but not always, there will be a prime or other superscript; $C_n{}'$ or $C_n{}^x$. Benzene (Fig. 1-1) has $C_6$, $C_3$, $C_2$ about its principal axis. $C_n{}^m$ is defined as $(C_n)^m$, so $C_6{}^2 = C_3$ and $C_6{}^3 = C_2$ but $C_3{}^2$ is a new symmetry operation. (It can be written as $C_3{}^2 = C_3^{-1}$ where $C_3^{-1}$ would be a rotation of $2\pi/3$ in the opposite sense but this will not be used in this book.) $C_6{}^5$ is also a new symmetry operation. Thus, about the z-axis of benzene there are six independent symmetry operations: E, $C_6$, $C_3$, $C_2$, $C_3{}^2$, $C_6{}^5$. In the plane of the benzene molecule there are three 2-fold rotations of the $C_2{}'$ type (through the origin and two carbon atoms) and three 2-fold rotations of the $C_2{}^d$ type (through the origin and the midpoint between carbon atoms). For example, $C_4(x, y, z) \rightarrow (y, -x, z)$. |
| $\sigma$ | **Reflection.** Reflection of the molecule in a plane. The particular plane of reflection will sometimes be specified by a subscript. Reflection in a plane means: transfer all of the points to the other side of the plane an equal distance along perpendiculars to the plane. So for reflection across the x-plane, which contains the yz-axis, we have $\sigma(x, y, z) \rightarrow (-x, y, z)$. |
| $\sigma_h$ | **Reflection in the horizontal plane.** The plane of reflection is perpendicular to the principal axis and contains the origin. Benzene has one $\sigma_h$. |
| $\sigma_v$ | **Reflection in a vertical plane.** The plane contains the principal axis. Benzene has three $\sigma_v$ through the z-axis and the three $C_2{}'$ axes. |
| $\sigma_d$ | **Reflection in a diagonal plane.** This plane also contains the principal axis and bisects the angle between the 2-fold axes perpendicular to the principal axis. Benzene has three $\sigma_d$ through z and the three $C_2{}'$ axes. (Actually for benzene the distinction between $\sigma_v$ and $\sigma_d$ is arbitrary.) |
| $S_n$ | **Improper rotation.** A rotation by $2\pi/n$ followed by a reflection in a horizontal plane, i.e., $S_n = \sigma_h C_n$. Again $S_n{}^m = (S_n)^m = (\sigma_h C_n)^m$. ($S_3{}^2 \neq S_3{}^{-1}$ but $S_3{}^2 = C_3{}^2$.) Benzene has $S_6$, $S_3$, $S_2$ (=i), $S_3{}^2$, $S_6{}^5$, E. (Note $S_2$ = i always.) A molecule possessing $S_n$ need not possess $\sigma_h$ and $C_n$. For example, methane $CH_4$. For $S_4$ we have $S_4(x, y, z) \rightarrow (y, -x, -z)$. |

one $\sigma_v$ (plane of oxygen and two hydrogens) and another $\sigma_v'$ perpendicular to the first. Clearly there are no other operations $C_n$, $S_n$, i, or $\sigma_h$. (Thus the molecule has four symmetry operations.)

**Example 2.** $NH_3$. The molecule has six symmetry operations E, $C_3$, $C_3^2$, and three $\sigma_v$.

**Example 3.** $PF_3Cl_2$. The three F atoms are in the xy-plane (every $2\pi/3$) and the Cl atoms are along the z-axis. The symmetry operations are E, $C_3$, $C_3^2$, $3C_2'$, $\sigma_h$, $3\sigma_v$, $S_3$, $S_3^5$. (Note, if one F is interchanged with one Cl, there are only two symmetry operations. If two F are interchanged with the two Cl so that in the xy-plane there are 2 Cl + 1 F, then the principal axis changes and there are four symmetry operations.)

At this stage it is worthwhile to deal with a point of **notation**. Proper or improper rotations about the principal axis have no superscripts apart from exponents when appropriate; rotations perpendicular to the principal axis are primed in the present examples or have some other superscript. In the next example the 2-fold rotations perpendicular to the principal axis are labeled $C_2^x$ or $C_2^y$ to denote rotations about the x- or y-axis, respectively. Similarly, rotations about the [110] and $[1\bar{1}0]$ axes are usually denoted by $C_2''$ or will be written as $C_2[110]$ and $C_2[1\bar{1}0]$ if such differentiation really should be needed. There are no hard and fast rules for these descriptive superscripts. As will be seen in the next chapter, the symmetry operations are often gathered together in classes where detailed superscripts are irrelevant. Similar descriptive superscripts sometimes will be used for $\sigma_v$ if it is necessary to differentiate among the various vertical planes. $\sigma_v^x$ means reflection across a plane perpendicular to the x-direction. Similarly, $\sigma[110]$ means reflection across a plane perpendicular to the [110] direction. (We will continue to use the Schoenflies notation for point symmetry operations until Chapter 11, where some of the other notations are discussed.)

Note the **origin** in the figures: For the $H_2O$ example we assumed the plane of the molecule was the $\sigma_v$ plane. Care must always be exercised with respect to this problem since other workers might use the $\sigma_v'$ for the plane of the molecule or put the origin at a different position in the molecule. These are not fundamental problems but it is important to keep them in mind. For $H_2O$ the origin can be anywhere on the z-axis.

**Example 4.** $PbTiO_3$. A unit cell of the crystal is shown in Fig. 1-2d, so we can think of 1/8 of a Pb ion at each corner, 1/2 of an oxygen at the center of each face, and the Ti ion at the center of the cell. In a

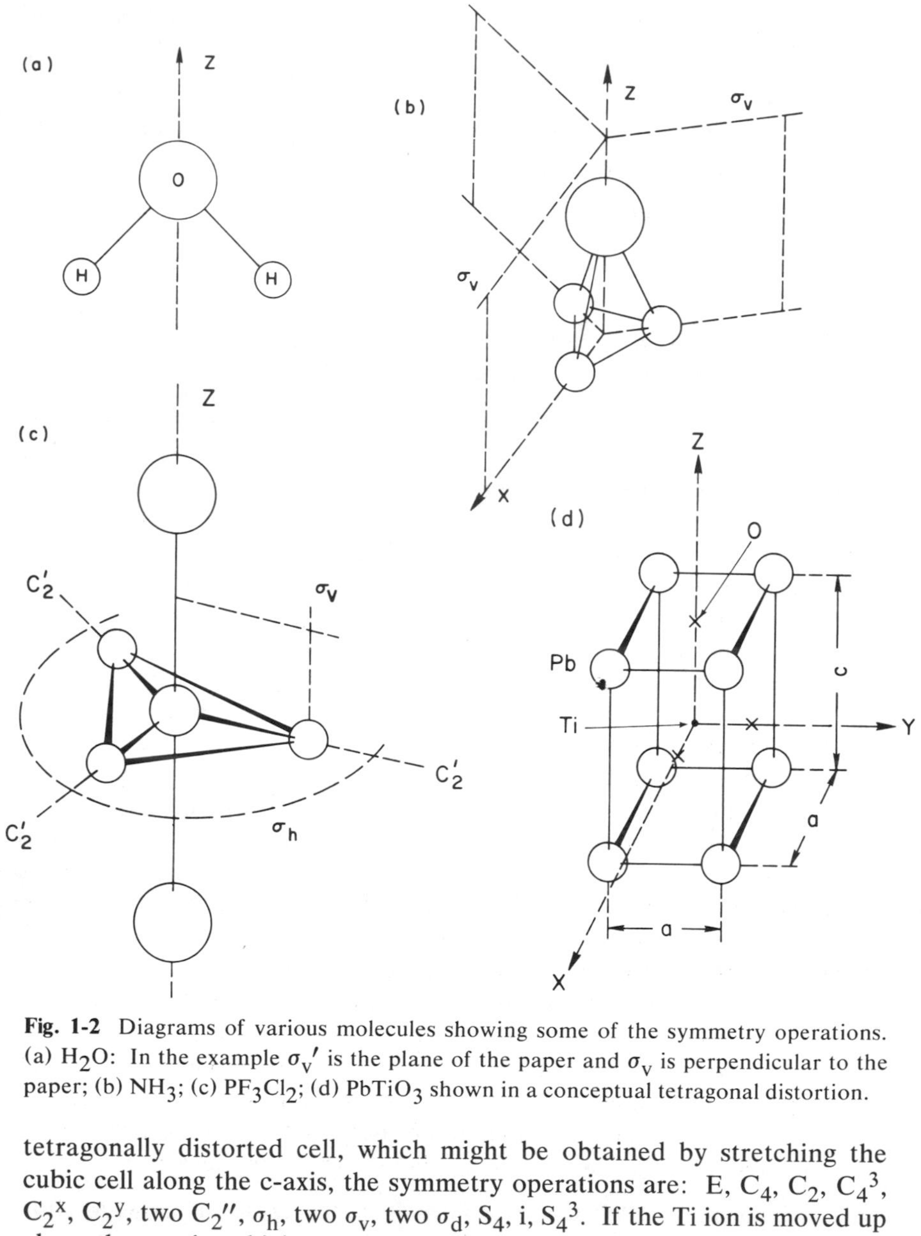


**Fig. 1-2** Diagrams of various molecules showing some of the symmetry operations. (a) $H_2O$: In the example $\sigma_v'$ is the plane of the paper and $\sigma_v$ is perpendicular to the paper; (b) $NH_3$; (c) $PF_3Cl_2$; (d) $PbTiO_3$ shown in a conceptual tetragonal distortion.

tetragonally distorted cell, which might be obtained by stretching the cubic cell along the c-axis, the symmetry operations are: E, $C_4$, $C_2$, $C_4^3$, $C_2^x$, $C_2^y$, two $C_2''$, $\sigma_h$, two $\sigma_v$, two $\sigma_d$, $S_4$, i, $S_4^3$. If the Ti ion is moved up along the z-axis, which symmetry operations are still appropriate? If the cell were cubic, not stretched along the z-axis, what are the symmetry operations? (48 operations — don't forget the four $C_3$ and four $C_3^2$.)

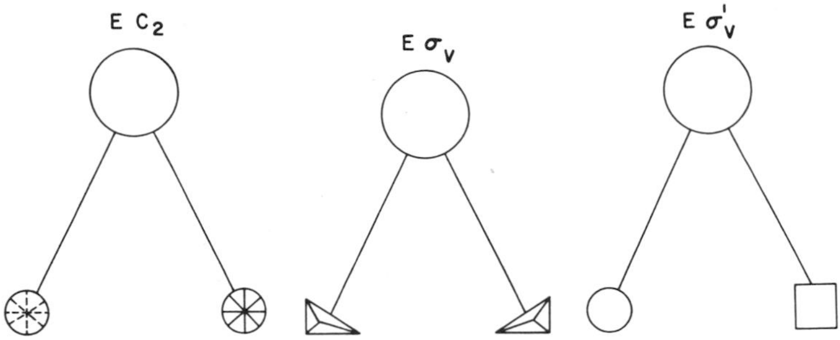


**Fig. 1-3** Various conceptual $H_2O$-like molecules that have lower symmetry than $H_2O$.

At this point we bring up a very important point. In $H_2O$, for example, only two symmetry operations are needed to transform the two H ions into themselves or each other. The two operations could be either E and $C_2$ or E and $\sigma_v'$. We might ask why we should say that $H_2O$ has four symmetry operations when it appears that two symmetry operations will suffice. The answer is simple. After we are convinced that the point symmetry operations listed in Table 1-1 are sufficient to describe any type of molecule, we want to find the **maximum number** of independent symmetry operations consistent with the molecule. If we describe the molecule with less than the maximum number of symmetry operations, we are really describing a molecule with lower symmetry. Figure 1-3 shows examples of similar molecules that have lower symmetry than $H_2O$.

### d. Inverse of symmetry operations

It should be noted in the examples given that, within any set of symmetry operations for a given molecule or crystal, the inverse of each symmetry element is also included in the set. The **inverse** of a symmetry operation A is a symmetry operation B such that the successive application of A and then B will return the molecule or crystal to exactly the same (not just an equivalent) position as if no operations were applied at all: BA = E. Thus, for any molecule or crystal in the examples given, if the set of symmetry operations is denoted by $\{A_i\}$ then for each $A_i$ in the set there is an $A_j$, also in the set, such that $A_jA_i = E$. (In the next chapter we show that a left inverse is also a right inverse, $A_jA_i = A_iA_j = E$.) Note that E, i, and any $\sigma$ are their own inverses. For the Examples: (1) $H_2O$: $C_2C_2 = E$, i.e., $C_2$ is also its own inverse; (2) $NH_3$: $C_3C_3^2 = E$ and $C_3^2C_3 = E$; (3) $PF_3Cl_2$: $S_3^5S_3 = E$ and $S_3S_3^5 = E$; etc. In general:

| Symmetry operation | Inverse | |
|---|---|---|
| $C_n^m$ | $C_n^{n-m}$ | |
| $S_n^m$ | $S_n^{n-m}$ | all m, n even |
| $S_n^m$ | $S_n^{2n-m}$ | m odd, n odd |

$S_n^m = C_n^m$ for n odd and m even and E, i, and $\sigma$ are their own inverses.

Note that for inverse operations it is not necessary to define which operation must be applied first, as mentioned above. However, if the symmetry operation B is applied to the molecule first, followed by A, by **convention** this is written as AB.

## 1-3 The Stereographic Projection

A stereographic projection is a useful and clear way of visualizing and understanding the effects of symmetry operations on molecules and crystals. The **stereographic projection** is usually defined as follows. A point in the +z hemisphere is projected on the xy-plane by determining the intersection of that plane and the line connecting the point with the south pole of the unit sphere. If the point to be projected is in the −z hemisphere, then the north pole is used. A stereogram is usually drawn with the z-axis (principal axis) projected onto the xy-plane. A **general point** or **general equivalent position** (this might be called an arbitrary point) on the unit sphere is projected onto the xy-plane and is labeled by a circle (o) or cross (×) if it is in the +z or −z hemisphere, respectively. A dot is often used instead of a cross.

Starting with a circle to denote a general point, we apply each independent symmetry operation to this first point, and map the resultant position with a circle or cross as appropriate, until all the independent symmetry operations have been applied. There will always be as many circles and crosses as independent symmetry operations because a general point has been used and the operations are independent. Figure 1-4 shows the stereograms for the four examples considered previously.

**Example 1.** $H_2O$. E, $C_2$, $\sigma_v$, $\sigma_v'$. Start with the circle at position 1 and apply E, resulting in 1; apply $C_2$ to 1, resulting in 2; apply $\sigma_v$ to 1, resulting in 3; apply the other $\sigma_v$ to 1 resulting in 4. Note that the order of the symmetry operations as well as the starting point is irrelevant. Note also the convention of thick lines representing the two $\sigma_v$ and the symbol in the center representing $C_2$.

**Example 2.** $NH_3$. E, $C_3$, $C_3^2$, $3\sigma_v$. E applied to 1 gives 1; $C_3$ applied to 1 gives 2; $C_3^2$ applied to 1 gives 3; each of the three $\sigma_v$ applied to 1 gives 4, 5, and 6.

**Example 3.** $PF_3Cl_2$. E, $C_3$, $C_3^2$, $3C_2'$, $\sigma_h$, $3\sigma_v$, $S_3$, $S_3^5$. Here, the results labeled with the first entry refer to the circle and with the second, to the cross. Note the convention on the diagram to show the three $C_2'$ axes and the thick line for the circle to designate $\sigma_h$.

**Example 4.** $PbTiO_3$. When stretched, as discussed, the symmetry operations are: E, i, $C_4$, $C_2$, $C_4^3$, $2C_2'$, $2C_2''$, $\sigma_h$, $2\sigma_v$, $2\sigma_d$, $S_4$, $S_4^3$. Note the obvious convention for $C_4$, $C_2$, $C_4^3$.

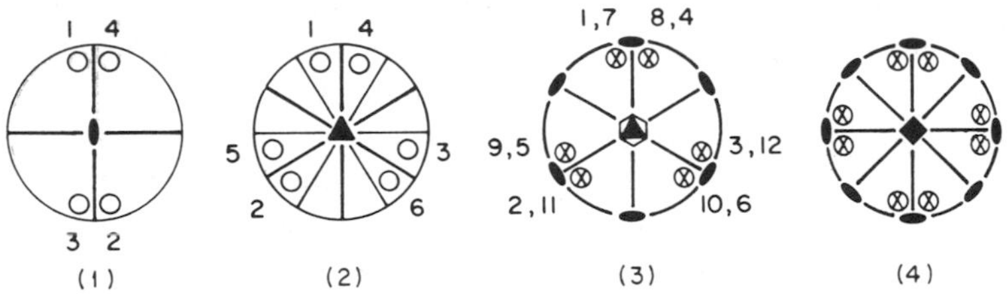


**Fig. 1-4** Stereograms for some of the examples.

It should be emphasized again that **general points** or **general equivalent positions** are pictured. If an atom is located on a symmetry element, then a symmetry operation will carry it into itself and fewer circles and crosses will be required on the stereogram to describe all the independent symmetry operations. For example, for $H_2O$, if an atom were located on one of the $\sigma_v$ planes, then all four symmetry operations would produce only two circles. If it were located on the $C_2$ axis, only one circle would result. Such positions are known as **special points**. However, general equivalent positions will always be shown on stereograms. This is consistent with the idea of describing the maximum number of point symmetry operations consistent with a given molecule. For example, compare the stereograms of the molecules in Fig. 1-3 with those for water in Fig. 1-4.

Stereograms for all 32 point groups are in Appendix 1 where two diagrams are shown for each point group. One shows the transformations of an arbitrary point and the other shows the symmetry elements.

## 1-4 The 32 Crystallographic Point Groups

For molecules there is no restriction on n in the $C_n$ rotations or $S_n$ improper rotations. Thus there is an infinite number of allowable point groups. For crystals however, the restriction that the translation of the unit cell must fill all of space, limits n to 1, 2, 3, 4, or 6. It turns out that only 32 distinct groups can be obtained (see Problem 9). We will always mean the point group with respect to the point of highest symmetry, i.e., the most symmetry operations. In crystals and molecules there is sometimes more than one such point.

We can visualize having a crystal with just one symmetry operation $E = C_1$; or two E, $C_2$; or three E, $C_3$, $C_3^2$; etc. The crystals would be said to possess **point symmetry** $C_1$, $C_2$, $C_3$, $C_4$, $C_6$ or transform as the $C_n$ **point group**. (Do not confuse the symmetry operation with the symbol used to designate the point group of the crystal, even though the symbols look alike for these particular point groups.) When a crystal is said to have point symmetry $C_n$, a collection of symmetry operations is meant. (For example, the point group $C_6$ immediately implies the six symmetry operations E, $C_6$, ..., $C_6^5$.) Five point groups have been enumerated so far. These correspond to molecules or crystals with very simple symmetry operations, namely proper rotations about the principal axis. The stereographic projections for these five point symmetries easily can be worked out and checked in Appendix 1.

From these five point groups, five more point groups can be obtained if the molecule or crystal possesses a $\sigma_h$ plane. This will double the number of symmetry operations for each $C_n$ group. The new point groups are called $C_{nh}$. For example, point group $C_3$ has three symmetry operations (E, $C_3$, $C_3^2$) and $C_{3h}$ has six symmetry operations (E, $C_3$, $C_3^2$, $\sigma_h = \sigma_h E$, $S_3 = \sigma_h C_3$, $S_3^5 = \sigma_h C_3^2$). As another example, consider $C_{4h}$. This point group has the four symmetry operations of the $C_4$ point group plus four more ($\sigma_h = \sigma_h E$, $S_4 = \sigma_h C_4$, $i = \sigma_h C_2$, $S_4^3 = \sigma_h C_4^3$). **Note** $i = S_2$ and $S_4 = iC_4^3$, $S_6 = iC_3^2$, $S_3 = iC_6^5$ which we list for completeness. The stereograms in Appendix 1 for these point groups should be checked. Notice how the stereograms indicate the number of symmetry operations for each point group. All point groups that have a $\sigma_h$ will also have an improper rotation.

By adding $\sigma_v$ symmetry operations to the $C_n$ point groups, four more point groups can be obtained which are different from those mentioned. These are called $C_{nv}$ (n = 2, 3, 4, 6) and also have twice as many symmetry operations as $C_n$. The reason n=1 is not included is that the symmetry operations (E, $\sigma_v$) are the same as those included in $C_{1h}$ since

$\sigma_v$ and $\sigma_h$ are the same if there is not principal axis. As an example of these point groups, consider $C_{3v}$. This point group has the three symmetry operations (E, $C_3$, $C_3^2$) as well as three others obtained by subsequently operating on the object with the three $\sigma_v$ planes. For most applications it is not necessary to specify one of these $\sigma_v$ planes as opposed to any other, so the symmetry operations are often written as $3\sigma_v$ to mean all three $\sigma_v$ planes. In fact, if $C_3$ is a symmetry operation, then $C_3^2$ must be one too. The six symmetry operations could be written for $C_{3v}$ in a shorthand notation as (E, $2C_3$, $3\sigma_v$). This type of arrangement will be useful in the next chapter where the term class will be defined and the symmetry operations are arranged according to these classes. The stereograms for the four $C_{nv}$ point groups can be checked in Appendix 1.

By allowing improper rotations $S_n$ about the principal axis, as well as proper rotations $C_n$, three new independent point groups (different from those discussed so far) can be obtained. These are point groups $S_n$, n = 2, 4, 6. The symmetry operations associated with the groups $S_n$ are $S_2$ (E,i); $S_4$ (E, $C_2$, $S_4$, $S_4^3$); $S_6$ (E, $C_3$, $C_3^2$, $S_6$, i, $S_6^5$). The stereograms are in Appendix 1. By convention $S_2$ is called $C_i$. Note that these groups never have $\sigma_h$ as an independent symmetry operation.

By taking the five $C_n$ point groups and adding a 2-fold symmetry operation perpendicular to the principal axis, $C_2'$ (which yields n 2-fold axes of the type $C_2'$ or $C_2'$ and $C_2''$), we obtain four new point groups $D_n$ (n = 2, 3, 4, 6) with twice as many symmetry operations.

From the point groups $C_{nh}$, a $C_2'$ will yield four new groups $D_{nh}$ (n = 2, 3, 4, 6).

By adding a $C_2'$ symmetry operation to the point groups $S_n$, two more new point groups can be obtained, $D_{nd}$ (n = 2, 3). These point groups can be checked in Appendix 1.

All the point groups discussed have a principal axis. In addition to these, there are five cubic point groups which are distinguished by possessing no unique principal axis and having four $C_3$ axes. The stereograms are in Appendix 1 and Fig. 2-1 shows some of the symmetry operations for the point group O. Appendix 2 lists the 32 point groups and all the symmetry operations for each. The five cubic groups are shown; they are T, $T_h$, $T_d$, O, $O_h$.

The 32 point groups discussed are appropriate for crystals and many molecules. The number is limited to 32 because only rotations with n = 1, 2, 3, 4, 6 are permitted. Molecules do not have this restriction so other rotations are allowed. (See the Notes.) Thus, molecules can belong to the point group $C_n$ where n is any integer. By the same procedure as outlined, stereograms and symmetry operations can be worked out for the point groups $C_{5h}$, $C_{5v}$, $D_5$, $D_{5h}$, $D_{5d}$, etc. For the 32 crystallographic point groups, there are at most 48 symmetry operations. (The point group $O_h$ has 48 symmetry operations.) Therefore there is a total of 48 points that can be generated for a general point for $O_h$ symmetry, as can be seen on the stereogram in Appendix 1.

We now show a method for classifying molecules into the appropriate point group. (The same method is appropriate for crystals and the 32 point groups with one important proviso as noted in Section 1-6.)

The problem to be addressed is: given a molecule, we would like to learn how to label it with a point group that will describe, to another scientist, the symmetry operations that are allowed. The problem is difficult if the molecule is cubic. It is very cumbersome to learn how to label systematically the five cubic groups to which a molecule may belong. It is usually obvious that a molecule does belong to one of the cubic groups. The cubic point groups are characterized by having no unique principal axis and having four $C_3$ axes. Then each molecule must be checked operation by operation to determine to which of the five cubic point groups it belongs. If the molecule does not belong to any of these special groups, then we look for the $C_n$ axis and proceed as in Fig. 1-5.

We illustrate the simple use of Fig. 1-5, using our four examples, before discussing its real significance.

**Example 1.** $H_2O$. Special groups? No! No $C_3$ at all. $C_n$? Yes, a $C_2$. $S_4$? No. $C_2'$? No $C_2'$ axis perpendicular to the z-axis. $\sigma_h$? No. $\sigma_v$? Yes, so the point group is $C_{2v}$.

**Example 2.** $NH_3$. $C_n$? Yes, $C_3$. $S_6$? No. $C_n'$? No. $\sigma_h$? No. $\sigma_v$? Yes, three of them. The result is $C_{3v}$.

**Example 3.** $PF_3Cl_2$. Result is $D_{3h}$.

**Example 4.** $PbTiO_3$. Result is $D_{4h}$.

We now can make use of one part of the character tables which are tabulated in Appendix 3. (The exact meaning of the character tables will be discussed in Chapter 4.) If we go through the procedure shown in Fig. 1-5 to determine the point group for any molecule, then by turning to Appendix 3 we can find the symmetry operations associated with that point symmetry. Several examples will be discussed. It was found, using Fig. 1-5, that $H_2O$ has point symmetry $C_{2v}$. In Appendix 3, finding $C_{2v}$ in the upper left hand corner, we can look across on the same line and see the four symmetry operations listed. Thus, if a molecule is known to have $C_{2v}$ point symmetry, the character table immediately shows the symmetry

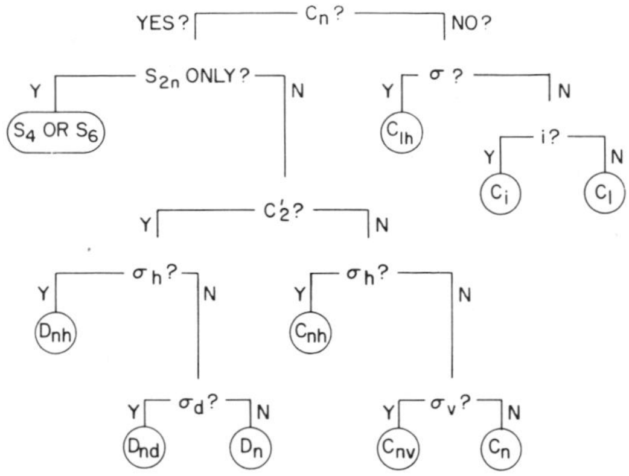


**Fig. 1-5** Chart to be used to determine the point group of a molecule.

operations allowable. For $NH_3$, $C_{3v}$ is obtained and in the Appendix we see E, $2C_3$, $3\sigma_v$; it is obvious that the $2C_3$ refers to $C_3$ and $C_3^2$, and the $3\sigma_v$ are the three vertical planes discussed previously. (Note that in the character tables the symmetry operations are collected into "classes." In Chapter 2 we will define the term class mathematically and see what it really means. However, it is obvious that similar symmetry operations are listed together and the number in front tells how many symmetry operations are included.) For $D_{3h}$, $2C_3$ refers to $C_3$ and $C_3^2$, $2S_3$ refers to $S_3$ and $S_3^5$. The symmetry operations of $D_{4h}$ are now clear.

In referring to the point symmetry of a molecule or crystal, remember that we always choose as the origin the point with the highest symmetry (most symmetry operations). There are always other points with lower symmetry. It is usually, but not always, obvious where the origin should be chosen. Also, it should be noted that an atom need not be located at the origin. A simple example of this would be $Cl_2$ treated in Section 1-5b or any triangular molecule.

## 1-5 Related Considerations

### a. Generating elements

It is already apparent that many of the symmetry operations can be written in terms of other operations. For example $C_3^2 \equiv C_3C_3 = C_6C_6C_3 = S_3S_3$ etc. For any particular one of the 32 point groups we can pick several of the symmetry operations and use these to express all of the symmetry operations for that point group. For example, for the point group $C_6$ the single symmetry operation $C_6$ can be used to generate all the point operations ($E = C_6{}^6$, etc.). Therefore we define the **generators of a point group** as the set of symmetry operations from which all the symmetry operations of the particular point group can be obtained by successive operations of the generators. The members of the set of generators for a particular point group are not necessarily unique. The following ten symmetry operations are a convenient set from which the generators of all the 32 point groups can be chosen: E, $C_4$, $C_3$, $C_2$, $C_2{}^y$, $C_3[111]$, $\sigma_v{}^y$, $\sigma_h$, i, $S_4{}^3$. In Appendix 2 a convenient set of generators for each point group is listed. The idea of generators will prove to be a useful one as will be discussed in Chapter 4.

### b. Other point symmetries

Section 1-4 discussed mainly the 32 point groups which are compatible with the translational symmetry of crystals. However, several closely related point groups such as $C_n$ where n is any integer, $C_{nv}$ where n is any integer, etc., were also mentioned since the extension from the 32 point groups to these other point groups is trivial. Other point groups are discussed here. Recall that the only requirement for a point group is that the symmetry operations be taken with respect to a fixed point.

A linear molecule, NaCl, can be rotated about its internuclear axis by any arbitrary amount and the molecule will be in an equivalent configuration. This rotation is a symmetry operation labeled $C_\infty$. However, the molecule has more symmetry operations. There is also an infinite number of $\sigma_v$ planes. The point symmetry for such a molecule is $C_{\infty v}$ which is an extension of the $C_{nv}$ notation.

Diatomic molecules such as $Cl_2$ also have an infinite number of $C_2'$ axes perpendicular to the prinicpal axis. The point group for such molecules is $D_{\infty v}$.

Some other point groups are occasionally discussed. For example, the icosahedral point group, which has twelve $C_5$ noncollinear axes among other unexpected operations, can be found in the character table in Appendix 3.

A very important point group consists of all the point symmetry operations of a sphere. This is usually called the full rotation group and consists of an arbitrary rotation about any axis through the origin as well as inversion. Atoms are examples of entities that transform into themselves under all these symmetry operations. Thus, the full rotation group is very important when the properties of atoms are discussed.

### c. Other symmetry operations

So far only point symmetry operations have been discussed. However, there are many other types of operations that leave the system in an equivalent position and thus are symmetry operations. We mention a few of them here.

A crystal is composed of an infinite array of primitive unit cells. If a unit cell can be described by three nonplanar vectors $\mathbf{a}_1$, $\mathbf{a}_2$, $\mathbf{a}_3$, then the position of any unit cell in the crystal can be denoted by the translation vector

$$\mathbf{t}_n = n_1\mathbf{a}_1 + n_2\mathbf{a}_2 + n_3\mathbf{a}_3 \tag{1-1}$$

where $n_i$ is any integer. $\mathbf{t}_n$ is a symmetry operation ($\mathbf{t}_n$ operating on the crystal will move it $n_i$ unit cells in the $\mathbf{a}_i$ direction). Before and after the operation the crystal will look exactly the same. This symmetry operation will be considered when the band theory of solids is discussed.

If the electrons in an atom or molecule or crystal are permutated (interchanged among themselves) the system will look exactly the same provided the electrons cannot be labeled in any manner. Then such a permutation operation is a symmetry operation and is important in many studies.

Most physical phenomena are unchanged if time is reversed. Thus time reversal is a symmetry operation. However, time reversal by itself is not a symmetry operation in the presence of an external magnetic field. In this case both time and the external magnetic field must be reversed to bring the system into a equivalent position.

Much of theoretical particle physics is concerned with finding the symmetry operations that the particles obey. For example, if a symmetry operation operating on a proton changes it to a neutron (the system is the same, to some degree of approximation, before and after the operation) a new quantum numbers can be obtained. In the "eightfold way" theory in particle physics we assume that the three quarks transform among themselves under all the symmetry operations of the problem. These quarks are combined in different ways to make up eight baryons (neutron, proton, $\phi^{\pm 0}$, $\Xi^{-,0}$, $\Lambda$) and their quantum numbers (isotopic spin and hypercharge) are predicted. By a slight extension of this theory, the symmetry is reduced and the differences in the rest masses of the eight particles can be calculated. This theory has been very successful (see the Notes in Chapter 13).

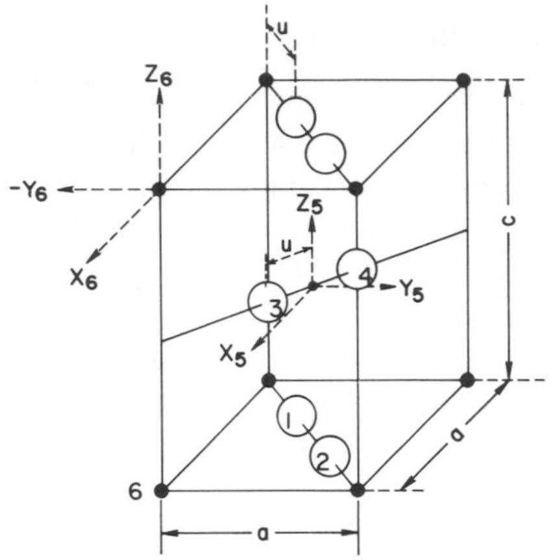


**Fig. 1-6** A unit cell of a rutile $TiO_2$ crystal. The oxygen atoms are the large open circles and the small solid circles are titanium atoms.

## 1-6 Space Group Example

Now that the point groups for molecules can be determined, we go on to crystals. Crystals can be more subtle than molecules if they have glide planes or screw axes. (See the Notes.) However, if the primitive unit cell does not have glide or screw axes, Fig. 1-5 can be applied directly to the crystal.

First consider the subtle aspects of the problem. Figure 1-6 shows a unit cell of a rutile $TiO_2$ crystal.

What is the point group? You probably will get $D_{2h}$. Strictly speaking, $D_{2h}$ is correct if the unit cell shown is thought of as a molecule. However, if it is a unit cell of a crystal with translational symmetry (Eq. 1-1), then there are other symmetry operations that the crystal possesses of a kind not mentioned yet. These new symmetry operations are not the point operations or the primitive cell translation operations (Eq. 1-1). Consider a $C_4$ rotation about the z-axis followed by a translation c/2 along the z-axis and a/2 along the x- and y-axes. The crystal will be indistinguishable before and after this operation (known as a screw operation). This symmetry operation is usually designated by the symbol

$$\{C_4 \mid \tau\} \qquad \tau = (1/2)\mathbf{a} + (1/2)\mathbf{a} + (1/2)\mathbf{c} \tag{1-2}$$

Note that the screw symmetry operation is a mixed operation involving a point rotation followed by a translation of some fraction of the primitive unit cell. A space group can be defined using the symbol introduced in

Eq. 1-2 (sometimes called the Seitz symbol). A **space group** is the set of symmetry operations $\{R \mid \mathbf{T}\}$ which transforms the crystal into itself. R can be any of the point group operations and **T** can be a unit cell translation vector (Eq. 1-1), $\mathbf{t}_n$, or a fraction of $\mathbf{t}_n$, $\boldsymbol{\tau}$, as in Eq. 1-2. Clearly there is an infinite number of symmetry operations since $\{E \mid \mathbf{t}_n\}$ exists. We might guess that the number of operations of the type $\{R \mid \boldsymbol{\tau}\}$ is finite. It is; there can be at most 48 such operations for a given crystal. It can be shown that all crystals can be classified according to 230 space groups. Further, we show here how to classify crystals (space groups) in terms of one of the 32 point groups. In Chapter 4 some of the usefulness and importance of such a classification will become apparent. First consider the classification of crystals that have no glides or screws.

No glide planes or screw axes are necessary to describe 73 of the space groups. These are called **symmorphic** space groups. The point group of such a crystal can be determined in the same manner as for molecules, via Fig. 1-5. Namely: find the principal axis; is there a $\sigma_h$; etc. As always, all these operations must be taken with respect to a fixed point. The $PbTiO_3$ example discussed several times in this chapter is an example of this procedure.

For crystals in which glide planes or screw axes are necessary to describe symmetry operations (**nonsymmorphic** crystals) the procedure is similar. Determine all the symmetry operations (point operations and glide and screw operations as in Eq. 1-2). Let all the $\boldsymbol{\tau}$ (Eq. 1-2) be zero. Then we are left with only point operations. The point group of this nonsymmorphic crystal is the point group that is determined with all $\boldsymbol{\tau}$ arbitrarily set to zero. The fundamental meaning of this procedure will become clear in Chapter 11 and the usefulness will be discussed in Chapter 4. As an example consider rutile in Fig. 1-6. The eight straightforward point symmetry operations are E, $C_2$, $2C_2''$, i, $\sigma_h$, $2\sigma_d$. There are eight screw symmetry operations involving $\boldsymbol{\tau}$ as in Eq. 1-2. They are $\{C_4 \mid \boldsymbol{\tau}\}$, $\{C_4^3 \mid \boldsymbol{\tau}\}$, $2\{C_2' \mid \boldsymbol{\tau}\}$, $\{S_4 \mid \boldsymbol{\tau}\}$, $\{S_4^3 \mid \boldsymbol{\tau}\}$, $2\{\sigma_v \mid \boldsymbol{\tau}\}$. If $\boldsymbol{\tau}$ is ignored, the 16 operations are identical to those found for the point group $D_{4h}$. Hence the nonsymmorphic crystal rutile can be classified as having the point group $D_{4h}$.

The 230 space groups are listed in Appendix 4 in several notations and the symmorphic groups are underlined. As can be seen in the Schoenflies notation, the point group that corresponds to each space group is obvious. Appendix 2 also lists the symbols for the 32 point groups in several notations although we will continue to use the Schoenflies notation.

## Notes

General references to group theory books can be found in the Bibliography. Several books which discuss the topics covered in this chapter are Bishop, Cotton, Schonland, and Shubnikov *et al.*

Schoenflies' discovery of the 230 space groups, in about 1890, had a great deal of influence on the development of structural crystallography. Actually, Fedorov discovered the 230 space groups several years earlier and Barlow a bit later. See E. S. Fedorov, "Symmetry of Crystals," translated by D. and K. Harker (American Crystallographic Association, 1971) and "Fifty Years of X-Ray Diffraction" by P. P. Ewald *et al.* (published for The International Union of Crystallography by N. V. A. Oosthoek's Uitgeversmaalschappij, Utrecht, The Netherlands, 1962).

Weinreich shows (in his Fig. 1.2) a nice set of "shapes" that have the symmetry of each of the 32 point groups.

(Some concepts used in this note will be discussed in the text in Chapters 2–4, so it can be read now for general content and should be reread at the end of Chapters 4 and 7.) Another more mathematical approach to the 32 point groups will be sketched (see Bhagavantam, Sections 5.5 and 5.6). First, we must still show that only proper rotations $C_n$ where n = 1, 2, 3, 4, 6 are compatible with translational symmetry. Then we show that there are only eleven crystallographic **pure rotational point groups**. These are point groups with only $C_n$ symmetry operations but rotation need not be about the prinicpal axis only. The resultant eleven point groups are listed in 1N-1.

$$C_1 \quad C_2 \quad C_3 \quad C_4 \quad C_6 \quad D_2 \quad D_3 \quad D_4 \quad D_6 \quad T \quad O \qquad (1N\text{-}1)$$

$$S_2 \quad C_{2h} \quad S_6 \quad C_{4h} \quad C_{6h} \quad D_{2h} \quad D_{3h} \quad D_{4h} \quad D_{6h} \quad T_h \quad O_h \qquad (1N\text{-}2)$$

In 1N-2 we have listed eleven other point groups each having twice the number of symmetry operations as the corresponding point group in 1N-1. The point groups in 1N-2 are obtained from the rotational point groups 1N-1 by multiplying each of the symmetry operations of a given point group by the symmetry operations E and then i. Symbolically if $C_i = \{E,i\}$, then, for example, $D_{2h} = D_2 \times C_i$. All of the point groups in 1N-2 have i as a symmetry operation and are said to be centrosymmetric. There are ten distinct subgroups of those listed in 1N-2 which do not contain the symmetry operation i explicitly and also are not the rotational point groups of 1N-1. These are:

$$C_{1h} \quad C_{2v} \quad S_4 \quad C_{4v} \quad D_{2d} \quad C_{3v} \quad C_{3h} \quad D_{3h} \quad C_{6v} \quad T_d \qquad (1N\text{-}3)$$

The 32 point groups are all listed in these three expressions. The 21 point groups in 1N-1 and 1N-3 are called noncentrosymmetry since they do not have i as a symmetry operation.

## Problems

**1.** Draw the stereograms for the symmetry operations (a) E, i; (b) E, $C_2$, $C_4$, $C_4^3$; (c) E, $C_4$, $C_2$, $C_4^3$, $2C_2'$, $2C_2''$; (d) E, i, $C_2$, $\sigma_h$.

**2.** Name the point group for each of the above. Check by looking in Appendix 2.

**3.** Answer the questions associated with $PbTiO_3$ in Section 1-2, Example.

**4.** What is the point group for: (a) the equilateral triangular molecular $A_3$. (b) Benzene in Fig. 1-1. (c) $CH_4$ (note the $4C_3$ and $4C_3^2$, i.e., $8C_3$). (d) The pyramids in Egypt. (e) The $(NiCl_4)^{2-}$ ion has $T_d$ point symmetry. What is the symmetry of $(NiCl_3Br)^{2-}$; of $(NiCl_2Br_2)^{2-}$; of $(NiClBr_3)^{2-}$; of $(NiFCl_2Br)^{2-}$?

**5.** Do Problem 3.9 in the 2nd edition of Cotton.

**6.** Show that the symmetry operations of the point group $C_{2h}$ are the same as those of the point group $C_2$ plus the two obtained by multiplying each of the operations in the point group $C_2$ by the symmetry operation i. (Symbolically $C_{2h} = C_2 \times C_i$.)

**7.** Show that the symmetry operations of $D_{4h}$ are those of $C_{4v}$ plus those obtained by multiplying each of the operations in $C_{4v}$ by the inverse operation i. (Symbolically $D_{4h} = C_{4v} \times C_i$.)

**8.** Which symmetry operations commute with each other?

**9.** (a) Prove that for the $C_n$ point groups only n = 1, 2, 3, 4, and 6 are consistent with translational symmetry of crystals. (Hint: See Bhagavantam, Falicov, Tinkham, etc.) (b) Derive the eleven crystallographic rotational point groups which are defined in the Notes.

**10.** Show that if a molecule possesses $S_n$ it either possesses $\sigma_h$ and $C_n$ or neither.

**11.** (a) Draw a figure that has $C_4$ symmetry and does not have higher symmetry. Do the same for $C_{4h}$ and $C_{4v}$. (b) Do the same for $D_{4h}$ and $D_{2d}$.

**12.** Work out the point group for $TiO_2$ as in Section 1-4. Check that the other operations, nonpoint operations, would indeed be symmetry operations of the crystal.

Chapter 2

# GROUP CONCEPTS

*Can there be any greater evil than discord and distraction and plurality where unity ought to reign? Or any greater good than the bond of unity?*

*Plato, "Republic"*

## 2-1 Introduction

In this chapter the basic group definitions and concepts are presented and discussed. The mathematical ideas are actually few and simple. We first discuss the definition of a group and a group multiplication table and give examples. Then we show that a collection of symmetry operations, as discussed in Chapter 1, forms a group in the mathematical sense. Various related group concepts such as subgroups and classes are discussed, as well as the meaning of isomorphism and homomorphism. Finally, a few special groups are presented and several more complicated but important group concepts are examined.

## 2-2 Definition of a Group

A **group** is defined as a **set**, or collection, of elements which obeys the four rules given below. The elements of the set are usually written as $\{A_i\}$; or A, B, C, ...; or $A_i$, $i = 1, ..., h$; whichever is most convenient. A binary operation called combination or product is defined which describes the operation by which two members of the set combine to form a third member of the set. This product, which is sometimes written as a small circle, i.e., A•B = C, could be ordinary multiplication, addition, operation as in a symmetry operation, matrix multiplication, etc. Here we will not write it explicitly. Thus, we will simply write AB = C.

(1) **Product or Closure.** The combination or product of any two elements in the set must also be an element of the set. $A_iA_j = A_k$.

(2) **Identity.** The set must contain as one of its elements the identity E, such that $EA_i = A_iE = A_i$ for all $A_i$.

(3) **Inverse.** Every element of the set must have an inverse or reciprocal that is also a member of the set, such that $A_i^{-1}A_i = E$ where $A_i^{-1} = A_j$ and $A_j$ is a member of the set. We can prove $E = A_iA_i^{-1}$ (see the Notes at the end of this chapter).

(4) **Associativity.** For any three elements in the set, (AB)C = A(BC). Thus the result can be written as ABC.

If the elements of a set obey these four rules, then they are said to form a group. Note that the group can be **finite** or **infinite**. For a finite group the number of elements in the set is called the **order of the group, h**.

### a. Examples

**Example 1.** The set {1, −1} forms a group under multiplication. The product is a member of the set, E = 1, each is its own inverse, and the associative law is obeyed.

**Example 2.** The set of all integers forms a group under addition, the identity is E = 0, and the negative of each is its inverse, i.e., −5 + (−(−5)) = E.

**Example 3.** The set of symmetry operations E, $C_2$, $\sigma_v{}^x$, $\sigma_v{}^y$ as in the example of Chapter 1 forms a group. It is easiest, at this point, to see this by using the stereogram in Fig. 1-4. $\sigma_v{}^x\sigma_v{}^y = C_2 = \sigma_v{}^y\sigma_v{}^x$, $C_2\sigma_v{}^x = \sigma_v{}^y$, etc. Each element is its own inverse. The associative law is axiomatic for symmetry operations, and true for matrix multiplications, ordinary multiplications, and ordinary additions. Associativity does not hold for mixed operations. For example, $10 + (10 \times 10) \neq (10 + 10) \times 10$.

### b. Multiplication table

An easy and important way to display the product of the group elements is by way of a multiplication table. Table 2-1 shows a table for the group E, $C_2$, $\sigma_v{}^x$, $\sigma_v{}^y$. We pick a **convention** such that the element in the left hand column comes to the left in multiplication of the element on the top row. An example of this convention is shown in the next paragraph. It is not needed for the particular case of the example in Table 2-1 but will be used in the next example and will be adhered to throughout.

**Example 1.** Do the following six matrices form a group (under matrix multiplication? (See Appendix 5 for the definition and properties of matrices.)

$$\mathrm{E} = \begin{bmatrix} 1 & 0 \\ 0 & 1 \end{bmatrix} \qquad \mathrm{A} = \tfrac{1}{2}\begin{bmatrix} -1 & \sqrt{3} \\ -\sqrt{3} & -1 \end{bmatrix} \qquad \mathrm{B} = \tfrac{1}{2}\begin{bmatrix} -1 & -\sqrt{3} \\ \sqrt{3} & -1 \end{bmatrix}$$

$$\mathrm{C} = \begin{bmatrix} -1 & 0 \\ 0 & 1 \end{bmatrix} \qquad \mathrm{D} = \tfrac{1}{2}\begin{bmatrix} 1 & -\sqrt{3} \\ -\sqrt{3} & -1 \end{bmatrix} \qquad \mathrm{F} = \tfrac{1}{2}\begin{bmatrix} 1 & \sqrt{3} \\ \sqrt{3} & -1 \end{bmatrix} \quad (2\text{-}1)$$

Table 2-2 shows the resultant multiplication table. (Ignore the dotted lines for the moment.) The convention is clear, i.e.,

$$\mathrm{CA} = \tfrac{1}{2}\begin{bmatrix} -1 & 0 \\ 0 & 1 \end{bmatrix}\begin{bmatrix} -1 & \sqrt{3} \\ -\sqrt{3} & -1 \end{bmatrix} = \tfrac{1}{2}\begin{bmatrix} 1 & -\sqrt{3} \\ -\sqrt{3} & -1 \end{bmatrix} = \mathrm{D} \quad (2\text{-}2)$$

From the multiplication table, the six matrices have closure. The identity matrix is E. Again from the multiplication table, all the elements have an inverse (C, D, F are their own inverses and AB = BA = E). Matrix multiplication is associative. Therefore the six matrices form a group.

**Example 2.** Do the set of symmetry operations E, $C_3$, $C_3^2$, $\sigma_v'$, $\sigma_v''$, $\sigma_v'''$ form a group? These operations were encountered in the examples in Section 1-3 where the stereogram, Fig. 1-4, is shown, and the point group is called $C_{3v}$. There is no need to show a new multiplication table for these elements because Table 2-2 is identically obtained if the correspondence is made between these elements and, respectively, E, A, B, etc., from Example 1. Perhaps Table 2-2 should be labeled with E, $C_3$, etc. Thus the symmetry operations of the point group $C_{3v}$ do indeed form a group. The point group $C_{3v}$ and the group of matrices Eq. 2-1 have identical multiplication tables, thus the two groups are said to be **isomorphic**. This will be discussed later in the chapter.

### c. Rearrangement theorem

Note that in the multiplication tables obtained so far, each element of the group appears once and only once in each row and in each column. This is always true and can be proved as follows. Suppose the element C appears twice so that AB = C and AD = C. Multiply these equations on

**Table 2-1** Multiplication table for a group of order 4

| | $E$ | $C_2$ | $\sigma_v^x$ | $\sigma_v^y$ |
|---|---|---|---|---|
| $E$ | $E$ | $C_2$ | $\sigma_v^x$ | $\sigma_v^y$ |
| $C_2$ | $C_2$ | $E$ | $\sigma_v^y$ | $\sigma_v^x$ |
| $\sigma_v^x$ | $\sigma_v^x$ | $\sigma_v^y$ | $E$ | $C_2$ |
| $\sigma_v^y$ | $\sigma_v^y$ | $\sigma_v^x$ | $C_2$ | $E$ |

**Table 2-2** Multiplication table for a group of order 6 and for the point group $C_{3v}$

| | E | A | B | C | D | F |
|---|---|---|---|---|---|---|
| E | E | A | B | C | D | F |
| A | A | B | E | F | C | D |
| B | B | E | A | D | F | C |
| C | C | D | F | E | A | B |
| D | D | F | C | B | E | A |
| F | F | C | D | A | B | E |

the left by $A^{-1}$ to obtain $A^{-1}AB = A^{-1}C = EB = B$. Thus $B = A^{-1}C$. Similarly, $D = A^{-1}C$ is obtained. By the definition of a group, all these operations are permissible and the results are also members of the group. Hence $B = D$, so they are the same element and the two entire columns in the table are identical. If the multiplication table ever does contain an element twice in the same row or column, two of the group elements are identical and one should be removed.

### d. Inverse of a product

We prove that the reciprocal of the product of several group elements is equal to the product of the reciprocals of the elements in the reverse order, i.e.,

$$(AB \ldots FG)^{-1} = G^{-1}F^{-1} \ldots B^{-1}A^{-1} . \qquad (2\text{-}3)$$

Consider $AB \ldots FG = X$ and start multiplying on the right by $G^{-1}$, eliminating $GG^{-1} = E$. Then, in succession multiply on the right by $F^{-1}$, ...,

$A^{-1}$. Thus,

$$\begin{aligned} E &= X(G^{-1}F^{-1} \ldots B^{-1}A^{-1}) \\ &= (AB \ldots FG)(G^{-1}F^{-1} \ldots B^{-1}A^{-1}) \\ (AB \ldots FB)^{-1} &= G^{-1}F^{-1} \ldots B^{-1}A^{-1}. \end{aligned} \tag{2-4}$$

## 2-3 Symmetry Operations Form a Group

As stated previously, most of this book will be concerned with group theory applied to a complete set of symmetry operations. We show in this section that a complete set of symmetry operationes does in fact obey the four group postulates and thus forms a group. What is surprising is that so much detailed, fundamental, and important understanding can result from such a simple concept. Part of the reason for this result is that group theory is a branch of mathematics and the results are exact.

A **complete set of symmetry operations** means that all the symmetry operations that transform a molecule into itself are included in the set. Thus, the product of any two operations is in the set because the product of two symmetry operations will transform the molecule into itself. Some other operation must also be able to make this same transformation or the set would, by definition, not be complete. Thus, the first group postulate is obeyed.

A complete set of symmetry operations contains the identity operation. The second postulate is obeyed.

All symmetry operations have an inverse as shown in Chapter 1. Each inverse must be included in the complete set by the definition of complete set, resulting in the fulfillment of the third postulate.

The fourth postulate, associativity, is axiomatically obeyed for symmetry operations. Note that this is different from commutivity, AB = BA, which in general is not true for symmetry operations.

Thus, the set of symmetry operations that transform the molecule into itself form a group in the mathematical sense.

## 2-4 Related Group Concepts

### a. Subgroups

The group S is a **subgroup** of the group G if all of the elements of S are also contained in a larger group G. It can be shown that if the order of the subgroup S is s and the order of G is h, then $h/s$ = integer.

We prove the above theorem and introduce a number of more subtle group concepts.

(a) A **complex** is a set of elements of a group. It is not necessarily a subgroup. If $\Omega$ and $\Delta$ are complexes, then the product $\Omega\Delta$ is the product of every element in $\Omega$ with every element in $\Delta$ where any product that occurs more than once is counted only once.

(b) Suppose p is an element of G not contained in the subgroup S. Then the complexes pS and Sp are called the **left coset** and the **right coset** of S. Cosets must be formed with respect to a subgroup S. Thus, cosets are never subgroups since p can not be E, because S contains E. If q is an element of G not contained in S or pS, then no element of qS is contained in S or pS by the rearrangement theorem.

(c) If G is a finite group, then it can be written as a **factored set** in terms of a finite number of cosets, $G = S + pS + qS + \ldots$ . (For a given S, the coset decomposition is unique as shown in Problem 10.)

(d) There are s nonequivalent elements in every complex S, pS, qS, etc. The number of cosets in the factored set is an integer n. So $h = ns$. QED

### b. Conjugate elements

A, B, C, ... are elements of a group. A and B are **conjugate elements** if they are related by a **similarity transformation** where X is any member of the group

$$A = X^{-1}BX \tag{2-5}$$

A number of simple theorems can be proved about conjugate elements:

(a) Every element is conjugate with itself. (Let $X = E$.)
(b) If A is conjugate with B, then B is conjugate with A.

**Proof.** $A = X^{-1}BX$. Multiply on the left by X and on the right by $X^{-1}$. Thus, $XAX^{-1} = B$. Since X can be any member of the group, $B = Y^{-1}AY$. QED

(c) If A is conjugate to B and A is conjugate to C, then B is conjugate to C. The proof is similar to that for (b).

### c. Class

A set of elements of a group which are conjugate elements are said to form a **class**. To find all the members in the same class as the element B, Eq. 2-5 should be applied successively for each X in the group. Thus, all elements conjugate to B can be found. Clearly, any member of one class will not be a member of any other class. Also, E always will be in a class by itself and will be the only class that is also a subgroup.

Some physical significance can be associated with different classes. For example, reflection operations will be in a different class from rotations and inversions. Rotations about the principal axis will be in different classes from rotations about an axis in the xy-plane. These results occur because, loosely speaking, from Eq. 2-5, $X^{-1}$ undoes what X did. Thus, A and B must be of a similar genre (i.e., rotations about the same axis, reflections containing the same axis, etc.). See Problem 9.

### d. Examples

**Example 1.** What subgroups of the group shown in Table 2-1? From the table it is clear that E and $C_2$ form a group so they are a subgroup. The other subgroups are E, $\sigma_v^x$ and E, $\sigma_v^y$ as well as the trivial subgroup E. Why is E, $\sigma_v^x$, $\sigma_v^y$ not a subgroup?

**Example 2.** What are the subgroups of the group shown in Table 2-2? It would help to label Table 2-2 as suggested in Section 2-2b. Obviously E, $C_3$, $C_3^2$ form a subgroup. Why is E, $3\sigma_v$ not a subgroup? ($6/4 \neq$ integer among other reasons.) What subgroups have order 2?

**Example 3.** There is no quick method for determining the classes of the symmetry operations of a group. Equation 2-5 must be applied to find all the elements conjugate to a given one as X is varied among all the group elements. For example, in the point group $C_{2v}$, what is conjugate to $C_2$? $\sigma_v^x C_2 \sigma_v^x = C_2$, $\sigma_v^y C_2 \sigma_v^y = C_2$, $C_2 C_2 C_2 = C_2$ so $C_2$ is **self-conjugate**. For this group, each element is conjugate only to itself; thus there are four classes. (Check by looking up the character table for the point group $C_{2v}$.)

**Example 4.** What are the classes of the group E, $C_3$, $C_3^2$, $\sigma_v'$, $\sigma_v''$, $\sigma_v'''$? The multiplication table is given in Table 2. Check by looking at the character table for the correct point group.

## 2-5 Isomorphism and Homomorphism

### a. Isomorphism

Two groups are **isomorphic** with each other if there is a one-to-one correspondence between the elements of the groups. Thus, the order h of the two groups must be the same and the group multiplication tables must be identical. Then, from the abstract group theoretical point of view, the two groups are identical even though the meaning of the elements in the two groups may be different. Herein lies the power of group theory, since the properties of isomorphic groups are the same from an abstract point of view. It will turn out that they have the same character tables. (Character tables will be discussed in Chapter 4.) Problem 1 shows several isomorphic and nonisomorphic groups of order four.

### b. Homomorphism

Two groups are said to be **homomorphic** if there is a many-to-one correspondence between the elements in the two groups. Thus the order h of the two groups is different. If there exist two groups $\{A_i\}$ and $\{B_j\}$, then the groups are homomorphic if $A_1$ corresponds to $B_1$, $B_2$, $B_3$, and $A_2$ corresponds to $B_4$, $B_5$, $B_6$, and so on, until all the elements in both groups are simultaneously exhausted. The homorphism is one-to-three. Thus, the multiplication tables of homomorphic groups are not identical (the order of the groups is different) but they are similar since the correspondence preserves products. For example, the group 1, $-1$ is homomorphic with the point group $C_{2v}$, the multiplication table of which is shown in Table 2-1. The correspondence is as follows: 1 corresponds to E, $C_2$ and $-1$ corresponds to $\sigma_v{}^x$, $\sigma_v{}^v$. Actually, between these two groups the correspondence can be made betweeen 1 and E along with any other element, and between $-1$ and the other two elements. Try it.

The point group $C_{3v}$, the multiplication table of which is Table 2-2, is also homomorphic with the group 1, $-1$. The correspondence or mapping is: 1 maps into E, $C_3$, $C_3{}^2$ and $-1$ maps into $3\sigma_v$.

Although it appears trivial, the most important homomorphism is between the group 1 (it is a group since it satisfies all the postulates) and every group.

## 2-6 Special Kinds of Groups

There are a number of special groups. We mention just a few here that will be used later.

### a. Abelian group

If all the elements of a group commute with each other (AB = BA), then it is called an **Abelian group**. It is clear that for an Abelian group each element is in a class by itself. (Using Eq. 2-5, $X^{-1}AX = X^{-1}XA = A$.) In general the point groups will not be Abelian, although some indeed are, such as $C_{2v}$.

### b. Cyclic group

If a group can be generated by repeated applications of one element, then it is called a **cyclic group**. For example, by repeated application of $C_3$ we can generate the group $C_3$, $C_3^2$ ($= C_3C_3$), E ($= C_3^3$). All cyclic groups are Abelian since each element commutes with itself. Let A be the generating element for the cyclic group of order h, $A^h = E$. Then the inverse of any element is given by $A^{h-n}A^n = E$. The conjugate to the element $A^q$ is, by Eq. 2-5,

$$A^{h-n}A^qA^n = A^{h-n}A^{q+n} = A^{h+q} = A^hA^q = A^q \qquad (2\text{-}6)$$

Therefore each element in a cyclic group is in a class by itself. This will be important in a later chapter when we consider the operations which translate a crystal from one unit cell to another as in Section 1-5c. These translation operations are clearly cyclic.

### c. Multiplication of groups

We define the **multiplication of two groups** or the direct product of two groups $\{a_i\}$ and $\{b_j\}$, which have only the identity element in common and $a_ib_j = b_ja_i$ in the set formed by all the elements $a_ib_j$. If the two groups are of order h and g, then the direct product consists of hg elements. It is easy to show that the set of elements formed by the direct product of two groups is indeed a group.

| | |
|---|---|
| Closure | $a_ib_ja_kb_l = a_ia_kb_jb_l = a_nb_m$ |
| Identity | $e_ie_j \equiv e$ |
| Inverse | $(a_ib_j)^{-1}\, a_ib_j = b_j^{-1}a_i^{-1}a_ib_j = e$ |
| Associativity | same as original groups |

Use of this concept is made in multiplying various point groups by the group $C_i$ consisting of elements E, i. Other point groups result. For example:

$$\begin{aligned} C_{2h} &= C_2 \times C_i, & C_{4h} &= C_4 \times C_i, \text{ etc.} \\ D_{3d} &= D_3 \times C_i, & D_{2h} &= D_2 \times C_i, \text{ etc.} \\ T_h &= T \times C_i, & O_h &= O \times C_i \end{aligned} \qquad (2\text{-}7)$$

These results may be confirmed by comparing the symmetry operations in the character tables. The classes of a direct product group are given by the direct product of the classes of the original groups as can be seen in the character tables and can be proved (see the Problems). It is also clear that each of the two groups which were multiplied together will be subgroups of the resultant group.

## 2-7 More Involved Group Concepts (including a Factor Group of a Space Group)

In Section 2-4 the concepts of complex, coset, and factored set were introduced. In this section several other more involved concepts will be introduced. This section could be skipped on a first reading or referred to when Chapter 11 on band theory is read.

### a. Invariant subgroup

A subgroup consisting only of complete classes is called an **invariant subgroup** or **normal divisor**. A subgroup, S, of a group, G, consists of complete classes if, for any element s of S and g of G, $g^{-1}sg$ is an element of S. From this we see that $g^{-1}Sg = S$ or $Sg = gS$. (S consists of complete classes.) Thus, for an invariant subgroup left and right cosets are the same.

For the point group $C_{2v}$ (Table 2-1) the subgroup E, $C_2$ is an invariant subgroup since E forms a complete class as does $C_2$. The other subgroups of this point group are also invariant as are the subgroups of the point group $C_{3v}$ (Table 2-2). Consider the cubic point group O. The classes are (from Appendix 3)

$$E \qquad 8C_3 \qquad 3C_2 \qquad 6C_4 \qquad 6C_2'$$

where $8C_3$ refers to four $C_3$ and four $C_3{}^2$; $3C_2$ refers to $C_2{}^x$, $C_2{}^y$, $C_2{}^z$, etc. (see Fig. 2-1). The subgroup E, $3C_2$ is invariant, but the subgroup E, $C_2{}^x$ is not.

### b. Factor group

A factor group is a set obtained by taking as elements an invariant subgroup and all of its cosets. Thus, if a group can be written in terms of a factored set $G = S + pS + qS + tS + \ldots$, where S is an invariant subgroup and p is not in S, and q is not in S or pS, etc., then the set with the elements S, pS, qS, ... is called a factor group with respect to the invariant subgroup S. S is the identity of the factor group. Clearly the elements of

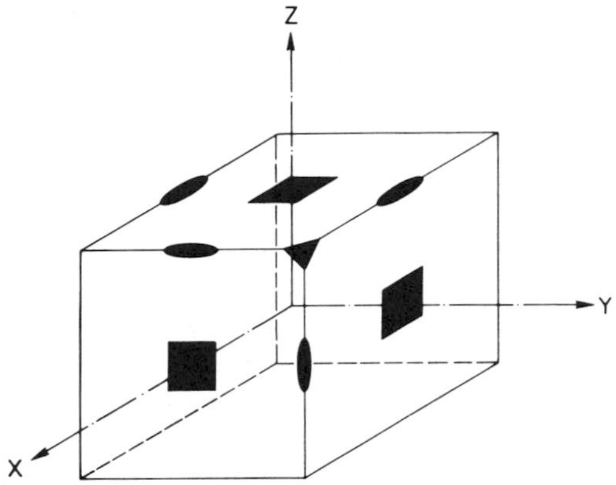


**Fig. 2-1** A cube schematically showing some of the 2-, 3-, and 4-fold symmetry operations.

the factor group will be different if a different invariant subgroup is used. However, given an S, there is no need to speak of right and left cosets because the complex pS = Sp when S is an invariant subgroup. We show that the set of elements forming a factor group indeed obeys the group definition. (However, note that the elements of a factor group are cosets and a factor group is not a subgroup of the original group.)

| | |
|---|---|
| Closure | (pS) (qS) = pSSq = pSq = (pq)S = tS |
| Identity | S(pS) = pSS = pS |
| Inverse | $(pS)^{-1}pS = S^{-1}p^{-1}pS = S^{-1}S = S$ |
| Associativity | same as the original group |

As can be seen, the invariant subgroup is the identity element and each coset is its own inverse.

As an example consider the point group $C_{2v}$ (Table 2-1). If the invariant subgroup is chosen as S = {E, $C_2$}, then the coset Q = $\sigma_v{}^xS$ = {$\sigma_v{}^x$, $\sigma_v{}^y$} exhausts all the group elements. The factored set of the group could be written $C_{2v}$ = S + Q, where the normal divisor is S = {E, $C_2$}. The factor group is S, Q which clearly forms a group. This group is homomorphic with the original group (it always must be). The group is isomorphic with {1, −1}. As another example consider the point group $C_{4v}$ = E, $C_2$, $2C_4$, $2\sigma_v$, $2\sigma_d$. Taking the invariant subgroup as S = {E, $C_2$} the factor group has four elements which are S, $C_4S$, $\sigma_vS$, $\sigma_dS$. This factor group is isomorphic with the point group $C_{2v}$.

In this book the factor group concept will be used for the discussion of crystal lattice vibration and other crystal or space group concepts.

A crystal has an infinite number of translational symmetry operations and a small number of operations on the unit cell which may be just the point operations (one point is fixed) or may contain a few glide planes or screw axes. This number of operations is 48 for $O_h$ point symmetry and is less for other point symmetries. The space group consists of all products of these operations. If the infinite translational group is taken as the invariant subgroup, then the factor group of the space group is a small group (48 or less elements). This factor group is isomorphic to one of the 32 point groups, and this is what is meant by **the point group of a crystal**. For crystals that have no glide planes or screw axes (symmorphic crystals) the factor group will be identical with the point group. However, for crystals that are nonsymmorphic the factor group will only be **isomorphic** with the point group. Thus a rutile crystal, which is nonsymmorphic (Section 1-6), has a point group $D_{4h}$ even though $D_{2h}$ would be obtained if only the point operations were considered. This will be discussed in detail in Chapter 11.

## Appendix to Chapter 2

In this Appendix we discuss a few simple aspects of the permutation group. Although most of the concepts to be used have already been covered in Chapter 2, the discussion on the irreducible representations should not be read until after Chapter 4.

The **permutation group** of n objects is usually written as $\mathbf{P}_n$ and is sometimes called the **symmetric group**. The theory of the irreducible representations of $\mathbf{P}_n$ is not simple (see Wigner, Chapter 13 or Hammermish, Chapter 7) and only a aspects will be stressed here. The connection with the Pauli principle will be discussed in the Appendix to Chapter 8.

Consider n boxes labeled from 1 to n, each containing one object. We can have a transformation of the objects among these labeled boxes producing a new arrangement, still with only one object in each box. This type of transformation is called a permutation and can be described as follows for six boxes;

$$A = \begin{pmatrix} 1 & 2 & 3 & 4 & 5 & 6 \\ 5 & 4 & 1 & 2 & 3 & 6 \end{pmatrix} = (153)(24)(6) \qquad \text{(2-A1)}$$

which means the object in box 1 goes into box 5, 2 into 4, 3 into 1, 4 into 2, 5 into 3, and 6 into 6. The first way of writing this permutation is straightforward. The second, shorter way, implies that within each bracket, the object in the last box in a bracket is put in the box with the first

**Table 2-A1** The multiplication table for the permutation group of order 3

| | (1)(2)(3) | (123) | (132) | (1)(23) | (2)(13) | (3)(12) |
|---|---|---|---|---|---|---|
| (1)(2)(3) | (1)(2)(3) | (123) | (132) | (1)(23) | (2)(13) | (3)(12) |
| (123) | (123) | (132) | (1)(2)(3) | (3)(12) | (1)(23) | (2)(13) |
| (132) | (132) | (1)(2)(3) | (123) | (2)(13) | (3)(12) | (1)(23) |
| (1)(23) | (1)(23) | (2)(13) | (3)(12) | (1)(2)(3) | (123) | (132) |
| (2)(13) | (2)(13) | (3)(12) | (1)(23) | (132) | (1)(2)(3) | (123) |
| (3)(12) | (3)(12) | (1)(23) | (2)(13) | (123) | (132) | (1)(2)(3) |

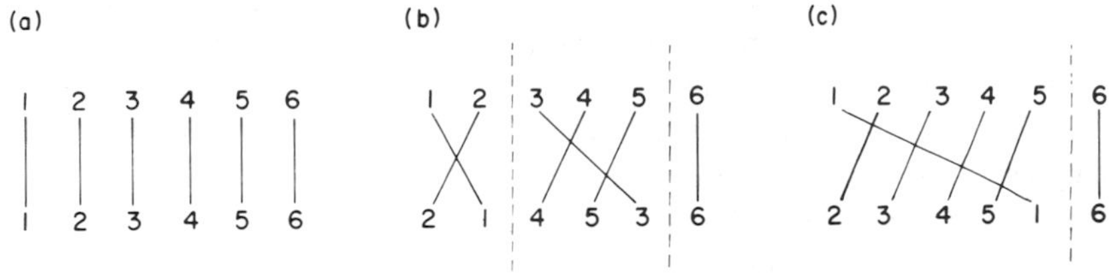


**Fig. 2-A1** A method for visualizing permuations: (a) no permutations, (b) an odd number of permutations, (c) an even number of permutations.

label within the bracket. The shorter way of writing the permutation shown in Eq. 2-A1, in terms of cycles, clearly shows which objects are interchanged among themselves. The ordering of the cycles is not important and within a bracket the ordering of the numbers can be changed as long as the cyclic order is maintained. Then A in Eq. 2-A1 can also be written as

$$A = (24)(153)(6) = (42)(315)(6),$$

etc. The product of two permutations AB = C is defined as the permutation C, which would give the same results as the permutation B applied first, followed by the permutaton A. Take A from Eq. 2-A1 and B = (164)(235); then

$$C = (153)(24)(6) \times (164)(235) = (162)(45)(3)$$

The permutations of n objects are closed under multiplication (although AB $\neq$ BA); each operation has an inverse [$B^{-1}$ = (461)(532)] obtained by writing the permutation in reverse order; there is an identity [E = (1)(2)(3)(4)(5)(6)]; associativity is obeyed [A(BC) = (AB)C].Therefore the permutations of n objects form a group. The order of the group is n!

Table 2-A1 shows the multiplication table for the group $\mathbf{P}_3$. Note that the multiplication table is the same as Table 2-2, so the two groups are isomorphic and they are both isomorphic with the point group $C_{3v}$.

As a last point concerning the permutation group, we show that two 1-dimensional irreducible representations, one antisymmetric and one symmetric, are included among the irreducible representations of the permutation group. Consider the rows of numbers in Fig. 2-A1a. Each number represents the label of a box. Every permutation within a cycle can be described by a series of interchanges of neighboring labels on the bottom row similar to the second term of Eq. 2-A1. Figure 2-A1b shows the result of the interchanges required to describe the permutation (12)(345)(6). Three interchanges were required which is necessarily the number of intersections of the lines. Figure 2-A1c shows the four interchanges required to obtain the permutation (12345)(6). Thus we can see that a permutation can be described as odd or even depending on the number of interchanges required to obtain it. A further interchange within a cycle can only add or remove one intersection. Consider a product AB of two permutations, where a and b interchanges are required to describe the permutations A and B, respectively. Then a + b interchanges are required to describe the product permutation AB. Thus, depending on the oddness or evenness of a and b, the product AB is even or odd, i.e.,

$$\begin{aligned} &\text{even} \times \text{even} = \text{even}, \qquad &&\text{even} \times \text{odd} = \text{odd} \\ &\text{odd} \times \text{odd} = \text{odd}, \qquad &&\text{odd} \times \text{odd} = \text{even} \end{aligned} \tag{2-A2}$$

so there must exist a 1-dimensional representation of the permutation group consisting of +1 for even permutations and −1 for odd permutations, which is called the antisymmetric representation. Of course there is always the symmetric representation consisting of all +1. If the order of the permutation group is greater than two, other higher-dimension irreducible representations can be found but these will not be needed in this book. For the example shown in Table 2-A1, an even number of interchanges is required to describe the identity permutations and (123) and (132), both of which are in the same class. An odd number of interchanges is required to describe the three other permutations which also are all in the same class.

## Notes

From the definition of a left inverse, $A_i^{-1}A_i = E$, we can prove that it is also a right inverse, i.e., $A_iA_i^{-1} = E$.

**Proof.** Multiply on the right by $A_i^{-1}$ so $A_i^{-1}A_iA_i^{-1} = EA_i^{-1} = A_i^{-1}$. Let B be the left inverse of $A_i^{-1}$ ($BA_i^{-1} = E$) and multiply on the left by B so $BA_i^{-1}A_iA_i^{-1} = BA_i^{-1}$ or $EA_iA_i^{-1} = E$ or $A_iA_i^{-1} = E$. QED

Wigner (Chapters 7 and 8), Cotton (Chapter 1), Tinkham (Chapter 2) cover material similar to that discussed in this chapter.

Permutation groups are discussed in many books. See also Heine in addition to those mentioned in the Appendix.

For a discussion of space groups see Burns and Glazer.

## Problems

**1.** Determine the multiplication table for the point groups: (a) $C_4$. (b) $D_2$; is it isomorphic with $C_4$? (c) $C_{2h}$, is it isomorphic with $C_4$ or $D_2$? (d) Show that the point group $C_4$ is isomorphic with the group $\{1, i, -1, -i\}$. What is the physical significance of the isomorphism?

**2.** Prove that if A is conjugate to B and A is conjugate to C, then B is conjugate to C.

**3.** Show that the classes listed for the point group $C_{3v}$ in Appendix 3 are correct. Note the physical similarity of the members of the same class. Does this physical similarity occur in the point groups $C_4$, $D_2$, $C_{2h}$, $C_{4v}$? Why?

**4.** Is there always a homomorphism between a group and its subgroups?

**5.** If the group G is homomorphic with the smaller group H, then prove that an invariant subgroup of G will be isomorphic with H.

**6.** Of the point groups $C_4$, $D_2$, and $C_{2h}$, which are Abelian? Cyclic?

**7.** For the point group O (Fig. 2-1), which subgroups are invariant?

**8.** Prove that the classes of a direct product group are given by the direct product of the classes of the original groups. (First make sure you know what the direct product of classes means.)

**9.** Prove that two rotations by the same angle, but about different axes, are in the same class if there is a third rotation that will rotate one axis into the other. Also prove that two reflections across different planes are in the same class if some other transformation in the group carries one plane into the other.

**10.** If S is a subgroup of a group G and p and q are elements of G, show that the two cosets pS and qS are identical or that $q^{-1}p$ is not in S and the two cosets have no element in common. (Thus the cosets decomposition in Section 2-4a is unique.)

**11.** Prove that a group with h = a prime number can only be a cyclic group.

**12.** Find the homomorphism between the cyclic groups of orders two and four.

**13.** Using only the multiplication table of the permutation group $\mathbf{P}_3$, find the class structure.

**14.** Do problem 7.3 given in Heine.

Chapter 3

# MATRIX REPRESENTATIONS OF FINITE GROUPS

*Among the many and varied literary and artistic studies upon which the natural talent of man is nourished, I think that those above all should be embraced and pursued with the greatest zeal which have to do with things that are very beautiful and very worthy of knowledge.*

*Copernicus, "On the Revolutions of the Celestial Spheres"*

## 3-1 Introduction

Most of the applications of symmetry to physics and chemistry are based on the methods developed in this and the next chapter. Much of the power of group theory comes from the fact that a set of square matrices can be found that behave just like the elements of the groups, that is, they are homomorphic with the group of symmetry operations. The characters (sum of the diagonal elements of the matrix) of these matrices are independent of the coordinate system. Once found, this group of square matrices, or their characters, can be used for most of the symmetry aspects of problems and they describe how the basis functions or wave functions of the molecule or crystal behave. The set of square matrices (representations) will be discussed in this chapter and their characters, in the next chapter.

## 3-2 Representations

We have already seen in Section 2-2 that a set of matrices can form a group. In that section, it was shown that the set of six 2 × 2 matrices forms a group isomorphic with the point group $C_{3v}$. This section will show that there are an infinite number of groups of six matrices that are isomorphic with $C_{3v}$ or any other group and that any of these groups of matrices are said to form a representation of the group $C_{3v}$.

### a. Matrix Representation

If there is a correspondence between a set of square matrices and the elements of a group, then the set of square matrices are said to form a **representation of the group**. That is, if the elements of the group G are E, A, B, C, ... and for each element there is a square matrix Γ(E), Γ(A), Γ(B),... such that the matrices have the same multiplication table (under matrix multiplication) as the elements,

$$\Gamma(A)\Gamma(B) = \Gamma(AB) \tag{3-1}$$

then the set of matrices form a representation of G.

If there is an isomorphism between the matrices and the group elements, then the representation is said to be **faithful**. The 2 × 2 matrices in Section 2-2 are a faithful representation of the point group $C_{3v}$. If there is only a homomorphism between the matrices and the group elements, the representation is **unfaithful**. For example, the set of six square (1 × 1) matrices 1, 1, 1, −1, −1, −1 is a representation of the group $C_{3v}$ (1 corresponds to E, $2C_3$ and −1 corresponds to $3\sigma_v$), but an unfaithful representation. The set of six matrices 1, 1, 1, 1, 1, 1, is also an unfaithful representation of the group. A set of 1 × 1 matrices composed of only +1 will always be a representation of every group.

### b. Transformation matrices as a representation

It is well to ask the origin of the six matrices in Section 2-2, and whether we can get more in the same manner. The six matrices came from a straightforward consideration of how the three Cartesian coordinates x, y, z transform among themselves under all the symmetry operations of the group.

Consider a right-handed set of coordinates x, y, z which we shall denote as $x_i$ with i = 1, 2, 3. After any orthogonal transformation, in particular a symmetry operation of the coordinate system, the coordinates are called $x_i'$. A fixed point on the fixed body can be described with respect to the $x_i$ or $x_i'$ system. The relation between the two coordinate systems is

$$x_i' = \Sigma_j \Gamma_{ij} x_j \qquad \begin{bmatrix} x' \\ y' \\ z' \end{bmatrix} = \begin{bmatrix} \Gamma_{11} & \Gamma_{12} & \Gamma_{13} \\ \Gamma_{21} & \Gamma_{22} & \Gamma_{23} \\ \Gamma_{31} & \Gamma_{32} & \Gamma_{33} \end{bmatrix} \begin{bmatrix} x \\ y \\ z \end{bmatrix} \qquad (3\text{-}2)$$

where $\Gamma_{ij}$ is a matrix.

Figure 3-1 shows an example of the effect of some of the symmetry operations of the point group $C_{3v}$. The matrices obtained for the symmetry operation E, $C_3$, and $C_3{}^2$ (not shown in Fig. 3-1) are

$$\Gamma(E) = \begin{bmatrix} 1 & 0 & 0 \\ 0 & 1 & 0 \\ 0 & 0 & 1 \end{bmatrix} \quad \Gamma(C_3) = \begin{bmatrix} -1/2 & \sqrt{3}/2 & 0 \\ -\sqrt{3}/2 & -1/2 & 0 \\ 0 & 0 & 1 \end{bmatrix} \quad \Gamma(C_3{}^2) = \begin{bmatrix} -1/2 & -\sqrt{3}/2 & 0 \\ \sqrt{3}/2 & -1/2 & 0 \\ 0 & 0 & 1 \end{bmatrix} \qquad (3\text{-}3)$$

Also, matrices for the other three symmetry operations, $3\sigma_v$, can be easily obtained. It should be noticed that while x and y transform between themselves under the symmetry operations of this group, z transforms only into itself. Thus, instead of the 3 × 3 matrices derived above, one could obtain 2 × 2 matrices by asking how x and y transform between themselves and the matrices obtained would be the six 2 × 2 matrices of Section 2-2. The first three, E, A, and B in Eq. 2-1 are in Eq. 3-3. All six matrices obey the fundamental requirement of a representation of the group, Eq. 3-1, since they are just the effect of the symmetry operations.

As a further example consider the point group $C_{4v}$. Again it is clear that z will transform into itself under all the symmetry operations of the group and thus, there is no need to deal with 3 × 3 matrices. We now consider how x and y transform between themselves. The resulting matrices form a representation of the $C_{4v}$ point group. The matrices are listed in the same order as their class groupings, i.e., E, $2C_4$, $C_2$, $2\sigma_v$, $2\sigma_d$.

$$\Gamma(E) = \begin{bmatrix} 1 & 0 \\ 0 & 1 \end{bmatrix} \quad \Gamma(C_4) = \begin{bmatrix} 0 & 1 \\ -1 & 0 \end{bmatrix} \quad \Gamma(C_4{}^3) = \begin{bmatrix} 0 & -1 \\ 1 & 0 \end{bmatrix} \quad \Gamma(C_2) = \begin{bmatrix} -1 & 0 \\ 0 & -1 \end{bmatrix}$$

$$\Gamma(\sigma_v{}^x) = \begin{bmatrix} -1 & 0 \\ 0 & 1 \end{bmatrix} \quad \Gamma(\sigma_v{}^y) = \begin{bmatrix} 1 & 0 \\ 0 & -1 \end{bmatrix} \quad \Gamma(\sigma_d{}') = \begin{bmatrix} 0 & 1 \\ 1 & 0 \end{bmatrix} \quad \Gamma(\sigma_d{}'') = \begin{bmatrix} 0 & -1 \\ -1 & 0 \end{bmatrix} \qquad (3\text{-}4)$$

It should be noted that there is nothing special about obtaining matrix representations from the three coordinates x, y, and z. Any functions that are operated on by the symmetry operators can be used. (Note that the magnitude of the radius vector $r = (x^2 + y^2 + z^2)^{1/2}$ will always

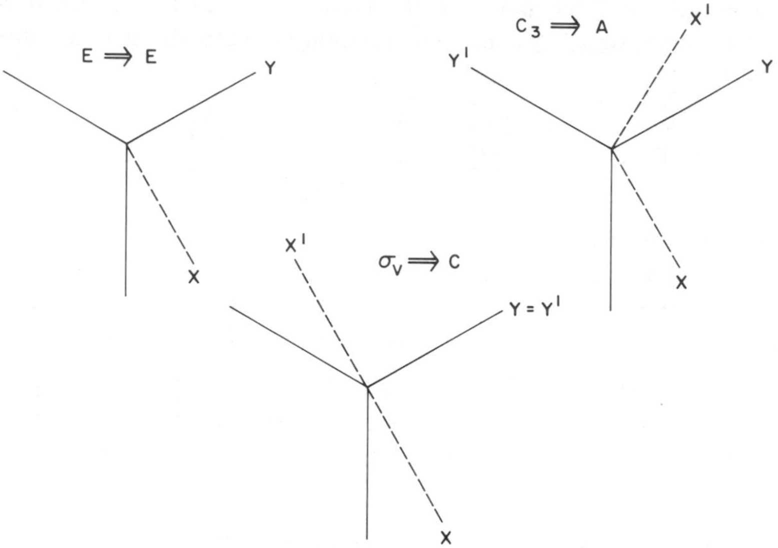


**Fig. 3-1** The effects of some symmetry operations of the point group $C_{3v}$.

transform into itself for all the point groups so that it can be ignored in any function or wave function containing it. This is a great convenience.) Consider the five products of the coordinates (d-type wave functions) $z^2$, $x^2 - y^2$, xy, xz, yz in $C_{4v}$ symmetry. The radial parts have not been included since they transform into themselves. Of these five products it is clear that $z^2$, like z, transforms into itself under all the symmetry operations of $C_{4v}$. The $1 \times 1$ matrix representation obtained from $z^2$ is the same as that obtained from z and consists of all 1's. The two functions xz and yz also transform between themselves in the same exact way as the functions x and y and the $2 \times 2$ matrix representation is al shown in Eq. 3-4. The function $x^2 - y^2$ transforms into plus or minus itself producing a new $1 \times 1$ representation. Writing the results in the same order as Eq. 3-4, one obtains 1, −1, −1, 1, 1, 1, −1, −1. Similarly, xy transforms only into itself but a new representation is obtained: 1, −1, −1, 1, −1, −1, 1, 1. (A careful look at the character table for $C_{4v}$ in the Appendix reveals these matrices. Remember, like the symmetry operations, they are grouped by classes.)

One could go on by taking f-type, g-type, etc., wave functions. However, helpful theorems exist and are associated with the so-called irreducible representations. Actually only the matrices of the **generators** need be obtained. The matrices of the other symmetry operations can always be determined by the appropriate matrix multiplication.

(Note that the set of square matrices obtained by considering how the coordinates transform are the matrix inverse of those obtained for the basis vectors transformations. See Appendix 5. As discussed in Chapter 7, basis functions transform as basis vectors. Therefore, actual representation tables conventionally will have the inverse matrices of those we determined from the coordinates and for an orthogonal basis, which we always use, the inverse equals the transpose.)

### c. Equivalent representations (similarity transformations)

Given an $l$-dimensional matrix representation of the group, then an **equivalent representation** $\Gamma'(A)$ can always be formed by

$$\Gamma'(A) = S^{-1}\Gamma(A)S \tag{3-5}$$

where S is a nonsingular $l \times l$ matrix (S is nonsingular if the determinant of $S \neq 0$. See Appendix 5.) To prove that the set of matrices $\Gamma'(A)$, $\Gamma'(B)$,... also form a representation of the group given $\Gamma(A)$, $\Gamma(B)$,... form a representation we must prove $\Gamma'(AB) = \Gamma'(A)\,\Gamma'(B)$. **Proof:**

$$\begin{aligned}\Gamma'(AB) &= S^{-1}\,\Gamma(AB)\,S = S^{-1}\,\Gamma(A)\,\Gamma(B)\,S\\ &= S^{-1}\,\Gamma(A)\,S\,S^{-1}\,\Gamma(B)\,S = \Gamma'(A)\,\Gamma'(B)\end{aligned} \tag{3-6}$$

Thus, 1-dimensional representations of a group are unique but for 2- or larger-dimensional representations, one can obtain an infinite number of equivalent representations by a similarity tranformation as in Eq. 3-5. The **dimension** of a representation consisting of $l \times l$ matrixes is defined as $l$. As will be seen later, it is most convenient to work with a representation composed of **unitary** matrices. (If the inverse of a matrix is equal to its complex conjugate transpose, then the matrix is unitary. See Appendix 5.) It can be shown that a nonsingular matrix exists that will transform any representation with nonvanishing determinants into a unitary representation. (See the Notes at the end of this chapter.)

## 3-3 Irreducible Representations

### a. General considerations

Consider a set of $l \times l$ matrices that form a representation of the group $\Gamma(R)$ where R can be any element of the group. Suppose a similarity transformation (Eq. 3-5) simultaneously transforms each $\Gamma(R)$ matrix into block form

$$\Gamma'(R) = \begin{bmatrix} \Gamma_1(R) & 0 \\ 0 & \Gamma_2(R) \end{bmatrix} \tag{3-7}$$

where $\Gamma_1(R)$ is an $l_1 \times l_1$ matrix and $\Gamma_2(R)$ is an $l_2 \times l_2$ matrix, and $l = l_1 + l_2$. Each representation $\Gamma_1(R)$ and $\Gamma_2(R)$ is said to be **irreducible representations** if neither one can be further transformed to a smaller-dimension block form. It is clear that $\Gamma_1(R)$ and $\Gamma_2(R)$ separately, are representations of the group since the matrix multiplications of the block form of Eq. 3-7 will never mix the $\Gamma_1(R)$ set of matrices with the $\Gamma_2(R)$ set. Namely,

$$\Gamma(A)\Gamma(B) = \begin{bmatrix} \Gamma_1(A) & 0 \\ 0 & \Gamma_2(A) \end{bmatrix} \begin{bmatrix} \Gamma_1(B) & 0 \\ 0 & \Gamma_2(B) \end{bmatrix}$$

$$= \begin{bmatrix} \Gamma_1(A)\Gamma_1(B) & 0 \\ 0 & \Gamma_2(A)\Gamma_2(B) \end{bmatrix} = \begin{bmatrix} \Gamma_1(AB) & 0 \\ 0 & \Gamma_2(AB) \end{bmatrix} = \Gamma(AB) \tag{3-8}$$

So $\Gamma'(R)$ in Eq. 3-7 is said to be a **reducible representation** and $\Gamma_1(R)$ and $\Gamma_2(R)$ are both said to be **irreducible representations**. Of course, the convenient block form of $\Gamma'(R)$ in Eq. 3-7 can always be obscured by a similarity transformation which will scramble $\Gamma_1$ and $\Gamma_2$. Even so, $\Gamma'(R)$ is still a reducible representation since some similarity transform exists that will put it into block form.

It is worthwhile to point out why irreducible representations are important. In Section 4-1 it will be shown that any arbitrary representation of the group can be decomposed, in a unique way, into irreducible representations of the group. Further, the number of irreducible representations of a finite group is small.

The expression in Eq. 3-7 can be written symbolically as $\Gamma'(R) = \Gamma_1(R) + \Gamma_2(R)$, but it is important to note that this is not ordinary matrix additions. ($\Gamma_1$ and $\Gamma_2$ in general do not have the same dimensions.) This denotes the result in Eq. 3-7 and in some other books, is written as $\Gamma_1(R) \oplus \Gamma_2(R)$ and called the **direct sum**.

An extremely important theorem about the character of a representation (irreducible or not) can be proved: **The character of each matrix in a representation is unaltered by a similarity transformation**. By **character**, (which is often called trace or sometimes spur) of a matrix one means the sum of diagonal elements.

$$\chi(R) = \Sigma_i \, \Gamma(R)_{ii} \tag{3-9}$$

where $\chi(R)$ is the character of the matrix $\Gamma(R)$. Before we prove this theorem, recall that if $\alpha$ is a $l \times m$ matrix and $\beta$ is a $m \times n$ matrix, they are conformable and can be multiplied together to form an $l \times n$ matrix, $\gamma = \alpha\beta$, where the ik element of $\gamma$ is given by

$$\gamma_{ik} = \Sigma_j \alpha_{ij} \beta_{jk} \tag{3-10}$$

We prove the theorem, namely, that the **characters of Γ′(R) and Γ(R) are identical** (in Eq. 3-5).

$$\begin{aligned}\chi[\Gamma'(R)] &= \chi[S^{-1}\Gamma(R)S] = \Sigma_i [S^{-1}\Gamma(R)S]_{ii} = \Sigma_{ijk} (S^{-1})_{ij} \Gamma(R)_{jk}(S)_{ki} \\ &= \Sigma_{ijk} (S)_{ki}(S^{-1})_{ij} \Gamma(R)_{jk} = \Sigma_{jk} (SS^{-1})_{kj} \Gamma(R)_{jk} \\ &= \Sigma_{jk} \delta_{kj} \Gamma(R)_{jk} = \Sigma_k \Gamma(R)_{kk} = \chi[\Gamma(R)]\end{aligned} \tag{3-11}$$

It is now clear why in Chapter 2 the concept of class was introduced. **Character of the matrices that represent symmetry operations in the same class are always identical.** This is the reason why in character tables, where only the characters of irreducible representations are listed, the symmetry operations are grouped together by class.

### b. Great orthogonality theorem (and other theorems)

Before discussing the very useful orthogonality theorem for irreducible representations, we discuss several other important theorems. (See the Notes at the end of this chapter for references to the proofs.) These theorems are needed to prove the orthogonality theorem and are used on other occasions.

**Theorem 1.** A nonsingular matrix exists that will transform any matrix representation into a unitary matrix representation. (This theorem has already been stated at the end of Section 3-2.)

**Theorem 2. (Schur's lemma)** If a matrix M exists such that $\Gamma(A_i)M = M\Gamma(A_i)$ for all i, then (a) if $\Gamma(A_i)$ is irreducible, M is a constant matrix (a constant times the unit matrix); (b) if M is not a constant matrix, then $\Gamma(A_i)$ is reducible.

**Theorem 3.** Given two irreducible representations of a group $\Gamma(A_i)$ and $\Gamma(B_i)$ of dimension p and q, respectively, and a matrix M of dimension p × q such that $M\Gamma(A_i) = \Gamma(B_i)M$ for all i, then (a) M = 0 if p ≠ q; (b) M = 0 or the determinant ≠ 0 if p = q. If the determinant ≠ 0, M has an inverse and the two representations are equivalent.

**Theorem 4.** This extremely important theorem for nonequivalent irreducible unitary representations describes the orthogonality among the different representations and the rows and columns within the matrices. This is the **Great Orthogonality Theorem** (GOT). If $\Gamma_i$ and $\Gamma_j$ are two nonequivalent irreducible unitary representations, then

$$\Sigma_R \Gamma_i(R)^*_{mn} \Gamma_j(R)_{op} = (h/l_i) \delta_{ij} \delta_{mo} \delta_{np} \tag{3-12}$$

The sum is over all the symmetry operations (not the classes but each operation), the * is the complex conjugate, h is the order of the group, and $l$ is the dimension of the matrix of the representation. By definition $\delta_{ij}$ is zero if $i \neq j$, or one if $i = j$. For the proof of this and the other theorems consult the Notes and see the Problems at the end of this chapter. We use the theorem by applying it to several examples.

### c. Examples of the great orthogonality theory

We already know three representations for the point group $C_{3v}$. Table 3-1 lists each representation according to the symmetry operation that the matrix represents. Each representation is denoted by $\Gamma_1, \Gamma_2, \Gamma_3$. On the right, functions that could be used to generate the representations are listed. Since all 1-dimensional representations are irreducible, $\Gamma_1$ and $\Gamma_2$ are irreducible. It is not obvious as yet but $\Gamma_3$ is also irreducible. First we check that the characters of the representation of symmetry operations in the same class are always identical in each irreducible representation. Recall that for the point group $C_{3v}$ the classes are E, $2C_3$, $3\sigma_v$, thus we obtain a shortened version of Table 3-1 by writing the characters instead of the complete representations:

| $C_{3v}$ | E | $2C_3$ | $3\sigma_v$ |
|---|---|---|---|
| $\Gamma_1$ | 1 | 1 | 1 |
| $\Gamma_2$ | 1 | 1 | −1 |
| $\Gamma_3$ | 2 | −1 | 0 |

(3-13)

Next we apply the GOT to the representations. Consider $\Gamma_i = \Gamma_2 = \Gamma_j$. Since $\Gamma_2$ consists of $1 \times 1$ matrices, the subscripts m, n, o, p are all 1 and can be ignored. For this group, $h = 6$ and for this representation, $l = 1$ and therefore, the right side of GOT is 6. On the left side when we sum R over, the six symmetry operations $1 + 1 + 1 + 1 + 1 + 1 = 6$ is obtained as hoped for. Obviously, the same would apply if $\Gamma_i = \Gamma_1 = \Gamma_j$. However, if $\Gamma_i = \Gamma_1$ and $\Gamma_j = \Gamma_2$, the right side of GOT is zero because $\delta_{12} = 0$. The left side gives +1 for each of the first three symmetry operations and −1 for the last three, therefore zero is obtained! Applying GOT to $\Gamma_3$ where $l = 2$ taking m, o, = 1 and n, p = 2, one obtains $0 + 3/4 + 3/4 + 0 + 3/4 + 3/4 = 3$ for the sum over R where the right side also gives $6/2 = 3$. Taking $\Gamma_i = \Gamma_3(R)_{12}$ and $\Gamma_j = \Gamma_3(R)_{21}$, one obtains $0 - 3/4 - 3/4 + 0 + 3/4 + 3/4 = 0$ as it should because $\delta_{12} = 0$. In this manner one can continue to use the GOT on all the elements (see also Problem 4).

**Table 3-1** Three representations for the point group $C_{3v}$. (Actually these are the only three irreducible representations for this point group.) (These are the conventional representations, i.e. see the note at the end of Section 3-2b.)

| $C_{3v}$ | E | $C_3$ | $C_3^2$ | $\sigma_v'$ | $\sigma_v''$ | $\sigma_v'''$ | |
|---|---|---|---|---|---|---|---|
| $\Gamma_1$ | 1 | 1 | 1 | 1 | 1 | 1 | Z OR $Z^2$ OR $X^2+Y^2$ |
| $\Gamma_2$ | 1 | 1 | 1 | -1 | -1 | -1 | |
| $\Gamma_3$ | $\begin{bmatrix}1 & 0\\ 0 & 1\end{bmatrix}$ | $\frac{1}{2}\begin{bmatrix}-1 & -\sqrt{3}\\ \sqrt{3} & -1\end{bmatrix}$ | $\frac{1}{2}\begin{bmatrix}-1 & \sqrt{3}\\ -\sqrt{3} & -1\end{bmatrix}$ | $\begin{bmatrix}-1 & 0\\ 0 & 1\end{bmatrix}$ | $\frac{1}{2}\begin{bmatrix}1 & -\sqrt{3}\\ -\sqrt{3} & -1\end{bmatrix}$ | $\frac{1}{2}\begin{bmatrix}1 & \sqrt{3}\\ \sqrt{3} & -1\end{bmatrix}$ | (X,Y) OR (XZ,YZ) |

Thus we see that the irreducible representations are not only orthogonal to each other but the various components of the matrices are also orthogonal to each other in the sense of Eq. 3-12.

## 3-4 Representations of a Factor Group

For the crystallographic point group $C_{3v}$, Eq. 3-13 or Table 3-1 shows that an invariant subgroup is $S = E, C_3, C_3^2$. S and its coset Q ($= \sigma_v S = \sigma'_v, \sigma_v'', \sigma_v'''$ for any $\sigma_v$) form a factor group $C_{3v}$ with respect to the invariant subgroup S. This factor group, of order 2, has two irreducible representations $\Gamma_1 = 1, 1$, and $\Gamma_2 = 1, -1$. With the correct correspondence, each irreducible representation of a factor group of a larger group G must always be an irreducible representation of the group G. For example, consider a function or functions that transform as one of the irreducible representations of the factor group in the sense of Eq. 3-2. Under all the symmetry operations in S, the transformation behavior of the function or functions is determined. Similarly, the transformation behavior for the function or functions is determined for each of the cosets, Q in this problem. Thus, when the full group is considered the transformation properties of the same function or functions are determined completely. Each square matrix in the irreducible representation of the factor group corresponds to every group element in the invariant subgroup or its appropriate coset. For $\Gamma_2$ of the factor group, 1 corresponds to E, $C_3$, $C_3^2$ (i.e., S), and $-1$ corresponds to Q. Then the corresponding irreducible representation of $C_{3v}$ is 1, 1, 1, $-1$, $-1$, $-1$ which in Table 3-1 is labeled $\Gamma_2$. The other irreducible representation $\Gamma_1$ of the factor group obviously corresponds to $\Gamma_1$ of $C_{3v}$. It is clear that $\Gamma_3$ cannot be obtained this way (see also Problem 10).

## Appendix to Chapter 3

Figure 3-A1 shows three mutually perpendicular unit vectors $\mathbf{a}_i$ and a point in space that can be described in terms of a position vector

$$\mathbf{r} = x_1\mathbf{a}_1 + x_2\mathbf{a}_2 + x_3\mathbf{a}_3$$

Thus, the coordinates of this point are ($x_1$, $x_2$, $x_3$) which are also referred to as the components of **r**. Consider the point to be held fixed and the unit vectors rotated in a right-hand screw sense (counterclockwise) about the $\mathbf{a}_3$ axis by an amount Δ, and this fixed point is described as **r′** and has coordinates ($x_1'$, $x_2'$, $x_3'$). Using the figure, we can describe the primed coordinates in terms of the unprimed ones as follows:

$$\begin{aligned} x_1' &= p\cos(\phi\text{-}\Delta) \\ &= p(\cos\phi\cos\Delta + \sin\phi\sin\Delta) \\ &= x_1\cos\Delta + x_2\sin\Delta \end{aligned}$$

$$\begin{aligned} x_2' &= p\sin(\phi\text{-}\Delta) \\ &= p(\sin\phi\cos\Delta - \cos\phi\sin\Delta) \\ &= -x_1\sin\Delta + x_2\cos\Delta \end{aligned}$$

or

$$\begin{bmatrix} x_1' \\ x_2' \\ x_3' \end{bmatrix} = \begin{bmatrix} \cos\Delta & \sin\Delta & 0 \\ -\sin\Delta & \cos\Delta & 0 \\ 0 & 0 & 1 \end{bmatrix} \begin{bmatrix} x_1 \\ x_2 \\ x_3 \end{bmatrix}$$

or

$$x_i' = \Sigma_j\, \Gamma_{ij}\, x_j \tag{3-A1}$$

where $\cos\phi = x_1/p$, $\sin\phi = x_2/p$, and Γ is defined from the last two equalities. Clearly the same results would be obtained if the unit vectors were considered to be held fixed and the point is rotated clockwise by an amount Δ. We will see in Chapters 4 and Appendix 5 that for functions of the coordinates $f_i$ we must have

$$f_i' = \Sigma_j\, f_j\, \Gamma_{ji} \tag{3-A2}$$

This same ordering of the subscripts, as in Eq. 3-A2, applies to the primed and unprimed unit vectors as shown in Appendix 5 while the ordering in Eq. 3-A1 applies the coordinates.

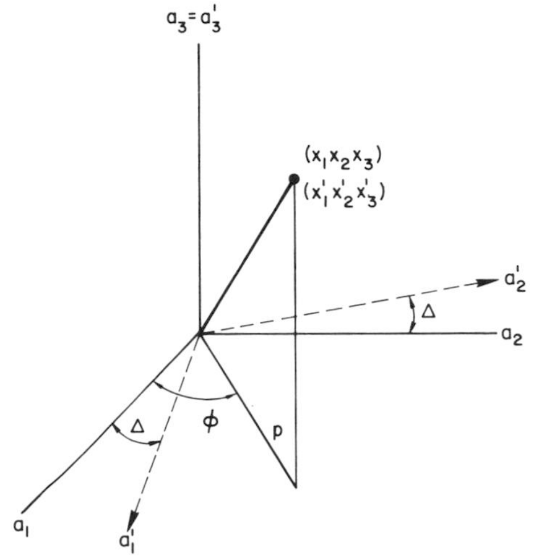


**Fig. 3A-1** A point in space before and after a rotation of the coordinate.

## Notes

The theorem stated at the end of Section 3-2 as well as theorems stated in Section 3-3b are proved in many books. See Wigner [Chapter 9], Falicov [pp. 221–236], or other books.

Bishop [Chapter 5] has a useful discussion about obtaining representations from coordinates and basis vectors as does Chisholm, Chapter 2.

The **regular representation** of a group is a matrix representation that is useful on certain occasions. To determine it, one writes the multiplication table rearranging the rows so that they correspond to the inverses of the columns. (See below.) Thus, E appears along the principal diagonal of the table. For a general group element R, the h × h matrix Γ(R) of the regular representation is obtained by taking the h × h multiplication table and replacing R by 1 and all the other group operations by 0. Then consider the resulting table as a matrix. This same procedure is done for each symmetry operation of the group. From the rearrangment theorem, each row and column consists of all zeros except for one entry which is a 1. The resulting h different matrices is called the regular representation of the group. It can be shown that this procedure, indeed, yields a representation of the group (Eq. 3-1), and that it contains each irreducible representation a number of times equal to the dimensionality of the irreducible

Chapter 4

# CHARACTERS OF MATRIX REPRESENTATIONS OF FINITE GROUPS

*In the series of things those which follow are always aptly fitted to those which have gone before;..., so the things which come into existence exhibit no mere succession, but a certain wonderful relationship.*

*Marcus Aurelius, "To Himself"*

In this chapter we continue the discussion of matrix representations in general and use examples from the 32 point groups. However, emphasis is on the characters of the matrix representations.

## 4-1 Properties of Characters of Irreducible Representations

A number of very important theorems concerning the properties of characters can be proved by using the GOT or considerations that follow from it. These theorems will be presented and discussed with examples. See the Notes of Chapter 3 for references to complete proofs.

(a) It has already been shown that **the characters of matrices in the same class are identical**. The only statement that must be added to Eq. 3-11 to complete the proof is that S is a matrix representation of any element of the group.

(b) **When summed over all the symmetry operations R, the character system of irreducible representations is orthogonal and normalized to the order of the group h.** Namely

$$\Sigma_R\ \chi_i(R)^*\ \chi_j(R) = h\ \delta_{ij} \tag{4-1}$$

This result follows from the GOT

$$\Sigma_R \Gamma_i(R)^*{}_{mm} \Gamma_j(R)_{pp} = (h/l_i)\, \delta_{ij}\, \delta_{mp}\, \delta_{mp}$$

$$\Sigma_{m,p,R} \Gamma_i(R)^*{}_{mm} \Gamma_j(R)_{pp} = (h/l_i)\, \delta_{ij}\, \Sigma_{m,p}\, \delta_{mp}\, \delta_{mp}$$

$$\Sigma_R \chi_i(R)^* \chi_j(R) = (h/l_i)\, \delta_{ij}\, l_i = h\, \delta_{ij}$$

For i = j, Eq. 4-1 is a necessary and sufficient condition for a representation to be irreducible. Thus, the use of Eq. 4-1 is a very simple test of irreducibility. As can be seen using the characters in Table 3-1 or Eq. 3-13, the three representations are indeed irreducible representations. For $\Gamma_1$: 6(1) = 6 is obtained for Eq. 4-1; for $\Gamma_2$: 6(1) = 6; and for $\Gamma_3$: $1(2)^2 + 2(1)^2 + 3(0) = 6$. In the same manner it can be seen that the sum for characters of the different irreducible representations yields zero as required by Eq. 4-1. (Always remember the sum is over all the symmetry operations, not just over the classes.) Equation 4-1 can be written in terms of characters of each class. Let $\chi_i(C_k)$ be the character of the ith irreducible representation of the kth class, which contains $N_k$ elements. Then Eq. 4-1 can be written as

$$\Sigma_k \chi_i(C_k)^* \chi_j(C_k) N_k = h\, \delta_{ij} \tag{4-1a}$$

(c) **A necessary and sufficient condition for two representations to be equivalent is that the characters of the representations must be equal.** The necessary condition for equivalent representations, Eq. 3-5, follows from Eq. 3-11. One important use would be in Problem 3-5 where two different two-dimensional representations $\Gamma'(R)$ and $\Gamma(R)$ are known, but finding a matrix S is not simple. The use of this theorem shows that the matrices are equivalent, making it unnecessary to find the similarity transformation. Another example of this theorem is its application to the representations in Table 3-1. As can be seen, the character systems of the three representations are not equal therefore the representations are not equivalent. From Eq. 4-1, we also see that **two inequivalent irreducible representations cannot have the same characters** and that **irreducible representations with equal characters are equivalent.**

(d) **If $l_i$ is the dimension of the ith irreducible representation of the group, then**

$$\Sigma_i\, l_i^2 = h \tag{4-2}$$

where the sum is over all the irreducible representations. Applying Eq. 4-2 to $C_{3v}$ irreducible representations in Table 3-1 or Eq. 3-13, where there are two 1-dimensional and one 2-dimensional representations, we obtain $(1)^2 + (1)^2 + (2)^2 = 6$. Since h = 6, it is clear that there are no more irreducible representations for $C_{3v}$. By realizing that $l_i = \chi_i(E)$, and substituting into Eq. 4-2, the theorem can be rewritten as

$$\Sigma_i [\chi_i(E)]^2 = h \tag{4-2}$$

Equation 4-2 can be proved from another orthogonality theorem of the characters of irreducible representations which differs from Eq. 4-1, namely

$$\Sigma_i \chi_i(C_m)^* \chi_j(C_n) N_m = h \delta_{mn} \tag{4-2a}$$

Note that here the sum is over all the irreducible representations, unlike Eq. 4-1a where the sum is over all the classes. To prove Eq. 4-2a form the matrices, $M_{im} = \chi_i(C_m)^*$ and $M_{nj}' = (N_n/h)\chi_j(C_n)$. Equation 4-1a shows

$$\Sigma_m (N_m/h)\chi_i(C_m)^* \chi_j(C_m) = \delta_{ij} = \Sigma_m M_{im} M_{mj}' = (MM')_{ij}$$

Thus, $MM' = I$, so $M' = M^{-1}$ and $M'M = I$ which is

$$(M'M)_{ij} = \delta_{ij} = \Sigma_m M'_{im} M_{mj} = \Sigma(N_i/h)\chi_m(C_i)(\chi_m(C_j)$$

which proves Eq. 4-2a. Actually, one must show that M and M′ are nonsingular (see the Notes).

(e) **The number of inequivalent irreducible representations equals the number of classes of the group.** This is a very useful theorem and can be proved by noting that the matrix M formed in proving Eq. 4-2a must be square (which can be seen by considering Eqs. 4-2a and 4-1a). For the point group $C_{3v}$, since there are only three classes, there can be only three irreducible representations. Since it is easy to compute the classes for any group, the number of irreducible representations can be quickly determined.

For any group that is Abelian (or cyclic), each symmetry operation is in a class by itself. Since the number of irreducible representations is equal to the order of the group, these groups have only 1-dimensional irreducible representations.

### a. Characters of some point groups

Using some of the theorems developed in this section, the characters for all the irreducible representations of the point groups can be easily obtained.

(i) All groups with two symmetry operations must have two classes, since E is in a class by itself. The trivial, but very important, 1-dimensional irreducible representation always consists entirely of 1's. For h = 2, it consists of 1, 1. In order to satisfy Eq. 4-2, the other irreducible representation can only be 1-dimensional and must be 1, −1. The only crystallographic point groups with h = 2 are $C_{1h}(E, \sigma_h)$, $S_2(E, i)$, and $C_2(E, C_2)$. Thus they all have the same characters and representations even though they contain different symmetry operations. These groups are isomorphic.

Figure 4-1a shows a character table for these three point groups. The two irreducible representations are arbitrarily labeled $\Gamma_1$ and $\Gamma_2$. For each point group, a few of the functions that transform according to each irreducible representation are listed for convenience. Consider the point group $C_2$. The function z transforms into itself under all the symmetry operations. However, x transforms into −x under the symmetry operation $C_2$, as does y, and each of the functions xz, yz, $x^3$, $y^3$, $zx^2y$, etc. These latter functions are not shown in the column under the point group $C_2$. Nevertheless these functions, along with x and y as shown, indeed transform as the $\Gamma_2$ irreducible representation of the point group $C_2$; in fact, there are an infinite number of functions that transform as the $\Gamma_2$ irreducible representation. There are also an infinite number of functions that transform as the $\Gamma_1$ irreducible representation of the point group $C_2$, z, $z^2$, $z^3$,..., xy, $x^2y^2$,...,xyz, $x^2y^2z$,..., etc.

For the point group $S_2$ in Figure 4-1a, only a few of the functions that transform as the $\Gamma_1$ and $\Gamma_2$ irreducible representations are shown. Other functions of the infinite number of functions, that transform as the two irreducible representations can be found. For the point group $C_{1h}$ only a few functions are listed, while others can be easily found.

(ii) The noncyclic crystallographic point groups of the order four ($D_2$, $C_{2v}$, $C_{2h}$) have four classes, and determining the four irreducible representations is easy. $\Gamma_1$ = 1, 1, 1, 1, as always. Each of the other irreducible representations must have $\chi(E) = 1$ and the other three characters must be 1, −1, −1. The arrangement of the characters must be consistent with Eq. 4-1. Figure 4-1b shows the four irreducible representations for the isomorphic point groups along with a few of the infinite number of functions that could have been used to generate the representation; also, the four representations are given arbitrary labels $\Gamma_1$ to $\Gamma_4$.

The products of functions transform under the symmetry operations just as one would expect. For example, in $C_{2h}$, z and y transform as the $\Gamma_3$ and $\Gamma_4$ irreducible representations. That is, under the symmetry operation $C_2$, z goes into +z, for i and $\sigma_h$, z → −z while y → y for $\sigma_h$, but y → −y for $C_2$ and i. This is all clearly expressed in the character table which for 1-dimensional representations is also a "representation table." Now consider the function yz; under any symmetry operation it transforms into ± yz depending on what y and z individually do. Thus, yz transforms as $\Gamma_2$. This is immediately obvious by multiplying the charac-

(a)

| | $C_{1h}$ | | | E | $\sigma_h$ |
|---|---|---|---|---|---|
| | | $S_2$ | | E | i |
| | | | $C_2$ | E | $C_2$ |
| $\Gamma_1$ | x,y | $z^2$, etc., xy, etc. | z | 1 | 1 |
| $\Gamma_2$ | z | x,y,z | x,y | 1 | $-1$ |

(b)

| | $D_2$ | | | E | $C_2^z$ | $C_2^y$ | $C_2^x$ |
|---|---|---|---|---|---|---|---|
| | | $C_{2v}$ | | E | $C_2$ | $\sigma_v$ | $\sigma_v$ |
| | | | $C_{2h}$ | E | $C_2$ | i | $\sigma_h$ |
| $\Gamma_1$ | $x^2,y^2,z^2$ | $x^2,y^2,z^2$ | $x^2,y^2,z^2,xy$ | 1 | 1 | 1 | 1 |
| $\Gamma_2$ | $y,xz,x^2y$ | x,xz | yz,xz | 1 | $-1$ | 1 | $-1$ |
| $\Gamma_3$ | $z,xy,y^2z$ | xy,xyz | $z,y^2z$ | 1 | 1 | $-1$ | $-1$ |
| $\Gamma_4$ | $x,yz,xy^2$ | y,yz | y,x | 1 | $-1$ | $-1$ | 1 |

**Fig. 4-1** (a) Character table for the three point groups $C_2$, $S_2$, and $C_{1h}$. (Clearly these point groups are of order 2 and are isomorphic.) (b) Character table for the point groups $D_2$, $C_{2v}$, $C_{2h}$.

ters of $\Gamma_3$ with $\Gamma_4$ and obtaining $\Gamma_2$. In fact, the product of any one of the infinite number of functions that transform as $\Gamma_4$ multiplied by a function that transforms as $\Gamma_3$ will transform as $\Gamma_2$, ($\Gamma_2 = \Gamma_3 \times \Gamma_4$). It is simple to determine the other products and even easier to determine the squares of any irreducible representation ($\Gamma_i \times \Gamma_i = \Gamma_1$ if $\Gamma_i$ is real). Hence, for 1-dimensional representations the characters express how the function transforms into itself.

(iii) For a point group such as $C_{3v}$ the character table can be easily obtained. With reference to Eq. 3-13, $\Gamma_1$ is immediately obtained, as it will always be. Since there are only three classes, Eq. 4-2 yields $l_1^2 + l_2^2 + l_3^2 = 6$ which can only be satisfied by representations of dimensions 1, 1, and 2. The only other 1-dimensional representation orthogonal (Eq. 4-1) to $\Gamma_1$ is $\Gamma_2$. To obtain $\Gamma_3$ we remember $\chi_3(E) = 2$ so the characters are, using a and b as unknowns for the characters of the $2C_3$ and $3\sigma_v$ classes: 2, a, a, b, b, b. Using the orthogonality with $\Gamma_1$ and $\Gamma_2$, one obtains from Eq. 4-1

$$2 + 2a + 3b = 0; \qquad 2 + 2a - 3b = 0$$

Therefore, a = $-1$ and b = 0 as shown in Eq. 3-13.

This procedure can be applied to the other point groups. (See the Notes at the end of the chapter for reference.)

### b. Cyclic groups

For a cyclic group, each symmetry element is in a class by itself therefore, there are as many 1-dimensional irreducible representations as there are group elements (see Section 2-6b). If A is the generating element and h is the order of the group, $A^h = E$. The representation and character (they are the same for 1-dimensional representations) are given by the roots of unity.

$$A = E^{1/h}, \qquad \chi(A) = \{\exp(2\pi im)\}^{1/h}, \qquad m = 1, 2, \ldots, h \quad (4\text{-}3)$$

for the h-irreducible representations for the symmetry operation A. Fig. 4-2a shows the construction of a character table for a cyclic group. For the symmetry operation A the characters are $\varepsilon$ ($= e^{2\pi i/h}$), $\varepsilon^2$, $\varepsilon^3$,... until m = h in Eq. 4-3, which corresponds to $\varepsilon^h = 1$ as can be seen in the table. For other operations

$$A^p = E^{p/h}, \qquad \chi(A^p) = \exp(2\pi imp/h), \qquad m = 1, \ldots, h \qquad (4\text{-}4)$$

For example, the characters of the symmetry operation $A^2$ are $\varepsilon^2$, $\varepsilon^4$,..., $\varepsilon^{2h} = 1$. This can be seen in Fig. 4-2a. The rest of the character table follows in the same manner until the last column which is the symmetry operation $A^h = E$. For this symmetry operation the characters are $\varepsilon^h$, $\varepsilon^{2h}$, etc., all of which are equal to one. Fig. 4-2b shows the identical character table but arranged in the more conventional manner. We discuss a simple example and then show why these 1-dimensional irreducible representations often are grouped in pairs.

For the cyclic point group $C_3$, Fig. 4-2c has the three irreducible representations. Note that the application of Eq. 4-1 for i = j = 2 gives $1^2 + \varepsilon * \varepsilon + (\varepsilon^2) * \varepsilon^2 = 3$ as expected; while with i = 1 and j = 2, we obtain $1 + \varepsilon + \varepsilon^2 = 0$; for i = 2 and j = 3, we obtain $1 + \varepsilon * \varepsilon^2 + (\varepsilon^2) * \varepsilon = 1 + \varepsilon + \varepsilon^2 = 0$. Thus the three irreducible representations (equal to the number of classes = h) behave properly with respect to orthogonality. However, each of the corresponding characters of $\Gamma_2$ and $\Gamma_3$ are complex conjugates of each other; hence the complex conjugate of a function that transforms as $\Gamma_2$, transforms as $\Gamma_3$. In Chapter 7, it will be shown that wave functions corresponding to different energy states transform as different irreducible representations. Also, as is well known, the wave function and its complex conjugate have the same energy in the absence of magnetic fields due to time reversal symmetry. Thus, when character tables are compiled, 1-dimensional representations in cyclic groups that

(a)

| $C_h$ | A | $A^2$ | ... | $A^h=E$ |
|---|---|---|---|---|
| | $\varepsilon$ | $\varepsilon^2$ | ... | $\varepsilon^h=1$ |
| | $\varepsilon^2$ | $\varepsilon^4$ | ... | $\varepsilon^{2h}=1$ |
| | $\varepsilon^3$ | $\varepsilon^6$ | ... | |
| | . | . | | . |
| | . | . | | . |
| | . | . | | . |
| | $\varepsilon^h=1$ | $\varepsilon^{2h}=1$ | | 1 |

(b)

| $C_h$ | E | A | $A^2$ | |
|---|---|---|---|---|
| $\Gamma_1$ | 1 | 1 | 1 | ... |
| $\Gamma_2$ | 1 | $\varepsilon$ | $\varepsilon^2$ | ... |
| $\Gamma_3$ | 1 | $\varepsilon^2$ | $\varepsilon^4$ | |
| . | . | . | . | |
| . | . | . | . | |
| . | . | . | . | |
| $\Gamma_h$ | 1 | $\varepsilon^{2h-2}$ | | ... |

(c)

| | $C_3$ | E | $C_3$ | $C_3^2$ | (where $\varepsilon=e^{2\pi i/3}$) |
|---|---|---|---|---|---|
| A | $\Gamma_1$ | 1 | 1 | 1 | $z^2, x^2+y^2$ |
| E | $\Gamma_2$ | 1 | $\varepsilon$ | $\varepsilon^2(=\varepsilon^*)$ | $(x \pm iy)$ |
| | $\Gamma_3$ | 1 | $\varepsilon^2(=\varepsilon^*)$ | $\varepsilon^4(=\varepsilon)$ | |

**Figure 4-2** (a) Character table for a cyclic group of order h. (b) Same character table but arranged in conventional order. (c) Character table of the cyclic group of order 3.

are the complex conjugate of each other, such as $\Gamma_2$ and $\Gamma_3$, are joined by a bracket to indicate that quantum mechanically they can be thought of as one 2-dimensional representation. However, this is just a convenience for quantum mechanical applications. From the mathematical point of view all the irreducible representations are 1-dimensional in agreement with the theorems in Section 4-1.

### c. Two-dimensional rotational groups $C_{\infty v}$ and $D_{\infty v}$

These groups, discussed in Section 1-5b, can be treated easily by considerations similar to those used for cyclic groups in the previous section.

The **group consisting of rotations about a fixed (z) axis by an amount $\phi$, $C(\phi)$, is called $C_\infty$**, which is an obvious extension of the point group $C_n$ notation. The group is Abelian since for any two rotations $C(\phi_1 + \phi_2) = C(\phi_1)C(\phi_2)$. Since the group is Abelian, all the irreducible representations are 1-dimensional. The rotation is continuous so the nth irreducible representation must be of the form $\Gamma_n(\phi) = \exp(in\phi)$. If we want the representation to be single valued, then $\Gamma(0) = \Gamma(2\pi)$ so that n can only be an integer. This is different from considerations in the last section where Eq. 4-3 has nonintegers in the exponent. The characters of $\Gamma_n$ and

$\Gamma_{-n}$ are complex conjugates of each other, so for quantum mechanical purposes, they are combined like the cyclic point groups in the previous section.

Now consider the $C_{\infty v}$ group. Here we note that the symmetry operation $\sigma_v$ changes the sign of $\phi$, hence $\sigma_v \exp(\pm in\phi) = \exp(\mp in\phi)$. For $n \neq 0$, the $\Gamma_n(\phi)$ and $\Gamma_n(-\phi)$ of $C_\infty$ must be combined to for a 2-dimensional irreducible representation. The representations of these 2-dimensional irreducible representation are

$$\Gamma_n[C(\phi)] = \begin{bmatrix} e^{-in\phi} & 0 \\ 0 & e^{-in\phi} \end{bmatrix} \qquad \Gamma_n[\sigma_v] = \begin{bmatrix} 0 & 1 \\ 1 & 0 \end{bmatrix} \tag{4-5}$$

The character table for $C_{\infty v}$ can be found in Appendix 3. $C(\phi)$ and $C(-\phi)$ are in the same class and the character is $e^{in\phi} + e^{-in\phi} = 2\cos n\phi$; all the $\sigma_v$ planes are also in the same class. The notation for the irreducible representations comes from the quantum mechanics of diatomic molecules where $\Sigma$-states have $n = 0$, $\pi$ states have $n = 1$, $\Delta$ have $n = 2$, $\Phi$ have $n = 3$, etc.

For the $D_{\infty h}$ group, $\sigma_h$ is a symmetry operation. The character table is also in Appendix 3 and has nothing new except that it has twice as many symmetry operations and irreducible representations with g and u subscripts as discussed in the next section.

## 4-2 Character Tables

Unfortunately, there are far too many different notations used in group theory. This has been a hindrance, often to the extent of causing the beginner to drop the entire study. In this book we continue the approach of ignoring the entire problem and just use the "best" notation. At a later time some of the other notations will be mentioned and the correspondence will be discussed. The object of putting off the discussion of the various notations is to learn as much about the subject of group theory before pointing out all the different notations. Hopefully this will be the least confusing approach. (If you really insist on looking at other notations see Chapter 11.)

We have been using the Schoenflies notation for the symmetry operations and the point groups. As you have noticed, in this chapter the irreducible representations have been labeled $\Gamma_1, \ldots, \Gamma_n$. This is Bethe's notation. It has the advantage of being simple and the disadvantage of being arbitrary. $\Gamma_1$ is always the representation containing all 1's, the **totally symmetric representation**. We shall use two notations for irreducible representations: Bethe's and Mulliken's. Bethe's will be used because its arbitrariness makes it easy to employ in equations. Mulliken's is very simple yet contains at a glance a considerable amount of information about the representation. These two notations are the most widely used. In Mulliken's notation the symbols A and B refer to 1-dimensional, E to 2-dimensional, and T to 3-dimensional irreducible representations. The explicit meaning ofthe subscripts 1 and 2 or superscript primes can be ignored. (The meaning is explained in Appendix 3.) If the crystal class has a center of inversion (i is a symmetry operation), then the functions that transform as the irreducible representations either transform into plus or minus themselves under inversion. The irreducible representations then have a subscript g (gerade) for even or plus under inversion, or the subscript u (ungerade) for odd or minus under inversion. Thus, the totally symmetric representation is either labeled A, $A_1$, or $A_{1g}$ depending on the point group.

We can restate the behavior of functions when a center of inversion is a symmetry operation in terms of **parity**. Even parity functions transform into themselves under inversion and thus transform as g-irreducible representations. Odd parity functions transform into minus themselves under inversion and thus transform as u-irreducible representations. Or one can say that parity is the eigenvalue of the inversion operation with eigenvalue $\pm 1$.

Appendix 3 contains the character tables for the 32 crystallographic point groups as well as other point groups. In the upper left, the symbol for the point group is shown. Across the top, the symmetry elements are listed and grouped by class. Down the left-hand side are listed, in Mulliken notation, the symbols for the irreducible representations. For each irreducible representation, the character for each class is listed. Down the right side of the table are listed some of the simple functions that transform as the different irreducible representations. These functions are x, y, z, the products of these two at a time, and the three components of an **axial vector** $R = \mathbf{A} \times \mathbf{B}$, i.e., $R_z = A_1B_2 - A_2B_1$, etc. For example, angular momentum transforms as an axial vector. For the 1-dimensional irreducible representation, various functions that transform as these representations are listed and separated by commas. For the E-irreducible representations, two functions transform between themselves under the symmetry operations. These two functions are enclosed in a bracket and separated by a comma. The three functions required for a T-irreducible representation are also enclosed in a bracket and separated by commas.

A surprisingly large amount of information can be extracted from the character tables. At the end of this chapter some simple but certainly nontrivial examples will be given. Actually, most of the remainder of the book is devoted to the use of these tables and the concepts developed in this chapter.

## 4-3 Reduction of a Reducible Representation

One will often have a matrix representation for a particular point group that might be reducible. The representation might obviously be reducible because its dimension is larger than the largest allowable for the particular point group. The maximum dimension of an irreducible representation for cubic crystallographic point groups is 3, for **axial point groups** (groups in which x and y transform between themselves) it is 2, and lower symmetry point groups have only 1-dimensional representations.

If the set of characters, symmetry operation by operation, is identical to one of the irreducible representations of the point group, then the representation in question is irreducible as in Section 4-1a. If not, we proceed to show how one can obtain the number of times a given reducible representation contains a particular irreducible representation. If the character of the reducible representation for each symmetry operation R is given by $\chi(R)$, and the reduction decomposes the arbitrary representation into $n_i$ irreducible representations with character $\chi_i(R)$, then the reduction is uniquely given by

$$\chi(R) = \Sigma_i \, n_i \, \chi_i(R) \qquad (4\text{-}6)$$

The sum is over all the irreducible representations $\chi_i(R)$ and $n_i$ is the number of times $\chi(R)$ contains $\chi_i(R)$. Multiplying Eq. 4-6 by $\chi_j(R)^*$, summing on R and using Eq. 4-1,

$$\Sigma_R \, \chi_j(R)^* \, \chi(R) = \Sigma_{i,R} \, n_i \, \chi_j(R)^* \, \chi_i(R) = h \, n_j$$

$$n_j = (1/h) \, \Sigma_R \, \chi_j(R)^* \, \chi(R) \qquad (4\text{-}7)$$

Always remember that R is summed over each symmetry operation, not over just the classes.

Suppose, for some problem in $C_{3v}$ symmetry, we have an obviously reducible representation whose characters listed in class groupings are $\chi(R) = 5, 2, -1$. Then, the number of times this representation contains

$A_1$ is $\quad [1(1)(5) + 2(1)(2) + 3(1)(-1)]/6 = 1,$

$A_2$, $\quad [1(1)(5) + 2(1)(2) + 3(-1)(-1)]/6 = 2,$

and E, $\quad [1(2)(5) + 2(-1)(2) + 3(0)(-1)]/6 = 1,$

where the first entry is the number of operations in the class. Thus, the reduction yields $A_1 + 2A_2 + E$. The results of Eq. 4-7 must always be an integer. Also the total dimensionality on both sides of Eq. 4-7 must be conserved.

Note that the reduction, Eq. 4-7, is unique. Often the reduction can be accomplished by direct observation of the character table, i.e., the sum of characters of irreducible representations gives the characters of the reducible representation. Try it on the given equations. Note that since the reduction is unique it is clear why two representations are equivalent if the character systems are equal (Section 4-1a).

## 4-4 Basis Functions

We have considered how the coordinates or products of coordinate transform among themselves under all the symmetry operations of the group. In Section 3-2 we had x, y, z transforming among themselves under the symmetry operations of $C_{3v}$. For the $C_{3v}$ problem, the resulting matrices of order 3 could easily be reduced by Eq. 4-7 to irreducible representations, although the result was intuitively obvious even then. In this section we consider the transformation of functions. It will turn out that wave functions have these same properties, hence these considerations are important. In Chapter 7 the transformation properties of functions will be discussed in greater detail.

For a given group consider a set of n linearly independent functions $\psi_1,\ldots,\psi_n$ such that a symmetry operation R will transform any $\psi_i$ into a linear combination of the others. By $R\psi$ we mean transform the coordinates and express the contours of the function that are fixed in space in terms of the new coordinates. Note that nothing is done to the functions, they are just expressed in terms of new coordinates. (A thorough understanding of this point is not necessary at this time but if desired see Section 7-2.) Thus

$$R\psi_i = \Sigma_j \psi_j \Gamma(R)_{ji} \tag{4-8}$$

Then the functions $\psi_j$ are said to form a **basis for a representation** $\Gamma(R)$ of the group. The representation could be reducible. Also, there are infinite sets of other functions that also are basis functions. It is clear that the matrices $\Gamma(R)$ form a representation of the group. Consider the symmetry operations AB = C, then

$$C\psi_i = AB\psi_i = A\,\Sigma_j\,\Gamma(B)_{ji}\,\psi_j = \Sigma_j\,\Gamma(B)_{ji}\,A\psi_j$$

$$\Sigma_k\,\Gamma(C)_{ki}\,\psi_k = \Sigma_{j,k}\,\Gamma(B)_{ji}\,\Gamma(A)_{kj}\,\psi_k$$

$$\Gamma(C)_{ki} = \Sigma_j\,\Gamma(A)_{kj}\,\Gamma(B)_{ji} = \Gamma(AB)_{ki} \qquad (4\text{-}9)$$

which is just the result required for Γ(R) to be a representation of the group (Eq. 3-1). In Appendix 5, we show that the order of the subscripts in Eq. 4-8 is natural and in agreement with our convention as in Eq. 3-1. If we had written $R\psi_i = \Sigma\,\Gamma(R)_{ij}\,\psi_j$ in place of Eq. 4-8, then the result in Eq. 4-9 would appear as $\Gamma(C)_{ki} = \Gamma(BA)_{ki}$ which would not be natural since C = AB.

We can prove that **the representations will be unitary if the basis functions are orthogonal and all normalized to the same constant,** usually picked to be 1. This is a very important theorem yet one that is often forgotten since we usually deal with orthonormal functions. Thus, the representations of these functions automatically are unitary. If the set of functions is not orthogonal, they can be altered to form an orthogonal set by the Schmidt orthogonalization process (see the Notes). The term **orthonormal functions** is used in the usual sense; the functions $\psi_i$ and $\psi_j$ are orthogonal if

$$c\,\delta_{ij} = \langle\psi_i\,|\,\psi_j\rangle \equiv \int \psi_i^*\,\psi_j\,d\tau \qquad (4\text{-}10)$$

where c is a constant and the integration is over all the space in which the function is defined. The proof of his theorem and the next rests on the fact that for two functions $\psi$ and $\phi$

$$\langle R\psi\,|\,R\phi\rangle = \langle\psi\,|\,\phi\rangle \qquad (4\text{-}11)$$

This follows from the meaning of $R\psi$ and $\psi$. Both of these functions have the same values at the same points in the space, expressed in terms of different coordinates (see Appendix 5). Now we can prove the theorem that if the basis functions are orthonormal, the representations will be unitary. Using Eq. 4-8

$$c\,\delta_{ij} = \langle\psi_i\,|\,\psi_j\rangle = \langle R\psi_i\,|\,R\psi_j\rangle$$
$$= \Sigma_{m,n}\Gamma(R)_{mi}^*\,\Gamma(R)_{nj}\langle\psi_m\,|\,\psi_n\rangle = c\,\Sigma_m\Gamma(R)_{mi}^*\,\Gamma(R)_{mj} \qquad (4\text{-}12)$$

which shows that the inverse of Γ(R) is indeed the complex conjugate of the transpose provided the constant c is independent of m and n. Thus, Γ(R) is unitary and the theorem is proved. In Section 7-2 this theorem is discussed again. However, for the rest of the book **we will only consider unitary representations and thus the orthonormal basis functions**. There is hardly ever any need to do anything but this.

Consider an irreducible representation $\Gamma_m$ of dimension $l$. Then if the functions $\psi_1{}^m,\ldots,\psi_l{}^m$ form a basis of the $\Gamma_m$ representation they are said to be **partners** of the $\Gamma_m$ irreducible representations. That is, the partners of the basis of an irreducible representation transform among themselves under all the symmetry operations of the group. For the example of $C_{4v}$, x and y are a basis for the E irreducible representation. It should be pointed out that there are an infinite number of sets of partners. For this case, xz and yz are also partners of the E irreducible representation as are $xz^2$ and $yz^2$, etc. In fact, any linear combination of different sets of partners are also partners. Again, for this case, x + cxz and y + cyz, where c is a constant, are also partners of the E irreducible representation. Notice that these partners of an irreducible representation are of mixed parity (odd and even functions, respectively). For a point group with a center of inversion, the odd functions transform as u-representations, and the even functions transform as g-representations so one cannot obtain partners of an irreducible representation with mixed parity.

The function $\psi_i{}^m$ is said to belong to the **ith row** of the mth irreducible representation. We now prove a theorem that has obvious quantum mechanical overtones. **Basis functions belonging to different irreducible representations or partners belonging to different rows of the same unitary irreducible representations are orthogonal.** Consider the integral $\langle\psi_i{}^m|\psi_j{}^n\rangle$. Then, as in Eq. 4-12,

$$\langle\psi_i{}^m|\psi_j{}^n\rangle = \langle R\psi_i{}^m|R\psi_j{}^n\rangle = \Sigma_{o,p}\,\Gamma(R)_{oi}{}^*\,\Gamma(R)_{pj}\,\langle\psi_o{}^m|\psi_p{}^n\rangle \quad (4\text{-}13)$$

Summing the right-hand side over all symmetry operations R and dividing by h does not alter the value. However, with the use of GOT (Eq. 3-12) on the right side, one obtains

$$\langle\psi_i{}^m|\psi_j{}^n\rangle = (1/l_m)\,\delta_{mn}\,\delta_{ij}\sum_{\rho=1}^{l_m}\langle\psi_\rho{}^m|\psi_\rho{}^m\rangle \qquad \text{QED} \qquad (4\text{-}14)$$

In addition to showing that the partners are orthogonal, Eq. 4-14 also shows that the partners of an irreducible representation are normalized the same way; that is, the result is independent of the index i.

An **invariant function** (or operator) f is defined as a function that has the property Rf = f, or f transforms as the $\Gamma_1$ irreducible representation. A very useful theorem states that **for the functions belonging to the ith row of the mth irreducible representation of dimension *l*** the expression $\Sigma_i|\psi_i{}^m|^2$ **is an invariant function** where i = 1 to $l$. To prove this, operate with a symmetry operation R on this function, and remember that representation is unitary. (We drop the irreducible representation superscript m for clarity.)

$$\begin{aligned} R\,\Sigma_i\,\psi_i^{m*}\,\psi_i^{m} &= \Sigma_i\,\Sigma_{j,k}\,\Gamma_m(R)_{ji}^{\ *}\,\psi_j^{\ *}\,\Gamma_m(R)_{ki}\,\psi_k \\ &= \Sigma_{j,k}\,\psi_j^{\ *}\,\psi_k\,\Sigma_i\,\Gamma_m(R)_{ki}\,\Gamma_m(R^{-1})_{ij} \\ &= \Sigma_{j,k}\,\psi_j^{\ *}\,\psi_k\,\Gamma_m(E)_{kj} \\ R\,\Sigma_i\,|\psi_i^{m}|^2 &= \Sigma_j\,|\psi_j^{m}|^2 \quad \text{QED} \end{aligned} \tag{4-15}$$

Examples of the use of this theorem would be: for cubic groups (x, y, z) transform as the partners of a T-irreducible representation so $x^2 + y^2 + z^2$ transforms as $\Gamma_1$. In an axial group (x, y) transform as an E-irreducible representation, thus $x^2 + y^2$ transforms as $\Gamma_1$.

## 4-5 Examples — Neumann Principle

The object of this section is to show the importance and the power of symmetry considerations in solids and at the same time to gain a deeper understanding of basis functions and character tables. While pursuing this object we cover much of the field of physical, or tensor, properties of crystals. This is done by realizing that a property of a material must transform into itself under all the symmetry operations of the material. We state this more clearly as **Neumann's principle** and then will apply Neumann's principle to a number of illustrative examples. The principle is: **Any physical property of a crystal or matter must possess at least the symmetry of the point group of the crystal.** Like many principles, after a little thought it becomes obvious, then self-evident, then trivial. However, it is very important. Restating it from a group theoretical point of view: **any physical property** (each tensor component separately) **of a crystal or matter transforms as the totally symmetric irreducible representation.** Thus, the meaning becomes more obvious: Namely, any symmetry operation applied to a physical property must transform it into +1 times itself.

After a brief review of the meaning of a tensor, we shall apply this principle to a number of important physical properties.

### a. Tensors

Most of the macroscopic properties of crystals are described by tensors. Tensors possess the proper behavior under symmetry transformations required by crystals. A tensor $T_{mno...}$ is defined by its behavior under a coordinate transformation. If, under a rotation of coordinates, a fixed point in the fixed body is described in the new ($x_i'$, i = 1, 2, 3) and original coordinate system ($x_i$) as in Eq. 3-2, $x_i' = \Sigma_j \Gamma_{ij} x_j$ where the direction cosines of a particular axis $x_i'$ with respect to $x_1, x_2, x_3$ are $\Gamma_{i1}$, $\Gamma_{i2}$, $\Gamma_{i3}$, etc., which have the usual property $\Sigma_j \Gamma_{ij} \Gamma_{kj} = \delta_{ik}$. Then a quantity is an r-rank tensor if

$$T_{ijk...}' = \Sigma_{m,n,o...} [\Gamma_{im} \Gamma_{jn} \Gamma_{ko}...] T_{mno...} \tag{4-16}$$

where there are a total of r subscripts on each of the T's. The tensor T is an r-rank tensor with $3^r$ components since each subscript has three values. For example, a second-rank tensor is defined as a nine-component object that behaves as

$$T_{ij}' = \Sigma_{mn} \Gamma_{im} \Gamma_{jn} T_{mn} \tag{4-17}$$

Examples of tensor properties of crystals are: 0-rank, temperature (it has only one component, and is called a scalar); 1-rank, electric polarization $P_i$ (three components and called a vector); 2-rank, strain $\varepsilon_{ij}$ (nine components although other nonsymmetry considerations can often be invoked which require $\varepsilon_{ij} = \varepsilon_{ji}$ thus reducing the number of components from nine to six resulting in a symmetric second-rank tensor); 3-rank, piezoelectricity $d_{ijk}$ (27 components but other considerations require $d_{ijk} = d_{ikj}$, thus reducing the number of components to 18); 4-rank, elasticity $c_{ijkl}$ (81 components but other considerations reduce this number to 21). For references see the Notes. It should be noted that polarization and strain can also be applied by the experimenter and are thus not always properties of the crystal itself. In our discussion in this section we will focus only on these terms as properties of the crystal.

To find the relations between the components of the tensors as well as the components that are zero, we use the direct inspection method (see the Notes). This method utilizes the fact that an r-rank tensor transforms as the Cartesian coordinates r at a time. That is, the second-rank tensor

$$a_{ij}' = \Sigma_{k,m} \Gamma_{ik} \Gamma_{jm} a_{km}$$

transforms as

$$x_i' x_j' = \Sigma(\Gamma_{ik} x_k)(\Gamma_{jm} x_m) = \Sigma \Gamma_{ik} \Gamma_{jm} x_k x_m$$

We also use the fact that the physical "constant" relating two 1-rank tensors is a 2-rank tensor. For example if $\mathbf{P} = \alpha\mathbf{E}$ where P and E are vectors, then $\alpha$ is a second-rank tensor. This is shown in the notes. The relation between an axial vector and a second-rank tensor is discussed in the Appendix to this chapter.

### b. Pyroelectricity

If the primitive cell of a crystal possesses a dipole moment, then it is called a pyroelectric crystal. (The dipole moment in a unit cell is nonzero if the center of mass of the positive charge is at a different position from that of the negative charge. $\boldsymbol{\mu} \equiv \Sigma\, e_i \mathbf{r}_i$ where the sum is over all the charges in the cell. Polarization $\equiv \boldsymbol{\mu}$/volume.) Since polarization is a vector, as is the position vector, its components will transform among themselves as Eq. 3-2. Using Neumann's principle, it is easy to see if any given crystallographic point group is pyroelectric. If x or y or z transforms as $\Gamma_1$ then, respectively, an x, y, or z component of polarization is allowable. For example, in the point group $C_{4v}$, z transforms as $A_1$ so crystals with this point group can possess a polarization along the z-axis and only this axis. (See Problem 3 in Chapter 1.) For the point group $C_{1h}$, x and y separately transform as A′ so a polarization in the x-y plane is allowable. There is no symmetry relation between the x and y components of the polarization so the vector can point anywhere in the plane but it must not have a z-component.

### c. Polarizability

The relation between the induced polarization $P_i$ and the applied electric field $E_j$ is given by the polarizability tensor $a_{ij}$ ($P_i = \Sigma_j\, a_{ij}\, E_j$) or

$$\begin{bmatrix} P_1 \\ P_2 \\ P_3 \end{bmatrix} = \begin{bmatrix} a_{11} & a_{12} & a_{13} \\ a_{21} & a_{22} & a_{23} \\ a_{31} & a_{32} & a_{33} \end{bmatrix} \begin{bmatrix} E_1 \\ E_2 \\ E_3 \end{bmatrix} \qquad (4\text{-}18)$$

$P_i$ and $E_j$ transform as vectors, and $a_{ij}$ transforms as a second-rank tensor. From energy considerations, it can be shown that $a_{ij} = a_{ji}$. Thus, there are only six independent components in Eq. 4-18. See the Notes for references where it is shown that $a_{ij} = a_{ji}$.

To further reduce the number of indpendent components compatible with symmetry, we use the direct inspection method. This method uses the fact that components of a second-rank tensor transform as the product of the orthogonal cartesian coordinates two at a time. The components of a third-rank tensor transform as the coordinates three at a time, $x_i x_k x_k$, etc., as in Eq. 4-16. If a symmetry operation shows $x_i x_j = -x_i' x_j'$, the the $\alpha_{ij}$ component must be zero because Neumann's principle demands that the component transform into +1 times itself. Also, the various symmetry operations require that various tensor components be equal to each other.

We illustrate the method with $C_{2h}$ {E, $C_2$, i, $\sigma_h$}. The six independent components of Eq. 4-18 transform in the same manner as $x^2$, $y^2$, $z^2$, xy, xz, yz. $C_2$ transforms x into −x, y into −y, and z into +z, etc.

$$
\begin{array}{llll}
C_2: & x^2 \to x^2 & xy \to +xy & \\
 & y^2 \to y^2 & xz \to -xz & \text{therefore } a_{13} = 0 \\
 & z^2 \to z^2 & yz \to -yz & \text{therefore } a_{23} = 0
\end{array}
\qquad (4\text{-}19)
$$

$$
\begin{array}{lll}
i: & x^2 \to x^2 & xy \to xy \\
 & y^2 \to y^2 & \\
 & z^2 \to z^2 &
\end{array}
\qquad (4\text{-}20)
$$

$$
\begin{array}{lll}
\sigma_h: & x^2 \to x^2 & xy \to xy \\
 & y^2 \to y^2 & \\
 & z^2 \to z^2 &
\end{array}
\qquad (4\text{-}21)
$$

Actually there is no need to use the $\sigma_h$ symmetry operation. All the restrictions on the tensor components can be found by considering the results from using the generators for each point group. (See Section 1-5a and Appendix 2 for the generators of the 32 point groups.) For this point group the generators are the symmetry operations $C_2$ and i. Thus, there are four independent tensor components of Eq. 4-18 for $C_{2h}$ symmetry:

$$
\begin{bmatrix} a_{11} & a_{12} & 0 \\ a_{12} & a_{22} & 0 \\ 0 & 0 & a_{33} \end{bmatrix} \qquad (4\text{-}22)
$$

For all of the cubic groups, the $C_3$ type of operation is a generator and will always result in $x^2 \to y^2 \to z^2$ and $xy \to yz \to zx$, so $x^2 = y^2 = z^2$ and independently, xy = xz = yz. Also, a $C_2[001]$ operation requires xz = −xz, i.e., $a_{12} = a_{13} = a_{23} = 0$ and $x^2 = x^2$, etc. The meaning of $x^2 \to y^2$ is nothing more than $a_{11}' \to a_{22}$ but the Neumann principle says that $a_{11}' = a_{11}$, therefore $a_{11} = a_{22}$. The other symmetry operations provide no further information. Thus, for the cubic groups the tensor is

$$
\begin{bmatrix} a_{11} & 0 & 0 \\ 0 & a_{11} & 0 \\ 0 & 0 & a_{11} \end{bmatrix} \qquad (4\text{-}23)
$$

with just one independent constant.

### d. Strain (seven crystal systems)

Usually strain is applied by a force in a certain direction. Thus, it is not a property of a crystal to be covered by Neumann's principle, but a property of an experiment. However, in this section we will consider a different kind of strain, a "self-strain" with reference to a cubic unit cell. For example, does the z-axis of the unit cell have exactly the same length as the x- or y-axis? Are the angles between it and the other axis $\pi/2$ or arbitrary? In other words, to which of the seven crystal systems does the point group belong. (The seven crystal systems are triclinic, monoclinic, orthorhombic, tetragonal, trigonal, hexagonal, and cubic. See Appendix 1 or Chapter 11.)

Strain $e_{ij}$, like polarizability, is a symmetric second-rank tensor with at most six independent components. If a crystal is squeezed along the z-axis, the $e_{33}$ strain is nonzero. If the crystal, at the same time, expands along the x-axis then $e_{11}$ is nonzero, and if the angle between the x- and z-axis changes then $e_{13}$ is nonzero. Similarly, if $z^2$ transforms into plus itself under all the symmetry operations of the crystal then the self-strain $e_{33}$ is nonzero and independent of what happens in the x and y directions. If $x^2$ and, independently, $y^2$ also transform into plus themselves, then the self-strains $e_{11}$and $e_{22}$ are independent of each other and nonzero. The crystal is then orthorhombic if all the other self-strains are zero.

Consider the point group $C_{2h}$. It is clear from the previous section that each of the four terms $x^2$, $y^2$, $z^2$, xy independently transforms as $\Gamma_1$ under all the symmetry operations of the group. Thus, the crystal must be monoclinic, namely, the a-, b-, and c-axes have different lengths and the angle between the a- and b-axes is not $\pi/2$.

The transformation properties of the six quadratic functions $x^2$, $y^2$, $z^2$, xy, xz, yz are of course included in the character tables in Appendix 3. Thus, $C_1$ and $S_2$ are triclinic; $C_{1h}$, $C_2$, and $C_{2h}$ are monoclinic, etc. The axial point groups have x and y transforming between themselves, so from Eq. 4-14 or the character tables, $x^2 + y^2$ transforms as $A_1$; thus the a- and b-axes are equal. The cubic groups have x, y, and z partners of an irreducible representation so the a-, b-, and c-axes are equal. It is only cubic point groups that have 3-dimensional irreducible representations; axial point groups have at most 2-dimensional irreducible representations and all the rest have only 1-dimensional irreducible representations (orthorhombic and lower symmetry). This will be most important in Chapter 7 where it is shown that wave functions transform as basis functions, wave functions that are partners of an irreducible representation are degenerate, and wave functions that transform as different irreducible representations have different energies. Therefore degeneracies of energy levels can be determined.

It is now easy to determine the independent tensor components of any symmetric second-rank tensor by inspection of the character tables.

Appendix 2 lists separately the point groups in the different crystal systems.

### e. Piezoelectric constants

An example of a third-rank tensor that describes a physical property of a crystal is the piezoelectric tensor $d_{ijk}$. This tensor relates an experimentally applied electric field $E_k$ to a resultant strain $e_{ij}$ as

$$e_{jk} = \Sigma_i \, d_{ijk} \, E_i \tag{4-24}$$

Since strain and electric field transform as second- and first-rank tensors, respectively the piezoelectric tensor must transform as a third-rank tensor. To find out which tensor components are zero, we consider how $x_i x_j x_k$ transforms under all the symmetry operations of the point group of interest. However, it is immediately obvious that for any point group with a center of inversion as a symmetry operation, $x_i x_j x_k = -x_i' x_j' x_k'$, so all the components must be zero because they cannot transform as $\Gamma_1$. It should also be pointed out that for this tensor $d_{ijk} = d_{ikj}$ since the strain tensor is symmetric. Thus, there are at most 18 independent tensor components.

We illustrate the method by obtaining the piezoelectric tensor for the point symmetry $D_2$ $\{E, C_2{}^x, C_2{}^y, C_2)\}$. The generating elements are $C_2$ and $C_2{}^y$

| | | |
|---|---|---|
| $C_2$: | $x^3 \rightarrow -x^3$, etc. | so $d_{111}=d_{222}=0$ |
| | $zx^2 \rightarrow zx^2$, etc. | |
| | $yx^2 \rightarrow -yx^2$, etc. | so $d_{211}=d_{133}=d_{233}$, etc.$=0$ |
| | $xyz \rightarrow xyz$, etc. | (4-25) |
| $C_2{}^y$: | $zy^2 \rightarrow -zy^2$, etc. | $d_{322}=d_{311}=d_{233}$, etc.$=0$ |
| | $xyz \rightarrow xyz$, etc. | (4-26) |

Thus, only the triple product xyz in various permutations transforms at $\Gamma_1$. This admits three nonzero independent tensor components for Eq. 4-24, $d_{123}$, $d_{312}$, $d_{231}$. For example, the term $d_{312}$ means that the application of an electric field along the z-axis will cause an xy-shear.

### f. Elastic coefficients

The properties of the crystal that relate an applied stress $t_{ij}$ to resulting strain $e_{mn}$ are the elastic coefficients $c_{ijmn}$:

$$t_{ij} = \Sigma_{m,n} \, c_{ijmn} \, e_{mn} \tag{4-27}$$

Since stress and strain are each symmetric second-rank tensors, the elastic coefficients transform as a fourth-rank tensor that is symmetric with respect to interchange of i and j, and separately, m and n. From energy considerations, it is also symmetric with respect to interchanges of ij with mn which reduces the maximum number of independent components from 81 to 21. Thus,

$$c_{ijmn} = c_{jimn} = c_{ijnm} = c_{jinm} = c_{mnij} = c_{nmij} = c_{mnji} = c_{nmji} \qquad (4\text{-}28)$$

Symmetry conditions can be applied as before. We again apply the ideas to point symmetry $D_2$, but this time we directly use the character table. It is immediately obvious that, for example, $(xy)(xy) = (x'y')(x'y')$, etc., under all the symmetry operations. Also, $(xy)(xz)$ will go into minus itself under $C_2$ and $C_2^y$ although it goes into plus itself under $C_2^x$. It is obvious that $x^4$, $x^2y^2$, $x^2y^2$, etc., independently transform into plus themselves under all symmetry operations. Thus, the independent coefficients are $c_{1111}$, $c_{2222}$, $c_{3333}$, $c_{1122}$, $c_{1133}$, $c_{2233}$, $c_{2323}$, $c_{1313}$, $c_{1212}$.

A **contracted notation** has been developed to make it easier to write out tensor components. For a set of symmetric components $x_ix_j$ there are only six, instead of nine, independent indices. Thus a single index that runs from 1 to 6 is all that is needed. The correspondence is

| | | | | | | |
|---|---|---|---|---|---|---|
| ij: | 11 | 22 | 33 | 23 | 13 | 12 |
| $x_ix_j$: | $x^2$ | $y^2$ | $z^2$ | yz | xz | xy |
| contracted index: | 1 | 2 | 3 | 4 | 5 | 6 |

(4-29)

Hence, the independent elastic coefficients written above in contracted notation would be written as $c_{11}$, $c_{22}$, $c_{33}$, $c_{12}$, $c_{13}$, $c_{23}$, $c_{44}$, $c_{55}$, $c_{66}$. The matrix would be written as

$$\begin{bmatrix} c_{11} & c_{12} & c_{13} & 0 & 0 & 0 \\ & c_{22} & c_{23} & 0 & 0 & 0 \\ & & c_{33} & 0 & 0 & 0 \\ & & & c_{44} & 0 & 0 \\ & & & & c_{55} & 0 \\ & & & & & c_{66} \end{bmatrix} \qquad (4\text{-}30)$$

Remember that in contracted notation the array in Eq. 4-30 is merely a matrix and should not be confused with a second-rank tensor. If we desire to study the transformation properties of the tensor components, it would be safest to use the full tensor notation.

In the contracted notation, the nonzero piezoelectric coefficients $d_{123}$, $d_{312}$, $d_{231}$ found for $D_2$ would be $d_{14}$, $d_{36}$, $d_{25}$. As a matrix:

$$\begin{bmatrix} 0 & 0 & 0 & d_{14} & 0 & 0 \\ 0 & 0 & 0 & 0 & d_{25} & 0 \\ 0 & 0 & 0 & 0 & 0 & d_{36} \end{bmatrix} \qquad (4\text{-}31)$$

## 4-6 Atomic Positions

The discussion on atomic positions has nothing to do with Neumann's principle. Nevertheless, we include it here as an important example of a property that transforms as $\Gamma_1$. After the application of one of the symmetry operations of the molecule or crystal, some of the atoms are interchanged and others are not. If there are n atoms and the position of the ith is given by $x_i$, $y_i$, $z_i$ with respect to a fixed x, y, z coordinate system, then when operated on by a symmetry operation R, $R(x_i, y_i, z_i) \rightarrow (x_j, y_j, z_j)$ for all i, where the operation implies that either the same atom stays at the same position or at most gets replaced by an atom of the same type (oxygen by oxygen, lead by lead, etc.). This is just a statement of the fact that the molecule or crystal looks exactly the same before and after the symmetry or covering operation. A function that expresses the internuclear distances explicitly $V = (x_1 - x_2)^2 + (x_1 - x_3)^2 + ... + (x_1 - x_n)^2 + (x_2 - x_3)^2 + ... + (y_1 - y_2)^2 + ...$ as would be required for a potential energy function also has the property RV = V.

Now consider a large molecule with $O_h$ symmetry (48 operations). There are at most 48 equivalent (from a chemical environment and symmetry point of view) atoms in the molecule no matter how large it is. For example, there can be at most 48 oxygen atoms that transform among themselves. Of course, there can be another set of 48 oxygen atoms that transform among themselves, but these two sets never interchange. The **exact** environment (neighbors, force constants, etc.) of each one of the 48 oxygens is equivalent to that of the other 47 oxygens. However, the exact environment of the 48 oxygens in one set is different from that of the other set. In general, a point group (or space group if you only consider a primitive unit cell) can have at most h equivalent atoms. The h equivalent points are called **general equivalent positions**. These are the points pictured on the stereograms in Appendix 1. The point symmetry of a general position (for an origin at the general point) is $C_1$. This is called the **site symmetry** of the particular point. (This will be discussed in detail in Chapter 11.) Atoms that lie on some of the symmetry elements have a smaller number of equivalent atoms and higher site symmetry. Finally if an atom lies at the origin (the point operations are taken with respect to a fixed origin), there are no equivalent atoms. Obviously all the h symmetry operations take the atom into itself and its site symmetry is the same as the point symmetry of the molecule.

Assume a large molecule has $O_h$ symmetry and 48 oxygen atoms correspond to one general point. Assume that one Li atom replaces one Na atom somewhere in the molecule and lowers the symmetry to $D_{4h}$. The original 48 oxygens no longer are equivalent. There can be three groups of 16 equivalent atoms. The 16 in one group will no longer see exactly the same environoment as those in each of the other two groups. The key word is "exactly." Group theory is a branch of mathematics and deals only with exact statements. However, it is clear that the new Li atom may not affect many experimental results of the original 48 oxygen atoms.

An entirely similar situation holds for crystals. Crystals with symmorphic space groups have one point in the unit cell that transforms into itself under all the point symmetry operations. If an atom is at this point, its site symmetry is the same as the point symmetry of the crystal. A general point in the unit cell still must refer to h equivalent points each having $C_1$ site symmetry. In nonsymmorphic space groups there is a slight difference. (Remember that the point group of a space group is obtained from the factor group of the sapce group taking the infinite number of unit cell translations as the invariant subgroup. For a symmorphic space group this factor group is identical to a point group. For a nonsymmorphic space group the factor group is only isomorphic to a point group because there are glide and screw symmetry opeations.) Within the unit cell, a general point still refers to h equivalent points even though they transform among themselves by glide and screw operations as well as point operations. However, there is no longer one point that transforms into itself under all the symmetry operations. Thus the point with the highest site symmetry in the unit cell no longer has symmetry as high (as many symmetry operations) as the point group of the crystal. In fact, there must be at least two equivalent positions in the unit cell that share the distinction of having the highest site symmetry. Site symmetry considerations are very important in many areas of physics and chemistry. For example, if the electron spin resonance of an impurity atom is measured, the site symmetry determines the form of the spin Hamiltonian. For crystals the symmetry can be lowered by external forces (strain, electric field, etc.) or internal changes (spontaneous polarization as in ferroelectricity, atom substitution, etc.).

Neumann's principle makes it physically clear why the point group of a nonsymmorphic space group is all that is required to determine the physical properties of crystals. Recall, from Section 1-6, how the point group of a nonsymmorphic space group is found. One finds all the space group operations $\{R \mid \tau\}$, where $\tau$ is not a primitive lattice vector, and the sets all $\tau = 0$. The resultant operations $\{R \mid 0\}$ are identical with one of the 32 point groups which is called the point group of the nonsymmorphic space group. It is perfectly reasonable that the macroscopic physical properties of the crystal will not be affected by a translation by an atomic dimension of the magnitude of $\tau$. So with respect to macroscopic properties, the definition of the point group of a nonsymmorphic space group is sensible. (In Chapter 11 we prove that the point group of a nonsymmorphic space group as defined here is indeed isomorphic to the factor group of the space group.)

## 4-7 The Hamiltonian

We discuss the Hamiltonian in this section and emphasize the fundamental point that it must transform into itself under all symmetry operations. The fact that the Hamiltonian must transform as $\Gamma_1$ is obvious. After all, if the system is unaltered by a symmetry operation, then the Schödinger equation must also be unaltered by the same operation.

To get acquainted with this principle, consider the Hamiltonian of an atom with q electrons. Omitting spin dependent terms, for a fixed nuclear mass of charge Ze, where m is the mass of the electron, the Hamiltonian is

$$H = -\frac{\hbar^2}{2m}\sum_{i=1}^{q}\nabla_i^2 - \Sigma_i \frac{Ze^2}{r_i} + \sum_{i<j}^{q}\sum^{q}\frac{e^2}{r_{ij}} \tag{4-32}$$

The rest of the symbols have their usual meaning. Consider a rotation about the z-axis in a right-handed sense by an arbitrary angle $\sigma$ so that the new $(x', y')$ and old $(x, y)$ coordinate systems are related in the usual way

$$\begin{bmatrix} x \\ y \end{bmatrix} = \begin{bmatrix} \cos\sigma & -\sin\sigma \\ \sin\sigma & \cos\sigma \end{bmatrix} \begin{bmatrix} x' \\ y' \end{bmatrix}$$

$$\begin{bmatrix} x' \\ y' \end{bmatrix} = \begin{bmatrix} \cos\sigma & \sin\sigma \\ -\sin\sigma & \cos\sigma \end{bmatrix} \begin{bmatrix} x \\ y \end{bmatrix} \tag{4-33}$$

Now consider each term in the Hamiltonian, Eq. 4-32.

The middle term in the Hamiltonian is the same in the primed and unprimed coordinate system since

$$\begin{aligned} r^2 &= x^2 + y^2 + z^2 \\ &= (x'\cos\sigma - y'\sin\sigma)^2 + (x'\sin\sigma + y'\cos\sigma)^2 + z'^2 \\ &= x'^2 + y'^2 + z'^2 \end{aligned} \tag{4-34}$$

The last term in the Hamiltonian is also the same in both coordinate systems since

$$\begin{aligned} r_{ij}^2 &= (x_i - x_j)^2 + (y_i - y_j)^2 + (z_i - z_j)^2 \\ &= (x_i' - x_j')^2 + (y_i' - y_j')^2 + (z_i' - z_j')^2 = r_{ij}'^2 \end{aligned} \tag{4-35}$$

The first term involving the kinetic energy is also the same function form in the primed and unprimed system. This is less obvious than the other two terms thus we show it here. Consider f to be any function. Then

$$\partial f/\partial x' = (\partial f/\partial x)(\partial x/\partial x') + (\partial f/\partial y)(\partial y/\partial x')$$

$$\partial^2 f/\partial x'^2 = \partial^2 f/\partial x^2\,(\partial x/\partial x')^2 + (\partial^2 f/\partial y^2)\,(\partial y/\partial x')^2 \tag{4-36}$$

with a similar relation for $\partial^2 f/\partial y'^2$. Thus

$$\begin{aligned} \nabla'^2 f &= [(\partial x/\partial x')^2 + (\partial x/\partial y')^2]\,\partial^2 f/\partial x^2 \\ &\quad + [(\partial y/\partial x')^2 + (\partial y/\partial y'^2]\,\partial^2 f/\partial y^2 \\ &= 1\,\partial^2 f/\partial x^2 + 1\,\partial^2 f/\partial y^2 = \nabla^2 f \end{aligned} \tag{4-37}$$

where the terms in the brackets [$\partial x/\partial x'$, etc.] are evaluted from Eq. 4-33 and are equal to 1 since $\cos^2\sigma + \sin^2\sigma = 1$. The extension to a rotation about all three axes is straightforward but a bit tedious.

We see that under a point symmetry transformation of the atom, the Hamiltonian is invariant to the transformation. For a hydrogen atom the Hamiltonian in Eq. 4-32 simplifies in that the last, interelectron repulsion term, is zero.

If this hydrogen atom were put in a box or at a point in a crystal that has cubic site symmetry, the potential energy in Eq. 4-32 is not complete. A solution of the wave equation with just the spherical potential will naturally just give the unperturbed hydrogen atom solution again. A cubic potential energy term of the type

$$C\,[x^4 + y^4 + z^4 - 3/5\,r^4] + \ldots \tag{4-38}$$

must be added to the Hamiltonian. This is the first term in an expansion for small r. Higher terms contain $x^6$, etc., and $x^4y^2$, etc. The expression in Eq. 4-38 transforms as $\Gamma_1$ under all point symmetry operations for the $O_h$ point group. The total potential energy of the electrons would then be $-Ze^2/r$ + Eq. 4-38. In fact the way to obtain the expression in Eq. 4-38 is to operate on the nine spherical harmonics of order four, $Y_4^l\Theta,\phi)$, with all the symmetry operations of $O_h$ to find out which transform into +1 times themselves. It is just the resulting expression that has the proper transformation properties to be a potential energy term in a Hamiltonian. (See the Notes.) Again if the crystal were strained along the z-axis, this potential energy would not completely describe the system. Axial terms of the type

$$A(3z^2 - r^2) + \ldots \tag{4-39}$$

must be added. The expressions in Eqs. 4-38 and 4-39 will transform into themselves under all the symmetry operations of any of the axial crystallographic point groups. However, Eq. 4-39 will not transform into itself under the four $C_3$ axes of the cubic point groups.

The set of transformation operations that leaves the Hamiltonian invariant is called the **group of the Schrödinger equation**. The coordinate transformations of this set are just the symmetry operations we have been discussing in this book so far and will continue to discuss. However, operations like time reversal will at times be included in the group of the Schrödinger equation. The relation between quantum mechanics and group theory will be discussed at length in Chapter 7. (See the Notes in Chapter 7 where we show that the set of symmetry operations that commute with the Hamiltonian form a group.)

## Appendix to Chapter 4

Tensors that transform as Eq. 4-16 are sometimes called true tensors or polar tensors. An **axial tensor** is defined as a quantity that transforms in the following manner

$$S_{ijk\ldots}{}' = \pm \Sigma [\Gamma_{im} \Gamma_{jn} \Gamma_{ko} \ldots] S_{mno\ldots} \tag{4-A1}$$

where the negative sign is used for a transformation, such as $\sigma$, i, $S_n$, that changes he coordinate system from a right- to left-hand one or vice versa. The positive sign is used for pure rotations. The most familiar axial tensor is the **axial vector C = ai + bj + ck** which can be thought of in terms of a vector cross product of two (polar) vectors. Thus, while the (polar) vector can be represented by a directed segment of length (an arrow), an axial vector has the directional properties of an element of area (a current loop) and the direction of the area must be defined in an arbitrary way such as a right-handed rule. If two (polar) vectors are given by **A** (= $A_1$**i** + $A_2$**j** + $A_3$**k**) and **B**, then the components of an axial vector **C** = **A** × **B** are $C_1 = A_2B_3 - A_3B_2$ and $C_2 = A_3B_1 - A_1B_3$, etc. Thus, an axial vector really contains the coordinates two at a time so we can write

$$C_1' = A_2'B_3' - A_3'B_2' = \Sigma \Gamma_{2i} \Gamma_{3j}(A_iB_j - A_jB_i)$$

which is not the tensor transformation law, Eq. 4-16. However, from the properties of the $\Gamma_{ij}$ matrix one can show that

$$C_i' = \pm \Sigma \, \Gamma_{im} \, C_m$$

for an axial vector in agreement with Eq. 4-A1. See the Notes for the appropriate references. As an example, look at the point group $C_{2v}$ in the character tables and note how a current loop in the xy,yz and xz-plane transform under the various symmetry operations. The result is quite different from a polar vector.

## Notes

For the proof and some discussion of Eq. 4-2a, see Falicov, p. 34. For the proofs of the other equations in Section 4-1 see Wigner, Chapter 9. Bethe shows how the character tables for the point groups can be obtained as in Section 4-1a; also see Falicov, Chapter 2.

**Theorem** If $P = \alpha E$ where P and E transform as vectors, then $\alpha$ transforms as a second-rank tensor. To prove this recall if $E_i' = \Sigma_j \Gamma_{ij} E_j$, then $E_j = \Sigma_i \Gamma_{ij} E_i'$ (i.e., $A = \Gamma B$ so $B = \Gamma^{-1} A$, but for real orthogonal coordinate transformations $\Gamma^{-1} = \Gamma$). Consider $P_i' = \Sigma \Gamma_{ij} P_j = \Sigma \Gamma_{ij} \alpha_{jk} E_k = \Sigma \Gamma_{jk} \alpha_{jk} \Gamma_{mk} E_m'$. But $P_i' = \Sigma \alpha_{im}' E_m'$ so $\alpha_{im}' = \Sigma_{j,k} \Gamma_{ij} \Gamma_{mk} \alpha_{jk}$. QED The generalization to higher order tensors is obvious.

For a general discussion of axial vectors and their transformation properties, see Bhagavantam, Section 2-5 and Birss, Sections 1-3 and 1-4.

For general references to tensor properties of crystals, see Nye or Bhagavantam. These books also show how energy considerations lead to polarizability and other tensors having the property $a_{ij} = a_{ji}$. Fumi's original paper [Acta Cryst. **5**, 44 (1952)] is worth reading. The direct inspection method is directly applicable only to crystal classes in which one can find Cartesian orthogonal coordinates that do not transform into linear combinations of themselves under the symmetry operations. This includes 22 of the 32 point groups.

## Problems

**1.** For $C_{2v}$, how can one easily tell which plane the $\sigma_v$ and $\sigma_v'$ refer to? (Hint: Consider how the functions transform.) In the same manner the meaning of $C_2'$ and $C_2''$ in $D_{4h}$ can be determined.

**2.** (a) For $C_{4v}$ point symmetry, find the reducible representation for the four functions m, n, o, p that resemble equal length arrows starting from the origin and pointing in the +x, +y, −x, −y directions.

(b) Reduce the representation by use of Eq. 4-7. If only part b were required note how much simpler the problem would be since only the character of the reducible representation is needed and it has a contribution only when the symmetry operation transforms a function into plus or minus itself.

**3.** Half of the symmetry operations of $C_{4v}$ are identical with those of $C_{2v}$. ($C_{2v}$ is a subgroup of $C_{4v}$.) Which 1-dimensional irreducible representations of $C_{4v}$ and $C_{2v}$ correspond to each other? (If the answer is not immediately obvious remember that a basis function transforms into plus or minus itself in the subgroup just as in the larger group.) The 2-dimensional irreducible representation of $C_{4v}$ is reducible in $C_{2v}$. Reduce it.

**4.** Show that $\Sigma_R\, \chi_i(R) = 0$ except for $\Gamma_1$.

**5.** What are the symmetry operations that exclude pyroelectricity. (Use the character tables.)

**6.** The character tables list basis functions of the type $x_i$ and $x_ix_j$. For the point group $C_2$, determine to which irreducible representation the functions of the type $x_ix_jx_k$ belong. In 20 sec what can you say about all functions of the $x_ix_jx_kx_l$ type?

**7.** Consider six atoms arranged octahedrally about an origin, i.e., $O_h$ symmetry. (See the six oxygen positions in Fig. 1-2, $PbTiO_3$.) Attached to each atom is an arrow (function) pointing toward the origin. Under which irreducible representations do these six functions transform? Why is it obvious that $A_{1g}$ must be part of the answer? If you can, write out the actual functions. (The problem that you just solved is finding the appropriate linear combination of ligand orbitals for $\sigma$-bonding to a central atom at the origin. It is also the same problem as finding the appropriate function or orbitals on the central atom for orbitals that point to the six surrounding ligands or octahedral hybridization.)

**8.** Consider cubic $PbTiO_3$ in Fig. 1-2. The point symmetry is $O_h$. (It is a symmorphic crystal so the Ti atom has $O_h$ point symmetry.)

(a) What is the point symmetry of oxygen atoms?

(b) A strain is applied along the z-axis. In the cubic phase all three oxygens transformed among themselves. In this case which transform among themselves? What is the point symmetry of a point anywhere along the z-axis?

(c) If in the cubic phase the Ti is moved up along the z-axis, the resulting point symmetry is $C_{4v}$ for the crystal. What is the point symmetry for the Ti site? Any point along the z-axis? How do the oxygen atoms transform?

**9.** In the point symmetry O how do the seven f-type wave functions transform, i.e., which are partners, etc.? In $C_i$ which irreducible representation do they transform? (The f-type wave functions are xyz, $y(x^2-z^2)$, $x(z^2-y^2)$, $z(y^2-x^2)$, $y^3-3yr^2/5$, $x^3-3xr^2/5$, $z^3-3zr^2/5$.)

**10.** The regular representation is defined in the Notes in Chapter 3. (a) Prove that it forms a representation as in Eq. 3-1. (b) Prove that it contains each irreducible representation a number of times equal to the dimensionality of the irreducible representation.

**11.** Prove that the components of an axial vector are elements of an antisymmetric second-rank tensor as in the Appendix to this chapter.

**12.** Obtain the character tables for the permutation groups $P_2$, $P_3$, and $P_4$. (These groups are discussed in the Appendix to Chapter 2.) Note that these groups are isomorphic with the point groups $C_2$, $C_{3v}$, and $T_d$, respectively.

Chapter 5

# VIBRATIONS OF MOLECULES AND CRYSTALS

*I cultivate them with my children, and the work keeps us from three great evils, boredom, vice, and poverty.*

*Voltaire, "Candide"*

While it might be more logical at this point to discuss the general implications of group theory in quantum mechanics, we will first apply group theory to atomic vibrations in molecules and solids. This is done for two reasons. First, atomic vibrations can be treated classically and the reader probably has some knowledge of vibrations in molecules. Second, an amazingly large amount of information can be obtained from very simple use of group theory. Thus we can see how powerful the apparatus is before going on. The following information can be obtained:

(a) The number of normal modes that transform as each irreducible representation of the symmetry group of the molecule or crystal.

(b) From these results, the infrared and Raman selection rules can be extracted very easily. [Results (a) and (b) are often all that must be established since the frequencies of the vibrations can be measured.]

(c) If only one normal mode transforms as a given irreducible representation, then with the use of projection operator techniques, the normal mode (or more exactly the eigenvector) can be determined. However, the frequency will not be determined. If more than one normal mode transforms as a given irreducible representation then projection operator techniques can be used to find symmetry-adapted vectors. Linear combinations of these symmetry-adapted vectors are the eigenvectors of the normal modes, but the appropriate linear combinations cannot be determined without completely solving the generalized Hooke's law problem.

(d) To calculate the actual frequencies for the normal modes, the forces between all the atoms must be known. This leads to the force constant matrix or dynamical matrix which, when diagonalized, gives the frequencies and eigenvectors of all the normal modes. Group theory is used to find symmetry-adapted vectors which partially diagonalize the dynamical matrix into blocks whose dimensions are equal to the number of normal modes that transform as a given irreducible representation. For example, if there are six normal modes that transform as the 3A + B + E irreducible representations, the $6 \times 6$ dynamical matrix can be reduced to a $3 \times 3$, a $1 \times 1$, and two identical $1 \times 1$ matrices. When diagonalized, the frequencies correspond to the three A-modes, the one B-mode, and the doubly degenerate E-mode, respectively.

## 5-1 3N Degrees of Freedom

For a molecule or crystal containing N atoms, there are 3N degrees of freedom. N is very large for a crystal ($\sim 10^{23}$) and we return to this problem at the end of this chapter to discuss only the infinite wavelength solutions. Not all of the 3N degrees of freedom of a molecule are internal vibrations. Three degrees of freedom are required to describe the center of mass of the molecule. These are the translational coordinates. Three degrees of freedom are also required to describe the rotational behavior of the molecule. These could be Euler angles with respect to an axis fixed along the principal moments of inertia (only two angles are required for a linear molecule). Thus, there are $3N - 6$ ($3N - 5$ for linear molecules) degrees of internal freedom, or true vibrations, where there is a change in the relative distance between atoms.

It should be pointed out that the ability to treat vibrations separately from rotations and electronic excited states follows from the fact that the characteristic energies of the excitations are very different (approximately 1 $cm^{-1}$, $10^2$ $cm^{-1}$, and $10^4$ $cm^{-1}$ for rotations, vibrations, and electron levels, respectively). Thus, the Hamiltonian can be written as a sum of separate terms involving the three processes and in most cases the small cross terms can be ignored or treated as a perturbation. Since the Hamiltonian is the sum of separate terms, the total wave function can be expressed as a product of the separate wave functions.

## 5-2 General Considerations

In this section, we show that arbitrary displacements from equilibrium that transform among themselves under all the symmetry operations of the point group from a reducible representation of the group. The normal modes do likewise but transform as the irreducible representations of the group. The number of normal modes that transform as each irreducible representation is given by the reduction of the reducible representation determined from the arbitrary displacements.

### a. Symmetry of normal modes

Consider a coordinate system fixed in the molecule with its origin at the center of mass. The N vectors to the equilibrium positions of the N atoms are given by $\mathbf{r}_i$, i = 1,..., N. The time varying small displacements of each atom are given by $\mathbf{u}_i$. Naturally, under all the symmetry operations R of the group, the $\mathbf{r}_i$ transform among themselves. Consider the effect of a particular $\mathbf{u}_i$ on the potential energy and some other $\mathbf{u}_n$ that is related to $\mathbf{u}_i$ by a symmetry operation. This is shown in Fig. 5-1a for a $H_2O$ molecule. Since the potential energy is the same after a symmetry operation, $\mathbf{u}_i$ and $\mathbf{u}_n$ will increase it by exactly the same amount. Both represent exactly equivalent bond distance and angle changes. For example, a displacement consisting of a certain set of $\mathbf{u}_i$ will have the same frequencies as the same displacement after a symmetry operation.

We try to make clear the effect of a symmetry operation on the instantaneous position $\mathbf{r}_i + \mathbf{u}_i$ of an atom identified by i. When operated on by a symmetry operation R, the i-atom will be moved to the former position of the j-atom.

$$R(\mathbf{r}_i + \mathbf{u}_i) = R\mathbf{r}_i + R\mathbf{u}_i = \mathbf{r}_j + R\mathbf{u}_i = \mathbf{r}_{R(i)} + R\mathbf{u}_{R^{-1}(j)} \qquad (5\text{-}1)$$

It is clear thar $R\mathbf{r}_i = \mathbf{r}_j$ since the vector to the i-atom, at the position of the j-atom, is equal to the vector to the j-atom before the symmetry operation. Then the meaning of $R\mathbf{r}_i = \mathbf{r}_{R(i)}$ is clear. The $\mathbf{r}_{R(i)}$ expression is the vector to the position of the i-atom after the symmetry operation. It is also clear that $R\mathbf{u}_i \neq \mathbf{u}_j$ because $u_i$ is an arbitrary displacement of the i-atom, and $\mathbf{u}_j$ can be quite different. $R\mathbf{u}_{R^{-1}(j)}$ means that R operates on the arbitrary displacement that will be at the site of the j-atom after the symmetry operation, or $R\mathbf{u}_{R^{-1}(j)} = R\mathbf{u}_i$. Thus, there is a subtle difference between equilibrium positions and arbitrary displacements. (See the Notes for a reference to Wigner.) A, B,... are point symmetry operations of the molecule. A symmetry operation on all 3N components of the

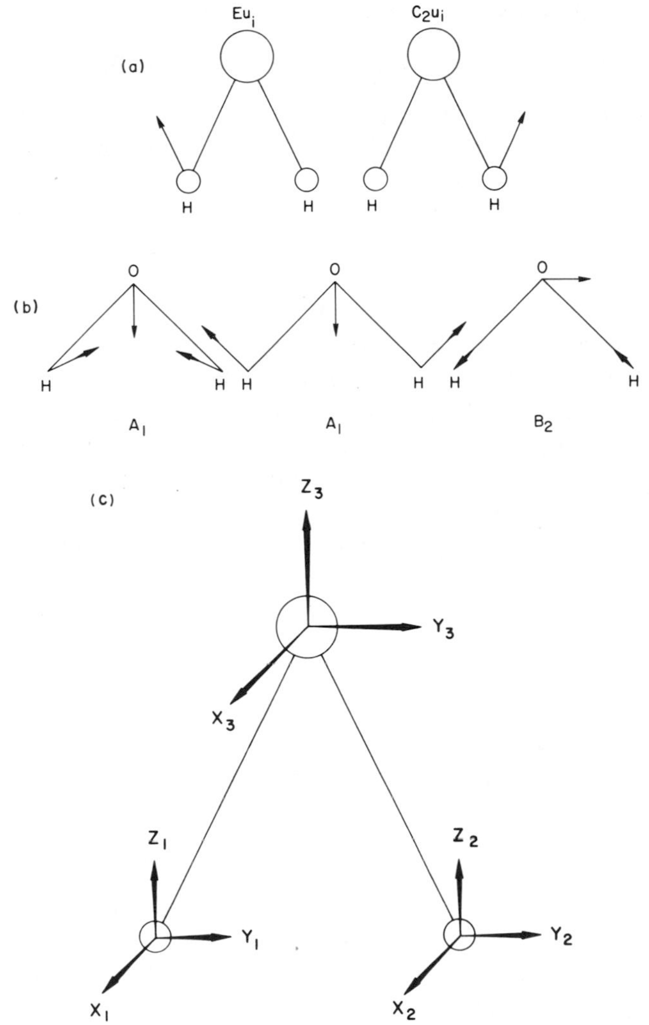


**Fig. 5-1** (a) The effect of symmetry operations on an arbitrary displacement of one of the hydrogen atoms. (b) The three normal modes for $H_2O$. (c) A coordinate system for $H_2O$.

arbitrary displacements of the N atoms, $Au_\alpha$, can be expressed in terms of the 3N components of the displacements:

$$Au_\alpha = \sum_{\beta=1}^{3N} u_\beta \, \Gamma(A)_{\beta\alpha} \tag{5-2}$$

If A is followed by another symmetry operation B, as in Eq. 4-9, then

$$BAu_\alpha = \Sigma_\gamma \Sigma_\beta \Gamma(B)_{\gamma\beta} \Gamma(A)_{\beta\alpha} u_\gamma = \Sigma_\gamma \Gamma(BA)_{\gamma\alpha} u_\gamma \qquad (5\text{-}3)$$

which is the requirement that Γ(R) be a representation of the group. This representation is reducible and when reduced will enumerate the irreducible representations under which normal modes transform. This will be shown in Section 5-2b. Examples of this procedure will be given in Section 5-3. However, in the next paragraph we show by example that normal modes do transform as irreducible representation and Section 5-2b shows that this is always to be expected.

Figure 5-1b shows the three normal mode displacements for $H_2O$. For $H_2O$ the point group is $C_{2v}$. Under all the four symmetry operations of this group, the set of three displacements on the left-hand molecule transforms directly into themselves. Therefore, that normal mode transforms as the $A_1$ irreducible representation of the point group $C_{2v}$. The normal mode shown in the middle molecule of the figure also transforms into itself under the symmetry operations E and $\sigma_v(yz)$, but under $C_2$ and $\sigma_v(yz)$ it transforms into $-1$ times itself. Thus, the mode transforms as the $B_2$ irreducible representation of the point group $C_{2v}$. Since $N = 3$, all $3N - 6$ modes are pictured in the figure.

**b. Dynamical matrix**

In this section a generalized Hooke's law for the forces between atoms is developed. To show that the normal modes of vibration transform as irreducible representations, one can skip to Eq. 5-11. (See the Notes.)

Let Φ be the effective potential function for the nuclei which is evaluated for a particular electronic state; $k = 1,2,\ldots,N$ for the N nuclei; $a,\beta = 1,2,3$ for the three rectuangular coordinates; $m_k$ = mass of the kth nuclei; $x_\alpha(k)$ = $\alpha$th component of the position of the kth nucleus; $u_\alpha(k)$ = $\alpha$th component of the displacement from equilibrium of the kth nuclei; and $j = 1,2,\ldots,3N$ for all the degrees of freedom.

The effective potential can be expanded for small displacements from equilibrium as follows

$$\Phi = \tfrac{1}{2} \sum_{k',k}^{N} \sum_{\alpha,\beta}^{3} \frac{\partial^2\Phi}{\partial x_\alpha(k)\, \partial x_\beta(k')} u_\alpha(k) u_\beta(k')$$

$$\equiv \tfrac{1}{2} \Sigma\, \Sigma\, \Phi_{\alpha\beta}(kk') u_\alpha(k) u_\beta(k') \qquad (5\text{-}4)$$

It is convenient to use reduced displacements and potentials

$$w_\alpha(k) = (m_k)^{1/2}\, u_\alpha(k), \qquad D_{\alpha\beta}(kk') = (m_\alpha m_\beta)^{-1/2}\, \Phi_{\alpha\beta}(kk') \tag{5-5}$$

The kinetic energy T and potential energy V can be written, using $p_\alpha(k) = -i\hbar\, \partial/\partial w_\alpha(k)$ the momentum canonical to $w_\alpha(k)$, as

$$H = \frac{1}{2}\, \Sigma_k \Sigma_\alpha\, p_\alpha(k)^2 + \frac{1}{2}\, \Sigma_{kk'} \Sigma_{\alpha\beta} D_{\alpha\beta}(kk') w_\alpha(k) w_\beta(k') \tag{5-6}$$

The Lagrangian is L = T − V. The equations of motion may be obtained from $\partial[\partial L/\partial\, p_\alpha(k)]/\partial t - \partial L/\partial w_\alpha(k) = 0$ which yields

$$w_\alpha(k) = -\, \Sigma_{k'} \Sigma_\beta\, D_{\alpha\beta}(kk') w_\beta(k') \tag{5-7}$$

This resultis just a generalized Hooke's law involving the forces between all the atoms. To solve Eq. 5-7, the form $w_\alpha(k) = C_j\, e_\alpha(k) \exp(i\omega t + \delta_j)$ is taken where j = 1,..., 3N and $C_j$ and $\delta_j$ are determined from the initial conditions of the molecule. With this form of $w_\alpha(k)$, Eq. 5-7 yeilds

$$\omega^2\, e_\alpha(k) = \Sigma_{k'} \Sigma_\beta\, D_{\alpha\beta}(kk') e_\beta(k') \tag{5-8}$$

This results in 3N simultaneous linear homogeneous equations for the 3N quantities $e_\alpha(k)$. The equations are solvable if the secular determinant is zero:

$$|\, D_{\alpha\beta}(kk') - \omega^2\, \delta_{kk'}\, \delta_{\alpha\beta}\,| = 0 \tag{5-9}$$

This equation will yield 3N solutions for $\omega^2$ denoted by $\omega_j^2$. Each $\omega_j^2$ substituted into Eq. 5-8 will yield a result for $e_\alpha(k)$, denoted by $e_\alpha(k\,|\,j)$. The $\omega_j^2$ are called **eigenvalues** and the $e_\alpha(k\,|\,j)$ are called **eigenvectors**. All the eigenvalues are real since the dynamic matrix is a real symmetric matrix. Six roots of Eq. 5-9 (five for a linear molecule) corresponding to three translational and three rotational degrees of freedom must be equal to zero because there is no restoring force for these motions.

The eigenvectors have a certain arbitrariness because multiplication by a constant will not change the solution, Eq. 5-8, i.e. the constant $C_j$ as in solving Eq. 5-7. There is an additional arbitrariness in the eigenvectors if some of the 3N − 6 $\omega_j$ are degenerate, as they will be for normal modes that transform as E or T-irreducible representations. However they can always be chosen such that

$$\Sigma_k\, \Sigma_\alpha\, e_\alpha(k\,|\,j) e_\alpha(k\,|\,j') = \delta_{jj'}$$

$$\Sigma_j\, e_\alpha(k\,|\,j) e_\beta(k'\,|\,j) = \delta_{\alpha\beta}\, \delta_{kk'} \tag{5-10}$$

A set of **normal coordinates** $q_j$, which will diagonalize the Hamiltonian, can be written in terms of eigenvectors as follows

$$q_j = \Sigma_k \Sigma_\alpha e_\alpha(k \mid j) w_\alpha(k) \tag{5-11a}$$

$$w_\beta(k') = \Sigma_j e_\beta(k' \mid j) q_j \tag{5-11b}$$

The simplicity of Eq. 5-11a cannot be overemphasized. It says that the jth normal mode is just made up of a specific linear combination of the 3N possible motions of the N atoms. Thus, the eigenvector gives the specific displacement of each of the atoms that is required for each of the j-normal modes. Using Eqs. 5-8, 5-10, and 5-11, the sum of the kinetic and potential energies of Eq. 5-6 can be written as

$$H = 1/2\, \Sigma_j P_j^2 + 1/2\, \Sigma_i \omega_i^2 q_i^2 \tag{5-12}$$

where $P_j = -i\hbar\, \partial/\partial q_j$. Solutions for this uncoupled Hamiltonian are the well-known simple harmonic oscillation solutions, each oscillator having a frequency $\omega$. Details of the solutions will be discussed in Chapter 6. If a photograph was taken of the motion of the molecule with one normal mode excited, the displacements of all the atoms from equilibrium are proportional to $e_\alpha(k \mid j)/(m_k)^{1/2}$, from Eq. 5-11. Thus, sometimes the word normal mode is used interchangeably with eigenvector.

From the form of the potential energy in Eq. 5-12, we can show that the $q_j$ transform as irreducible representations of the point group of the molecule. The potential energy transforms into itself under all the symmetry operations of the molecule as do the frequencies $\omega_i$. If $\omega_i$ is not equal to any other $\omega_j$, then $q_i^2$ must transform into itself under all the symmetry operations; $q_i$ must transform into $+1$ or $-1$ times itself under all the symmetry operations, or $Rq_i = \pm q_i$. Hence, $q_i$ transforms as a 1-dimensional irreducible representation. If, on the other hand, three frequencies are equal $\omega_1 = \omega_2 = \omega_3$, the potential energy will have a term $\omega_1^2\ (q_1^2 + q_2^2 + q_3^2)$. Then $q_1$, $q_2$, $q_3$ transform among themselves under all the symmetry operations of the group and are partners of a 3-dimensional irreducible representation. Equation 4-15 shows that the sum of the squares of partners transform into exactly itself. If only, say, $q_1$ and $q_2$ transform between themselves, but $\omega_1 \approx \omega_3$ we have an accidental degeneracy. This offers no problems. It is very unusual and is discussed in Chapter 7.

## 5-3 Number and Type of Normal Modes for Molecules

The application of Eq. 5-2 to a molecule results in a $3N \times 3N$ matrix for each symmetry operation of the group. There are h matrices of $3N \times 3N$ dimensions which form a representation of the group as shown by Eq. 5-3. One need only reduce this reducible representation to determine how many times each irreducible representation is contained in it. This result reveals how many normal modes transform as a given irreducible representation. Note that in order to reduce a reducible representation it is not necessary to find a similarity transformation as in Eq. 3-5. Rather, the reduction can be accomplished very quickly by considering only the characters of the reducible representation and using Eq. 4-7.

To fix these ideas firmly we apply them to the $H_2O$ molecule. Figure 5-1c shows this molecule with a separate coordinate system fixed to each atom. Each coordinate system is used to describe each one of the 3N $u_\alpha$ arbitrary displacements discussed in the last section. (See Eq. 5-2.) We now find the 3N × 3N matrix $\Gamma(R)$ that describes the transformation properties of the arbitrary displacements under the four symmetry operations of the point group $C_{2v}$. For each example under a $C_2$ and $\sigma_v(yz)$ symmetry operation, the results are shown in Eq. 5-13.

$$
C_2 \begin{bmatrix} x_1 \\ y_1 \\ z_1 \\ x_2 \\ y_2 \\ z_2 \\ x_3 \\ y_3 \\ z_3 \end{bmatrix} = \left[\begin{array}{ccc|ccc|ccc} & & & -1 & 0 & 0 & & & \\ & 0 & & 0 & -1 & 0 & & 0 & \\ & & & 0 & 0 & 1 & & & \\ \hline -1 & 0 & 0 & & & & & & \\ 0 & -1 & 0 & & 0 & & & 0 & \\ 0 & 0 & 1 & & & & & & \\ \hline & & & & & & -1 & 0 & 0 \\ & 0 & & & 0 & & 0 & -1 & 0 \\ & & & & & & 0 & 0 & 1 \end{array}\right] \begin{bmatrix} x_1 \\ y_1 \\ z_1 \\ x_2 \\ y_2 \\ z_2 \\ x_3 \\ y_3 \\ z_3 \end{bmatrix}
$$

$$
\Gamma(\sigma_v(yz)) = \left[\begin{array}{ccc|ccc|ccc} -1 & 0 & 0 & & & & & & \\ 0 & 1 & 0 & & 0 & & & 0 & \\ 0 & 0 & 1 & & & & & & \\ \hline & & & -1 & 0 & 0 & & & \\ & 0 & & 0 & 1 & 0 & & 0 & \\ & & & 0 & 0 & 1 & & & \\ \hline & & & & & & -1 & 0 & 0 \\ & 0 & & & 0 & & 0 & 1 & 0 \\ & & & & & & 0 & 0 & 1 \end{array}\right] \qquad (5\text{-}13)
$$

The partitioning is helpful for keeping track of what is going where. The 9 × 9 matrix on top is $\Gamma(C_2)$. $\Gamma(C_2)$ contains mostly zeros, as is typical; it can have at most three nonzero entries in each row and column. The character of this matrix is $-1$ and in Table 5-1a it is listed under the $C_2$ operation in the $\chi[\Gamma(R)]$ row. Equation 5-13 also shows the matrix

**Table 5-1a** Character table for the point group $C_{2v}$ and the characters of the reducible representations for the vibrational mode problem

| $C_{2v}$ | E | $C_2$ | $\sigma_v(xz)$ | $\sigma_v(yz)$ |
|---|---|---|---|---|
| $A_1$ | 1 | 1 | 1 | 1 |
| $A_2$ | 1 | 1 | −1 | −1 |
| $B_1$ | 1 | −1 | 1 | −1 |
| $B_2$ | 1 | −1 | −1 | 1 |
| $\chi[\Gamma(R)]$ | 9 | −1 | 1 | 3 |
| a | 2 | 0 | 0 | 2 |
| b | 1 | 1 | 1 | 1 |

representation of the $\sigma_v(yz)$ symmetry operation. Its character 3 is also listed in the appropriate column in Table 5-1a. The matrix representation of the E symmetry operation obviously consists of 1's along the diagonal and zeros in every other position. Thus, the character is 9 as shown in Table 5-1a. (Clearly the character of the E symmetry operation will always be 3N.) The representation of the remaining symmetry operation can be easily worked out and the character is 1, as shown. We only need the character of the representation since the representation will be reduced by use of Eq. 4-7 and not by a similarity transformation, Eq. 3-5. Thus, there is no reason to write out or consider the representation itself in any detail. In fact, **only those atoms that transform into themselves under a symmetry operation need be considered,** since a contribution to the character can come only from such coordinate systems. (See Problem 4-2.) Reducing the reducible representation, the characters of which are given in Table 5-1, by Eq. 4-7 gives

$$\Gamma = 3A_1 + A_2 + 2B_1 + 3B_2 \tag{5-14}$$

One could then write, for each of the four symmetry operations of $C_{2v}$ symmetry, a 9 × 9 matrix for the representation in block form expressed by Eq. 5-14. This is the meaning of Eq. 5-14. Of course, dimensionality is conserved in Eq. 5-14 and we have all nine degrees of freedom. However, only the three internal degrees of freedom are desired. The three translational degrees of freedom transform as x, y, and z, or for this point group ($C_{2v}$) as $A_1 + B_1 + B_2$. The three rotational degrees of freedom transform as $R_x$, $R_y$, and $R_z$, or for this point group $A_2 + B_1 + B_2$.

**Table 5-1b** Characters of the reducible representation for $SF_6$ and $PbTiO_3$

| $O_h$ | E | $8C_3$ | $6C_2$ | $6C_4$ | $3C_2$ | i | $6S_4$ | $8S_6$ | $3\sigma_h$ | $6\sigma_d$ |
|---|---|---|---|---|---|---|---|---|---|---|
| $\chi[\Gamma(R)]$ | 21 | 0 | −1 | 3 | −3 | −3 | −1 | 0 | 5 | 3 |
| a | 6 | 0 | 0 | 2 | 2 | 0 | 0 | 0 | 4 | 2 |
| b | 12 | 0 | 2 | 0 | 0 | 0 | 0 | 0 | 4 | 2 |
| Pb | 3 | 0 | −1 | 1 | −1 | −3 | −1 | 0 | 1 | 1 |
| Ti | 3 | 0 | −1 | 1 | −1 | −3 | −1 | 0 | 1 | 1 |
| 3O | 9 | 0 | −1 | 1 | −3 | −9 | −1 | 0 | 3 | 1 |

Subtracting these six degrees of freedom from Eq. 5-14, one obtains for the internal degrees of freedom

$$\Gamma_{vib} = 2A_1 + B_2 \tag{5-15}$$

This agrees with the result in Fig. 5-1b. Again we remark that had we taken the xz instead of the yz plane to be the plane of the molecule, the labeling of the normal modes would be different. (See the Problems.) Thus it is important to state the coordinate system with respect to the symmetry operations in the character table.

For practice we work a few examples. First consider a molecule $AB_4$ with the atoms atthe corners of a square prism (see Fig. 5-2a). The point group is $C_{4v}$. For the character of the reducible representation for each of the symmetry operations one should obtain

| $C_{4v}$ | E | $2C_4$ | $C_2$ | $2\sigma_v$ | $2\sigma_d$ |
|---|---|---|---|---|---|
| $\chi[\Gamma(R)]$ | 15 | 1 | −1 | 3 | 1 |

(5-16)

This reduces to $\Gamma = 3A_1 + A_2 + 2B_1 + B_2 + 4E$. Subtracting the three translational degrees of freedom $(A_1 + E)$ and the three rotational ones $(A_2 + E)$, one finds that the nine normal modes of vibration transform as $2A_1 + 2B_1 + B_2 + 2E$.

For the tetrahedral $CH_4$ (Fig. 5-2b) with $T_d$ point symmetry one should obtain

| $T_d$ | E | $8C_3$ | $3C_2$ | $6S_4$ | $6\sigma_d$ |
|---|---|---|---|---|---|
| $\chi[\Gamma(R)]$ | 15 | 0 | −1 | −1 | 3 |

(5-17)

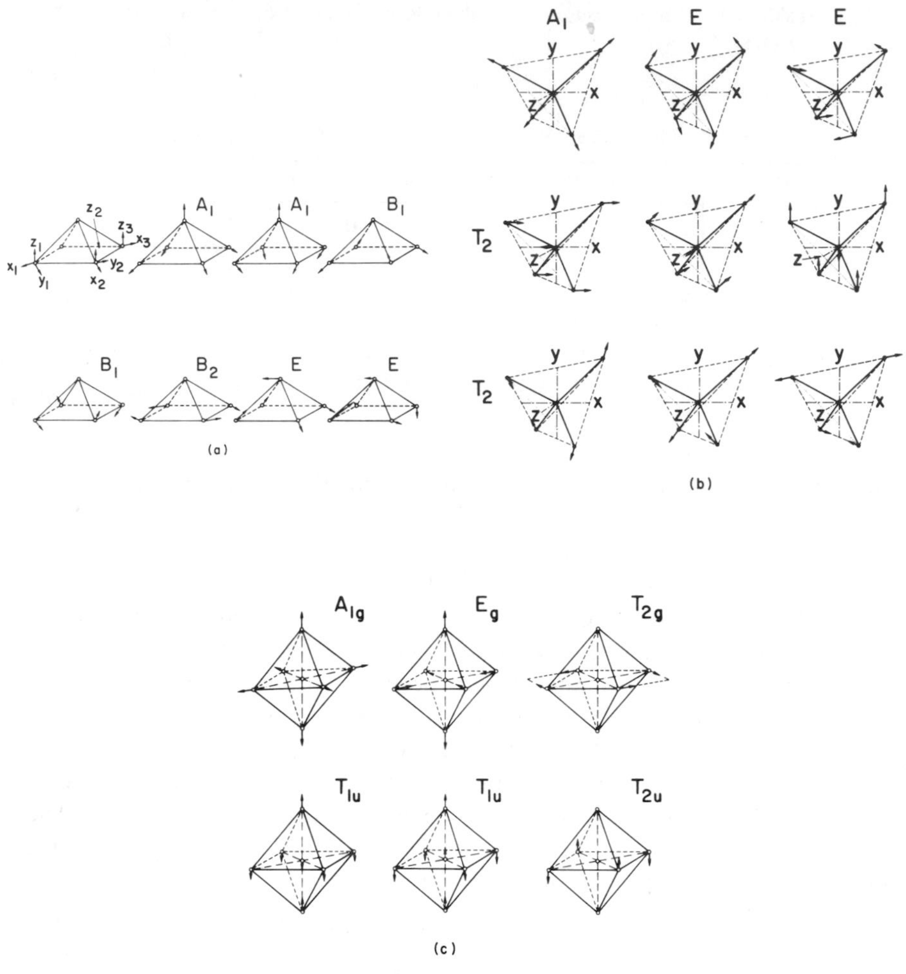


**Fig. 5-2** (a) A coordinate system for an $AB_4$ molecule. For clarity the coordinate system is left off the B-atom in the back and the apex-atom. The various normal modes are shown. (b) The normal modes of a tetrahedral molecule such as $CH_4$. (c) The normal modes of an octahedral molecule such as $SF_6$.

which reduces to $\Gamma = A_1 + E + T_1 + 3T_2$. Subtracting the six nonvibrational modes one obtains the normal modes of vibration transform as $A_1 + E + 2T_2$.

For a seven-atom molecule with octahedral symmetry $O_h$, such as $SF_6$, Fig. 5-2c, one finds the results shown in Table 5-1b. Reducing it and subtracting the translations $T_{1u}$ and the rotations $T_{1g}$, the true internal normal modes transform as $A_{1g} + E_g + T_{2g} + 2T_{1u} + T_{2u}$.

Note, in all of these examples one could immediately subtract from the character of the reducible representation, $\chi[\Gamma(R)]$, the characters of the three translational and three rotational degrees of freedom. The resulting set of characters would contain only the true vibrational modes and would be easier to reduce by Eq. 4-7 or by inspection.

The character of an atom that is not displaced by the action of a symmetry operation can be determined in general. If the coordinates x, y, z describe an arbitrary displacement from equilibrium, then under a $C_n^{\,m}$ proper rotation

$$C_n^{\,m}\begin{bmatrix} x \\ y \\ z \end{bmatrix} = \begin{bmatrix} \cos 2\pi m/n & -\sin 2\pi m/n & 0 \\ \sin 2\pi m/n & \cos 2\pi m/n & 0 \\ 0 & 0 & 1 \end{bmatrix}\begin{bmatrix} x \\ y \\ z \end{bmatrix}$$

$$\chi(C_n^{\,m}) = 1 + 2\cos 2\pi m/n \qquad (5\text{-}18)$$

For $S_n$ point operation a $-1$ instead of $+1$ is obtained in this equation for the character because the symmetry operation inverts the z-component of the coordinate system.

## 5-4 Internal Coordinates

In the previous section all 3N degrees of freedom were obtained in each example. Since we know the irreducible representation under which the uniform rotations and translations transform, we might like to simplify the problem to obtain only the 3N − 6 internal vibrations. Often we can do this by using only internal coordinates instead of all 3N coordinates and insight into the actual motion or eigenvector can be obtained.

As **internal coordinates we will consider bond lengths and bond angles.** For the water molecule in Fig. 5-1c, consider the stretching of the bond length between hydrogen and oxygen 1 as bond length 1 and the other hydrogen–oxygen as bond length 2. The increase of the angle between these bonds is the third internal coordinate. In this simple problem there are three natural independent internal coordinates. The three vibrational modes should then be obtained directly and the uniform rotations and translations should never enter the calculation. In Table 5-1a lines a and b have the characters of the representation for the two bond lengths and the one bond angle, respectively. The character for the transformation of the bond lengths can be found separately from the bond angle since the bond lengths transform only between themselves. When such separations occur it is always best to determine the character separately, eliminating the possibiity of mistakes, and reduce each set of characters separately by inspection or by Eq. 4-7. Under the E symmetry operation, the bond lengths transform into themselves so the character of the 2-dimensional representational is 2; under $C_2$ they transform into each other resulting in a character of the representation of zero; likewise, under $\sigma_v(xz)$; under $\sigma_v(yz)$ they again transform into themselves. An increase in the bond angle will transform into inself under all the symmetry operations of the group. Reducing the representation, the characters of which are shown in line a, by inspection of Eq. 4-7 yields $A_1 + B_2$. Line b yields $A_1$ thus the previous result is obtained but with less work and insight into normal modes is obtained. Increases in bond length will take place in normal modes that transform as $A_1$ and $B_2$ irreducible representations but changes in the bond angle will only take place in $A_1$ vibrations. This agrees with the normal modes in Fig. 5-1b.

For the octahedral molecule $SF_6$ we take the internal coordinates to be the six bond lengths and the twelve bond angles. This amounts to 18 coordinates yet there are only 15 true internal coordinates. The three extra coordinates derive from the fact that while the six bond lengths are independent of one another, the twelve bond angles are not. For example, they cannot all increase at the same time. Neglecting this aspect of the problem for a moment, in Table 5-1b lines a and b have the characters of the representation for the lengths and angles, respectively. Reducing the representations yields $a = A_{1g} + E_g + T_{1u}$ and $b = A_{1g} + E_g + T_{2g} + T_{1u} + T_{2u}$. We immediately know that the $A_{1g}$ in the b-reduction is not an internal mode since a mode with an increase in all twelve angles would transform just as $A_{1g}$ would. Since there are just two more modes to eliminate, they must be the partners in the $E_g$ mode. (Actually it is difficult to picture the overdetermination of the angles for this mode. See the Notes.) The 15 true internal modes then agree with what was previously obtained. Thus, bond length changes do not occur in the normal modes that transform as $T_{2g}$ and $T_{2u}$ irreducible representations.

## 5-5 Crystals

For crystals the situation is much more complicated as might be expected since 3N is a large number. We will consider just the infinite wavelength or zero wave vector vibrations (wave vector $k = 2\pi/\lambda$). This condition implies that the displacements of one unit cell are identical to the displacements of all the other cells. Thus, the translational symmetry is preserved. To consider the opposite extreme, imagine the motions in one unit cell to be $\pi$ out of phase with its neighbor (i.e., vibrating in the opposite direction). This repeats throughout the crystal. Thus, starting at one unit cell (origin) the next cell is out of phase, the next is in phase with the origin, then out of phase, etc. If the unit cell is of length $l$, then the wavelength of the vibration is clearly $2l$ and $k = \pi/l$ which is just the Brillouin zone boundary. Wavelengths between these two extremes clearly can be obtained and each of the allowed wavelengths corresponds to the allowed points in the Brillouin zone. This will be discussed in Chapter 12 in detail when electronic states are considered. It should be pointed out however,that the symmetry considerations for the electron states in a crystal and the normal modes of vibration of the crystal are the same.

Consideration of the $k \approx 0$ vibrations is very important besides being easily within our grasp. It is just these very long wavelength vibrations that are measured in infrared and Raman spectroscopy. Thus, we proceed to consider the $k = 0$ vibrations in solids.

If n is the number of atoms in a primitive unit cell of a crystal (i.e., the smallest cell that still has the translational properties of the crystal), then there are 3n degrees of freedom for any one wavelength including $k = 0$. Of these 3n modes, three are called acoustic modes involving the positive and negative charge moving in the same direction. The frequency of this acoustic mode is zero for $k = 0$ (i.e., the restoring force is zero). (In fact, since the motion in all the unit cells is in phase, this motion represents the uniform translation of the entire crystal and is one of the three degrees of translational freedom subtracted from the 3N modes of the crystal.) Again the acoustic modes transform as the position coordinates x, y, and z, and can be subtracted from the result. Thus, there are $3n - 3$ optic normal modes. It is these $3n - 3$ optic modes that can be infrared or Raman active.

**a. Symmorphic crystals**

(Recall, from Section 1-6, that a symmorphic crystal is one which has no glide planes or screw axes as symmetry operations.) Consider a unit cell of $O_h$ symmetry that consists only of octahedral linked as in Fig. 5-2c. The cubic crystal $WO_3$ would be an example and the figure of $PbTiO_3$, Fig. 1-2d would be appropriate if the Pb ions were ignored. One can consider each oxygen ion as shared by two unit cells. For crystalline $WO_3$ there are only nine $k \approx 0$ optic vibrational $(4 \times 3-3)$ modes com-

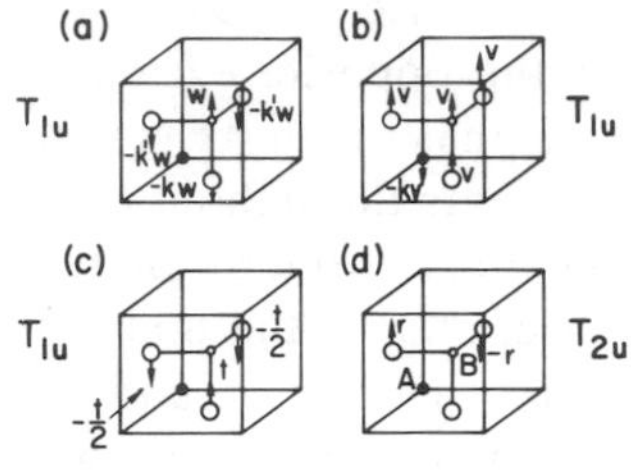


**Fig. 5-3** Four normal modes at k ≈ 0 for a $PbTiO_3$ crystal.

pared to 15 for $SF_6$. Without doing any work, it is clear that certain of the normal modes pictured in Fig. 5-2c cannot be k = 0 modes, for such modes require that oxygen ions on opposite faces have identical motion to satisfy translational symmetry. The $T_{2u}$ mode in Fig. 5-2c does have the appropriate oxygen ions moving identically. The $T_{1u}$ modes also satisfy the k = 0 criterion. However, the other modes in Fig. 5-2c, as pictured, cannot be k = 0 modes. The same general considerations apply to all crystals. Consider the similar crystal $PbTiO_3$ as shown in Fig. 1-2d. Let us solve the problem the usual way by attaching a coordinate system to each of the ions and determining the character of the representation for each symmetry operation. The characters are listed separately for each type of ion in Table 5-1b. (Note that one type of ion can transform only into like ions so there is no need to treat the entire cell as one entity. This is helpful and always can be done in molecules as well as crystals. See Problem 1.) The reduction is Pb = $T_{1u}$; Ti = $T_{1u}$; 3O = $2T_{1u} + T_{2u}$. The result for the optic normal modes at k = 0 is obtained by adding up the three results and subtracting the acoustic modes that transform as $T_{1u}$. Thus we obtain $3T_{1u} + T_{2u}$ for the optic normal modes. In the discussion in the above paragraph these types of modes were shown to be particularly favorable for k = 0 vibrations. The result has a unique $T_{2u}$ mode but there are three 3-fold degenerate $T_{1u}$ modes. Figure 5-3 shows these four types of optic modes. The $T_{2u}$ mode is just the same as shown in Fig. 5-2c for the $SF_6$ octahedral. It is clear that since Pb and Ti ions do not transform as a $T_{2u}$ mode, they cannot be involved in the $T_{2u}$ motions. The other three $T_{1u}$ modes are not unique since any appropriate linear combination of these modes also transforms as $T_{1u}$. The particular modes shown were all chosen somewhat arbitrarily but on physical grounds. They could be obtained exactly with knowledge of the force constants. Mode (c) involves mostly relative motion of oxygen ions against each other. In the $PbTiO_3$ crystal the force constants for this motion is expected to be the largest so this mode should have the highest frequency.

Mode (b) involves an approximately rigid $TiO_6$ octahedron vibrating against the Pb ions. This should be of intermediate frequency. Mode (a) involves for the most part the Ti ion vibrating against a rigid oxgyen octahedron. Based on understanding of the ferroelectric phenomena, a mode mostly composed of this motion should be the lowest in frequency (smallest restoring force).

A $WO_3$ crystal has the same cubic structure as $PbTiO_3$ in Fig. 1-2d with the Pb ion site empty. Thus the optic modes transform as $2T_{1u}$ + $T_{2u}$. Our normal modes in Fig. 5-3 would be appropriate if, for example, we leave out the (b) mode.

### b. Nonsymmorphic crystals

(Recall, from Section 1-6, that a nonsymmorphic crystal is one which does have glide planes and/or screw axes as symmetry operations.) The $k = 0$ modes of nonsymmorphic crystals present no new difficulties and one only has to become familiar with the symmetry operations. References to the symmetry operations for all 230 space groups can be found in the Notes at the end of this chapter. Although the space groups in general will be discussed in detail later, we can proceed now for $k = 0$ modes.

Table 5-2 lists all the symmetry operations appropriate to the $D_{4h}{}^{14}$ space group which describes the rutile form of $TiO_2$ (see Fig. 1-6). Also, above the normal headings, the point group operations which are isomorphic to the symmetry operations of the space group are listed. $\boldsymbol{\tau} = (\mathbf{a}, \mathbf{b}, \mathbf{c})/2$, i.e., a translation along the x-axis by a/2, the y-axis by a/2, and the z-axis by c/2, where a and c are the usual tetragonal unit cell lengths as pictured. Then the operation $\{C_4 | \boldsymbol{\tau}\}$ is a $C_4$ rotation followed by a translation by an amount $\boldsymbol{\tau}$. If the atom is moved out of the unit cell shown, it can always be moved back by a primitive cell distance, namely, an amount **a** in the x- or y-direction or **c** in the z-direction. Then $\{C_4 | \boldsymbol{\tau}\}$ takes atom 1 into atom 4, etc. ($1 \rightarrow 4$; $2 \rightarrow 3$; $4 \rightarrow 2$; $3 \rightarrow 1$; $5 \rightarrow 6$). Note that all eight Ti atoms at the corners of the unit cell are equivalent by a unit cell translation **a** or **c**, just as the oxgyen atoms on the top face are equivalent to the oxygen atoms 1 and 2 on the bottom face. Since the symmetry operation $\{C_4 | \boldsymbol{\tau}\}$ interchanges all atoms, the character of the representation of the 3n arbitrary displacements is zero for both the Ti and O ions as listed in Table 5-2. The operation $\{C_2{}^y | \boldsymbol{\tau}\}$ takes $5 \rightarrow 6$; $1 \rightarrow 4$; $2 \rightarrow 3$; $3 \rightarrow 2$; $4 \rightarrow 1$; again giving zero character; $\{S_4 | \boldsymbol{\tau}\}$ takes $5 \rightarrow 6$; $1 \rightarrow 4$; $2 \rightarrow 3$; $4 \rightarrow 2$; $3 \rightarrow 1$; and $\{\sigma_v{}^y | \boldsymbol{\tau}\}$ takes $5 \rightarrow 6$; $1 \rightarrow 4$; $2 \rightarrow 3$; $3 \rightarrow 2$; $4 \rightarrow 1$. For all the

**Table 5-2** Character of the reducible representation for the k $\approx$ 0 normal mode problem for the rutile form of $TiO_2$ (space group $D_{4h}{}^{14}$). (The isomorphic symmetry operations of the point group $D_{4h}$ are shown in the appropriate places.)

| | $E$ | $2C_4$ | $C_2$ | $2C_2{}^x$ | $2C_2''$ | $i$ | $2S_4$ | $\sigma_h$ | $2\sigma_v$ | $2\sigma_d$ |
|---|---|---|---|---|---|---|---|---|---|---|
| $D_{4h}$ | $E$ | $2\{C_4 \mid \tau\}$ | $C_2$ | $2\{C_2{}^x \mid \tau\}$ | $2C_2''$ | $i$ | $2\{S_4 \mid \tau\}$ | $\sigma_h$ | $2\{\sigma_v \mid \tau\}$ | $2\sigma_d$ |
| $T_i$ | 6 | 0 | −2 | 0 | −2 | −6 | 0 | 2 | 0 | 2 |
| O | 12 | 0 | 0 | 0 | −2 | −6 | 0 | 4 | 0 | 2 |

operations involving $\tau$ the character is zero because no atom transforms into itself. Although this is often true, it is not always true. The symmetry operations that do not involve the nonprimitive translation $\tau$ are the usual point operations and the results in Table 5-2 can be checked in the usual manner. Reducing the reducible representations, one obtains Ti = $A_{2u} + B_{1u} + 2E_u$ and O = $A_{1g} + A_{2g} + B_{1g} + B_{2g} + E_g + A_{2u} + B_{1u} + 2E_u$. The acoustic modes transform as $A_{2u} + E_u$. Thus the optic modes transform as $A_{1g} + A_{2g} + B_{1g} + B_{2g} + E_g + A_{2u} + 2B_{1u} + 3E_u$. Only oxygen motions are involved in the g-modes, but both oxygen and titanium ions are involved in the u-modes.

## 5-6 Eigenvectors and Symmetry Adapted Vectors

So far we have considered how all of the normal modes of vibration transform under all the symmetry operations of the point group, i.e., according to which irreducible representation they transform. However, we have not determined the normal modes or eigenvectors in terms of the displacements of the atoms. In this section we make some general comments about the use of symmetry in considerably simplifying the problem and show how some eigenvectors can be obtained. In the next section projection operator techniques are introduced. These techniques enable one to attack systematically this problem.

Figures 5-1b and 5-2 show these normal modes or a snap shot of the motion. Group theory will never give any information about the frequency of the normal mode of vibration. To calculate a frequency the secular equation, Eq. 5-9, must be solved. However, if only one normal mode transforms as a given irreducible representation then projection operator techniques can be used to determine the normal mode, or more exactly the eigenvector Eq. 5-11. From the knowledge of the eigenvector, the dynamical matrix can be put in block form, directly yielding the frequency of this particular normal mode in terms of the force constants. If a number of normal modes transform as a given irreducible representation, then projection operator techniques can only determine a linear combination of the eigenvectors involved, and these are called **symmetry-adapted vectors**. If M normal modes transform as a given irreducible representation, then the symmetry-adapted vectors can be used to obtain an M × M block in the dynamical matrix. Within this block, the dynamical matrix must be solved algebraically to determine the frequencies and the eigenvectors. These more formal aspects of the vibrational problem will only be mentioned briefly in this chapter. There are several references in the Notes to entire books devoted to the vibrational problem. The eigenvector for a mode that transform as 1-dimensional irreducible representation can be found with a little imagination.

All molecules will have a **uniform breathing mode** which must transform as $\Gamma_1$. The motions of this mode only involve displacement away (or toward) the origin (the center of mass). It is clear from Section 4-6 that such a motion is equivalent to a change of the atomic positions, retaining the point symmetry of the molecule, and transforms as $\Gamma_1$.

Other 1-dimensional modes are often easy to determine. For the $B_2$ mode in $H_2O$, it is clear that the vibration must be in the plane of the molecule because under $\sigma_v(yz)$ it must transform into itself. Also the motions must be oppositely directed because of the $-1$ character under $C_2$ and $\sigma_v(xz)$. (Remember the vibration must not rotate the molecule as a whole.)

For the $B_1$ mode in the $AB_4$ molecule with $C_{4v}$ symmetry, the $+1$ character under $\sigma_v$ (the character table is given in Appendix 3) fixes the motions to the $\sigma_v$ planes. The character for $C_2$ fixes the motions of opposite B atoms. The character for the $2C_4$ fixes the opposite-directed motions on alternating B atoms. Similarly for $B_2$ the $-1$ character of $\sigma_v$ fixes the motions to be in the basal plane, etc.

Another general approach to modes is to use the fundamental properties of basis functions. If a normal mode transforms as x, y, or z then the motion of the normal mode must have an induced polarization as in Section 4-5b. Thus, for such modes the center of gravity of the positive and negative charge must separate. For example, in Fig. 5-2c one can see that a polarization results from the motion of the $T_{1u}$ mode in $O_h$. Also, one can see in Fig. 5-2, that modes that do not transform as $T_{1u}$ (x,y,z) do not have a polarization resulting from their motion. In the same manner it can be shown that the motion of a normal mode that transforms as any of the six components of polarizability, $x_i x_j$, must induce a change in polarizability in the molecule. However, in practice this approach is difficult to visualize. These points will be brought up again when the selection rules for infrared and Raman spectra are discussed in the next chapter.

Consider $SF_6$ with $O_h$ symmetry. It should be clear that Fig. 5-2 shows only one partner of the T-modes. For the $T_{1u}$ modes the motions along the z-axis are shown. The other partners are identical motions along the x- and y-axes. Given one partner of a T-mode, the symmetry operation $C_3$ will quickly determine the other two T-mode partners.

A last, but most important, point should be made again. When two or more modes transform as one irreducible representation, any linear combinations of the two orthogonally chosen modes wiil satisfy all the symmetry requirements. For example, the two $A_1$ modes for $AB_4$ in Fig. 5-2a are orthogonal and each has $A_1$ symmetry. In such cases the force constants between the atoms are required to solve the complete problem as previously mentioned. Then two different $A_1$ modes, often with widely separated frequencies, will be determined. The same considerations apply to the two $T_{1u}$ modes in Fig. 5-2c.

## 5-7 Projection Operators

The basis function that belong to the ith row of the mth unitary irreducible representation is $\psi_i{}^m$ as in Section 4-4. When operated on by a symmetry operation the result can be expressed as a linear combination of its $l$ partners,

$$R\psi_i{}^m = \sum_{j=1}^{l} \Gamma_m(R)_{ji}\, \psi_j{}^m \tag{5-19}$$

as in Eqs. 4-8 and 4-9. Multiplying both sides by $\Gamma_n(R)_{op}{}^*$, summing over R, and using GOT (Eq. 3-12), we obtain

$$\begin{aligned} \Sigma_R\, \Gamma_n(R)_{op}{}^*\, R\, \psi_i{}^m &= \Sigma_j\, (h/l_m)\, \delta_{nm}\, \delta_{oj}\, \delta_{pi}\, \psi_j{}^m \\ &= (h/l_m)\, \delta_{nm}\, \delta_{pi}\, \psi_o{}^m \end{aligned} \tag{5-20}$$

On rearranging we define the **transfer projection operator** $V_{oi}{}^m$

$$\psi_o{}^m = \{(l_m/h)\, \Sigma_R\, \Gamma_m(R)_{oi}{}^*\, R\}\psi_i{}^m \equiv \{V_{oi}{}^m\}\, \psi_i{}^m \tag{5-21}$$

The transfer projection operator $V_{oi}{}^m$ operating on a function that belongs to the ith row of the mth irreducible representation will yield its orthogonal partner belonging to the oth row. Given one partner, Eq. 5-21 can be used to determine the other partners. More important, the transfer projection operator operating on an arbitrary function F will either yield zero or the basis function $\psi_o^m$, provided of course that the operation RF is defined. This is a very useful result that quickly enables basis functions (normal modes or symmetry-adapted vectors in this chapter) to be obtained. Thus we wish to show that Eq. 5-19 is satisfied for the function

$$\psi_o^m = V_{oi}^m F \tag{5-22}$$

i.e., that the function on the left obtained by the operations on the right does indeed transform as a basis function. To prove this we operate on Eq. 5-22 by another symmetry operation S such that T = SR

$$\begin{aligned} S\psi_o^m &= (l_m/h)\,\Sigma_R\,\Gamma_m(R)_{oi}^*\,SRF = (l_m/h)\,\Sigma_T\,\Gamma_m(S^{-1}T)_{oi}^*\,TF \\ &= (l_m/h)\,\Sigma_T\,\Sigma_p\,\Gamma_m(S^{-1})_{op}^*\Gamma_m(T)_{pi}^*\,TF \\ &= \Sigma_p\,\Gamma_m(S^{-1})_{op}^*\{V_{pi}^m F\} \\ &= \Sigma_p\,\Gamma_m(S)_{po}\{V_{pi}^m F\} = \Sigma_p\Gamma_m(S)_{po}\,\psi_p^m \end{aligned} \tag{5-23}$$

Thus, if a function results from the operation $V_{pi}^m$ F, it will have the properties required of a basis function Eq. 5-19. (Note the unitary property is used in the above proof.) The quantities $l_m/h$ in the definition of the transfer projection operator do not normalize the basis function when they are obtained from Eq. 5-22. Functions that result from using this equation should be separately normalized in the way appropriate for the particular problem.

Using just the diagonal elements of the representation, Eq. 5-21 yields the projection operator

$$\psi_i^m = V_{ii}^m\,\psi_i^m \tag{5-24}$$

This is not very useful by itself but when the diagonal components of the projection operator operate on an arbitrary function, either zero or a basis function (unnormalized in general) is still obtained, similar to Eq. 5-22

$$\psi_i^m = V_{ii}^m F \tag{5-25}$$

The use of the projection operators to obtain basis functions from an arbitrary function, Eqs. 5-22 and 5-25, requires the knowledge of the matrix representation. Information can also be obtained from the characters. From the definition of the transfer projection operator in Eq. 5-21, we can define the character projection operator

$$V^m = \sum_{i=1}^{l_m} V_{ii}^m = (l_m/h)\,\Sigma_R\,\chi_m(R)^*\,R \tag{5-26}$$

When operating on an arbitrary function F, this operator will project out of F the sum of all parts transforming as the rows of the mth irreducible representation. The result will be zero if F contains no part that transforms as the mth irreducible representation and therefore, it would be useful to apply Eq. 5-27 to F before applying Eq. 5-22 or 5-25.

$$\psi^m = V^m F \tag{5-27}$$

If the result of the character projection operator is nonzero then the resulting function can be taken as a basis function. This is true because any linear combination of basis functions is also a basis function of an equivalent representation. For 1-dimensional irreducible representations, Eqs. 5-22, 5-25 and 5-27 reduce to the same equation and the use of projection operator techniques is very simple and almost intuitive.

It should be pointed out that any arbitrary function F, that can be operated on by the symmetry operations can be decomposed into the sum of functions that transform as the $l_m$-partners of the m-irreducible representation as

$$F = \Sigma_m \sum_{i=1}^{l_m} \psi_i^m \tag{5-28}$$

This is proven in Wigner, Chapter 12. The result is intuitive since any function, $x^4y^2z$, for example, when operated on by a symmetry operation, transforms into some other similar looking function. Perhaps $f = x^4y^2z - y^4x^2z$ and $g = x^4y^2z + y^4x^2z$ where f and g are basis functions of two particular irreducible representations. Then $x^4y^2z = (f + g)/2$. (See Problem 9.)

## 5-8 Projection Operators Applied to Normal Coordinates

We now apply the techniques of the last section to describe the eigenvectors or symmetry-adapted vectors of vibrations of molecules and crystals. Recall that symmetry-adapted vectors (sometimes called symmetry coordinates in other books) are linear combinations of eigenvectors that transform as the same irreducible representations. If more than one eigenvector transforms as a given irreducible representation, then projection operator techniques are not able to separate the eigenvectors. All that can be obtained is a set of symmetry-adapted vectors. The dynamical matrix must be diagonalized numerically to determine the appropriate linear combinations of symmetry-adapted vectors which yield eigenvectors.

Consider molecules with $C_{2v}$ point symmetry. The projection operators for the four 1-dimensional irreducible representations of $C_{2v}$ are using Eq. 5-26

$$\begin{aligned} V^{A1} &= E + C_2 + \sigma_v(xz) + \sigma_v(yz) \\ V^{A2} &= E + C_2 - \sigma_v(xz) - \sigma_v(yz) \\ V^{B1} &= E - C_2 + \sigma_v(xz) - \sigma_v(yz) \\ V^{B2} &= E - C_2 - \sigma_v(xz) + \sigma_v(yz) \end{aligned} \tag{5-29}$$

The factor of 1/4 has been left out since it is irrelevant because the function that results from projection operators must be separately normalized. We apply these operators to the $H_2O$ molecule which has this symmetry. The coordinates are shown in Fig. 5-1c and there are normal vibrational modes of $A_1$ and $B_2$ symmetry only. Treating the $A_1$ mode first, the relative motions along the x-axis are determined by Applying $V^{A1}(x_1) = x_1 - x_2 + x_2 - x_1 = 0$ and $V^{A1}(x_3) = 0$. Thus, there is no motion along the x-axis for modes with $A_1$ symmetry. Next we determine relative motions along the y-axis: $V^{A1}(y_1) = y_1 - y_2 - y_2 + y_1 = 2(y_1 - y_2)$ and $V^{A1}(y_3) = 0$. Thus, for both $A_1$ modes the oxygen does not move in the y-direction, but both hydrogen ions are vibrating in a manner which first brings them toward each other, then away from each other. Immediately, we also know that when the hydrogen ions are moving toward the center of mass, with the condition $y_1 = -y_2$, the oxygen must move in phase along the z-axis to keep the center of mass stationary. We know that projection operator techniques on the z-coordinates will not be too helpful since uniform motion of the entire molecule along the z-axis also transforms as $A_1$. Nevertheless we proceed, $V^{A1}(z_1) = 2(z_1 + z_2)$ and $V^{A}{}_1(z_3) = 4z_3$. This result can be interpreted as uniform motion of the entire molecule or, the two hydrogen atoms moving in phase along the z-axis (i.e., $z_1 + z_2$ transform as $A_1$) against the oxygen atom to keep the center of mass fixed. Actually, we know that this latter motion must occur since there are $2A_1$ modes. Figure 5-1b shows the two normal modes meeting the conditions determined here.

(Rotations with angular momentum along the z-axis $R_z$ can easily be determined. $R_z$ transforms as $A_2$ and $V^{A2}(x_1) = 2(x_1 - x_2)$, which is just the required rotation.)

Now consider the $B_2$ mode for $H_2O$. Again, there are no motions along the x-axis. $V^{B2}(z_1) = 2(z_1 - z_2)$ and $V^{B2}\, z_3 = 0$, so there is no oxgyen motion along the z-axis but the two hydrogen atoms move with opposite phase along the z-axis. Clearly $y_3$ must be negative to keep the molecule from rotating when $z_1$ is positive and $z_2$ is negative. Again the vibrational motions along the y-axis will be difficult to determine by projection operator alone since the y-vibrational motion will be mixed with the uniform translation which transforms like $B_2$. We already know from Section 5-4 that there are no bond angle changes for $B_2$ modes, thus $y_1 + y_2$ must be the vibrational motion.

One can appreciate that while projection operator techniques can give unique results, this is not always the situation. The particular operator $V_{oi}{}^{m}$ will treat equivalently all functions that transform as the mth irreducible representation. Thus, not only internal modes but motions associated with uniform translation of the center of mass and rotation about the center of mass will be projected out, as encountered in the $H_2O$ problem. Another aspect of the use of projection operators bears mentioning. If there is more than one internal vibration mode that transforms as the mth irreducible representation, the projection operator can not separate the motion even when there is no interference from uniform translation or rotation coordinates. The two $A_1$ modes in the $H_2O$ example showed this problem. Thus, projection operators can determine which functions transform as $A_1$ but can not separate the functions into its two parts since any linear combination of functions that individually transform as the mth irreducible representation individually also transform the same way. To find the two orthogonal vibrational modes, both of which transform as $A_1$, we must know the internuclear forces and solve for the eigenfunctions and eigenvalues. Actually, if the frequency of vibration of even the single $B_2$ mode in $H_2O$ is to be calculated the force constant problem still must be solved. However, if the frequency of the $B_2$ mode is to be experimentally measured and only the eigenvector must be determined then one can say that group theory completely describes the mode.

As another example of unique results that can be obtained by the use of projection operators, we consider the $B_2$ mode of the $AB_4$ molecule having $C_{4v}$ symmetry. For $C_{4v}$ symmetry none of the uniform translation or rotation modes transform as $B_2$ hence there is no interference from those modes. Also there is just one vibrational mode of $B_2$ symmetry. The projection operator is

$$\begin{aligned} V^{B_2} &= E - C_4 - C_4{}^3 + C_2 - \sigma_v{}^x - \sigma_v{}^y + \sigma_d{}' + \sigma_d{}'' \\ V^{B_2}(z_1) &= z_1 - z_2 - z_4 + z_3 - z_1 - z_3 + z_2 + z_4 = 0 \\ V^{B_2}(x_1) &= 0 \end{aligned} \tag{5-30}$$

zero is also obtained for $x_5$, $y_5$, and $z_5$; the only nonzero term comes from $V^{B_2}(y_1) = 2(y_1 - y_2 + y_3 - y_4)$ which is just the mode pictured in Fig.

5-2a. The only thing group theory does not determine about this mode is the frequency of vibration. However, the eigenvector will diagonalize the dynamic matrix to a 1 × 1 part which immediately yields the frequency of the $B_2$ mode.

We apply the projection oprator techniques to nonsymmorphic space groups. The example of $TiO_2$ discussed at the end of the last chapter will be used. We again emphasize that nonsymmorphic space groups for k = 0 offer no more difficulty than the point operations. One must only become familiar with the operations and know the isomorphism between the factor group of the space group and point group. Knowing the correspondence between the operations, as in Table 5-2 the character table for $D_{4h}$ in Appendix 3 can be used. For example, $V^{A_{2u}} = E + \{C_4|\tau\} + \{C_4^3|\tau\} + C_2 - \{C_2^x|\tau\} - \{C_2^y|\tau\}...$, $V^{A_{2u}}(x_5) = 0$ and $V^{A_{2u}}(y_5) = 0$. Thus, while the Ti-ion motion is involved in an $A_{2u}$ normal mode, it has no component of motion along the x- and y-axes. Since the oxygen ions are also involved in the $A_{2u}$ normal mode and there is only one such optic mode, the mode consists of the rigid Ti-ion sublattice vibrating against the rigid oxygen-ion sublattice each with opposite phases along the z-axis so that the center of mass remains fixed.

As an example of the use of projection operators for a degenerate mode we treat the doubly degenerate $E_u$ mode for the Ti-ion motion in $TiO_2$. In this case, Eqs. 5-21, 5-24, and 5-26 are no longer equivalent. Obviously the easiest to use is Eq. 5-26, but it will also provide the least information, therefore, we start by using Eq. 5-24. It is rather easy to determine the diagonal elements of the $E_u$ irreducible representation since x and y are partners of this representation for the $D_{4h}$ point group. Thus we determine the representation from the point operations, but for the projection operator we use the symmetry operations appropriate to the nonsymmorphic space group. For example, under $C_2$, $C_2^y$, i, and $\sigma_v^x$, −1 is obtained for the 11-term in the $E_u$ representation, and for $2C_4$, $2C_2''$, $2S_4$, and $2\sigma_d$, zero is obtained. Similarly, the 22-term can be obtained for all the point symmetry operations. Then, the diagonal projection operators for the space group operations are

$$\begin{aligned} V_{11}^{E_u} &= E - C_2 + \{C_2^x|\tau\} - \{C_2^y|\tau\} - i + \sigma_h - \{\sigma_v^x|\tau\} + \{\sigma_v^y|\tau\} \\ V_{22}^{E_u} &= E - C_2 - \{C_2^x|\tau\} + \{C_2^y|\tau\} - i + \sigma_h + \{\sigma_v^x|\tau\} - \sigma_v^y|\tau\} \end{aligned} \tag{5-31}$$

Applying thse operations to $z_5$, zero is obtained, $V_{11}^{E_u}(x_5) = 4(x_5 + x_6)$, so the acoustic mode is projected out since it transforms as x. $V_{11}^{E_u}(y_5) = y_5 + y_5 - y_6 + y_5 + y_5 - y_6 - y_6 = 4(y_5 - y_6)$. Similar results are obtained for $V_{22}^{E_u}$. The only interesting result is $V_{22}^{E_u}(x_5) = 4(x_5 -$

$x_6$). Thus, for the optic $E_u$ modes the Ti ions do not move along the z-axis but the two ions move in opposite phase (against each other) along the x-axis and similarly along the y-axis. These two motions are partners of the $E_u$ irreducible representation. As an added check, the transfer projection operator Eq. 5-21 can be calculated. This is done using the functions x and y to determine the representation. The result is

$$V_{12}^{E_u} = -\{C_4 | \tau\} + \{C_4^3 | \tau\} + C_2[110] - C_2[\bar{1}10] - \{S_4 | \tau\} + \{S_4^3 | \tau\} - \sigma_d[110] + \sigma_d[\bar{1}10] \qquad (5\text{-}32)$$

and $V_{12}^{E_u}(x_5 - x_6) = 4(y_5 - y_6)$ which is what was obtained previously. The oxygen-ion motion must now be calculated. However, as in the molecular case, group theory will not give unique answers to this problem since there are three optic $E_u$ modes.

As a last example of projection operators we apply the character projection operator, Eq. 5-26, to basis functions of a doubly degenerate irreducible representation. Consider the $E_g$ irreducible representation for the point group $D_{4h}$. From the character table

$$V^{E_g} = 2(E - C_2 + i - \sigma_h) \qquad (5\text{-}33)$$

Suppose we are interested in the d-wave functions. $V^{E_g}(xz) = 0$ so xy does not transform as $E_g$. $V^{E_g}(xz) = 8xz$, hence xz transforms as $E_g$, but its partner is not known and the other d-functions must be tried, one at a time, to determine the partner. One way around this is to operate on xz with each symmetry operation, one at a time. For example, $C_4xz = yz$, yz is a partner of xz since the symmetry operation will only transform a function into a linear combination of its partners. The trouble with this method is that the functions obtained are not necessarily orthogonal. This is not a fundamental problem since if they are not orthogonal they can always be orthogonalized by the Schmitt process.

## Notes

Wigner's 1930 fundamental paper on vibrations appears in English in Cracknell as well as in Knox and Gold. It is readable and discusses the differences between the equilibrium position and a displacement. The rest of this short paper covers material similar to that found in Sections 5-2a and 5-3. A reprint of an extensive paper describing properties of normal modes of certain molecules by E. Bright Wilson, Jr. is also included in Cracknell. An early review paper [J. E. Rosenthal and G. M. Murphy,

Rev. Mod. Phys. **8**, 317 (1936)] is also worth looking at for a clear account of material similar to that found in this chapter.

Section 5-2b is very similar to Born and Huang "Dynamical Theory of Crystal Lattices" (Oxford, 1954), Section 15. In Section 38 of this same classic book, the normal mode problem in solids is treated for the case where the wavelength of the vibration can be of the order of the unit cell dimension.

The problem of internal coordinates and redundancy is treated by Wilson, Decius, and Cross. S. J. Cyvin, "Molecular Vibrations and Mean Square Amplitudes" (Elsevier Publishing Co., 1968) might also be consulted. These books, as well as many others, treat the general vibrational problem. Solving the dynamical matrix, usually by the Wilson FG-matrix method, is also treated in these references.

Bishop, Section 5-4 and 9-4, has a nicely worked out example of the reducible representations of an $A_3$ molecule. Leech and Newman have a fairly complete solution for the $CH_3Cl$ vibration problem including writing the dynamic matrix.

The symmetry operations of all 230 space groups can be found in Kovalev or Zak *et al.* These references are discussed in Chapter 11.

## Problems

**1.** For the molecule $H_2O$ determine the normal modes by considering how the two hydrogen ions transform between themselves and reduce the representation and then how the oxygen transforms and reduces the representation. Check the results with the text.

**2.** For the molecule DHO how do the vibrational modes transform? What does the result mean?

**3.** Consider a square planar molecule $B_4$. How do the normal modes of vibration transform? What do some of the modes look like?

**4.** Consider a square planar molecule $A_2B_2$ with $D_{2h}$ symmetry. Determine under which irreducible representations the normal modes transform. How do the normal modes compare to those in Problem 3?

**5.** Consider the $C_{4v}$ phase of $PbTiO_3$ (move the Ti ion along the z-axis).

(a) Determine the $k = 0$ normal modes of vibration. Keep the appropriate groups of ions separate in analogy to Fig. 5-3, determine the normal modes.

(b) Instead of the above approach determine the irreducible representation, modes, etc., by realizing the $C_{4v}$ is a subgroup of $O_h$ and reduce the representation of $O_h$ to representations of $C_{4v}$.

**6.** For the $H_2O$ molecule vibrational problem, take the plane of the molecule as the xz-plane and determine the irreducible representations that describe the normal modes. Why is the difference between this result and that in Eq. 5-15 clear?

**7.** For planar cis-$PtCl_2Br_2$ determine to which irreducible representations the modes belong. Do the same for trans-$PtCl_2Br_2$.

**8.** For an $X_4$ molecule with $T_d$ symmetry guess the normal modes and eigenvectors, given the answer for $CH_4$ and Fig. 5-2b.

**9.** Prove for transfer projection operators that $V_{oi}{}^m V_{pj}{}^n = \delta_{mn} \delta_{ip} V_{oj}{}^m$. Thus, $V_{ii}{}^m V_{ii}{}^m = V_{ii}{}^m$ and operators with this property (all powers are equal) are called itempotent. Show that $(V_{ij}{}^k)^\dagger = V_{ji}{}^k$ so the transfer projection operator is not Hermitian. (In Quantum Mechanics the term projection operators is usually reserved for operators that are Hermitian and itempotent. Thus, $V_{ij}{}^k$ is called a transfer projection operator, while $V_{ii}{}^k$ is, indeed, a projection operator.)

**10.** (a) An f-atomic function transforms like $xz^2$. For $D_3$ point symmetry, under which irreducible representation does $xz^2$ transform? Using orthogonal functions find the unitary irreducible representation and check that its characters are correct. Using $F = xz^2$ in Eq. 5-28 determine the right side of this equation. (b) Do the same thing for the rest of the f-atomic functions.

**11.** For a water molecule using the three internal coordinates determine the potential and kinetic energies. By transforming to symmetry coordinates (see Fig. 5-1b) determine the partially factored Dynamical Matrix. What is the frequency of the $B_2$ mode? (Hint: See Wilson, Decius, and Cross, Section 6-3.)

**12.** For the tetrahedral $CH_4$, using internal coordinates, determine the irreducible representation that describes the vibrational normal modes. Determine the normal modes or, if not possible, the symmetry coordinates.

**13.** For the octahedral $SF_6$ molecule, using internal coordinates, determine the normal modes that transform as the $A_{1g}$, $E_g$, and $T_{2g}$ irreducible representations.

Chapter 6

# NORMAL MODES (DIRECT PRODUCT AND SELECTION RULES)

*Now you shall hear how a man may become perfect, if he devotes himself to the work which is natural to him.*

*"Bhagavad-Gita"*

In this chapter we continue to discuss vibrations of molecules and crystals but with the emphasis on selection rules in infrared and Raman spectroscopies. In order to do this the important concept of the direct product of irreducible representations is introduced. This concept is applied immediately to the normal mode problem. However, it is an important general concept that will be used throughout the remainder of the book.

In finding and using the selection rules, the only fact that must be known is the number of normal modes that transform as the different irreducible representations. The actual eigenvectors need not be known. Thus, the determination of the number of modes that will be observed in various polarizations with infrared and Raman techniques is very simple. We proceed by first considering the direct product of irreducible representations.

## 6-1 Direct Product of Irreducible Representations

### a. General

In Section 2-6 we find what is meant by the multiplication of two groups. In this section we define what is meant by the direct product of two irreducible representations. This is a concept that is used often since it is intimately related to products of wave functions.

One is often concerned with the products of basis functions (wave functions) that transform as different or the same irreducible representations. Consider a set of basis functions $A_1, A_2, \ldots, A_p$ that transform as a p-dimensional irreducible representation and another set of basis functions $B_1, B_2, \ldots, B_q$ for a q-dimensional irreducible representation of a group of order h. A symmetry operation R operating on any function, will yield a linear combination of its partners in the usual way

$$RA_i = \sum_{j=1}^{p} A_j\, a_{ji}$$

$$RB_m = \sum_{n=1}^{q} B_n\, b_{nm} \tag{6-1}$$

where $a_{ji}$ and $b_{nm}$ are appropriate irreducible representations of the group. The complete set of all possible products of functions $A_iB_m$ contains pq functions and forms a basis of a representation of the group. This representation, called the **direct product** (or Kronecker product or outer product) of the representations is of dimension pq. To define the direct product more clearly, let $\mathbf{a}^\alpha$ and $\mathbf{b}^\beta$, where $\alpha, \beta = 1,2,\ldots,h$, and h is the order of the group, be the irreducible representations. Then the set of h square matrices $\mathbf{c} = \mathbf{a}^\alpha \times \mathbf{b}^\alpha$ comprises this direct product representation, each matrix being of dimension pq. The method of obtaining the h matrices, each called **c**, is described below. (Not used in this book but included for completeness: An outer Kronecker product of a set of matrices is formed by $\mathbf{a}^\alpha \times \mathbf{b}^\beta$. The dimension of each matrix in the set is pq, but the set contains hh′ matrices where $\alpha = 1, 2, \ldots, h$ and $\beta = 1, 2, \ldots, h'$.) As in Eq. 6-1, a symmetry operation operating on any one of the pq functions will be expressed as a linear combination of all the functions,

$$RA_i\, B_m = \sum_{j=1}^{p} \sum_{n=1}^{q} a_{ji}\, b_{nm}\, A_j\, B_n = \Sigma_j\, \Sigma_n\, c_{jn,im}\, A_j\, B_n \tag{6-2}$$

Each matrix **c** is the **direct product** of the matrices **a** and **b** ($\mathbf{c} = \mathbf{a} \times \mathbf{b}$). (This is written in some books as $\mathbf{c} = \mathbf{a} \otimes \mathbf{b}$.) Direct product should not be confused with the usual matrix product where the matrices **a** and **b** would have to be compatible. To see more clearly what Eq. 6-2 means, consider that **a** and **b** are each 2 × 2 matrices

$$\mathbf{c} = \mathbf{a} \times \mathbf{b} = \begin{bmatrix} a_{11} & a_{12} \\ a_{21} & a_{22} \end{bmatrix} \times \begin{bmatrix} b_{11} & b_{12} \\ b_{21} & b_{22} \end{bmatrix} = \begin{bmatrix} a_{11}\mathbf{b} & a_{12}\mathbf{b} \\ a_{21}\mathbf{b} & a_{22}\mathbf{b} \end{bmatrix}$$

$$= \begin{bmatrix} a_{11}b_{11} & a_{11}b_{12} & a_{12}b_{11} & a_{12}b_{12} \\ a_{11}b_{21} & a_{11}b_{22} & a_{12}b_{21} & a_{12}b_{22} \\ a_{21}b_{11} & a_{21}b_{12} & a_{22}b_{11} & a_{22}b_{12} \\ a_{21}b_{21} & a_{21}b_{22} & a_{22}b_{21} & a_{22}b_{22} \end{bmatrix} \tag{6-3}$$

Thus, for each **a** × **b**, Eq. 6-2 defines **c**, which is a representation (not necessarily irreducible) of the group. For the matrix **c**, we have the notation in which two numbers denote columns and two other numbers denote rows with a comma separating the first two from the second two. In Eq. 6-3 the first two numbers indicate each of the blocks and the second two numbers, a row and column within a block. (Wigner, Chapter 2, arranges the matrix **c** differently, but the arrangement in Eq. 6-3 is more often used.)

As is usually the case, a simplification can be made if the characters of this representation are considered

$$\chi(c) = \Sigma_j \Sigma_n c_{jn,jn} = \Sigma_j \Sigma_n a_{jj} b_{nn} = [\chi(a)][\chi(b)] \qquad (6\text{-}4)$$

Thus, the character of a direct product representation is the product of the characters.

As an example, consider some direct products of the irreducible representations of the point group $D_{4h}$. By inspection of the character table, using Eq. 6-4, $B_{1g} \times B_{2g} = A_{2g}$. Thus the function $(x^2 - y^2)xy = x^3y - xy^3$ transforms as the $A_{2g}$ irreducible representation. Also, $B_{2g} \times E_g = E_g$. Hence, for example, the partners of the $E_g$ irreducible representation are also $(x^2yz, xy^2z)$. On the other hand, the direct product $E_u \times E_u$ is clearly reducible. Reducing it by Eq. 4-7 or by inspection yields $A_{1g} + A_{2g} + B_{1g} + B_{2g}$. We would conclude that the four functions $x^2$, xy, yx, $y^2$ transform as these four irreducible representations. That conclusion would be true if x and y did not commute. Since they do commute the result is incorrect and we must consider the symmetric and antisymmetric nature of the problem more carefully. However, if we are dealing with two sets of functions, each transforming as $E_u$, which do not commute, i.e., $(\alpha,\beta) \times (\gamma,\Delta) = \alpha\gamma, \alpha\Delta, \beta\gamma, \beta\Delta$, then the above result is true. This is discussed at the end of the following section.

### b. Symmetric and antisymmetric direct products

When both sets of functions transform as the same irreducible representation, $a_{ij} = b_{ij}$ and $p = q$. For this case we show that there are symmetrized and antisymmetrized products with respect to the interchange of the two sets of functions. We work out the relevant equations for a 2-dimensional irreducible representation. However, the results can be easily generalized. For a 2-dimensional irreducible representation, we take two sets of partners as the basis function, $(A_1,A_2)$ and $(B_1,B_2)$. There are four direct product functions $A_1B_1$, $A_1B_2$, $A_2B_1$, $A_2B_2$. For these functions, Eq. 6-2 written in matrix form is

$$R[A_1B_1 \quad A_1B_2 \quad A_2B_1 \quad A_2B_2] = [A_1B_1 \quad A_1B_2 \quad A_2B_1 \quad A_2B_2] \begin{bmatrix} a_{11}a_{11} & a_{11}a_{12} & a_{12}a_{11} & a_{12}a_{12} \\ a_{11}a_{21} & a_{11}a_{22} & a_{12}a_{21} & a_{12}a_{22} \\ a_{21}a_{11} & a_{21}a_{12} & a_{22}a_{11} & a_{22}a_{12} \\ a_{21}a_{21} & a_{21}a_{22} & a_{22}a_{21} & a_{22}a_{22} \end{bmatrix} \quad (6\text{-}5)$$

As can be seen, the right side of Eq. 6-5 is normal matrix multiplication. Equation 6-5 can also be rewritten in a simpler form if the product functions are taken to be the symmetric functions $A_1B_1$, $(A_1B_2 + A_2B_1)$, $A_2B_2$, and the antisymmetric function $(A_1B_2 - A_2B_1)$:

$$R[A_1B_1 \quad (A_1B_2+A_2B_1) \quad A_2B_2 \quad (A_1B_2-A_2B_1)] = [A_1B_1 \quad (A_1B_2+A_2B_1) \quad A_2B_2 \quad (A_1B_2-A_2B_1)] \times \begin{bmatrix} a_{11}a_{11} & 2a_{11}a_{12} & a_{12}a_{12} & 0 \\ a_{11}a_{21} & a_{11}a_{22} + a_{12}a_{21} & a_{12}a_{22} & 0 \\ a_{21}a_{21} & 2a_{21}a_{22} & a_{22}a_{22} & 0 \\ 0 & 0 & 0 & a_{11}a_{22} - a_{12}a_{21} \end{bmatrix} \quad (6\text{-}6)$$

The symmetric parts of the product transform among themselves and the antisymmetric product transforms into itself. Thus, there are two separate representations. In a similar manner one can show in general that the direct product of a representation $\Gamma_n(R)$ with itself, can always be written in terms of a symmetric and antisymmetric part

$$\Gamma_n(R) \times \Gamma_n(R) = [\Gamma_n(R) \times \Gamma_n(R)]_S + \{\Gamma_n(R) \times \Gamma_n(R)\}_{AS} \quad (6\text{-}7)$$

We consider the characters of the terms in Eq. 6-7 to simplify reducing the representations on the right side. Previously such considerations led to Eq. 6-3 for direct products of two different irreducible representations. For the example of the 2-dimensional irreducible representation, the character of the symmetric product can be written as

$$[\chi(R)\chi(R)]_S = a_{11}{}^2 + a_{11}a_{22} + a_{12}a_{21} + a_{22}{}^2$$
$$= \tfrac{1}{2}[a_{11} + a_{22}]^2 + \tfrac{1}{2}[a_{11}{}^2 + a_{12}a_{21} + a_{21}a_{12} + a_{22}{}^2] \quad (6\text{-}8)$$

The two bracketed quantities on the right side of Eq. 6-8 can easily be reduced. The first term is

$$[\chi(R)]^2 = [\Sigma_i\, \Gamma(R)_{ii}]^2 = [a_{11} + a_{22}]^2 \quad (6\text{-}9)$$

The second term is the effect of a symmetry operation applied twice

$$RR\psi_i = R\, \Sigma_j\, a_{ji}\, \psi_j = \Sigma_j\, \Sigma_k\, a_{kj}\, a_{ji}\, \psi_k \quad (6\text{-}10)$$

The sum over j on the right side of Eq. 6-10 is an ordinary matrix multiplication. Thus, using matrix multiplication (not a direct product)

$$\mathrm{TR}[\Gamma(R)\Gamma(R)] = \mathrm{Tr}\begin{bmatrix} a_{11} & a_{12} \\ a_{21} & a_{22} \end{bmatrix}\begin{bmatrix} a_{11} & a_{12} \\ a_{21} & a_{22} \end{bmatrix}$$
$$= a_{11}^2 + a_{12}a_{21} + a_{21}a_{12} + a_{22}^2 \qquad (6\text{-}11)$$

Thus, Eq. 6-8 reduces to

$$[\chi(R) \times \chi(R)]_S = 1/2\,[[\chi(R)]^2 + \chi(R^2)] \qquad (6\text{-}12)$$

In a similar manner one can show for the antisymmetric part of the direct product representation

$$\{\chi(R) \times \chi(R)\}_{AS} = 1/2\,\{[\chi(R)]^2 - \chi(R^2)\} \qquad (6\text{-}13)$$

Equations 6-12 and 6-13 are easy to use. They just involve, for a given symmetry operation, the character squared and the character of the operation that results when the symmetry operation is applied twice. The result that the direct product of an irreducible representation with itself can be expressed as a symmetric and antisymmetric part as in Eq. 6-7 is a general one. The expressions for the parts, Eqs. 6-12 and 6-13, are also general and not just applicable to 2-dimensional representations. If there are n-basis functions in the representation, then there are $n(n+1)/2$ symmetric functions $A_iB_j + A_jB_i$ and $n(n-1)/2$ antisymmetric functions $A_iB_j - A_jB_i$ for $i \neq j$ when a direct product of the representation with itself is considered. (See the Notes for other references.)

As an example, we apply Eqs. 6-12 and 6-13 to $E_u \times E_u$ in $D_{4h}$ symmetry. For the second term in Eqs. 6-12 or 6-13, $C_2C_2 = ii = \sigma\sigma = E$, but $C_4C_4 = S_4S_4 = C_2$. Table 6-1 shows the result in ten (not 16) bracketed numbers, one for each R in each class as in the character table with the first and second numbers in the bracket referring, respectively, to the first and second terms on the right of the equation. The character of the resulting symmetric and antisymmetric representation in Fig. 1 can be reduced by inspection or by Eq. 4-7. The result is $[\chi(R) \times \chi(R)]_S = A_{1g} + B_{1g} + B_{2g}$ and $\{\chi(R) \times \chi(R)\}_{AS} = A_{2g}$. Now consider basis functions that can be associated with the symmetric and antisymmetric irreducible representations in this example. Consider the two sets of partners that separately transform as the $E_u$ representation $(A_1, A_2)$ and $(B_1, B_2)$. The antisymmetric function $(A_1B_2 - A_2B_1)$ is unique. This indeed transforms as $A_{2g}$ under all the symmetry operation of $D_{4h}$. It is also a familiar function; namely, the z-component of an axial vector (like angular momentum). In fact, the character table shows thar $R_z$ transforms as $A_{2g}$. The functions that transform as the symmetric irreducible representations are somewhat more subtle. From Eq. 6-6, the three functions $A_1B_1$, $A_1B_2 + A_2B_1$, and $A_2B_2$ would be chosen. It is clear that while $A_1B_2 + A_2B_1$ transforms as $B_{2g}$, the other functions as they stand are not basis functions of the other two irreducible representations. However, linear combinations of them, $A_1B_1 + A_2B_2$ and $A_1B_1 - A_2B_2$ are basis functions of the $A_{1g}$ and $B_{1g}$ irreducible representations, respectively. This result for the symmetric basis functions is clear from a careful observation of Eq. 6-6. In this equation, for $D_{4h}$, $a_{11}a_{12} = 0 = a_{11}a_{21}$ for all R such that the symmetric function $A_1B_2 + A_2B_1$ transforms only into itself. However, the terms $a_{12}a_{12}$ are not always zero, so the functions $A_1B_1$ and $A_2B_2$ will transform between themselves for some symmetry operations of the group. Thus, when the symmetric representation is reduced, some linear combination of these functions must be found so that basis functions of the irreducible representation can be determined.

**Table 6-1** Characters of symmetric and antisymmetric product representations of $E_u \times E_u$ in $D_{4h}$

| R | E | $C_4$ | $C_2$ | $C_2'$ | $C_2''$ | i | $S_4$ | $\sigma_h$ | $\sigma_v$ | $\sigma_d$ |
|---|---|---|---|---|---|---|---|---|---|---|
| $\{\chi(R) \times \chi(R)\}_S$ | $\frac{(4+2)}{2}$ | $\frac{(0-2)}{2}$ | $\frac{(4+2)}{2}$ | $\frac{(0+2)}{2}$ | $\frac{(0+2)}{2}$ | $\frac{(4+2)}{2}$ | $\frac{(0-2)}{2}$ | $\frac{(4+2)}{2}$ | $\frac{(0+2)}{2}$ | $\frac{(0+2)}{2}$ |
| $\{\chi(R) \times \chi(R)\}_{AS}$ | $\frac{(4-2)}{2}$ | $\frac{(0+2)}{2}$ | $\frac{(4-2)}{2}$ | $\frac{(0-2)}{2}$ | $\frac{(0-2)}{2}$ | $\frac{(4-2)}{2}$ | $\frac{(0+2)}{2}$ | $\frac{(4-2)}{2}$ | $\frac{(0-2)}{2}$ | $\frac{(0-2)}{2}$ |

Note that if the direct product of an irreducible representation with itself is being considered, and the basis functions are also identical (in the notation of this section $a_{ij} = b_{ij}$ and $A_i = B_i$), then the antisymmetric product in Eq. 6-6 and the following equations, is zero. This is perfectly clear in the preceding example. The antisymmetric basis function $A_1B_2 - A_2B_1$ is zero if $(A_1A_2) = (B_1B_2)$. This is an important considerationfor wave functions when selection rules are being considered.

In Appendix 6 we list the direct product for all the irreducible representations for several groups. Also listed are the symmetric and antisymmetric products for several groups.

## 6-2 Vibrational Wave Function

The wave functions for the normal modes will be needed to determine selection rules for infrared and Raman absorptions in the next section. As will be seen, the symmetry of the wave functions is particularly simple, so the selection rules will be rather easy to determine.

The Hamiltonian of Eqs. 5-6 and 5-12 can be written in terms of a sum of Hamiltonians for individual normal modes

$$\mathbf{H} = \sum_{i=1}^{3N\text{-}6} \mathbf{H}_i \qquad \mathbf{H}_i = 1/2\,(-\hbar^2\,(\partial^2/\partial q_i^{\,2}) + \omega_i^{\,2}\,q_i^{\,2} \qquad (6\text{-}14)$$

Thus, the wave function for the vibrational normal mode $q_i$ is

$$\mathbf{H}_i\,\psi_n(q_i) = E_n\,\psi_n(q_i) \qquad (6\text{-}15)$$

As is well known, the energy and wave function resulting from Eq. 6-15 can be written in terms of Hermite polynomials, $h_n$

$$E_n = \hbar\,\omega_i(n + 1/2), \qquad (E_{n+1} - E_n)/\hbar = \omega_i \qquad (6\text{-}16)$$

$$\psi_n(q_i) = K_i\,[\exp(-\alpha_i^{\,2} q_i^{\,2}][h_n(\alpha_i q_i)] \qquad (6\text{-}17)$$

where $\alpha_i$ and the normalization are

$$\alpha_i = (\omega_i/\hbar)^{1/2}$$
$$K_i = [(\omega_i/\hbar)^{1/2}/2^n\,n!\,\pi^{1/2}]^{1/2}$$

The Hermite polynomials are simple functions of $\alpha_i q_i$ and for the lowest values of n, these function are

$$h_0 = 1, \qquad h_1 = 2(\alpha q)$$
$$h_2 = 4(\alpha q)^2 - 2, \qquad h_3 = 8(\alpha q)^3 - 12(\alpha q) \qquad (6\text{-}18)$$

The **vibrational quantum number n** in the energy and wave function, Eqs. 6-16 and 6-17, describes the excitation of the oscillator. If the oscillator is in the ground state, $n = 0$, the energy has the zero point value $\hbar\omega/2$. If the oscillator absorbs one quantum of energy it goes into its first excited state, it has energy $3\hbar\omega/2$, and the wave function is proportional to $h_1$. The amount of energy absorbed is $\hbar\omega_i$ with frequency $\omega_i$ corresponding to the ith normal mode. The $n = 2$ state is the second excitated state. A transition from the ground state to the $n = 2$ state would occur if the energy absorbed is $2\hbar\omega_i$.

Since the Hamiltonian of the ensemble of oscillators can be written as a sum of Hamiltonians for individual oscillators, the total wave function can be written as the product of wave functions, $\Psi = \Pi\,\psi_n(q_i)$. We condense the notation with a subscript on the principal quantum number for the ith mode, so $n_i$ would be the nth excited state of the ith mode and the wave function in Eq. 6-17 could be written as $\psi_n(q_i) = \psi_{n_i}$. The total wave function of the ensemble is ($i = 1$ to $3N - 6$)

$$\Psi(n_1, n_2, ..., n_{3N-6}) = [\Pi_i\,K_i]\,[\exp(-\Sigma_i\,\alpha_i^{\,2}\,q_i^{\,2})][\Pi_i\,h_{n_i}(\alpha_i q_i)] \qquad (6\text{-}19)$$

Normally one is interested in low excitation of the system of oscillators. The ground state of the system has all $n_i = 0$ hence all the Hermite polynomials in the wave function are one and the wave function is given by the first two bracketed terms in Eq. 6-19. The system with one normal mode excited is given by all $n_i = 0$ except $n_j = 1$. Then the wave function is

$$\Psi(0_1,\ldots, 1_j, 0_{j+1},\ldots) = [\Pi_i K_i]\,[\exp(-\Sigma_i\, \alpha_i^2 q_i^2)][2\alpha_j\, q_j] \qquad (6\text{-}20)$$

In a similar manner the wave function for other excited states can be determined.

We now consider the effect of transformations in Eq. 6-19. Under a symmetry operation the normalization (first term within the square brackets) clearly transforms into itself; the second term within the square brackets transforms just like the potential energy, thus it transforms into itself; the transformation of the third term depends on which Hermite polynomials are involved. If all $n_i = 0$, then the third square bracketed term in Eq. 6-19 is one and the term clearly transforms into itself under all the symmetry operations of the molecule. If all $n_i = 0$ except $n_j = 1$, as in Eq. 6-20, then the third term transforms as $q_i$, i.e., it transforms in the same way as the normal mode. If all $n_i = 0$ except $n_j = 1$ and $n_k = 1$, then the term transforms as the direct product $q_j q_k$ and can be handled in a straightforward manner as in Section 6-1. Thus, we have arrived at the general result that **in the ground vibrational state (all $n_i = 0$), the total vibration wave function transforms as the totally symmetric irreducible representation $\Gamma_1$. For the one-phonon state where all $n_i = 0$ except $n_j = 1$, then the total vibrational wave function transforms as the same irreducible representations as does the normal coordinate $q_j$.** The extension to the two-phonon state (all $n_i = 0$ except $n_n = 1 = n_k$ or all $n_i = 0$ except $n_j = 2$) simply involves the direct product considerations as already discussed.

These extremely simple symmetry transformations enable general infrared and Raman selection rules to be derived.

## 6-3 Selection Rules — Infrared and Raman

### a. Selection rules — general

We are interested in integrals of the type

$$<\psi_i^m \,|\, O_a^b \,|\, \psi_v^n> \equiv \int (\psi_i^m)^*\, O_a^b\, \psi_v^n\, d\tau \qquad (6\text{-}21)$$

where $\psi_i^m$ is a wave function transforming as the ith row of the mth irreducible representation and $O_a^b$ is an operator that transforms as the ath row of the bth irreducible representation. The operator we have in mind is perhaps a component of the electric dipole, the quadrupole or the magnetic dipole operator that transforms as an irreducible representation. (If the total operator does not transform as an irreducible representation then it may be decomposed into parts that do as in Eq. 5-28.) If the integral is nonzero the operator can cause transitions between the two states. Selection rules tell us when the integral is indeed nonzero. It should be appreciated that the operator in Eq. 6-21 often is made up of several parts, each part transforming as an irreducible representation. Thus, the one integral could be replaced by several terms adding no fundamental difficulty in determining the selection rules. If Eq. 6-21 is zero then a transition induced by the particular operator between the levels represented by the two wave functions is forbidden.

The appreciation of whether or not the integral in Eq. 6-21 is zero can be obtained by recalling the results in Section 4-4. Namely that basis functions of irreducible unitary representations or partners belonging to different rows of the same irreducible unitary representation are orthogonal. Then integrals of orthogonal functions are zero. See Problem 11. Thus, the transition is forbidden if none of the direct products $O_a^b \psi_j^n$ or linear combinations of these products belong to the mth irreducible representation. This result is usually expressed in several ways in terms of direct products of irreducible representations as in Section 6-1. The transition is forbidden unless: $m \times b \times n$ contains $\Gamma_1$; $b \times n$ contains m; $m \times n$ contains b. Although all these statements are equivalent, at times the last one makes more intuitive physical sense. This is because if $m = n$ and the two wave functions are the same functions, only the symmetric direct product and not the direct product should be considered since the antisymmetric product is zero as discussed in Section 6-1. Examples of this occurrance are given below.

### b. Infrared absorption

When the interaction of electromagnetic radiation with matter is considered, the perturbing term that causes transitions is the term involving the vector potential **A** and momentum of the electrons **p**. As shown in quantum mechanics texts this perturbation for a single electron is $H_1 = (e/mc)\mathbf{A}\bullet\mathbf{p}$. The vector potential involves components such as $\exp(\pm i\mathbf{k}\bullet\mathbf{r})$ where $k = 2\pi/\text{wavelength}$. Since the wavelength of the radiation is generally much larger than the dimensions of the molecule or unit cell, we can expane $H_1$ for small kr, $\exp(\pm i\,\mathbf{k}\bullet\mathbf{r})\mathbf{p} \approx \mathbf{p} \pm i(\mathbf{k}\bullet\mathbf{r})\mathbf{p}$. The first term leads to the electric dipole approximation while the second terms gives both the magnetic dipole and electric quadrupole perturbation. (See the

Notes for references to the electric dipole, etc. expansion, and the interaction of radiation with matter.) General selection rules for these various approximations will be given in the following chapter. Here we just discuss infrared absorption and the electric dipole moment operator.

The dipole moment operator of a molecule is

$$\boldsymbol{\mu} = \Sigma_i\, c_i\, \mathbf{r}_i \qquad (6\text{-}22)$$

where the sum is over all the charge $c_i$ at position $\mathbf{r}_i$. Only the electronic ground state is involved since there is no change in electronic state when infrared absorption involving a vibrational normal mode occurs. The coordinates of $\boldsymbol{\mu}$ are fixed to the x, y, z coordinates of the molecule. The dipole moment in the laboratory coordinates $\mathbf{U}$ is related to $\boldsymbol{\mu}$ by the direction cosines of the molecular axis with respect to the laboratory axes $U_\alpha = \Sigma_i\, l_{\alpha i}\, \mu_i$ where $\alpha$ and i = 1, 2, 3. Consider the infrared radiation polarized along the laboratory 1-axis. The selection rule is determined by substituting $U_1$ for the operator in Eq. 6-21 and considering each state to be represented by a rotational and vibrational ground and excited state wave function $\psi_{rot_g}\, \psi_{vib_g}$ and $\psi_{rot_e}\, \psi_{vib_e}$, respectively. The infrared absorption is proportional to the following matrix element squared

$$<e\,|\,U_1\,|\,g> = \sum_{i=1}^{3} <\psi_{rot_e}\,|l_{1i}\,|\,\psi_{rot_g}><\psi_{vib_e}\,|\,\mu_i\,|\,\psi_{vib_g}> \qquad (6\text{-}23)$$

The energies associated with the rotational part of this result, as mentioned earlier, are very small compared to the vibrational energies. In infrared spectra the rotational part at most gives rise to some fine structures and is not usually seen at all in liquids and solids. A transition involving no change in rotational quantum number is allowed, so the integral will not vanish due to the rotational part. So we ignore the rotational part of Eq. 6-23 and consider only how the infrared selection rules are determined by the vibrational integral. The components of the dipole moment defined in Eq. 6-22 transform in the same way as the Cartesian coordinates x, y, z as discussed in Section 4-5b. Thus, we can state the selection rule for infrared absorption. **Infrared absorption between two states g and e is allowed if the reduction of the direct product of the irreducible representations $e^* \times g$ transforms as the same irreducible representation of one or more of the Cartesian coordinates.**

This result is even simpler in the usual experimental situation. The usual infrared (one-phonon) absorption experiment excites the vibrational system from the ground state, which transforms as $\Gamma_1$, to an excited vibrational state (all $n_i = 0$ except $n_j = 1$), which transforms as $q_j$, as discussed. Thus, **a one-phonon infrared absorption at frequency $\omega_j$ is allowed when the normal mode $q_j$ transforms as a Cartesian coordinate.**

Before we go on to examples, more physical insight can be obtained by expanding the dipole moment of a molecule in terms of normal mode coordinates

$$\mu_i = \mu_{i0} + \Sigma_j (\partial\mu_i/\partial q_j)_0 q_j + \Sigma_{k,j} (\partial^2\mu_i/\partial q_k \partial q_j)_0 q_k q_j + \ldots \quad (6\text{-}24)$$

The subscript zero means the term should be evaluated in the ground state for zero displacement. The first term on the right side of Eq. 6-24 is the permanent dipole moment of the molecule. It is easy to determine if this term is nonzero as shown in Section 4-5b. The second term of Eq. 6-24 has a frequency component at $\omega_j$ and can be measured in the infrared region of the spectrum. The coefficient $(\partial\mu_i/\partial q_j)_0$ must transform as $\Gamma_1$ since it is a material property of the molecule. (This is a slight generalization of Neumann's principle, Section 4-5, and follows because an equation like Eq. 6-24 can always be written for any material property. The derivative of this property will give a change as in Eq. 6-24 which transforms as $\Gamma_1$.) Thus, $q_j$ must transform as $\mu_i$ otherwise $(\partial\mu_i/\partial q_j)_0$ must be zero. If $q_j$ transforms as $\mu_i$ then one-phonon infrared absorption is allowed and $(\partial\mu_i/\partial q_j)_0$ will be different from zero. The third term on the right side of Eq. 6-24 allows two-phonon infrared absorption. The selection rules for this absorption in molecules can easily be worked out and were touched upon at the end of Section 6-2.

### c. Infrared absorption — examples

In Chapter 5 we found that an $H_2O$ molecule with $C_{2v}$ symmetry has three vibrational normal modes that transform as $2A_1 + B_2$. Consider the fundamental absorption to the first excited $A_1$ state. Then $\psi_g$, $\mu_z$, and $\psi_e$ transform as $A_1$, $A_1$, $A_1$ so that the direct product contains $A_1$ and the transition is allowed. Note that only the z-component of the dipole operator is effective in the infrared absorption to the first excited $A_1$ state. For an absorption of the ground state to the first excited state of a $B_2$ normal mode, only the y-component of the dipole operator will be effective since it transforms as $B_2$, that is, $A_1 \times B_2 \times B_2 = A_1$. Thus the transition is allowed. The z- and x-components of the dipole operator transform as $A_1$ and $B_1$, respectively. An absorption of a $B_2$ mode by these components is not allowed because the relevant matrix element Eq. 6-23 gives $A_1 \times A_1 \times B_2 = B_2$ and $A_1 \times B_1 \times B_2 = A_2$; neither result contains $A_1$. A gas composed of $H_2O$ molecules has all three modes allowed in fundamental (one-phonon) infrared absorption.

Actually there is no reason to look at the details of the vibrational modes for a particular molecule. Given the point group $C_{2v}$, a general statement can be made. It is clear from the character table for $C_{2v}$ that the z-component of the dipole operator transforms as the $A_1$ irreducible representation, similarly, the x-component as $B_1$ and the y-component as $B_2$. If the molecule has a normal mode that transforms as $A_1$, $B_1$, or $B_2$ irreducible representation, then that particular component of the dipole operator can cause first-order infrared absorption at an infrared frequency corresponding to the $\omega_i$ of the normal coordinate.

Going on to the $AB_4$ example with $C_{4v}$ symmetry in Chapter 5, it is clear from the character table that the z-component of the dipole moment transforms as $A_1$ and the x- and y-components transform as E. For this point group in general, fundamental infrared absorption is allowed only for modes that have $A_1$ and E symmetry. The other normal modes, $2B_1$ and $B_2$, found for this molecule cannot be seen in one-phonon infrared absorption spectra. However, these modes are allowed for two-phonon infrared absorption. For example, in a two-phonon absorption a $B_1$ mode can be excited to its n = 2 state (Hermite polynomial $h_2$ in Eq. 6-18). The total vibrational wave function, Eq. 6-19, transforms as $A_1$ and the z-component of the dipole moment can excite the vibrational system from its $A_1$-ground state to this two-phonon state. This process is called **first overtone absorption**. A two-phonon absorption to a state where one $B_1$ mode and one $B_2$ mode are excited would be called a **combination band.** Note in this case since $B_1 \times B_2 = A_2$ this combination band is not infrared active. Usually the intensity of the overtone and combination bands is considerably smaller than the intensity of the fundamental (one-phonon) absorption. The energy of the overtone and combination bands is still given by Eq. 6-16. For a first overtone the energy separation from the ground state (n = 0) is twice the energy separation for the one-phonon state. (A detailed discussion of overtone and combination bands can be found in Wilson, Decius, and Cross.)

For $T_d$ symmetry, it is clear from the character table that all three components of the dipole moment transform as $T_2$. Thus only modes that transform as $T_2$ can be observed in fundamental infrared absorption spectra. Recall in the example in Chapter 5, the tetrahedral molecule $CH_4$ has two vibrational modes that transform at $T_2$. Therefore these modes are infrared active.

**The selection rules for infrared absorption in solids** are governed by the same simple considerations as for molecules except that more information can be obtained. For solids the crystal and laboratory axes can be aligned uniquely so the angular integral and the sum in Eq. 6-23 do not enter the expression. The electric field of the infrared radiation can be made parallel to each crystallographic axis in turn by rotating the crystal. For example, in any crystal with $C_{2v}$ symmetry, fundamental infrared absorption of only $A_1$ modes will be observed when the electric field of the radiation is parallel to the z-axis. When the electric field is parallel to the x-axis only modes with $B_1$ symmetry can be observed. These modes will be observed if the crystal has normal modes with this symmetry. When the electric field is parallel to the y-axis only modes with $B_2$ symmetry will be observed. Thus,for solids the various types of modes can be separately measured if infrared work is done with single crystals. If the measurements are carried with powders the situation is just like a liquid or gas in that all the modes will be simultaneously observed.

The appearance and splitting of modes associated with phase transitions in solids is an interesting point. The high temperature phase of $PbTiO_3$, treated in Chapter 1, has $O_h$ symmetry. The 12 vibrational normal modes transform as $3T_{1u} + T_{2u}$. From the character table it is clear that x, y, and z transform as $T_{1u}$, hence the three $T_{1u}$ modes can be observed separately in fundamental infrared absorption, but the $T_{2u}$ mode cannot be observed. This crystal has a phase transition and below the transition temperature the crystal is ferroelectric with $C_{4v}$ symmetry. It can be seen from the correlation table in Appendix 7 or standard subgroup considerations that for this transition $T_{1u} \rightarrow A_1 + E$ and $T_{2u} \rightarrow B_1 + E$. In the $C_{4v}$ phase, radiation polarized along the z-axis will detect $A_1$ modes while radiation polarized along the x- or y-axis will detect E modes. These "pairs" of modes will now be observed at different energies in $C_{4v}$ while they were degenerate in the $O_h$ phase. In fact, the study of the splitting will give information about the phase transition. Also, the E mode arising from the $T_{2u}$ mode in the cubic phase will now be observable.

Consideration of Eq. 6-24 in conjunction with the actual normal modes yield considerable insight into the physical reason that particular modes are infrared-active for particular polarizations. Consider the normal modes pictured in Fig. 5-1 for $H_2O$. Note that for the $A_1$ modes the dipole moment of the molecule decreases as the hydrogens move up and the oxygens move down. Therefore, the second term on the right of Eq. 6-24 is nonzero for $A_1$ modes and obviously for the time dependence of just the frequency of the particular $A_1$ mode under consideration. Note also that this particular oscillation will cause a time-dependent dipole moment only along the z-axis, not the x- or y-axis. On the other hand, the $B_2$ mode, Fig. 5-1, produces a time varying dipole moment only along the y-axis.

For the modes for $C_{4v}$ in Fig. 5-2a, it is again clear that the motions of the $A_1$ normal mode causes a time varying dipole moment only along the z-axis. Similarly, the motions involved in the $B_1$ and $B_2$ normal modes do not produce a time varying dipole moment in any direction. The E modes have dipole moments in the x- and y-directions.

It is clear from Fig. 5-3 for $PbTiO_3$ that the three partners of the $T_{1u}$ representation produce time varying dipole moments along the x-, y-, and z-directions, whereas the $T_{2u}$ mode does not produce a dipole moment since as many oxygen atoms move up as move down.

### d. Raman absorption

In a Raman experiment, light from an intense, spectrally narrow frequency $\omega_{in}$ is incident on a sample. (A laser or mercury arc is usually used.) One usually observes the frequency of the light scattered from the sample $\omega_{sc}$ at lower frequencies (Stokes's radiation). The difference between the two frequencies is due to the fact that a phonon vibrational state has been raised from n to n + 1. The energy is given by Eq. 6-16. Therefore, $\omega_{in} = \omega_{sc} + (E_{n+1} - E_n)/\hbar$. Different vibrational modes often can be observed in Raman and infrared experiments.

Infrared spectra are due to the interaction of light with the modulation of the dipole moment of the molecule at the normal mode frequency as in Eq. 6-24. We show in a simple way that Raman spectra are due to the modulation of the polarizability of a molecule or solid by the normal mode. The polarizability relates an electric field to a polarization as in Section 4-5c. $P_i = \alpha_{ij}E_j$ summed on j. Similar to Eq. 6-24 one has

$$\alpha_{ij} = [\alpha_{oj}]_0 + [\partial\alpha_{ij}/\partial q_k]_0\, q_k + [\partial^2 \alpha_{ij}/\partial q_l\, \partial q_k]_0\, q_k\, q_l + \ldots \qquad (6\text{-}25)$$

Neglecting the tensor notation and writing the time dependence of the phonon normal mode as cos $\omega$t, this equation for the first two terms is $\alpha = \alpha_o + A \cos \omega t$. Now consider an electric field oscillating at $\omega_{in}$. We have for the polarization

$$\begin{aligned} P &= (\alpha_0 + A \cos \omega t)(E_o \cos \omega_{in} t) \\ &= \alpha_o E_o \cos \omega_{in} t + 1/2\, AE_o\, [\cos(\omega_{in} + \omega)t + \cos(\omega_{in} - \omega)t] \end{aligned} \qquad (6\text{-}26)$$

The time varying polarization in Eq. 6-26 can cause transitions in the quantum mechanical sense. The first term on the right of Eq. 6-26 causes emission at just the incident light frequency and is called Rayleigh scattering. The two terms in the brackets will cause transitions and result in the emission of scattered light a phonon frequency above and below the incident light. This is known as anti-Stokes and Stokes radiation, respectively. We can see that the one-phonon Raman transition is caused by the second term on the right of Eq. 6-25. Again, since the derivative evaluated at zero normal mode displacement transforms as $\Gamma_1$ only, those normal modes that transform as one of the six components of the polarizability (the symmetric terms of $x_ix_j$ for i, j = 1, 2, 3) will have a nonzero $(\partial\alpha_{ij}/\partial q_k)_0$. The selection rules for Raman activity can be determined in a similar manner to the infrared selection rules. The matrix element, the square of which gives the intensity, for the $\alpha_{12}$ laboratory component of the polarizability is given by

$$<e|\alpha_{12}|g> = \sum_{i,j=1}^{3} <\psi_{rot_e}|l_{1i}\ l_{2j}|\ \psi_{rot_g}><\psi_{vib_e}|\alpha_{ij}|\psi_{vib_g}> \quad (6\text{-}27)$$

analogous to Eq. 6-23. Again the rotational part in gases can be ignored. Thus, we immediately see the general selection rule for Raman activity. **A Raman line will be observed corresponding to the separation between two vibrational states g and e if the reduction of the direct product of the irreducible representations $e^* \times g$ transforms as the same irreducible representation as one or more of the Cartesian coordinates taken two at a time.**

Most experimental Raman spectra correspond to an excitation from the ground vibrational state, which transforms as $\Gamma_1$, to a one-phonon state which transforms in the same way as the normal mode $q_i$. For this process the Raman shift for a normal mode $q_i$ with frequency $\omega_i$ is given by $\omega_{in} - \omega_{sc} = \omega_i$ for the Stokes radiation. **Thus, a Raman line of frequency $\omega_i$ will be allowed whenever the normal modes transform as one or more of the six symmetric terms $x_i x_j$ where i,j = 1, 2, or 3.**

### e. Raman spectra — examples

For $C_{2v}$ point symmetry at least one of the six components of the polarizability transform as each of the irreducible representations.Therefore, all the normal modes can be observed in Raman spectra. For a gas of $H_2O$ molecules the frequencies of the normal modes of vibration can be observed by infrared or Raman techniques.

For $C_{4v}$ symmetry only normal modes that transform as $A_2$ will be unobservable in a Raman experiment. For the $AB_4$ example discussed in Chapter 5 the vibrational modes were found to transform as $2A_1 + 2B_1 + B_2 + 2E$. All of these modes are Raman active while only the $2A_1 + 2E$ are infrared active.

For solids the same considerations govern the selection rules. However, as in the infrared case, more information can be obtained since the electric field direction of the electromagnetic incident and scattered radiation can be uniquely determined. The modes that transform as each separate irreducible representation can be measured separately.

### f. Exclusion principle

It should be noted that in centrosymmetric molecules or cyrstals, an infrared active mode cannot be Raman active and vice versa. This is known as the exclusion principle. In centrosymmetric point groups the irreducible representations are divided into g- and u-irreducible representations (gerade or ungerade). If the basis functions transform as even (gerade) functions under inversion, then they transform as a g-irreducible representation. If the basis functions transform into minus themselves under inversion then they transform as an odd (ungerade) representation. Clearly the dipole moment which transforms as the components of the Cartesian coordinates $x_i$ transforms as a u-irreducible representation. The polarizability, on the other hand, transforms as $x_i x_j$, hence normal modes that transform as it does clearly transform as a g-irreducible representation. Thus, the Raman active normal modes that transform as a g-irreducible representation cannot transform appropriately for infrared activity. Modes that are neither infrared or Raman active are said to be silent modes.

It should be remarked that noncentrosymmetric molecules and crystals can have normal modes that are both infrared and Raman active. In fact, in piezoelectric crystals (Section 4-5e) there must be at least one normal mode that is infrared and Raman active.

## 6-4 Molecular Approximations (Site Symmetry and Davydov Splitting)

In crystals that contain some tightly bound chemical complexes it is often useful to make a molecular approximation in which they are treated as a "molecule" in free space and calculate its properties. We then take into account the rest of the crystal with a perturbation approach to see how the "molecular" properties are affected. We discuss vibrations in this section but any property of a molecule such as the electronic energy levels will be affected in a similar manner. Using this molecular approximation we discuss two perturbations in particular that can lift degeneracies and shift levels from the free space properties. First, the levels of a single molecule can be affected by the site symmetry. Second, Davydov splittings will occur for two or more molecules that are at equivalent lattice sites (sites that transform into one another under a symmetry operation of the crystal other than a primitive lattice translation).

### a. Site symmetry

Consider a complex such as an $(SO_4)^{2-}$ ion in a crystal. In calculating the normal modes of the crystal these five atoms could be treated just as the other atoms in the unit cell. This is perfectly correct but will usually add to the difficulty of comparing the normal mode results to experiment. To a good approximation the $(SO_4)^{2-}$ molecule can be treated as if the symmetry were $T_d$. This has been done in Section 5-3 with the result that the internal vibrational modes transform as $A_1 + E + 2T_2$. The four frequencies for the appropriate modes are well known and have been studied in aqueous solutions, the force constants are well known, etc. This is the reason why the molecular approximation is useful, namely, a great deal is known about the complex independent of the details of the particular crystal lattice. The $(SO_4)^{2-}$ complex in a crystal usually has a symmetry lower than $T_d$ so the degeneracies can split. The center of mass of the ion (the sulfur atom in this case) has a certain point symmetry in the crystal (site symmetry as discussed in Section 4-6 and in Chapter 11). This site symmetry is fixed by the rest of the atoms in the crystal. Then, the splittings of each of the four molecular modes can be determined by straightforward subgroup considerations or use of the correlation tables.

For example, consider the site symmetry of the sulfur atom to be $D_{2d}$. Then the mode that has E symmetry in the molecule will split into two modes in the solid. Using correlation diagrams these modes transform as the $A_1 + B_1$ irreducible representations of $D_{2d}$. However, the frequencies of these two modes will be close to each other and to the molecular E mode. If this is not so, the molecular approximation is a poor one and the problem must start again without this approximation. Note that in the crystal the $A_1$ and $B_1$ modes are both Raman active and they can be observed separately since they transform as different components of the polarizability. Each molecular $T_2$ mode will split into a $B_2$ + E mode for the $D_{2d}$ site symmetry. Again the $B_2$ and E modes can be determined separately by polarized infrared or Raman measurements. Note that for $T_d$ symmetry only the normal mode that transforms as the $T_2$ irreducible representation is infrared active. By lowering the symmetry of the complex the other modes can be made observable.

A study of the splitting and shifts from the molecular results as a function of the surroundings can be interesting in itself.

### b. Davydov splitting

Besides Davydov splitings this effect is sometimes called factor group or correlation field splittings. If two complexes in the same unit cell transform between themselves under some of the symmetry operations of the crystal, then we can observe a splitting of the eigenstates due to the coupling between the two complexes introduced by a phase relationship between the vibrations. We briefly discuss this effect for vibrations.

For example, consider two $(SO_4)^{2-}$ complexes. Suppose the molecular approximation is quite good and site symmetry considerations are applied to each complex lowering the symmetry to $D_{2d}$. The uniform breathing mode ($A_1$ of $T_d$ symmetry) will still transform as $A_1$ in $D_{2d}$ and Raman techniques can be used to measure its frequency. However, if some symmetry operation of the crystal transforms one $(SO_4)^{2-}$ into the other then the wave functions and motions are no longer independent but are coupled. The symmetry operations of the crystal cause the coupled motion to transform appropriately as two irreducible representations in the true point group of the crystal. Thus, the frequencies and selection rules will be different from the uncoupled frequencies. Let us call these two modes M and N. The transformation properties of the basis functions of mode, M, might be such that both $(SO_4)^{2-}$ ions breathe out (radially expand) at the same time. The other normal mode, N, has one ion breathing out while the other breathes in (radially contracts). It is reasonable to expect the frequency of mode M to be slightly larger than that of mode N because the restoring force for M will be slightly larger since the unit cell must expand and contract while for N the unit cell can remain at the same volume.

It is important to realize that this is the first section in which we have discussed some gross approximations. Group theory is a branch of mathematics and is exact. In fact, the usefulness of group theory is that it treats the exact symmetry aspects of the problem exactly! Sometimes, in various branches of science treating a problem exactly, obscures it by bringing in unnecessary complications.

## Notes

Wigner, Chapter 16, discusses direct products.

Discussion of symmetric and antisymmetric direct products with examples can be found in: Bhagavantam and Venkatarayudu, pp. 51ff and pp. 93ff; Wilson, Decius, and Cross, pp. 152ff where formulas for the symmetric product of a 2-and 3-dimensional irreducible representation to the vth power are given; Heine pp. 258ff; Hamermesh, Section 5-2.

Any standard quantum mechanics textbook discusses the interaction of radiation with matter. DiBartolo, Chapter 14, discusses this problem with a group theory slant including the expansion of the vector potential to give electric dipole, etc. radiation. Also see Heine Section 24.

Bishop, Chapter 9, and Schonland, Chapter 8, should be consulted for further readings on molecular vibrations.

In crystals the optic normal modes of vibrations that are infrared active are split into transverse optic modes (TO) and longitudinal optic modes (LO). The splitting is due to the strong electric fields of these modes. The splitting is observed for a nonzero wave vector, **k** (= $2\pi$/wavelength), and is an additional symmetry restriction to the calculations performed on crystals in this chapter, which is for **k** = 0. For example, consider a crystal with $C_{4v}$ symmetry and assume the result is one mode with the symmetry of each of the irreducible representations. For the $A_2$, $B_1$, and $B_2$ modes there will only be one frequency of vibration independent of the direction of **k** because these modes have no oscillating dipole moment. However, for the $A_1$ and E modes this is not true. For these modes, for small but nonzero **k** ($10^4$ to $10^5 cm^{-1}$ which is typical experimentally) one can observe an $A_1$(TO) and $A_1$(LO) when **k** is perpendicular to the z-axis (direction of the oscillating dipole moment for $A_1$ modes) or parallel to the z-axis, respectively. These modes occur at different frequencies with the LO always higher than the TO. Similarly, an E(TO) and E(LO) can be observed. There still are $3n - 3$ optic modes for any one direction of **k**. This complicated subject is discussed in Born and Huang "Dynamical Theory of Crystal Lattices" (Oxford, 1954), Sections 7 and 9. Also see: Burns and Scott, Phys. Rev. B7, 3088 (1973). Section 8 of Born and Huang also discusses the important topic of polaritons. Polariton effects occur when **k** $\approx 10^3$ $cm^{-1}$.

## Problems

**1.** For the molecule $AB_3$ with $D_{3h}$ symmetry determine under which irreducible representations the normal mode transforms. (Give a little thought to the coordinate system.) Which modes are infrared active? Raman active? Determine the symmetry-adapted vectors of the modes.

**2.** For the molecule $AB_3$ with $C_{3v}$ symmetry do the same as in Problem 1 but use as much information from that problem as possible to help with this one.

**3.** Under which irreducible representations of $D_{6h}$ do the vibrations of benzene transform? Which of these vibrations can be observed by infared and which by Raman techniques? What can you say about these vibrations with respect to in-plane vibrations, which involve C–H stretching (presumably high frequency) and which involve C–C stretching?

**4.** For trans-$N_2F_2$ ($C_{2h}$ point symmetry) find the irreducible representation spanned by the normal modes of vibration. Which modes can be observed by infared and which by Raman techniques? Which vibration will involve an out of plane deformation? For the other five modes find a set of five internal coordinates and solve the vibration problem using these coordinates. What can you say about the order of the frequencies of the vibratons? Can the cis and trans isomers of $N_2F_2$ be distinguished by infrared and Raman techniques?

**5.** For the seven carbon atoms that form the frame work of the trivinyl-methyl radical determine under which irreducible representations the normal modes transform. Which modes involve in-plane motion?

**6.** An atom is in a site symmetry of $D_3$. There is one p-electron in an eigenfunction that transforms as the E-irreducible representation (x, y). Another p-electron on the atom also transforms as the E-irreducible representation (x′, y′). If there is a small coupling between the electrons how do the resultant eigenfunctions transform?

**7.** For a linear molecular of n atoms with $C_{\infty v}$ symmetry, show that there are $(n-1)\Sigma^+ + (n-2)\Pi$ optic modes.

**8.** For an $A_2O$ molecule with $D_{\infty h}$ symmetry find the irreducible representations that describe the normal modes. Show the correlation of these modes with those obtained if the molecule has $C_{2v}$ symmetry as in $H_2O$.

**9.** How do the triple products of the coordinates $x_ix_jx_k$ (i, j, k = 1, 2, 3) transform in the point group $D_2$? Which elements of the piezoelectric tensor are nonzero?

**10.** For the point group $O_h$, (x,y,z) are partners of the $T_{1u}$ irreducible representation. Using these three functions write the basis functions for the representation formed by $T_{1u} \times T_{1u}$.

**11.** With reference to Eq. 6-21, show that the matrix element is zero unless $\Gamma_m \times \Gamma_b \times \Gamma_n$ contains $\Gamma_1$. (Hint. Since the matrix element is invariant to the symmetry operations, as in Section 4-4, operate with a symmetry operation R, sum over all R and use orthogonality properties of the irreducible representations. See Knox and Gold or Falicov.)

**12.** Prove that piezoelectric crystals that have at least one normal mode that is infrared and Raman active. (This is why crystals with the point symmetry O are not piezoelectric even though they lack a center of inversion.)

Chapter 7

# QUANTUM MECHANICS

*Columbus found a world, and had no chart,*
*Save one that faith deciphered in the skies:*
*To trust the soul's invincible surmise*
*Was all his science and his only art.*

*G. Santayana, "O World"*

In this chapter we bring together and clarify some of the points previously discussed and show how group theory is used in quantum mechanics.

## 7-1 Atomic Wave Functions

For solutions to the wave equation with a central field $V = V(|\mathbf{r}|)$, the angular parts of the wave functions are the spherical harmonics $Y_m^l(\theta,\phi)$ where $\theta$ and $\phi$ are defined in the usual way as in Fig. 7-1. The radial part of the wave function can be very complicated and might have to be solved by some self-consistent procedure. However, an eigenfunction can in principle be found for each eigenstate, and the energy of each state can be determined by a principal quantum number n and the orbital quantum number $l$. The radial eigenfunction is a function of n and distance $\rho(n,r)$. Under a rotational symmetry operation, $\rho(n,r)$ transforms into itself since radial distance is unaltered. However, the various angular parts of the wave function (the spherical harmonics) transform among themselves. Thus, here we need only concern ourselves with the angular parts of the wave functions unless the actual energies are to be calculated. For very complicated systems the many electron wave functions will often be written in terms of single particle wave functions. Thus the function will still have these simple angular dependences.

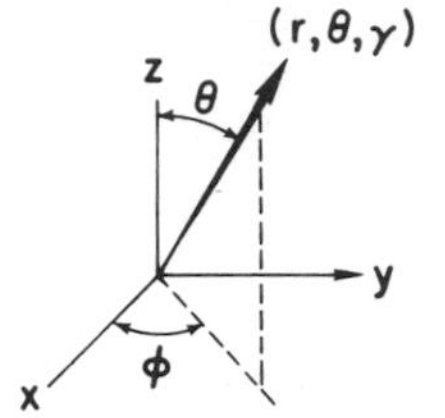


**Fig. 7-1** The definition of the angles in spherical coordinates.

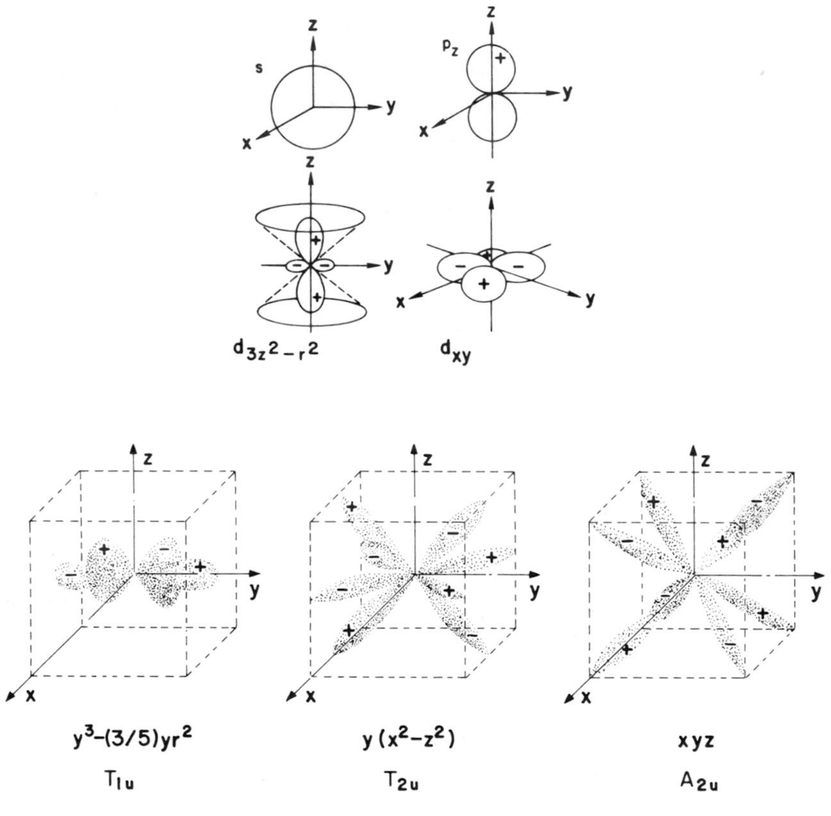


**Fig. 7-2** A schematic of some of the orbitals.

As is well known the spherical harmonics are written in terms of two quantum numbers, the orbital quantum number $l$ and the azimuthal quantum number m, $Y_m^l$. For a central potential with no external fields the eigenvalue is independent of m. For $l$ = 0, 1, 2, 3,... we have functions that conventionally are called s, p, d, f,... respectively. Table 7-1 lists the angular parts of the s, p, d, and f wave functions and Fig. 7-2 shows some of the orbitals.

## 7-2 Transformation of Functions

In previous chapters we have operated on a coordinate system to give a new coordinate system where we have expressed the position of a fixed point in space in terms of the new (primed) coordinate system which is related to the old (unprimed) system. The symmetry transformation is given by the symbol R. The 3 × 3 matrix that describes the transformation of coordinates, written as **R**, is given by

$$x_i' = \sum_{j=1}^{3} R_{ij} x_j \qquad \text{or} \qquad \begin{aligned} x' &= Rx \\ x &= R^{-1}x' \end{aligned} \tag{7-1}$$

For proper rotations $|\mathbf{R}| = 1$ (meaning determinant of R) while for improper rotations $|\mathbf{R}| = -1$. These are all **real orthogonal transformations** so $R^{-1} = \tilde{R}$ or $(R^{-1})_{ij} = R_{ji}$. (See Appendix 5 for the definition of matrix operations.) Thus, a representation derived from these transformations is unitary. The appendix at the end of this chapter shows explicitly how the matrices for the proper and improper real orthogonal transformations can be written. Also various other characteristics of **R** are derived.

Now we want to define formally what we mean when we operate on functions with our symmetry transformation R. The resulting matrices from this group of transformations will form a group that is isomorphic with the group of coordinate transformations. Other books often define an operator $P_R$ which operates on functions rather than coordinates but we will use R for simplicity. When R is followed by a function, operating on that function is implied, and when R is followed by coordinates, operating on coordinates is implied.

When considering the transformation properties of scalar wave functions we have in mind a scalar quantity that at every point gives a definite value and this value is independent of coordinate system. Thus, we expect the scalar quantity to have different functional forms in different coordinate systems. The result of a transformation of coordinates, as in Eq. 7-1, will give a new function $\psi'$ evaluated at the point $x'$ which gives the same value as the function $\psi$ evaluated at the point x, i.e. $\psi'(x') = \psi(x)$. The effect of a function operator on $\psi$ is $R\psi = \psi'$ and if $x' = Rx$ (coordinates) then the effect of the function operator is $R\psi(x') \equiv \psi'(x')$

**Table 7-1** Angular parts of wave functions for $l = 0$ to 3

| Orbital Type | Normalized orbital $A(\theta, \phi)$ $\left[\int_0^{2\pi}\int_0^{\pi} [A(\theta, \phi)]^2 \sin\theta \, d\theta \, d\phi = 1\right]$ |
|---|---|
| s | $\sqrt{1/\pi}/2$ |
| $p_z$ | $(\sqrt{3/\pi}/2) \cos\theta$ |
| $p_x$ | $(\sqrt{3/\pi}/2) \sin\theta \cos\phi$ |
| $p_y$ | $(\sqrt{3/\pi}/2) \sin\theta \cos\phi$ |
| $d_{z^2}$ | $(\sqrt{5/\pi}/4)(3\cos^2\theta - 1)$ |
| $d_{xz}$ | $(\sqrt{15/\pi}/2) \sin\theta \cos\theta \cos\phi$ |
| $d_{yz}$ | $(\sqrt{15/\pi}/2) \sin\theta \cos\theta \sin\phi$ |
| $d_{x^2-y^2}$ | $(\sqrt{15/\pi}/4) \sin^2\theta \cos 2\phi$ |
| $d_{xy}$ | $(\sqrt{15/\pi}/4) \sin^2\theta \sin 2\phi$ |
| $f_{xyz}$ | $(\sqrt{105/\pi}/4) \sin^2\theta \cos\theta \sin 2\phi$ |
| $f_{x(5x^2-3r^2)}$ | $(\sqrt{7/\pi}/4) \sin\theta \cos\phi (5\sin^2\theta \cos^2\phi - 3)$ |
| $f_{y(5y^2-3r^2)}$ | $(\sqrt{7/\pi}/4) \sin\theta \sin\phi (5\sin^2 \upsilon \sin^2\phi - 3)$ |
| $f_{z(5z^2-3r^2)}$ | $(\sqrt{7/\pi}/4)(5\cos^3\theta - 3\cos\theta)$ |
| $f_{x(z^2-y^2)}$ | $(\sqrt{105/\pi}/4) \sin\theta \cos\phi (\cos^2\theta - \sin^2\theta \sin^2\phi)$ |
| $f_{y(z^2-x^2)}$ | $(\sqrt{105/\pi}/4) \sin\theta \sin\phi (\cos^2\theta - \sin^2\theta \cos^2\phi)$ |
| $f_{z(x^2-y^2)}$ | $(\sqrt{105/\pi}/4) \sin^2\theta \cos\theta \cos 2\phi$ |

$= \psi(x)$. Thus, the function operator R changes the functional form in such a way as to compenstate for the change of variable. We have

$$\psi'(x') = R\psi(Rx) = \psi(x) \quad \text{or} \quad R\psi(x) = \psi(R^{-1} x) \qquad (7\text{-}2)$$

The latter result is just $R\psi(x') = \psi(R^{-1}x')$ where the primes are dropped. This latter form is particularly simple to use and it says that the function after rotation, $R\psi(x)$, has the same value as the unrotated function $\psi(x)$, but evaluated with the coordinates inversely rotated, i.e., $\psi(R^{-1}x)$.

As an example consider the effect of an $S_4$ symmetry operation in which the new, primed, coordinates are $x' = y$, $y' = -x$, $z' = -z$. Thus,

$$\begin{bmatrix} x' \\ y' \\ z' \end{bmatrix} = \begin{bmatrix} 0 & 1 & 0 \\ -1 & 0 & 0 \\ 0 & 0 & -1 \end{bmatrix} \begin{bmatrix} x \\ y \\ z \end{bmatrix} = R \begin{bmatrix} x \\ y \\ z \end{bmatrix}, \quad \text{so } R^{-1} \begin{bmatrix} x \\ y \\ z \end{bmatrix} = \begin{bmatrix} -y \\ x \\ -z \end{bmatrix} \tag{7-3}$$

Thus, $R\psi(x, y, z) = \psi(-y, x, -z)$ or in the function operated on by the function operator one must replace x with $-y$, y with x, and z with $-z$. If the functions being operated on were p-type wave functions xf(r), yf(r), zf(r), then one would obtain $Rxf(r) = -yf(r)$; $Ryf(r) = xf(r)$; $Rxf(r) = -zf(r)$. The matrix representation could be given by

$$R \begin{bmatrix} p_x \\ p_y \\ p_z \end{bmatrix} = \begin{bmatrix} p_x & p_y & p_z \end{bmatrix} \begin{bmatrix} 0 & 1 & 0 \\ -1 & 0 & 0 \\ 0 & 0 & -1 \end{bmatrix} \tag{7-4}$$

for the functions, treating the basis functions on the right side as a row matrix. Figure 7-3 shows a xf(r) orbital; after the symmetry operation $S_4$, this is $-yf(r)$ as noted. (Thus, while the x and y coordinates were rotated counterclockwise by the operation, one could treat the problem, alternately, as though the contours of the function were rotated clockwise.) Remember that in our definition of the function operation of R (Eq. 7-2) the contours of the function remain fixed. Thus, to describe these same contours the functional form changes as described in Eq. 7-2 and shown in the example.

The function operator R, operating on one basis function will give a linar combination of its partners

$$R\psi_i = \Sigma_j \psi_j \Gamma(R)_{ji} \tag{7-5}$$

as in Section 4-4 and Eq. 4-8 in particular. Equation 4-9 proves that the matrices involved are indeed representations of the group. Also see the discussion in Appendix 5. Note that in Eq. 7-2 the operator acts directly on the coordinates. This can be seen in Eq. 4-9 and is important to remember. Thus the effect of two symmetry operations is

$$AB\psi(x) = A\psi(B^{-1} x) = \psi(B^{-1} A^{-1} x) = \psi([AB]^{-1} x) \tag{7-6}$$

which is in the proper order. The function operator is a linear operator so that $R[f(x) + g(x)] = Rf(x) + Rg(x)$. (See Appendix 5.)

The other important property of the function operator is that it is unitary: (See Appendix 5.)

$$\langle R\psi \mid R\phi \rangle = \langle \psi \mid \phi \rangle \tag{7-7}$$

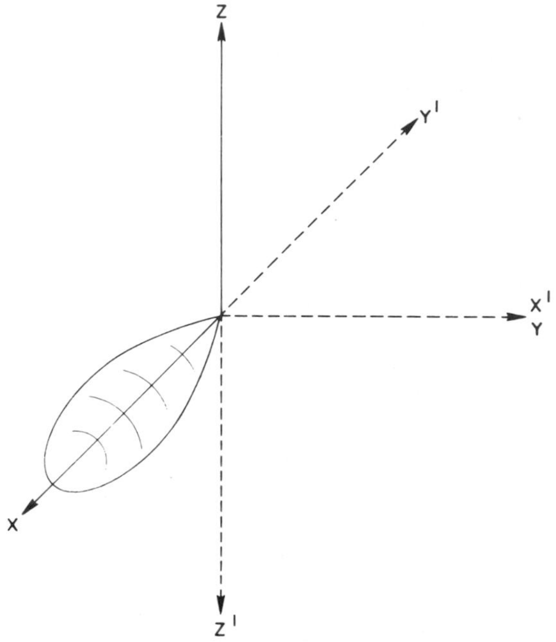


**Fig. 7-3** Part of a p-orbital before and after a symmetry operation.

which follows from the meaning of the operation, i.e., before or after the operation the function or resulting functions have the same values at the same point in space. See also Appendix 5. This was seen in the above examples and discussed briefly in Section 4-4, where we also proved the theorem that if n-basis functions are orthogonal the representation is unitary, and we only deal with unitary representations in this book. Consider Γ an n × n unitary representation with n-basis function $\psi_i$ and

$$\phi_j = \sum_{i=1}^{n} A_{ij}\,\psi_i \qquad \text{for } j = 1 \text{ to } n \tag{7-8a}$$

Then it is straightforward to prove that

$$\Gamma' = A^{-1}\,\Gamma\,A \tag{7-8b}$$

where Γ′ is also an n × n unitary representation with basis functions $\phi_j$ and A is an n × n matrix whose pq element is given by $A_{pq}$. Thus, we see how different linear combinations of basis functions induce a similarity transformation of the representation and vice versa. However, since the character of a matrix is unaltered by a similarity transformation (Eq. 3-11), Γ′ and Γ have the same character.

It should also be remembered that there are an infinite number of sets of basis functions for a representation that can not be written as linear combination of one another. For example the p- and d-wave functions are not linear combinations of one another. Thus, for any irreducible representation there are an infinite number of different sets of basis functions.

## 7-3 Eigenfunctions as Basis Functions

In this section we prove two very important theorems that relate group theory to quantum mechanics. These considerations use the fact that the Hamiltonian of a system is **invariant** under all symmetry operations. (See Section 4-7.) Consider what is meant in general by the effect of rotations on an operator O. Suppose before rotation

$$f = Og \tag{7-9}$$

After rotation a new operator will be obtained and $f' = Rf$ and $g' = Rg$, Eq. 7-2. The transformed operator O′ still relates the functions

$$f' = O' g' \tag{7-10}$$

Thus, we can relate O to the transformed operator

$$f' = Rf = ROg = RO(R^{-1} R)g = (ROR^{-1})g' \tag{7-11}$$

which gives the desired relation

$$O' = ROR^{-1} \tag{7-12}$$

**Invariant operators** are those for which O′ and O are indistinguishable. Thus, for an invariant operator H

$$H' = RHR^{-1} = H \qquad \text{or} \qquad RH = HR \tag{7-13}$$

The latter relation states that the invariant operators commute with all symmetry operations. It is not immediately obvious how Eq. 7-12 is to be applied to some specific operator, for example if $O = x$ or the various angular momentum operators $S_x$ or $J_z$. In Section 7-4 this will be discussed in detail where the relation between rotations and angular momentum is discussed.

For a Hamiltonian we have, from Eq. 7-13, $RH = HR$. So $RH(x) = H(R^{-1} x) = H(x)$ as discussed in Section 4-7, which is a fundamental property of a Hamiltonian. In the Notes we show that the set of symmetry operations that commute with the Hamiltonian indeed form a group.

Thus we have a group G of symmetry operations, R or A,B,..., under which the Hamiltonian is invariant and we have eigenfunctions $\psi_i$ and eigenvalues $E_i$ for the Hamiltonian.

**Theorem 1. Eigenfunctions that have the same eigenvalue form a basis of a representation of G.**

Let $\psi_i$, i = 1,2,..., *l*, be all the orthonormal eigenfunctions with eigenvalue E. Let a linear combination of these degenerate eigenfunctions be $\Psi = \Sigma_i a_i \psi_i$ where the $a_i$ are picked so that $<\Psi | \Psi> = 1$ or $\Sigma a_i^* a_j = 1$

$$H\Psi = E\Psi \qquad \text{and} \qquad H\psi_i = E\psi_i \tag{7-14}$$

Operating on the equation with a symmetry operation R

$$RH\Psi = H(R\Psi) = E(R\Psi) \qquad \text{and} \qquad H(R\psi_i) = E(R\psi_i) \tag{7-15}$$

So $R\psi_i$ is also an eigenfunction with the same eigenvalue as $\psi_i$. In general $R\psi_i$ must be a linear combination of the *l* original functions. Thus, a symmetry operation operating on one of the eigenfunctions must transform it into a linear combination of the *l*-values. The matrix describing the linear combination is given by

$$R\psi_i = \sum_{j=1}^{l} \psi_j \, \Gamma(R)_{ji} \tag{7-16}$$

By applying each of the symmetry operations of the group, a set of square matrices $\Gamma(A)$, $\Gamma(B)$, ... is obtained. As discussed in the last section and in Section 4-4 these matrices form a representation of the group G.

**Theorem 2. Eigenfunctions that are partners of an irreducible representation have the same eigenvalue.**

A symmetry operation operating on a partner of a basis of an irreducible unitary representation in general will result in a linear combination of the orthogonal partners. Thus, all the partners of the irreducible representation must be included in Eq. 7-16 and each is an eigenfunction with the same energy. In fact, $R\psi_i$ will result in a linear combination of functions of only the partners of some irreducible representation of $\psi_i$. Thus, while there is no restrictuion that eigenvalues that transform as the kth irreducible representation must have a different energy from those that transform as the mth, the probability that the two have the same energy is very small. If these two eigenvalues were the same it would be called an accidental degeneracy. **Accidental degeneracy** is defined as a degeneracy that is not a consequence of symmetry considerations.

As an example of Theorem 2, consider a hydrogen atom in a crystal at a site of $C_{4v}$ symmetry. Assume that the extra potential energy of the electron on the hydrogen atom due to the crystal can be written as a simple crystal field expansion $V = a(x^2 + y^2) + bz^2$ where x, y, and z are the usual coordinates with the origin at the center of the hydrogen atom and a and b are coefficients. This is the most general Hamiltonian (in quadratic form) that transforms as the $A_1$ irreducible representation in $C_{4v}$. In first-order perturbation theory the energy of the s-wave function will shift from its free atom value when it is in the crystal by an amount $\langle s|V|s\rangle$. This result will depend on the values of the coefficients a and b. In the free atom all three p-wave functions have the same energy. However, in the crystal since the coefficients in the Hamiltonian are not equal $\langle p_x|V|p_x\rangle = \langle p_y|V|p_y\rangle \neq \langle p_z|V|p_z\rangle$. Thus, the $p_x$- and $p_y$-wave functions have the same energy, which is different from the energy of the $p_z$-wave function. This is in agreement with our theorem since $p_x$ and $p_y$ are partners of the E-irreducible representation in $C_{4v}$ symmetry. Also we can see that $p_z$ transforms as a different irreducible representation and has in general a different energy, so there is little chance of accidental degeneracy. In a similar manner for the d-wave functions the xz and yz functions which are partners of the E-irreducible representation, will have the same energy. This energy will be different from that of the $z^2$ and the $x^2 - y^2$ and the xy wave functions. These latter three wave functions will also have energies that will all be different from each other. Notice that there is no symmetry reason to prevent, for particular values of the coefficients a and b, thc energy in the crystal of the $p_x$- and $p_y$-wave function from having the same value as that of the s-wave function. If these were to happen the degeneracy would be of an accidental kind. On the other hand, we will see in the next section that, in general, symmetry will prevent the s- and $z^2$-wave functions from being degenerate since they transform as the same irreducible representation. Similarly, the $(p_x, p_y)$ eigenfunctions in general will not be degenerate with the (xz, yz) eigenfunctions.

As another example of the use of these very important theorems consider an isolated atom with three p-wave functions, e.g., nitrogen with an outer electron configuration $2s^2p^3$. These three p-wave functions have the same energy. If the atom is put in a crystal at a site with cubic site symmetry, then all three p-wave functions still have equal energy since they are partners of an irreducible representation. Of course, the energy in the crystal will be different from that in the free state. If the symmetry is lowered from cubic, then the energy levels must split because only cubic point groups have 3-dimensional irreducible representations. As a more detailed example consider Fig. 7-4 where the five d-wave functions are shown to be degenerate in the free ion. If the symmetry is lowered to $O_h$ then the 5-fold degeneracy must split since cubic symmetry will allow at most a 3-fold degeneracy. The wave functions split as shown into a 2-fold degenerate level that in $O_h$ transforms as the $E_g$ irreducible representation and a 3-fold level that in $O_h$ transforms as the $T_{2g}$ irreducible representation. (These are sometimes called the d$\varepsilon$ and d$\gamma$ levels.) It is

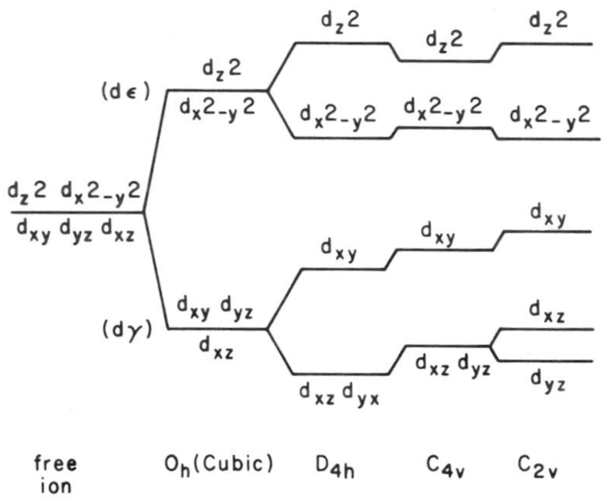


**Fig. 7-4** Free ion d-wave functions in several lower symmetries.

very important to realize that group theory does not determine which of the two levels lies higher in energy. It shows only that they have different energies. On lowering the symmetry to $D_{4h}$, the levels split further with only the wave functions that transform as the $E_g$ irreducible representation in $D_{4h}$ symmetry remaining degenerate. If the symmetry is further lowered to $C_{4v}$ no new splittings result, although in general there will be shifts in the levels due to different values of the radial integrals. To break the last degeneracy the symmetry must be reduced to orthorhombic or lower symmetry. Orthorhombic point groups have only 1-dimensional irreducible representations.

Thus, we see one of the very important uses of group theory in quantum mechanics. The irreducible representations can be used to label the eigenfunctions and eigenvalues. Further, the degeneracies and splittings are clear from the labels of the irreducible representation, and splitting resulting from perturbations are straightforward to handle.

## 7-4 Proper Rotations and Angular Momentum

The connection between proper rotations and angular momentum is very important in quantum mechanics. Here we make the connection between the rotation operator and angular momentum. Then we show that the angular momentum has the usual commutation relations. Thus, our previous knowledge derived from the commutation relations in quantum mechanics is apropos.

Notice that we will be considering infinitesimal rotations and not necessarily finite rotation as one does for the 32 point groups. After a rotation the new wave function is given by $\psi' = R\psi$, Eq. 7-2. Consider a rotation about the z-axis by a small angle $\varepsilon$, such that $\sin\varepsilon \approx \varepsilon$ and $\cos\varepsilon \approx 1$. Then

$$\begin{bmatrix} x' \\ y' \\ z' \end{bmatrix} = \begin{bmatrix} 1 & \varepsilon & 0 \\ -\varepsilon & 1 & 0 \\ 0 & 0 & 1 \end{bmatrix} \begin{bmatrix} x \\ y \\ z \end{bmatrix} \tag{7-17}$$

Remembering $R\psi(x) = \psi(R^{-1}x)$ from Eq. 7-2, to first order in $\varepsilon$ in a Taylor series

$$\psi' = \psi(x - \varepsilon y, \varepsilon x + y, z) \approx \psi(x, y, z) + \varepsilon\,[y\,(\partial\psi/\partial x) - x\,(\partial\psi/\partial y)] \tag{7-18a}$$

$$R_z = [1 + (i/\hbar)\varepsilon J_z] \tag{7-18b}$$

where $J_z$ is the usual orbital angular momentum

$$\mathbf{J} = \mathbf{r} \times \mathbf{p} = \begin{vmatrix} \mathbf{i} & \mathbf{i} & \mathbf{k} \\ x & y & z \\ p_x & p_y & p_z \end{vmatrix} \qquad J_z = x[-i\hbar\,(\partial/\partial y)] - y[-i\hbar\,(\partial/\partial x)] \tag{7-19}$$

Equation 7-18b shows explicitly what $R\psi$ means. The same result is obtained for R as in Eq. 7-18 in terms of $J_x$ or $J_y$ if the rotation were performed about the x- or y-axis. R for a finite rotation $\varepsilon$ can be directly obtained by considering the rotation to be performed n-times by an infinitesimal amount $\varepsilon/n$, i.e.,

$$R_z(\varepsilon) = \lim_{n\to\infty} [1 + (i/\hbar)\varepsilon/n]^n$$

$$= 1 + (i/\hbar)\,\varepsilon J_z + (1/2!)[(i/\hbar)\varepsilon\, J_z]^2 + \ldots = \exp[(i/\hbar)\varepsilon J_z] \tag{7-20}$$

It should always be remembered that the meaning of an exponential operator is just the expansion $e^x = 1 + x + (x^2)/2 + (x^3)/3! + \ldots$.

If the infinitesimal rotation takes place about an axis whose direction cosines are $n_x$, $n_y$, $n_z$ then the rotation can be expressed as

$$\begin{aligned} R_{\mathbf{n}}(\varepsilon) &= [1 + (i\varepsilon/\hbar)n_x J_x][1 + (i\varepsilon/\hbar)n_y J_y][1 + (i\varepsilon/\hbar)n_z J_z] \\ &= [1 + (i/\hbar)\,\varepsilon\,\mathbf{n} \cdot \mathbf{J}] \end{aligned} \tag{7-21}$$

Thus for a finite rotation the result, which is clearly unitary, is

$$R_{\mathbf{n}}(\varepsilon) = \exp[(i/\hbar)\,\varepsilon\,\mathbf{n} \cdot \mathbf{J}] \tag{7-22}$$

So far the rotation operator $R_{\mathbf{n}}(\varepsilon)$ that describes the function with the coordinates rotated as $\psi' = R\psi$ has been expressed in terms of **J**, the orbital angular momentum (Eqs. 7-18 and 7-19). This is the only type of angular momentum that can enter the problem because only functions of real space have been considered. We now give a generalized definition of the angular momentum J in Eq. 7-22. We will show that such a definition immediately leads to the commutation conditions that are expected for angular momentum. (See also Rose, Section 7.) The rotation operator R in Eq. 7-22, which contains the angular momentum operator, can then be used to operate on spinor wavefunctions which are two-component wave functions (or three-component wave functions if SU(3) symmetry is being considered, etc.) Consider a small rotation $\varepsilon$ about an axis **n** given by $n_x = 1$, $n_y = \phi$, $n_z = 0$ where $\phi$ is very small. Then

$$R(\varepsilon) = 1 + (i/\hbar)\,(J_x + \phi J_y)\varepsilon \tag{7-23}$$

Another way to obtain this same rotation is to first rotate about the z-axis an amount $\phi$, then rotate about the x-axis an amount $\varepsilon$, and then finally rotate the coordinates back by rotating about the z-axis an amount $-\phi$. Thus

$$\begin{aligned} &R_z(-\phi)R_x(\varepsilon)R_z(\phi) = \\ &\quad [1 - (i/\hbar)\phi J_z][1 + (i/\hbar)\varepsilon J_x][1 + (i/\hbar)\phi J_z] \end{aligned} \tag{7-24}$$

By equating terms in Eqs. 7-23 and 7-24, $J_xJ_z - J_zJ_x = -i\hbar J_y$ is immediately obtained. Similar results are obtained by rotating about other axes. Thus, with the usual definitions of a commutator, we have

$$\begin{aligned} &[J_x, J_y] \equiv J_xJ_y - J_yJ_x = i\hbar J_z \\ &[J_y, J_z] = i\hbar J_x \qquad [J_z, J_x] = i\hbar J_y \end{aligned} \tag{7-25}$$

It should be recalled from quantum mechanics (see the Notes) that from the commutation relations (Eq. 7-25), and the fact that the Hamiltonian **H** is invariant to rotations, one can derive the usual properties of angular momentum. Namely,

$$J^2 \psi_m{}^J = J(J+1)\,\psi_m{}^J \qquad\qquad J_z \psi_m{}^J = m\psi_m{}^J$$

$$J_+ \psi_m{}^J \equiv (J_x + iJ_y)\,\psi_m{}^J = [J(J+1) - m(m+1)]^{1/2}\,\psi_{m+1}{}^J$$

$$J_- \psi_m{}^J \equiv (J_x - iJ_y)\,\psi_m{}^J = [J(J+1) - m(m-1)]^{1/2}\,\psi_{m-1}{}^J \qquad (7\text{-}26)$$

where $m = -J, J-1, \ldots, -J$; $J = 0, 1/2, 1, 3/2, 2, \ldots, \infty$; the raising and lowering operators $J_+$ and $J_-$ have the usual definition; and $\psi_m{}^J$ is an eigenvector of $\mathbf{J}^2$ and $J_z$ as in Eq. 7-26.

As an example of the use of R in terms of angular momentum in connection with the rotation of operators (Eq. 7-12) consider an operator $\mathbf{f} \cdot \mathbf{S}$, where $\mathbf{f}$ is the external magnetic field. Consider the magnetic field along the z-axis of a crystal, thus the only operator is $f_z S_z$. If the crystal is rotated an amount $\varepsilon$ about the y-axis how is the spin operator changed? For small $\varepsilon$, clearly $S_z' = (1 + i\varepsilon S_y/\hbar)S_z(1 - i\varepsilon S_y/\hbar)$. Using the commutation relations Eq. 7-25 immediately gives $S_z' = S_z - \varepsilon S_x$ to first order in $\varepsilon$.

## 7-5 Perturbations

In this section we consider the group theory aspects of perturbation theory. The total Hamiltonian is taken as $\mathbf{H} = H_0 + V$ where $H_0$ is the unperturbed part for which the solutions already are known. For the $H_0$ part we have orthonormal eigenfunctions which are partners of a basis for a unitary representation that is irreducible (Theorem 2). $H_0$ transforms as $\Gamma_1$ of group G. V is a Hamiltonian that also must transform as the total symmetry irreducible representation of a group g. Now g can have at least the symmetry of G; an example would be the application of hydrostatic stress to a cubic crystal. Or g can have lower symmetry than G and an example would be a perturbation consisting of uniaxial stress on a cubic crystal. If the uniaxial stress were applied along one of the 4-fold axis of a crystal with $O_h$ symmetry, V would have $D_{4h}$ symmetry. These two different types of perturbations are treated separately.

### a. No reduction in symmetry

All the eigenfunctions of $H_0$ are of the form $\psi_i{}^m$ which transform as the ith partner of the mth irreducible representation. If V has at least the symmetry of $H_0$ then it also transforms as $\Gamma_1$ of the group G. Its effect can only be to shift the energy levels of $H_0$ and there will be no further splittings. There can be no splitting since the direct product of V with the basis functions will yield new basis functions that transform as the same irreducible representation, i.e., $\Gamma_1 \times \Gamma_m = \Gamma_m$. However, there can be a shift of the energy levels since the diagonal matrix elements will be nonzero, $<\psi_i^m | V | \psi_j^m> \propto \delta_{ij}$. Since the energies of the eigenstates of $H_0$ are shifted due to V it is possible that accidental degeneracies will occur or be removed.

If there is more than one set of eigenfunctions of $H_0$ that transform as the same irreducible representation, then V will result in matrix elements between basis functions that transform as the same row of the two sets of basis functions. Thus, the l-functions $\phi_i$ are also eigenfunctions of $H_0$ that transform as the same m-irreducible representation. The eigenvalues of $\phi_i$ are of course different from those of $\psi_i$. Now the perturbation V will shift the eigenvalues associated with $\psi_i$ by a different amount. Moreover, cross terms involving $<\psi_i | V | \phi_i>$ will also be nonzero. This is a simple problem to deal with since only a 2 × 2 matrix must be diagonalized if only two energy levels are considered. An example of this would be a hydrogen atom as an impurity at a site of $O_h$ symmetry in a crystal. The three 2p and 3p type eigenfunctions will have eigenvalues that are different from each other. Hydrostatic pressure is applied to the crystal resulting in a perturbation V. Then there is a shift of the 2p energy $E_{2p}$ given by $<2p_x | V | 2p_x>$, and a shift of a different amount of the 3p energy $E_{3p}$ given by $<3p_x | V | 3p_x>$. Using second order perturbation theory, the result from the off-diagonal terms will cause an increase of the energy separation of the two levels by $2 | <2p_x | V | 3p_x> |^2 / (E_{3p} - E_{2p})$.

A similar situation would be encountered in the problem of the hydrogen atom in $C_{4v}$ symmetry discussed in the last section. We work out this easy 2 × 2 matrix problem and discuss the very important no crossing rule.

### b. No crossing rule

Consider two eigenfunctions $\psi_1$ and $\psi_2$ of $H_0$ that have different eigenvalues. They can transform as the same 1-dimensional irreducible representation or they can transform as the same row of a larger-dimensional representation. Assume that V has the same symmetry as $H_0$. Take

$$H_{11} = <\psi_1 | H_0 + V | \psi_1> \qquad H_{12} = <\psi_1 | V | \psi_2> = H_{21}$$
$$H_{22} = <\psi_2 | H_0 + V | \psi_2> \qquad (7\text{-}27)$$

Note that $H_{12} = 0$, unless $\psi_1$ transforms as the same irreducible representation as $\psi_2$, remembering that V transforms as $\Gamma_1$. Then the secular equation is

$$\begin{vmatrix} H_{11} - E & H_{12} \\ H_{12} & H_{22} - E \end{vmatrix} = 0 \qquad (7\text{-}28)$$

Solving for the energy one obtains

$$E = (H_{11} + H_{22})/2 \pm 1/2\,[(H_{11} - H_{22})^2 + 4H_{12}{}^2]^{1/2} \qquad (7\text{-}29)$$

This can be put in simple form if $H_{12} << H_{11} - H_{22}$

$$E_+ = H_{11} + 2H_{12}{}^2/(H_{11} - H_{22})$$

$$E_- = H_{22} - 2H_{12}{}^2/(H_{11} - H_{22}) \qquad (7\text{-}30)$$

These results lead to the very general and important conclusion that energy levels that interact (have a nonzero $H_{12}$) repel each other. **So energy levels belonging to eigenfunctions that are not partners of each other but transform as the same irreducible representation repel each other if there is any interaction.** In Chapter 10 we will use this result when we consider the interaction of a central ion with its neighbors resulting in low lying energy levels that are depressed and called bonding levels and higher lying ones that are raised and called antibonding levels.

We can consider the supposed crossing of these levels which we now know repel each other. Suppose $H_{11}$ is lower than $H_{22}$ but V is a perturbation that raises $H_{11}$ and lowers $H_{22}$ such that for a large enough perturbation $H_{11} = H_{22}$. If just the diagonal terms of the secular equation Eq. 7-28 were considered, the levels would indeed cross. However, there is an allowed interaction, or mixing, between the two levels given by $H_{12}$. As can be seen from the exact expression for the energy, Eq. 7-29, the two levels at $H_{11} = H_{22}$ never get closer than $2\,|H_{12}|$. **Thus, interacting levels never cross.** Figure 7-5 shows this result for our example of a hydrogen atom in $C_{4v}$ symmetry. The $(p_x, p_y)$ and $(d_{xz}, d_{yz})$ functions transform as th E-irreducible representation. As the perturbation is increased they approach each other due to the diagonal terms but the off-diagonal term causes the two levels to repulse each other. The $p_z$-eigenfunction can cross these levels since it transforms as the $A_1$-irreducible representation and has no interaction term with functions that transform as other irreducible representations.

The new eigenfunctions can be obtained. They are linear combinations of $\psi_1$ and $\psi_2$. Associating (+) or (−) with the plus or minus sign in Eq. 7-29, the new eigenfunctions are obtained by using the following

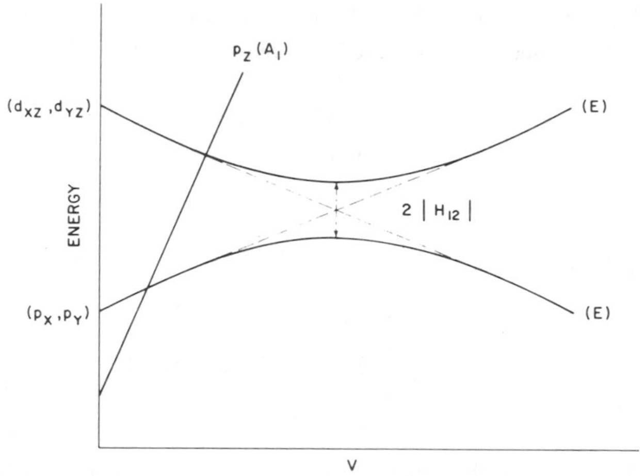


**Fig. 7-5** A conceptual diagram of eigenvalues versus perturbation potential V. The potential has $C_{4v}$ symmetry. Crossing and noncrossing is shown as the potential is increased.

orthonormal functions and solving for x from the equation for $E_+$

$$\psi_+ = (\psi_1 + x\psi_2)/(1+x^2)^{1/2}$$

$$\psi_- = (\psi_2 - x\psi_1)/(1+x^2)^{1/2}$$

$$E_+ = \langle \psi_+ | H_0 + V | \psi_+ \rangle \qquad (7\text{-}31)$$

From Eqs. 7-31 and 7-29 one determines that $x = 1$ for the value of V that causes $H_{11} = H_{22}$. Thus for this value of V the eigenfunctions are made up equally of the two originally separate eigenfunctions. In Fig. 7-5, the pure $p_x$ eigenfunction at $V = 0$ admixes more $d_{xz}$-like character as V is increased. At large V, this curve, which is now lowest, exhibits mostly $d_{xz}$ character with very little $p_x$ admixture.

### c. Reduction of symmetry

The perturbation V can lower the symmetry of the Hamiltonian from group G to group g, where g is a subgroup of G. In this case, although energy shifts can occur, the most interesting effects of V usually will be the removal of degeneracies that existed under group G. It is straightforward to calulate which representations of G correspond to those of g, although this can be determined by the use of the Correlation Table in Appendix 7 as discussed previously. The group g has some of the symmetry operations of G but not all. Let R be a symmetry operation of the group g and $\chi_i(R)$ be the character of the ith irreducible representation of the group g. $\chi(R)$ is the character of a possibly reducible representation obtained from an irreducible representation of the group G, but remember there are less R's in g than there are in the group G. To see if the irreducible representation of G is reduced in g and, if so, to accomplish the reduction, we ask how many times $\chi(R)$ is contained in the ith irreducible representation of g

$$n_i = (1/h)\ \Sigma_R\ \chi_i(R)^*\ \chi(R) \qquad (7\text{-}32)$$

where h is the order of group g and the sum is over the h symmetry operations of g. This is of course just Eq. 4-7 and we have previously used these considerations. For example, $PbTiO_3$ with $O_h$ symmetry has $T_{1u}$ normal modes. When the symmetry is lowered to $C_{4v}$ these normal modes split into modes that transform as the $A_1$ and E irreducible representations of $C_{4v}$. The total dimension remains three as it must. Each of the infinite sets of three basis functions of the $T_{1u}$ representation must be basis functions of the $A_1$ and E in $C_{4v}$. For example, (x, y, z) are basis functions of $T_{1u}$ and z and (x, y) are basis functions of $A_1$ and E, respectively. Of course, for E in $C_{4v}$ (x + iy, x − iy) could also be used for convenience.

Since the lowering of symmetry and the use of correlation tables have been discussed and used in several previous chapters we need not dwell on it further here. However, we mention that considerations of symmetry lowering associated with phase transitions in solids are of considerable interest. The lowering of the symmetry, besides causing splittings, can reduce selection rule restrictions. See Section 6-3c for these considerations with respect to splitting and selection rules changes for the modes in $PbTiO_3$. Also Fig. 7-4 and the discussion associated with it in Section 7-3 is a good example of the effects of symmetry reduction.

## 7-6 Matrix Elements (Selection Rules)

### a. General

As discussed in Section 6-3 we are concerned with calculating terms, or matrix elements of the form

$$M = \langle \psi_i{}^m \,|\, O_a{}^b \,|\, \psi_j{}^n \rangle \qquad (7\text{-}33)$$

where $\psi_i^m$ is a function transforming as the ith row of the mth irreducible representation and $O_a^b$ is a function or operator transforming as the ath row of the bth irreducible representation. Terms such as these enter in secular determinants, transition probabilities, etc. The transition rate between two states g and e is given by Fermi's golden rule where $\psi_i^m$ would be the eigenfunction of the g-state and $\psi_j^n$ of the e-state in Eq. 7-33. The M in Eq. 7-33 would be given a superscript $M^{eg}$

$$W_{eg} = (2\pi/\hbar)\,|\,M^{eg}\,|^2\,\rho(e) \tag{7-34}$$

where $\rho(e)$ is the density of final states. If the matrix element is zero, the transition is forbidden. As discussed in Section 6-3a the **matrix element is zero if the reduction of the irreducible representations m × n does not contain b.**

It should also be recalled from the discussion on direct products of irreducible representations (Section 6-1) that complications can arise when m = n. For this case, if $\psi_i^m$ and $\psi_i^n$ are not independent sets of partners, one must form only the symmetric products, the antisymmetric products being identically zero. Thus, for the case of sets of partners which are not independent the matrix element would be zero if the symmetric directproduct $[m \times n]_S$ does not contain the b irreducible representation.

We will proceed with several general examples.

### b. Electric dipole transitions

The electric dipole operator transforms as the coordinates x, y, z as discussed in Section 4-5b. The transition probability will be zero unless the reduction of the direct product m × n contains an irreducible representation in which x or y or z transforms. In the case thatone can uniquely fix the axis, as in a crystal, general selection rules as well as polarization selection rules can be determined. This occurred in the examples of Section 6-3c.

We consider an example of states that transform as the irreducible representations of the point group $C_{4v}$. Table 7-2 shows a few of inifinite number of basis functions and their transformation properties. Also shown are most of the direct products. The products not shown are the very simple ones; $\Gamma_1 \times \Gamma_n = \Gamma_n$; if $\Gamma_a$ is real and 1-dimensional, $\Gamma_a \times \Gamma_a = \Gamma_1$. If a transition from a state that transforms as the $B_1$ irreducible representation to one that transforms as E is considered, $B_1 \times E = E$. Such a transition is permitted by the x- and y-components of the dipole

**Table 7-2** Some basis functions and the direct products of the irreducible representation for point group $C_{4v}$

| $\Gamma_m$ | Basis Functions | | |
|---|---|---|---|
| $A_1$ | z | $x^2 + y^2$ | $z^2$ |
| $A_2$ | $R_z$ | | |
| $B_1$ | | $x^2 - y^2$ | |
| $B_2$ | | xy | |
| E | (x, y) | $(R_x, R_y)$ | (xz, yz) |

Direct products

$A_2 \times B_1 = B_2$ $\qquad$ $B_1 \times B_2 = A_2$

$A_2 \times B_2 = B_1$ $\qquad$ $B_1 \times E = E$

$A_2 \times E = E$ $\qquad$ $B_2 \times E = E$

$E \times E = A_1 + A_2 + B_1 + B_2$

$[E \times E]_S = A_1 + B_1 + B_2$ $\qquad$ $\{E \times E\}_{AS} = A_2$

operator but not the z-component. Table 7-3 lists this as $\mu_\perp$. The other transitions that are allowed by $\mu_\perp$ are also listed as well as those that can occur due to the z-component of the dipole operator, $\mu_\parallel$. For this example, no problem arises from $\{E \times E\}_{AS}$ because none of the components of the dipole operator transform as the $A_2$ irreducible representation. Note that there are a number of states between which electric dipole transitions are not allowed.

### c. Magnetic dipole transitions

The magnetic dipole transforms as an axial vector, i.e., $R_x$, $R_y$, $R_z$ in the character tables. That is, R transforms as the cross product of two vectors **A** × **B**,

$$R_x=(A_yB_z-A_zB_y), \qquad R_y=(A_zB_x-A_xB_z), \qquad R_z=(A_xB_y-A_yB_x) \tag{7-35}$$

Thus, R does not transform as a vector but as an antisymmetric second-rank tensor, or is called axial vector. (See the Notes to Chapter 4.) Pictorially an axial vector can be thought of as a current loop. $R_z$ would be a current loop in the right-handed sense (counterclockwise if looking down the z-axis). Similarly for $R_x$ and $R_y$. For rotations about the principle axis, $R_z$ transforms into plus itself. Similarly, $\sigma_h R_z = R_z$, $iR_z =$

**Table 7-3** Allowed electric dipole ($\mu$) and magnetic dipole (R) transitions

| | $A_1$ | $A_2$ | $B_1$ | $B_2$ | E |
|---|---|---|---|---|---|
| $A_1$ | $\mu_{\parallel}$ | $R_{\parallel}$ | — | — | $\mu_{\perp}R_{\perp}$ |
| $A_2$ | | $\mu_{\parallel}$ | — | — | $\mu_{\perp}R_{\perp}$ |
| $B_1$ | | | $\mu_{\parallel}$ | $R_{\parallel}$ | $\mu_{\perp}R_{\perp}$ |
| $B_2$ | | | | $\mu_{\parallel}$ | $\mu_{\perp}R_{\perp}$ |
| E | | | | | $\mu_{\parallel}$ |

$R_z$, but $\sigma_v R_z = -R_z$, $C_2^x R_z = -R_z$, etc. These results are tabulated in the character tables.

The determination of the allowed transitions, for our example of the states that transform as the irreducible representation of $C_{4v}$ symmetry, is straightforward. For example, transitions from $A_2$ to $B_1$ states are not allowed since the direct product $A_2 \times B_1 = B_2$ and none of the components of R transform as $B_2$. The allowed transitions are shown in Table 7-3 along with the polarization selection rules. The only result that needs a comment is the transition involving independent states, both of E-symmetry. There is an antisymmetric function that transforms as $\{E \times E\}_{AS} = A_2$ and the transition is allowed. However, if the two E states are not independent there is no function that transforms as $A_2$ thus the transition probability is zero.

### d. Electric quadrupole transitions

The electric quadrupole operator transforms as a symmetric second-rank tensor. It is just like the polarizability of Section 4-5c with the additional restriction that its trace is zero; i.e. instead of six independent components for a second-rank tensor there are only five since $Q_{xx} + Q_{yy} + Q_{zz} = 0$. Thus, the components of the quadrupole operator transform as the symmetric combinations of the coordinates two at a time, except one takes $x^2 + y^2 + z^2 = 0$. Since there are five components rather than three for the electric and magnetic dipole operators, more transitions usually are allowed.

The only transitions not allowed for our example of states that have $C_{4v}$ symmetry occur when the product of the states transforms as $A_2$. Thus, transitions $A_1$ to $A_2$ and $B_1$ to $B_2$ are not allowed for quadrupole radiation. All others are allowed by some component of the quadrupole operator. For real irreducible representation, the direct product $\Gamma_a \times \Gamma_a = \Gamma_1$ for 1-dimensional representations. Such a transition is not allowed for cubic or higher symmetry by the quadrupole operator since the restriction is $x^2 + y^2 + z^2 = 0$. However, for $C_{4v}$ symmetry $(x^2 + y^2)$ transforms as $A_1$ as does $z^2$ but the two functions are independent of each other so the transition is allowed for symmetries lower than cubic.

### e. Transitions for centrosymmetric point groups

For centrosymmetric point groups one can make some generalizations regarding forbidden transitions. For these point groups the irreducible representations are either even parity (g) or odd parity (u) under inversion symmetry.

The electric dipole operator transforms as a u-irreducible representation. Thus, transitions of states g to g are not allowed (g × g = g). Similarly, u to u are not allowed (u × u = g).

The magnetic dipole operator, since it is a cross product of two vectors, transforms as the g-irreducible representations. Thus, transitions of states g to u are not allowed (g × u = u). This same condition holds for the electric quadrupole operator which also transforms as the g-irreducible representation. For the case of infrared and Raman selection rules of normal modes these generalities reduce to the so-called exclusion principle (Section 6-3f).

### f. Vibronic coupling

For molecules and solids that are centrosymmetric we often find weak electric-dipole transitions where, at first glance, they are not allowed. This happens frequently in centrosymmetric complexes that contain d-wave functions from transition metal ions (Ti, V, Cr,...,Cu). The d-wave functions transforms as g-irreducible representations. Since g × g = g it follows that d to d transitions are forbidden by an electric dipole mechanism.

The mechanism that weakly breaks the electronic selection rules is vibronic coupling. The wave functions that enter Eq. 7-33 are the complete wave functions, not just the electronic part. Thus, for a solid or molecule the wave function should be a product of the electronic and vibrational wave functions as in Section 6-3. Denoting the ground state vibrational and electron wave function as $\psi_v \psi_e$ and the excited state with primes, the dipole matrix element for vibronic coupling is

$$\langle \psi_v' \psi_e' | \mu_i | \psi_v \psi_e \rangle \tag{7-36}$$

We have already discussed the symmetry properties of the vibrational wave functions. $\psi_v$ transforms as $\Gamma_1$, and for a one-phonon process $\psi_v'$ transforms as the same irreducible representation as the excited normal mode. Thus, the problem is reduced to the direct product of one more irreducible representation, that of the normal modes. Of course the molecule must have normal modes of the required symmetry for the vibrationally induced transition to occur.

We examine a generalized example. Consider the centrosymmetric point group $C_{2h}$. All the electronic g to u transitions are allowed by some component of the dipole operator. However, pure electronic g to g or u to u dipole transitions are not allowed. Nevertheless an $A_g$ to $A_g$ electronic transition would be allowed with the z-component of the dipole operator if a phonon of $A_u$ symmetry was involved. This is because of the direct product $\psi_e \times \psi_v' \times \psi_e' = A_g \times A_u \times A_g = A_u$. Similarly, for the same electronic transition, a $B_u$ phonon would make the transition allowed in x- or y-polarization. These and the general results are shown in Table 7-4.

Energy is still conserved in these processes. At 0°K, the absorption of electromagnetic radiation will take place at an energy corresponding to the separation of the electronic levels plus the vibrational energy of the phonon. Thus, a phonon is created. If the electronic states can be observed to emit light (photoluminescence or fluorescence), the emitted electromagnetic radiation will correspond to the energy separation of the electronic levels minus the vibrational energy (phonon created). The difference in energy between the absorption and fluorescence is twice the energy of the normal mode of vibration involved.

(For solids it should be remarked that the normal modes mentioned in this section are not necessarily those discussed in Chapters 5 and 6. In those chapters we were discussing the normal modes of solids that had very large wavelength. In this section, if the electronic transition is of a localized kind, the normal modes that can be coupled may have very small wavelength. In fact, they can be modes with wavelength of the order of the lattice distance and are therefore modes near the Brillouin zone edge. For free molecules such problems do not exist.)

Vibrationally assisted transitions can occur in complexes that lack a center of symmetry. In our example of $C_{4v}$ symmetry, Tables 7-2 and 7-3, electric dipole transitions for $A_1$ to $B_1$ were not allowed. However, if the complex has a normal mode with $B_1$ symmetry then the transition could be observed with z-polarization. If the complex has a normal mode

**Table 7-4** Symmetries of vibrations that will allow various electric dipole transitions for point group $C_{2h}$

| | $A_g$ | $B_g$ | $A_u$ | $B_u$ |
|---|---|---|---|---|
| $A_g$ | ($A_u$ - z<br>$B_u$ - x,y) | $A_u$ - x,y<br>$B_u$ - z | ALLOWED | |
| $B_g$ | | ($A_u$ - z<br>$B_u$ - x,y) | | |
| $A_u$ | | | ($A_u$ - z<br>$B_u$ - x,y) | $A_u$ - x,y<br>$B_u$ - z |
| $B_u$ | | | | ($A_u$ - z<br>$B_u$ - x,y) |

with E-symmetry the transition could be observed in x- and y-polarization.

### g. Noncentrosymmetric d-d transitions

Transitions between various d-orbitals of the transition metal series (Ti, V, Cr,...,Cu) are of considerable interest. We might, if we are not thinking, say that the d-states always transform as g-irreducible representations. Thus, g to g transitions will somehow not be allowed in the electric dipole approximation. However this is not correct since if the transition metal is at a site with no center of symmetry, then the d-orbitals transform as an irreducible representation with no g or u label. The example will clarify this point. (Remember there are an infinite number of functions in a centrosymmetric point group that transform as u-irreducible representations. When the corresponding noncentrosymmetric point group is used, by eliminating i as a symmetry operation, these transform in the same way as the d-states. The noncentrosymmetric Hamiltonian will have just the correct symmetry to mix these states which are even and odd in the centrosymmetric case.)

In our $C_{4v}$ problem the d-states (xz, yz) and p-orbitals (x, y) transform as E. Hence, the transition metal "d-states" will be a linear combination of the 3d-atomic-like states with some 4p-atomic states admixed. Thus ($xz + \alpha x$, $yz + \alpha y$), where $\alpha$ is a constant, also has E-symmetry and perhaps is a better way to write the "d-state." Actually, such a function is only an approximation. Coupled into the $3d_{xz}$-state is the $4p_x$, $5p_x$, etc., and the infinite number of other states that transform as the E-irreducible representation. As discussed in Section 7-4 it is precisely these states that have the correct symmetry to mix. Similarly, $p_z$

can mix with the d-states that transform as $A_1$. There are no p-states that transform as $B_1$ and $B_2$; however, there are f-states that transform as $B_1$ and $B_2$. Clearly, $z(x^2 - y^2)$ and zxy transform as $B_1$ and $B_2$, respectively. We often see a statement to the effect: "The d-d transitions in noncentrosymmetric complexes borrow their intensity from higher p- and f-states." The meaning of this statement is now clear.

## 7-7 General Secular Equation Problem

There are very few problems in quantum mechanics that are exactly soluble. When we treat molecules or solids, symmetry must be used as much as possible. Even then, there are an infinite number of terms that transform as a given irreducible representation. Naturally the number actually used in the approximate solution of the problem must be severely restricted. In this section we discuss some aspects of the solutions.

Suppose we are forming a molecule from several atoms. The Hamiltonian **H** in the wave equation transforms as $\Gamma_1$ in the group G. Each nondegenerate eigenfunction of the molecule $\psi$ transforms as a 1-dimensional representation of G and degenerate eigenfunctions transform as partners of a higher dimension irreducible representation of G. Here, we ignore accidental degeneracies. The molecule transforms into itself under all the symmetry operations of G. For the isolated atoms that make up the molecule, the eigenfunctions are given by $\phi$. Naturally the $\phi$'s do not transform as irreducible representations of G. However, all the infinite number of $\phi$'s form a complete set and the molecular wave function can be expanded in terms of the complete set of functions.

$$\psi = \sum_{i=1}^{\infty} a_i \phi_i \tag{7-37}$$

It can already be appreciated that the problem is being approached as though a great deal is known about the eigenfunctions of the isolated atoms. The wave equation for the molecule is now obtained by substituting Eq. 7-37 into Eq. 7-14.

$$\Sigma_i a_i [H\phi_i - E\phi_i] = 0 \tag{7-38}$$

Multiplying on the left by $\phi_j$ and integrating over all space, one obtains for each j

$$\sum_{i=1}^{\infty} a_i [\langle\phi_j | H | \phi_i\rangle - E\langle\phi_j | \phi_i\rangle] = 0 \qquad j = 1, 2, \ldots \tag{7-39}$$

This infinite set of equations will have a solution for all $a_i$ if the determinant of the coefficients, called the **secular determinant or secular equation**, equals zero.

$$| <\phi_j | H | \phi_i> - E<\phi_j | \phi_i> | = 0 \tag{7-40}$$

This is usually written as

$$\begin{vmatrix} H_{11} - ES_{11} & H_{12} - ES_{12} & \cdots \\ H_{21} - ES_{21} & H_{22} - ES_{22} & \cdots \\ \vdots & \vdots & \vdots \end{vmatrix} = 0$$

$$H_{ij} = <\phi_i | H | \phi_j> \qquad S_{ij} = <\phi_i | \phi_j> \tag{7-41}$$

Note that the overlap $S_{ij}$ of wave functions of different atoms in general will not be zero which leads to E appearing in the off-diagonal terms. The solution of this secular equation with an infinite number of rows and columns will yield the infinite number of E's of the molecule. For each E, we can obtain the $a_i$ by substituting the particular E into Eq. 7-39 for j = 1, 2,... There will be as many equations as unknown $a_i$'s, thus in principle the $a_i$'s can be determined.

Faced with this, in general, unsolvable problem, we make approximations and try to determine only the lower lying eigenvalues and eigenfunctions of the molecule. We take only the low lying $\phi_i$'s in Eq. 7-37 and try to neglect as many of the overlap functions as possible.

For example, if six equivalent atoms make up the molecule we would take for $\phi$ the lowest atomic orbital of each. The secular equation would then be a 6 × 6 matrix to be solved for six values of E. There would be many off-diagonal elements and the problem would be difficult. The problem can be simplified considerably if the symmetry aspects of the molecule are considered. Since the atoms transform among themselves under all the symmetry operations of the group G, so must the $\phi$'s. The resulting representation can be reduced and linear combinations of the $\phi$'s can be found that transform as the irreducible representations of G, i.e., $\phi^m$ as the wave functions. If the problem has six equivalent atoms then there will be six $\phi^m$ each composed of some or all of the $\phi$'s. Now we may apply to this problem all that has been discussed so far in the chapter, since now we are dealing with functions that transform irreducibly under the operations of the group G. In particular, there is no overlap or energy terms between orbitals that transform as different irreducible representations $<\phi^m | H | \phi^n>$ and also $<\phi^m | \phi^n> = \delta_{mn}$. If only one of the six $\phi^m$ transforms as the k-irreducible representation then its energy can be determined immediately $E = \langle\phi^k | H | \phi^k\rangle$, and the remaining secular determinant will be at most 5 × 5. If two of the $\phi^m$ transform as the j-irreducible representation, then a 2 × 2 determinant can be factored out leaving at most a 3 × 3 determinant, etc.

The molecular orbitals that are formed, $\phi^m$, are linear combinations of the atomic orbitals $\phi$'s, hence the name given to this approach **linear combination of atomic orbitals – molecular orbital, LCAO-MO**. We will return to this in greater detail in Chapters 9 and 10.

The ideas discussed above are very general. We expanded the eigenfunctions of a given complicated problem in terms of a complete set of functions (Eq. 7-37), but then we made the approximation by using a severely truncated set. In order to solve the secular equation easily, we found the appropriate linear combinations of functions that transform irreducibly under the group operations.

We can see that formally the above discussion is just like the normal mode problem of Chapters 5 and 6. The normal modes are expanded in terms of a complete set of atomic displacements. The complete set contains only 3N arbitrary displacements instead of an infinite number thus, no approximation is required. Then, appropriate linear combinations of the 3N arbitrary displacements are determined (symmetry-adapted vectors) that factor the dynamical matrix. Again, if only one symmetry-adapted vector transforms as the k-irreducible representation, the problem is completely solved.

We discuss a related but somewhat different problem of a hydrogen atom in free space. There are an infinite number of s-eigenfunctions, each one separately transforming into itself under all the symmetry operation of the full rotation group which will be discussed in Chapter 8. There are an infinite number of p-eigenfunctions, each 3-fold degenerate set consists of partners of a 3-dimensional irreducible representation. Similarly, for the five d-eigenfunctions which are partners of a 5-dimensional irreducible representation of the full rotation goup; and also for f, g, h, i,...-eigenfunctions. The secular determinant for these orthonormal eigenfunctions is completely diagonal. Now consider what happens if the hydrogen atom is put into a site of $O_h$ symmetry, i.e., a perturbing Hamiltonian V has $O_h$ symmetry. There are only 10 irreducible representations in $O_h$ symmetry so many hydrogen eigenfunctions now transform as the same irreducible representations. For example, the three p-wave functions transform as $T_{1u}$. Of the seven formally degenerate f-functions, three are now partners of the $T_{1u}$ representation. Off-diagonal terms such as $\langle p | V | f\rangle$ are nonzero and the p-function will now have some f-function admixed and vice versa. The same reduction occurs with g-functions which mix with s- and d-functions. (Also, as discussed in Section 7-5g, if there is no center of symmetry d-functions can mix with p-functions, etc.) Under the full rotation symmetry we diagonalized the secular equation by obtaining s, p, d,... eigenfunctions. When diagonalized the secular equation will have no off-diagonal elements In $O_h$ symmetry the eigenfunctions of the diagonalized problem will be mixed functions p + $\alpha$f, etc. as discussed above. Further lowering of the symmetry allows new off-diagonal elements, $H_{ij}$ and/or $ES_{ij}$ terms. The new eigenvalues and eigenfunctions that diagonalized the secular equation must be found. However, if the perturbations are small enough it is possible that the mixing of the original s, p, d eigenfunctions is small enough so that their labeling can be maintained.

## Appendix to Chapter 7

In this appendix we show explicitly how real orthogonal transformations can be written and how these transformations represent a proper or an improper rotation.

Consider a 3 × 3 real orthogonal matrix **R** as in Eq. 7-1. Real implies all elements are real; orthogonal implies $\mathbf{R}^{-1} = \tilde{\mathbf{R}} = \mathbf{R}^{\dagger}$, therefore $\mathbf{R}\tilde{\mathbf{R}} = \mathbf{1}$. The matrix **R** effects a linear orthogonal transformation of real space as implied by Eq. 7-1 (i.e., $x' = \mathbf{R}x$). The linearity is obvious. The orthogonality can be seen by showing that the distance between two points $(x_1, y_1, z_1)$ and $(x_2, y_2, z_2)$ is unaltered by the transformation. Let the row matrix $\tilde{\Delta} = (x_1 - x_2, y_1 - y_2, z_1 - z_2)$. The square of the distance between these two points is $\tilde{\Delta}\Delta$, which is the same after transformation, i.e.,

$$\tilde{\Delta}\Delta = \tilde{\Delta}\tilde{\mathbf{R}}\mathbf{R}\Delta = (\widetilde{\mathbf{R}\Delta})(\mathbf{R}\Delta) = (\tilde{\Delta})'(\Delta)' \tag{7-A1}$$

The determinant of **R**, $|\mathrm{R}|$, is $\pm 1$. This is immediately obvious by recalling a theorem on determinants: the determinant of a product of matricies is equal to the product of determinants of the individual matrices. Thus $1 = |\tilde{\mathbf{R}}\mathbf{R}| = >\tilde{\mathbf{R}}|\ |\mathbf{R}| = (|\mathbf{R}|)^2$, therefore $|\mathbf{R}| = \pm 1$. We formally will shown +1 corresponds to a proper rotation and −1, to an improper rotation.

Consider a proper rotation about the z-axis by an amount $\theta$, $R_z(\theta)$,

$$R_z(\theta) = \begin{bmatrix} \cos\theta & \sin\theta & 0 \\ -\sin\theta & \cos\theta & 0 \\ 0 & 0 & 1 \end{bmatrix} \tag{7-A2}$$

(Note that $|R_z(\theta)| = 1$ as required for a proper rotation.) The three eigenvalues of this equation, determined in the standard way $|R_{ij} - \lambda\delta_{ij}| = 0$, are $\lambda = 1, e^{\pm i\theta}$. Thus the unitary matrix **U** that diagonalizes the rotation in Eq. 7-A2 gives

$$\mathbf{U}^{\dagger}\mathrm{R}_z(\theta)\mathbf{U} = \mathbf{U}^{-1}\mathrm{R}_z(\theta)\mathbf{U} = \Lambda = \begin{bmatrix} e^{i\theta} & 0 & 0 \\ 0 & e^{-i\theta} & 0 \\ 0 & 0 & 1 \end{bmatrix} \tag{7-A3}$$

or $\mathrm{R}_z(\theta)\mathbf{U} = \mathbf{U}\Lambda$ as in a standard eigenvalue equation. Note: $\mathrm{Tr}\,\mathrm{R}_z(\theta) = \mathrm{Tr}\,\Lambda = 1 + 2\cos\theta$. The matrix **U** can be determined in the usual way resulting in

$$\mathbf{U} = \begin{bmatrix} 1/\sqrt{2} & 1/\sqrt{2} & 0 \\ -i/\sqrt{2} & i/\sqrt{2} & 0 \\ 0 & 0 & 1 \end{bmatrix} \tag{7-A4}$$

from which Eq. 7-A3 can be checked.

The rotation in Eq. 7-A2 is general in the sence that other orthogonal rotations can be used to bring the z-axis into the direction of the new rotation axis. This is explicitly shown in Section 8-1 in terms of Eulerian angles. The rotation matrix Eq. 7-A2 is a proper rotation, but a rotation followed by a reflection in the xy-plane can be represented by the product of two matrices.

$$\sigma_h \mathrm{R}_z(\theta) = \begin{bmatrix} 1 & 0 & 0 \\ 0 & 1 & 0 \\ 0 & 0 & -1 \end{bmatrix} \begin{bmatrix} \cos\theta & \sin\theta & 0 \\ -\sin\theta & \cos\theta & 0 \\ 0 & 0 & 1 \end{bmatrix}$$

$$= \begin{bmatrix} \cos\theta & \sin\theta & 0 \\ -\sin\theta & \cos\theta & 0 \\ 0 & 0 & -1 \end{bmatrix} \tag{7-A5}$$

The trace of this improper rotation is $-1 + 2\cos\theta$. The improper rotation can also be represented by a rotation by $\theta + \pi$ followed by an inversion through the origin

$$i\mathrm{R}_z(\theta + \pi) = \begin{bmatrix} -1 & 0 & 0 \\ 0 & -1 & 0 \\ 0 & 0 & -1 \end{bmatrix} \begin{bmatrix} \cos(\theta+\pi) & \sin(\theta+\pi) & 0 \\ -\sin(\theta+\pi) & \cos(\theta+\pi) & 0 \\ 0 & 0 & 1 \end{bmatrix}$$

$$= \begin{bmatrix} \cos\theta & \sin\theta & 0 \\ -\sin\theta & \cos\theta & 0 \\ 0 & 0 & -1 \end{bmatrix} \tag{7-A6}$$

**In summary** A proper rotation can be described by a matrix Eq. 7-A2 with trace $1 + 2\cos\theta$. An improper rotation can be described by a matrix Eq. 7-A5 (rotation and then reflection through a plane perpendicular) with trace $-1 + 2\cos\theta$. However, an improper rotation can also be described by Eq. 7-A6 (rotation $\theta + \pi$ followed by an inversion through the origin). This latter approach was discussed in the Notes of Chapter 1.

The set of $3 \times 3$ matrices **R**, discussed here, form a group (all the group postulates are satisfied). This group of **orthogonal transformations in three dimensions** is known as O(3) (or in some books $R_3$ or $D_3$). The determinant of the representation is $\pm 1$. This group can be expressed as the direct product of the subgroup $O^+(3)$, consisting of proper rotations, and the subgroup $C_i$, consisting of {E, i} with determinant $-1$.

We list some rotation matrices besides $C_2^z$ (Eq. 7-A2 with $\theta = \pi$) and i (in Eq. 7-A6)

$$\begin{matrix} C_2^y & C_3^z & C_4^z & C_6^z \\ \begin{bmatrix} -1 & 0 & 0 \\ 0 & 1 & 0 \\ 0 & 0 & -1 \end{bmatrix} & \begin{bmatrix} -1/2 & \sqrt{3}/2 & 0 \\ -\sqrt{3}/2 & -1/2 & 0 \\ 0 & 0 & 0 \end{bmatrix} & \begin{bmatrix} 0 & 1 & 0 \\ -1 & 0 & 0 \\ 0 & 0 & 1 \end{bmatrix} & \begin{bmatrix} 1/2 & \sqrt{3}/2 & 0 \\ -\sqrt{3}/2 & 1/2 & 0 \\ 0 & 0 & 1 \end{bmatrix} \end{matrix}$$

Matrices for the improper rotations can be obtained by multiplying the appropriate proper rotation by i. For example, $\sigma_z = iC_2^z$, $\sigma_y = iC_2^y$, $S_4^z = i(C_4^2)^3$, etc.

## Notes

The transformation of functions is leisurely and nicely discussed by McWeeny, Chapter 7 and Bishop, Chapter 5.

Lyubarskii has a very complete discussion of L. D. Landau's theory of second-order phase transitions (Chapter 7).

The electric dipole, etc. selection rules are discussed in Hamermech, Chapter 6; DiBartolo, Chapter 14; Wigner, Chapter 18; Heine, Section 13; as well as many other places. The expansion that leads to the various approximations can be found in any quantum mechanics textbook or see DiBartolo, Chapter 14.

Angular momentum operators and their commutation rules are discussed in Rose, Sections 5–7, as well as most quantum mechanics books. See the Notes to Chapter 13.

An English translation of von Neumann and Wigner's paper [Physikalische Zeitschrift **30**, 467 (1929)] on the no crossing rule can be found in Knox and Gold.

We show that the set of symmetry operations, A, B, C,..., that commute with the Hamiltonian **H** form a group. Symmetry operations are axiomatically associative and clearly the identity is included in the set. We know that $\mathbf{H} = \mathbf{A}\mathbf{H}\mathbf{A}^{-1} = \mathbf{B}\mathbf{H}\mathbf{B}^{-1} =$ etc. Thus, $\mathbf{A}^{-1}\mathbf{H}\mathbf{A} = \mathbf{H}$, so the inverse of each operation is a member of the set. Finally we show that the set is closed.

$$\mathbf{H} = \mathbf{A}\mathbf{H}\mathbf{A}^{-1} = \mathbf{A}(\mathbf{B}\mathbf{H}\mathbf{B}^{-1})\mathbf{A}^{-1} = (\mathbf{A}\mathbf{B})\mathbf{H}(\mathbf{B}^{-1}\mathbf{A}^{-1}) = (\mathbf{A}\mathbf{B})\mathbf{H}(\mathbf{A}\mathbf{B})^{-1}$$

Thus, the product AB also commutes with **H** so must be included in the set. Since the set satisfies all the group postulates it is a group.

## Problems

**1.** From the use of Eqs. 7-2 (7-3 and 7-4) find irreducible representations of the point group $C_{4v}$ using the five d-functions. Check that Eq. 7-2 does indeed keep the contours of the functions fixed.

**2.** For the E representation of the point group $C_{4v}$ write a representation using the functions (x, y) as basis functions. Do the same thing if (x + iy, x − iy) are basis functions. Find a similarity transformation between these two representations. (Note the physical significance of the similarity transformation. See Problem 3-5.)

**3.** Show results similar to Fig. 7-4 for f-wave functions.

**4.** A hydrogen atom is in a crystal at a site of symmetry $D_{4h}$. An electron is in the $2p_x$ eigenstate, and electromagnetic radiation polarized along the x-axis is present. To what d-states can we have electric dipole transitions?

**5.** What are the selection rules for electric dipole, magnetic dipole, and electric quadrupole transitions for eigenstates that transform as the irreducible representations of T point symmetry?

**6.** Given the character table for $D_{4h}$, make a correlation table for $D_{4h}$ to $C_{4v}$. What would you say if the request were for $D_{4h}$ to $C_{3v}$?

**7.** g-wave functions have a part that transforms as $y^4f(r)$. Find the functions that transform as irreducible representations of the O point group, in the sense of Eq. 5-28, that contain $y^4 f(r)$. What is the result for $O_h$ symmetry?

**8.** Consider four identical spherical charge distributions at the corners of a square ($D_{4h}$), i.e., a $B_4$ type, molecule. Write out the secular determinant taking the nearest neighbor overlap and energy to be S and V, respectively, and neglecting second nearest neighbor effects. Can you factor this secular determinant? These four charge distributions transform among themselves under all the symmetry operations of the group and give a reducible representation. Reduce the representation and find the proper linear combination of these four charge distributions that transform as the irreducible representations. (Use projection operators if you cannot guess.) Now write out the secular determinant using these new wave functions that transform as irreducible representations. Is the result not beautiful? What are the three eigenvalues? (Parts of this problem are similar to Problems 4-2, 4-7.)

**9.** For eigenfunctions that transform as the irreducible representations of the point groups O, work out the selection rules for (a) electric dipole, (b) magnetic dipole, and (c) electric quadrupole transitions between all possible eigenstates.

**10.** Consider f-atomic orbitals. Write out the angular parts of the seven wave functions in terms of sin $m\phi$ and cos $m\phi$ by taking the appropriate linear combinations. Compare these results with those in Table 7-1. The results in Table 7-1 are f-orbitals that transform irreducible under the symmetry operations of $O_h$. Find the appropriate linear combinations of the functions that transform irreducible under the full rotation that give the results shown in Table 7-1.

**11.** (a) For the E-irreducible representation of the $C_{3v}$ point group find the representation for the basis functions xy and $x^2 - y^2$. Is this irreducible reducible representation unitary? (b) Compare this last unitary representation with the unitary representation obtained from the basis functions xz and yz. Again alter the functions in part a of this problem so that the representation is the same as obtained for xz and yz.

Chapter 8

# CRYSTAL FIELD THEORY (AND ATOMIC PHYSICS)

*A well-constructed plot should, therefore, be single in its issue, rather than double as some maintain.*

*Aristotle, "Poetics"*

The basic work in crystal field theory was written in 1929 by Bethe. In this long paper, he showed how the symmetry of the electric field, or potential energy, due to the crystalline environment will cause the electronic levels of the central ion to split. The splitting of the degeneracies can be determined completely by symmetry although, as always, the magnitude of the splittings will depend on other details of the problem. Early history of the work on crystal field theory is reviewed by Ballhausen's book. (See the Notes.)

In this introduction, we would also like to make clear the difference between the terms crystal field theory, ligand field theory, and molecular orbital theory although they are often used loosely and sometimes interchangeably. All three approaches treat the symmetry of the problem correctly. The term **crystal field theory** is usually reserved for the problem in the original Bethe sense. Namely, the only role of the entire crystal that surrounds the central ion is to produce a static crystal field of the correct symmetry. The magnitude of the crystal field is calculated using the electrostatic approximation; that is, the atoms surrounding the central ion are treated as point changes, point dipoles, etc. In **ligand field theory** the electrostatic approximation is dropped; rather, the fact that the radial part of the central ion wave function overlaps with the ligands (neighbors) is recognized. The ionic radial wave functions lose some of their physical significance since they are mixed with the ligands. Thus, in ligand field theory, the radial wave function, interelectronic repulsion parameters, etc. are taken as adjustable parameters that are varied to fit the experimental data. **Molecular orbital theory** takes as a starting point the fact that overlap is important. When this is done, electronic wave functions of the central ion and the ligands including excited states, are obtained.

In this chapter, we talk in terms of crystal field theory. However, since the symmetry aspects of all these theories are the same, the labeling of the eigenfunctions in terms of irreducible representation is the same for all the theories.

## 8-1 Rotations in Terms of Euler Angles

In order to describe the rotation of a sphere with fixed origin, three parameters are needed.

To specify a rotation quantitatively we will use the Euler angles. Two angles are required to specify the direction of the axis of rotation and one, the magnitude of the rotation. There are various conventions for these angles. We follow Rose (see the Notes). $R_w(\theta)$ means a rotation about the w-direction by an amount $\theta$ in the right-handed screw sense. Then the rotation can be thought of as taking place (see Fig. 8-1):

$$\begin{aligned} R_z(\alpha) &\rightarrow (x', y', z') \\ R_{y'}(\beta) &\rightarrow (x'', y'', z'') \\ R_{z''}(\gamma) &\rightarrow (x''', y''', z''') \end{aligned} \qquad (8\text{-}1a)$$

This means a rotation by an amount $\alpha$ about the z-axis yields the coordinate $(x', y', z')$. This is followed by a rotation by an amount $\beta$ about the $y'$-axis, etc. The complete symmetry operation $R(\alpha, \beta, \gamma)$ which describes a general rotation in terms of Euler angles is written as

$$R(\alpha, \beta, \gamma) = R_{z''}(\gamma)R_{y'}(\beta)R_z(\alpha) \qquad (8\text{-}1b)$$

The identical rotation can be written in terms of rotations about the fixed (x,y,z) axis. (See Problem 8-1.)

$$R(\alpha, \beta, \gamma) = R_z(\alpha)R_y(\beta)R_z(\gamma) \qquad (8\text{-}2)$$

Note that the order of the rotations is reversed in Eq. 8-2 compared to Eq. 8-1. The expression for the symmetry operation $R(\alpha, \beta, \gamma)$ in Eq. 8-2 will be used here and in Chapter 13 in discussions of the full rotation group because it is more convenient than its equivalent expression in Eq. 8-1. We also note in passing that Eq. 8-2 can be expressed in terms of angular momentum operations, as in Section 7-4

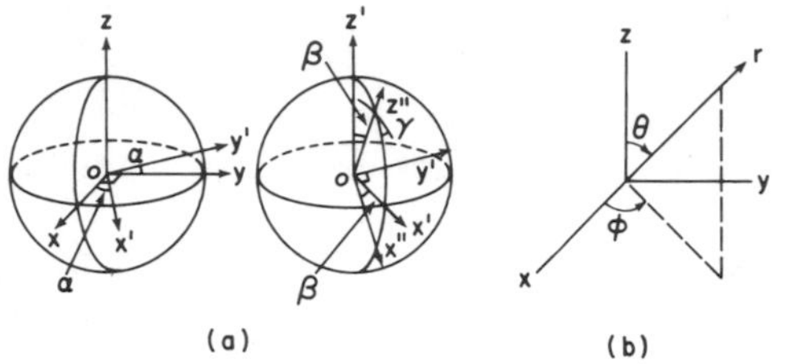


**Fig. 8-1** (a) Rotations that specify the Euler angles. (b) Definition of $\theta$ and $\phi$ that are used throughout.

$$R(\alpha, \beta, \gamma) = \exp(i\alpha J_z/\hbar) \exp(i\beta J_y/\hbar) \exp(i\gamma J_z/\hbar).$$

To make totally clear the meaning of Eqs. 8-1 and 8-2 we could write out explicitly $(x''', y''', z''')$ in terms of $(x, y, z)$. The relevant matrices required in Eq. 8-1 are

$$R_z(\alpha) \begin{bmatrix} x \\ y \\ z \end{bmatrix} = \begin{bmatrix} x' \\ y' \\ z' \end{bmatrix} = \begin{bmatrix} \cos\alpha & \sin\alpha & 0 \\ -\sin\alpha & \cos\alpha & 0 \\ 0 & 0 & 1 \end{bmatrix} \begin{bmatrix} x \\ y \\ z \end{bmatrix} \tag{8-3}$$

$$R_{y'}(\beta) \begin{bmatrix} x' \\ y' \\ z' \end{bmatrix} = \begin{bmatrix} x'' \\ y'' \\ z'' \end{bmatrix} = \begin{bmatrix} \cos\beta & 0 & -\sin\beta \\ 0 & 1 & 0 \\ \cos\beta & 0 & \sin\beta \end{bmatrix} \begin{bmatrix} x' \\ y' \\ z' \end{bmatrix} \tag{8-4}$$

The matrix for $R_{z''}(\gamma)$ is the same for in Eq. 8-3 but with $\gamma$ instead of $\alpha$. Clearly $R(\alpha, \beta, \gamma)$ is the $3 \times 3$ matrix formed by the matrix product of these three $3 \times 3$ matrices. Instead of writing out this resultant $3 \times 3$ matrix, which is straightforward, we write out a closely related, but more interesting one. Consider the three functions $Y_1 = -(x+iy)$, $Y_0 = \sqrt{2}z$, $Y_{-1} = (x-iy)$. Exactly the same definitions apply for the primed, double primed, and triple primed, i.e., $Y_1' = -(x'+iy')$, ..., $Y_{-1}''' = (x''' - iy''')$. Then by straightforward substitution from Eq. 8-3 one obtains

$$R_z(\alpha) \begin{bmatrix} Y_1 \\ Y_0 \\ Y_{-1} \end{bmatrix} = \begin{bmatrix} Y_1' \\ Y_0' \\ Y_{-1}' \end{bmatrix} = \begin{bmatrix} e^{-i\alpha} & 0 & 0 \\ 0 & 1 & 0 \\ 0 & 0 & e^{-i\alpha} \end{bmatrix} \begin{bmatrix} Y_1 \\ Y_0 \\ Y_{-1} \end{bmatrix} \tag{8-5}$$

This result is the same as that found in the Appendix to Chapter 7 for Eq. 7-A3. By straightforward substitution from Eq. 8-4

$$\begin{bmatrix} Y_1'' \\ Y_0'' \\ Y_{-1}'' \end{bmatrix} = \begin{bmatrix} (1+\cos\beta)/2 & \sin\beta/\sqrt{2} & (1-\cos\beta)/2 \\ -\sin\beta/\sqrt{2} & \cos\beta & \sin\beta/\sqrt{2} \\ (1-\cos\beta)/2 & -\sin\beta/\sqrt{2} & (1+\cos\beta)/2 \end{bmatrix} \begin{bmatrix} Y_1' \\ Y_0' \\ Y_{-1}' \end{bmatrix} \tag{8-6}$$

The third matrix for $R_z''(\gamma)$ is again the same as $3 \times 3$ matrix in Eq. 8-5 but with $\alpha$ replaced by $\gamma$. The resultant matrix product of the three $3 \times 3$ matrices is in Eq. 8-7. This result gives the representation of $R(\alpha, \beta, \gamma)$ as in Eq. 8-1.

$$\begin{bmatrix} Y_1''' \\ Y_0''' \\ Y_{-1}''' \end{bmatrix} = R(\alpha, \beta, \gamma) \begin{bmatrix} Y_1 \\ Y_0 \\ Y_{-1} \end{bmatrix}$$

$$= \begin{bmatrix} e^{-i(\alpha+\gamma)}(1+\cos\beta)/2 & e^{-i\gamma}\sin\beta/\sqrt{2} & e^{i(\alpha-\gamma)}(1-\cos\beta)/2 \\ -e^{-i\alpha}\sin\beta/\sqrt{2} & \cos\beta & e^{i\alpha}\sin\beta/\sqrt{2} \\ e^{-i(\alpha-\gamma)}(1-\cos\beta)/2 & -e^{i\alpha}\sin\beta/\sqrt{2} & e^{i(\alpha+\gamma)}(1+\cos\beta)/2 \end{bmatrix} \begin{bmatrix} Y_1 \\ Y_0 \\ Y_{-1} \end{bmatrix} \quad (8\text{-}7)$$

This representation is interesting because the three functions $Y_1$, $Y_0$, $Y_{-1}$ are proportional to the spherical harmonics of order one. As we will soon discuss, the spherical harmonics of each order $l$ ($l = 0, 1, 2, 3, \ldots$) form the basis of a $2l + 1$ dimension irreducible representation of the full rotation group. Thus, the transpose of the matrix that appears in Eq. 8-7 is just the irreducible representation of the $l = 1$ irreducible representation. It is the transpose because of the way we are defining our irreducible representations, Eq. 7-5. We can see how the $l = 2, 3, \ldots$ irreducible representations can be generated.

## 8-2 Representations of the Full Rotation Group

The **full rotation group** is the point group containing the symmetry operations of a sphere. It is called O(3) since it contains all the real orthogonal transformations about a fixed point in 3-dimensional space. (These transformations consist of all the transformations with real coefficients which leave the sphere $x^2 + y^2 + z^2$ invariant.) The group that contains only the proper rotations about a point is called the rotation group $O(3)^+$. If the group containing only {E,i} is called $C_i$ then $O(3) = O(3)^+ \times C_i$, where group multiplication is defined in Section 2-6c. Then once the irreducible representations of $O(3)^+$ are obtained, the irreducible representations of O(3) can be obtained immediately and have even (g) and odd (u) parity. This is done in Section 8-3 and in Table 8-3.

In this section we first find and discuss the basis functions of the odd dimension irreducible representations of $O(3)^+$. Although it may appear that all the irreducible are accounted for, actually the even dimension ones will be discussed in Chapter 13 when the SU(2) group is discussed. However, in the next section we use Bethe's simplified approach and show how we can use the same formulas for the character of the odd and even dimensional irreducible representations. The resulting combined character tables when specialized to the 32 point groups are called the **double group character tables**.

Basis functions for the rotation group can be found by considering the spherical harmonics, as in Section 8-1. Consider Laplace's equation

$$(\partial^2\psi/\partial x^2) + (\partial^2\psi/\partial y^2) + (\partial^2\psi/\partial z^2) = \nabla^2\psi = 0 \,. \qquad (8\text{-}8)$$

We have already shown in Section 4-6 that any rotational symmetry operation acting on $\nabla^2$ leaves it invariant or $R(\alpha, \beta, \gamma)\nabla^2 = \nabla^2 R(\alpha, \beta, \gamma)$. Thus, if $\psi$ is a solution of Eq. 8-8 then $R(\alpha, \beta, \gamma)\,\psi$ is also a solution. Writing Eq. 8-8 in polar coordinates, we recognize that solutions are $r^l Y_m{}^l(\theta\phi)$, where $Y_m{}^l(\theta\phi)$ are the well-known spherical harmonics. For a given $l$ there are $(2l + 1)$ linearly independent solutions with $m = l, l - 1, \ldots, -l$, i.e., $Y_l{}^l, Y_{l-1}{}^l, \ldots, Y_{-l+1}{}^l, Y_{-l}{}^l$. The spherical harmonics are completely defined in Appendix 8 where some are tabulated. The general idea is shown in Eq. 8-9.

$$\begin{aligned}
Y_0{}^0 &= (1/4\pi)^{1/2} \\
Y_1{}^1 &= -(3/8\pi)^{1/2}\,[(x + iy)/r] \\
Y_0{}^1 &= [3/(4\pi)^{1/2}][z/r] \\
Y_{-1}{}^1 &= (3/8\pi)^{1/2}\,[(x - iy)/r] \\
Y_2{}^2 &= (15/32\pi)^{1/2} \sin^2\theta \exp(+2i\phi) \\
Y_1{}^2 &= (15/8\pi)^{1/2} \sin\theta\cos\theta \exp(+i\phi), \text{ etc.}
\end{aligned} \qquad (8\text{-}9)$$

The spherical harmonics are polynomials of the $l$th degree, the most general form of which is

$$\Sigma_{a,b}\, N_{ab}(x + iy)^a\,(x - iy)^b\, z^{l-a-b} \qquad (8\text{-}10)$$

From Eq. 8-10 and the recursion relationship for $N_{ab}$, it can be seen that $(a - b)$ can take on the $2l + 1$ values from $+l$ to $-l$ as is already known for the spherical harmonics. In Chapter 13 the polynomial approach will be used for SU(2).

Under an arbitrary rotation $R(\alpha, \beta, \gamma)$, we know that a given $Y_m{}^l$ can be expressed in terms of a linear combination of the other spherical harmonics with the same $l$ but different m's. This is also clear because

$2l + 1$ polynomials of order $l$ can transform only among themselves under all the symmetry operations of the rotation group. Thus

$$R(\alpha, \beta, \gamma)Y_m{}^l(\theta, \phi) = \sum_{m'=-l}^{+l} \mathbf{D}^l(\alpha, \beta, \gamma)_{m'm} Y_{m'}{}^l(\theta, \phi) \tag{8-11}$$

or the spherical harmonics are basis functions of a $(2l + 1)$ dimensional irreducible representation of $\mathbf{D}^l$ of the rotation group. (**D** comes from the German Darstellung.) Since these are in infinite number of classes for the rotation group, there are an infinite number of irreducible representations. Also note that all rotations through the same angle are in the same class no matter what axis the rotation is about, because some other rotation connects the two rotation axes. This is a very useful result because it is easy to calculate the character of $\mathbf{D}^l$ for a rotation about a convenient axis. We show this for the group O(3). Let A be a rotation about the axis OA, B a rotation of the same amount but about the axis OB, and let R be a transformation of O(3), a rotation reflection or inversion, which carries OA into OB. Some rotation will always carry OA into OB and this rotation must be a member of the group O(3). We want to show that $B = RAR^{-1}$. Consider the right side. $R^{-1}$ carries OB into OA, A rotates the object by an angle, then R rotates back to OB without changing the angle of rotation that was performed about the OA axis. This is the same result as B so both sides of the equation give the same result, and indeed $B = RAR^{-1}$; therefore, B and A are in the same class.

Using this theorem we consider a rotation about the z-axis by an amount $\gamma$. This is a convenient axis to consider

$$R(0, 0, \gamma)Y_m{}^l(\theta, \phi) = Y_m{}^l(\theta, \phi - \gamma) = e^{-im\gamma}\, Y_m{}^l(\theta, \phi) \tag{8-12}$$

from the $\exp(im\phi)$ part in Eq. 8-9. Thus, from Eq. 8-11

$$\mathbf{D}^l(0, 0, \gamma)_{m'm} = e^{-im\gamma}\, \delta_{mm'} \tag{8-13}$$

which is just a diagonal matrix. The character of this representation is the sum of all these terms from $l$ to $-l$

$$\begin{aligned}
\chi^l(\gamma) &= \sum_{m=-l}^{l} e^{-im\gamma} = e^{-il\gamma} \sum_{j=0}^{2l} (e^{i\gamma})^j = e^{-il\gamma}\, \frac{e^{i(2l+1)\gamma} - 1}{e^{i\gamma} - 1} \\
&= \frac{\exp[i(l + 1/2)\gamma] - \exp[-i(l + 1/2)\gamma]}{\exp(i\gamma/2) - \exp(-i\gamma/2)} \\
&= \frac{\sin(l + 1/2)\gamma}{\sin(\gamma/2)}
\end{aligned} \tag{8-14}$$

This is a very simple result that is appropriate for the character of a rotation by an amount $\gamma$ about any axis. This equation will be very useful when we want to consider how the s, p, d, f,... atomic-like functions split when in a crystal with symmetry rotations $C_2$, $C_3$, $C_4$, and $C_6$.

Before we consider lowering the symmetry from the rotation symmetry, we show that there are no other irreducible representations besides $\mathbf{D}^l$. If there will be another irreducible representation its character must be orthogonal to the characters found in Eq. 8-14 for all $l$-values. Then it must be orthogonal to the difference between two successive characters $\chi^{l+1} - \chi^l$. The difference is $2 \cos l\gamma$ for $\gamma = 0$ to $\pi$. Thus the character of this other irreducible representation must be orthogonal to a cosine Fourier series, which is not possible. From this we conclude that the spherical harmonics as basis functions determine all the odd dimensional irreducible representation of the proper rotation group. We will see when we consider the character table for the double groups that there are even dimensional irreducible representations that are possible when the domain of $\gamma$ is extended from $\pi$ to $2\pi$. This seemingly nonphysical extension is required for the spinor functions of half-integer angular momentum. However, the basis functions of these representations will not be discussed until Chapter 13 when the group SU(2) is considered.

## 8-3 Reduction of Symmetry

So far we have considered the rotation group $O(3)^+$ [sometimes O(3) is called $R_3$ or $D_3$] and the $2l + 1$ functions $Y_m{}^l$ that form bases of the odd dimensional irreducible representations of this group. We also know that, with the central field approximation, the electronic states in atoms have this symmetry and the wave functions can be given in terms of the spherical harmonics.

Now we would like to consider how the irreducible representations of O(3) split when the symmetry is lowered to that of one of the 32 point groups. (Again we consider $O(3)^+$ and we will show how the extension to O(3) is very simple.) This problem is encountered when an atom is placed at a particular site in a crystal where the atom experiences the site symmetry. Since the highest degeneracy of the 32 point groups is three (for cubic point groups) and the degeneracy of the irreducible representations $\mathbf{D}^l$ of O(3) is $2l + 1$ with $l = 0, 1, 2, 3, ..., \infty$(s, p, d, f, ... wave functions), it is clear that $\mathbf{D}^l$ will split when the symmetry is lowered from O(3) to one of the 32 point groups. The solution to this problem is straightforward. For a given $l$, Eq. 8-14 can be used for each type of proper rotational operation $C_2$, $C_3$, $C_4$, $C_6$, $C_3{}^2$,... to determine the character. Some of the results are given in Table 8-1. Since all of the 32 point groups are subgroups of either $O_h$ or $D_{6h}$ or both, it is only necessary to consider how the s, p, d,... wave functions split for the point symmetric O and $D_6$, from which immediately we get the results for $O_h$ and $D_{6h}$. Table 8-2 shows, for a given *l*-value, the characters of the representation of the full rotation group under the symmetry operations of O and $D_6$. As can be seen, most of the representations are reducible and the resultant reduction is shown to the right of the table.

As an example consider a d-wave function. For a free atom all five orbitals, $Y_2^2$, $Y_1^2$, ..., $Y_{-2}^2$, are degenerate. When the symmetry of the atom is lowered to O symmetry, the energy level is split since the oribitals now transform as $E + T_2$. Thus the d-electrons transform as a doubly degenerate and triply degenerate level, and these two levels have different energies. The basis functions of $E + T_2$ are linear combinations of the five $Y_m^2$. Projection operator techniques can be used to obtain the basis functions. For a p-wave function in O symmetry, it is clear that there is no splitting; thus the three orbital functions maintain their degeneracy. However, a p-function in $D_6$ symmetry transforms as the $A_2 + E_1$ irreducible representatives of $D_6$. The basis functions immediately can be written as $A_2$: $Y_0^1 \sim z$; $E_1$: $(Y_1^1, Y_{-1}^1)$ or any linear combination of these. An f-state in cubic symmetry splits into three energy levels, one singly degenerate and two triply degenerate ones.

To go from irreducible representations in O and $D_6$ to those in $O_h$ and $D_{6h}$ is trivial. Recall that the group $C_i$ has symmetry operations E, i. $O_h = O \times C_i$ where the group multiplication was defined in Section 2-6c. If $R_1$, $R_2$, ..., $R_{24}$ are the symmetry operations of O, the then 48 symmetry operations are $R_1$, $R_2$, ..., $R_{24}$, $iR_1$, $iR_2$, ..., $iR_{24}$. The character table $O_h$ is then given in Table 8-3 where M is just the character table of O. Thus, the basis functions of the point group O that are an even power polynomials, Eq. 8-9 or 8-10, transform as the g-irreducible representations of $O_h$. Similarly, odd power polynomials transform as the u-irreducible representations of $O_h$. For $O_h$ these results are indicated in Table 8-2 on the extreme right. The identical considerations apply to $D_6$ and $D_{6h}$. Since all of the 32 point groups are subgroups of $O_h$ or $D_{6h}$ or both, the irreducible representations and splittings of all the free ion wave functions can be determined by standard subgroup considerations or by the use of the correlation tables in Appendix 7. Eigenfunctions which transform as g-irreducible representations are said to have **even parity** (or to be even) and those that transform as u-irreducible representations have **odd parity** (or are odd). Since the electric dipole operator has odd parity, we say that electric dipole transitions only occur between levels of differ-

**Table 8-1** Characters of the lowest order irreducible representations of the full rotation group for certain angles

| $l$ | $\chi^l(\pi)$ | $\chi^l(2\pi/3)$ | $\chi^l(\pi/2)$ | $\chi^l(2\pi/6)$ |
|---|---|---|---|---|
| 0 | 1 | 1 | 1 | 1 |
| 1 | −1 | 0 | 1 | 2 |
| 2 | 1 | −1 | −1 | 1 |
| 3 | −1 | 1 | −1 | −1 |
| 4 | 1 | 0 | 1 | −2 |
| 5 | −1 | −1 | 1 | −1 |

**Table 8-2** Characters of the irreducible representation of the full rotation group for the symmetry operations of the point group O and $D_6$, and the reduction of the various irreducible representations of O(3) to O and $D_6$ (as well as to $O_h$).

| $l$ | Symmetry Operations in O<br>E | <br>$8C_3$ | <br>$3C_2$ | <br>$6C_4$ | <br>$6C_2'$ | Reduction | Reduction for $O_h$ |
|---|---|---|---|---|---|---|---|
| 0(s) | 1 | 1 | 1 | 1 | 1 | $A_1$ | $A_{1g}$ |
| 1(p) | 3 | 0 | −1 | 1 | −1 | $T_1$ | $T_{1u}$ |
| 2(d) | 5 | −1 | 1 | −1 | 1 | $E + T_2$ | $E_g + T_{2g}$ |
| 3(f) | 7 | 1 | −1 | −1 | −1 | $A_2 + T_1 + T_2$ | $A_{2u} + T_{1u} + T_{2u}$ |
| 4(g) | 9 | 0 | 1 | 1 | 1 | $A_1 + E + T_1 + T_2$ | $A_{1g} + E_g + T_{1g} + T_{2g}$ |

| $l$ | Symmetry Operations in $D_6$<br>E | <br>$2C_6$ | <br>$2C_3$ | <br>$C_2$ | <br>$3C_2'$ | <br>$3C_2''$ | Reduction |
|---|---|---|---|---|---|---|---|
| 0(s) | 1 | 1 | 1 | 1 | 1 | 1 | $A_1$ |
| 1(p) | 3 | 2 | 0 | −1 | −1 | −1 | $A_2 + E_1$ |
| 2(d) | 5 | 1 | −1 | 1 | 1 | 1 | $A_1 + E_1 + E_2$ |
| 3(f) | 7 | −1 | 1 | −1 | −1 | −1 | $A_2 + B_1 + B_2 + E_1 + E_2$ |
| 4(g) | 9 | −2 | 0 | 1 | 1 | 1 | $A_1 + B_1 + B_2 + E_1 + 2E_2$ |

ent parity. This is known as Laporte's rule. This **selection rule** is rigorously obeyed in atoms but can be broken in solids by vibronically assisted transitions as discussed in Section 7-6f. Since the electric quadruple operator and the magnetic dipole operator both have even parity, these operators will cause transitions between states of the same parity.

**Table 8-3** Character table of $O_h$ in terms of O

| $O_h$ | | $R_1 \ldots R_{24}$ | $iR_1 \ldots iR_{24}$ |
|---|---|---|---|
| $A_{1g}$ | $\Gamma_1$ | | |
| . | . | | |
| . | . | M | M |
| . | . | | |
| $T_{2g}$ | $\Gamma_5$ | | |
| $A_{1u}$ | $\Gamma_6$ | | |
| . | . | | |
| . | . | M | −M |
| . | . | | |
| $T_{2u}$ | $\Gamma_{10}$ | | |

We should note that in the same trivial manner we go from the point group O to the point group $O_h$ in obtaining irreducible representation characters (Table 8-3) and basis functions, we proceed from the rotation group $O(3)^+$ to the full rotation group $O(3)$ because of the group multiplication $O(3) = C_i \times O(3)^+$. Thus, for the even irreducible representations $\mathbf{D}^l(iR)_g = +\mathbf{D}^l(R)_g$, while for the odd irreducible representations $\mathbf{D}^l(iR)_u = -\mathbf{D}^l(R)_u$ just as in Table 8-3 for the point group $O_h$. The parity g or u of the eigenfunctions of spinless particles that transform as the $\mathbf{D}^l$ irreducible representation is given by $(-1)^l$. Thus, eigenfunctions that transform as $l = 0, 2, 4, \ldots$ are g-functions, and those with $l = 1, 3, \ldots$ are u-functions. For n spinless particles the parity is given by $\Pi(-1)^{l_k}$ where $l$ is the orbital angular momentum of the kth particle and the product is taken from k = 1 to n. In Chapter 13 the addition of spin will be considered. The double value of the spin representation somewhat obscures the simple distinction between even and odd representations. However, the parity of an atomic state is determined by the orbital part of the wave function (angular momentum transforms as an axial vector and therefore into itself under inversion) and remains important in selection rules.

Figure 8-2 shows schematically the energy level diagram for a d-electron in various crystal fields. The 5-fold degenerate orbital state $Y_m^2$ is found for the free atom. When the atom experiences $O_h$ symmetry, the degeneracy is lifted and wave functions that transform as the $E_g$ and $T_{2g}$ irreducible representations are separated in energy by an amount

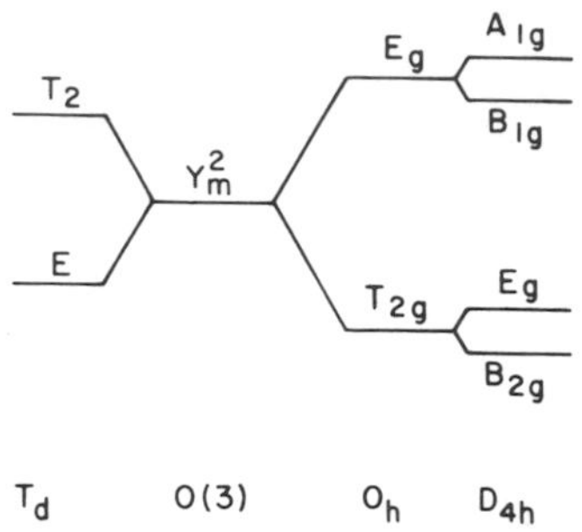


**Fig. 8-2** The energy levels of a d-electron in various crystal fields.

called 10Dq. These levels are 2- and 3-fold orbitally degenerate.

To examine the origin of the crystal field in more detail, consider an expansion of the potential $V = e/r^{1/2}$ due to a charge at $x = -a$, $y = z = 0$. Thus, $V = e[(a + x)^2 + y^2 + z^2]^{-1/2} = (e/a)\,[1 - (x/a) + \ldots]$. Adding the potential from this charge and five others at (a, 0, 0)(0, ±a, 0), (0, 0, ±a), the dipole, quadrupole, etc. terms cancel. The first nonzero crystal field potential after the spherical term due to these six charges (which have $O_h$ symmetry about the origin) is

$$\begin{aligned} V_c &= (35/4)\,(e/a^5)\,(x^4 + y^4 + z^4 - (3/5)\,r^4) \\ &= 22\sqrt{\pi}\,b_4\,[Y_0{}^4 + (5/14)^{1/2}\,(Y_4{}^4 + Y_{-4}{}^4)] \end{aligned} \quad (8\text{-}15)$$

where the second equality is a reminder that $V_c$ can be written in terms of spherical harmonics with $l = 4$. ($b_4 = 28e <r^4> /264$, where $<r^4>$ is the integral of $r^4$ over the wave function centered at the origin. For example, if dealing with an ion at the origin with 3d electrons, the 3d wave function will be used. This form will be discussed again in Section 8-7a.) Thus, in terms of the wave functions that transform as the $E_g$ and $T_{2g}$ irreducible representations,

$$10Dq = <E_g \,|\, V_c \,|\, E_g> - <T_{2g} \,|\, V_c \,|\, T_{2g}> \quad (8\text{-}16)$$

In Figure 8-2 it is implied that the $T_{2g}$ eigenvalue is lower than that of the $E_g$. This is usually the situation for $O_h$ symmetry due to six neighbors about a central ion. The reason for this is that the $T_{2g}$ wave functions have an angular dependence like xy, yz, xz (see Fig. 7-2). These three functions have their maximum electron density in between the six neighboring atoms at (±a, 0, 0), etc. Thus, they avoid overlapping the neighboring ions to a larger extent compared to the $E_g$ wave functions. On the other hand, if $T_d$ symmetry is formed by four neighbors as in a tetrahedron (Fig. 5-2b), the $E_g$ level will lie lower than the $T_{2g}$ level for the same reason. Consider neighboring ions at a fixed distance for different symmetries and a different number of neighbors. The $V_c$ (Eq. 8-15) for an $O_h$ 6-neighbor situation is $-9/4$ larger than the $T_d$ 4-neighbor case. Thus, $V_c(O_h \text{ 6-nb}) = (-9/4)V_c(T_d \text{ 4-nb}) = (-9/8)V_c(O_h \text{ 8-nb}) = (-2)V_c(O_h \text{ 12-nb})$, where 8-nb refers to eight atoms at the corners of a cube and 12-nb refers to 12 atoms each in the center of the 12 edges of a cube.

Symmetries lower than cubic can be obtained in a number of ways and are found in molecules and crystals. For example, in the 6-neighbor case the two atoms at $(0, 0, \pm a)$ can be expanded to $[0, 0, \pm(a + \Delta)]$ keeping the other four atoms fixed. This will lead to a term in addition to Eq. 8-15 proportional to $(\Delta^2/a^2)(e/a^3)(3z^2 - r^2) \propto Y_0{}^2$. Clearly this will lower the symmetry from $O_h$ to $D_{4h}$ and cause splittings of the type shown in Fig. 8-2.

## 8-4 Energy Level Diagrams (Correlation Diagrams)

We would like to consider the energy level diagrams when an atom with more than one electron is embedded in a crystal field. We consider the electrons that are not in closed shells to have certain quantum numbers. Two limiting cases are apparent. First, there is the **strong crystal field case** where the interaction of the individual electrons with the crystal field is much larger than the electron–electron interaction. In this case, the individual electrons have their angular momentum orientated by the crystal field and then the electrons are considered to interact between themselves while preserving their orientation with respect to the crystal axis. The second case is the **weak crystal field case**. The interaction between the electrons on the free atom must be considered first. (The angular momentum of the individual electrons is combined into a total angular momentum.) The total angular momentum is split by the crystal field. A **correlation diagram** is an energy level diagram that shows how the levels vary as one goes from one extreme case to the other. Symmetry can be used extensively to show which states from one extreme must join (correlate) with states on the other extreme.

For a many-electron atom (q electrons), the orbital part of the Hamiltonian can be written as Eq. 8-17a where the symbols have their usual meaning:

$$\mathbf{H} = (-\hbar^2/2m) \sum_{i=1}^{g} \nabla_i^2 - \Sigma\, Ze^2/r_i + \sum_{i}^{g} \sum_{<j}^{g} e^2/r_{ij} \tag{8-17a}$$

$$H_i \approx -(\hbar^2/2m)\, \nabla_i^2 + V_i(r_i) \tag{8-17b}$$

Equation 8-17b is the type of approximate Hamiltonian that is often used to solve the eigenvalue problem since it is much simpler to use than Eq. 8-17a. The effective one-electron potential $V_i(r_i)$ is of the Hartree or Hartree–Fock type. To obtain this term, the last term in Eq. 8-17a, which is the electron–electron interaction, is averaged over all the electrons and made spherically symmetric by averaging over angles. The one-electron Hamiltonian in Eq. 8-17b is usually solved in a self-consistent manner. That is, from $H_i\psi_i = E_i\psi_i$, $\psi_i$ is determined for all i. Then $V_i(r_i)$ is redetermined with these eigenfunctions. The process is repeated until there is no variation in the resultant $\psi_i$ for all i to the degree of accuracy desired. Since the potential is spherical symmetric [$H_i(\mathbf{r}_i)$ transforms into itself under all the symmetry operations of the full rotation groups] for each i-electron, the eigenfunction is $\psi_i = R_{n_i l_i}(r_i)\, Y_{m_i}^{l_i}(\theta_i, \phi_i)$ where R is a radial part that depends on the principle quantum number n and orbital quantum number $l$. $Y_m^l$ is a function of angles only and is spherical harmonic as in Section 8-2. Obviously, $Y_m^l$ contains the symmetry information.

For the rest of this chapter we will use the **convention** of small letters for orbital or spin angular momentum and irreducible representations of individual electrons and capital letters for whole atoms.

### a. Nonequivalent electrons

We proceed to determine a correlation diagram for a simple example. We take the case of two d-electrons with different principal quantum numbers (ndmd). The different principal quantum numbers frees one of the restrictions of the Pauli exclusion principle. (In Section 8-4b, the case of $nd^2$ will be covered.) A single d-electron in a free atom transforms as the $\mathbf{D}^2$ irreducible representation of the full rotation groups.

First consider the strong crystal field approximation for an averaged spherically symmetric potential as in Eq. 8-17b. For the free atom (no crystal field $V_c$), the energy level for the ndmd configuration has an orbital degeneracy of $(2l_1 + 1)(2l_2 + 1)$ which in this case, $l_1 = 2 = l_2$, gives a degeneracy of 25. The deviation from a spherical potential will split this degeneracy. The true orbital Hamiltonian Eq. 8-17a does not transform into itself under the full rotation group for each electron being rotated separately since it contains internuclear distances. However, the true Hamiltonian does transform into itself it at the same time all the electrons are rotated through the same angle. As we shall see, this will have important effects. Now back to the crystal field problem. As shown in Section 8-3, a cubic crystal field of O-symmetry splits the 5-fold irreducible representation into $E + T_2$. The right side of Fig. 8-3 shows the energy levels of ndmd for zero interaction between the electrons, or if the averaged spherical potential in Eq. 8-17b is used, or it is assumed that the crystal field is infinitely large ($\infty V_c$) or at least very much larger than the nonspherically symmetric part of the potential in the true orbital Hamilto-

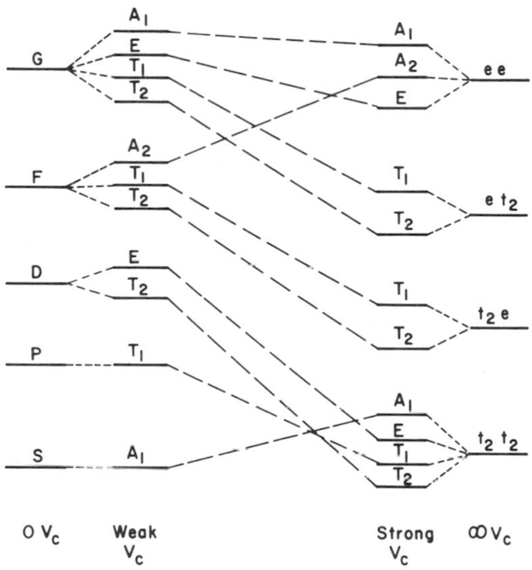


**Fig. 8-3** The correlation diagram for the two-electron system ndmd from the free atom to the infinite crystal field case.

nian Eq. 8-17a. The electrons in the four energy levels transform as the irreducible representations of the point group O as shown. When the electrons are allowed to interact (strong $V_c$ but not $\infty V_c$), the wave function transforms as the irreducible components of the product representation ($e \times e = A_1 + A_2 + E$; $e \times t_2 = T_1 + T_2$; $t_2 \times t_2 = A_1 + E + T_1 + T_2$). This is shown under "strong $V_c$" in the figure. The fact that the direct product is used is shown in Section 8-4b. The ordering is arbitrary since a detailed evaluation of the appropriate radial integrals would be required to determine the proper order. However, we have the $(t_2)^2$ level lowest because an octahedral field of $O_h$ symmetry, the $t_2$ orbitals should be lower than the e orbitals as discussed in Section 8-3. The separation between the $t_2$ and e orbitals would be 10Dq, which in this case has two different values.

Now consider the weak crystal field side of the correlation diagram where the angular momentum of the two electrons is first coupled together. Each d-electron transforms as the $\mathbf{D}^2$ irreducible representation of the full rotation groups. The two d-electrons transform as the product $\mathbf{D}^2 \times \mathbf{D}^2$, which we show in the next part of this section (Section 8-4b). The product can be decomposed into $\mathbf{D}^0 + \mathbf{D}^1 + \mathbf{D}^2 + \mathbf{D}^3 + \mathbf{D}^4$ and thus, the free atom can be in an S, P, D, F, G angular momentum state (angular momentum is 0, 1, 2, 3, 4 respectively). We use the **convention** of capital letters for entire systems and small letters, s, p, d, ... for individual electrons. These levels are listed on the left side of Fig. 8-3; the order is arbitrary. We already know, from Section 8-3, how eigenfunctions that transform as irreducible representations of the full rotation split where a crystal field is applied. The splitting is shown under the label "Weak V".

The correlation between the weak and strong crystal field case can now be completed. Notice that there are as many levels of the same irreducible representations on both sides of the figure as there should be. The point is that it does not matter how large or small $V_c$ is, what is important is the fact that the symmetry is the same. As $V_c$ is increased going from left to right there can be no new splitting, and eigenfunctions that transform as the same irreducible representations join each other. There can be no crossing of eigenfunctions that transform as the same irreducible representation although different irreducible representations can cross as discussed in Section 7-5b.

### b. Decomposition of angular momentum (vector model)

In elementary atomic physics the vector model of the addition of angular momentum is used. This model shows that when two angular momenta $j_1$ and $j_2$ are coupled on an atom, the resultant total angular momentum can vary from $j_1 + j_2$, $j_1 + j_2 - 1$, ..., $|j_1 - j_2|$. That is, it varies from $j_1 + j_2$ to $|j_1 - j_2|$ with all integer values in between. (If $j_1 = 3/2$ and $j_2 = 1/2$ one obtains a total angular momentum of 5/2, 3/2 and 1/2.) In this section we prove this result.

In Sections 8-2 and 7-4, it was shown that if an electron in an atom has the symmetry of the full rotation group O(3) its angular momentum is a label of an irreducible representation of O(3). This angular momentum is a good quantum number. We can also say the Hamiltonian is invariant under the operations of O(3).

If two electrons are considered but there is no term in the Hamiltonian that allows for an interaction between them, then the Hamiltonian is still invariant under O(3) applied separately to both electrons. This is the situation for the Hamiltonian in Eq. 8-17b, where the potential energy is averaged over all electrons and angles. Let the $2j_1 + 1$, $u_{m1}$ functions ($m_1 = j_1, ..., -j_1$) be eigenfunctions with angular momentum $j_1$. Similarly the functions $v_{m2}$ ($m_s = j_2, ..., -j_2$) have angular momentum $j_2$. If there is no interaction between the electrons then the eigenvalues are sums of the individualized systems and the eigenfunctions are products. The Hamiltonian will be invariant to the symmetry operation $R\underline{R}$ where R operates only on the coordinates of the function $u_{m1}$ and $\underline{R}$, only on those of $v_{m2}$. Clearly R and $\underline{R}$ commute. The Hamiltonian is invariant to the symmetry operation $R\underline{S}$ which corresponds to simultaneous but different rotations of the coordinate systems of the two functions u and v. Thus, for the Hamiltonian without the electron–electron coupling, the resultant **electronic configuration** has a degeneracy equal to $(2j_1 + 1)(2j_2 + 1)$. The eigenfunctions for this electron configuration are $u_{j1}\, v_{m2}$ ($m_2 = j_2, ..., -j_2$), $u_{j1-1}\, v_{m2}$ ($m_2 = j_2, ..., -j_2$), etc.

In the correct total orbital Hamiltonian Eq. 8-17a there are terms that depend explicitly on the distances between the electrons ($r_{ij}$). Thus, the Hamiltonian containing the coupling term will not be invariant to separate rotations of the two electrons. We now allow for the coupling so that the interelectron repulsion in the Hamiltonian is treated properly from a symmetry point of view. Thus, the Hamiltonian is invariant to the symmetry operation $R\underline{R}$ which corresponds to the same rotations for u and v. Proper linear combinations of the $(2j_1 + 1)(2j_2 + 1)$ functions give eigenfunctions of the coupled problem and transform as the direct product of the uncoupled eigenfunctions as in Section 6-1. Since the direct product is generally reducible, reduction will be done below.

As is usually the case, it is easiest to work with the characters of the representations. The character of the direct product $\mathbf{D}^{j1} \times \mathbf{D}^{j2} = \Sigma\, a(J)\, \mathbf{D}^J$ where the sum is $J = 0, ...\infty$ and the problem is to determine all the a(J). From Eq. 8-14 the character of the direct product is

$$\chi^{j1}(\gamma_1)\chi^{j2}(\gamma_2) = \sum_{m_1=-j_1}^{j_1} \exp(-im_1\gamma_1) \sum_{m_2=-j_2}^{j_2} \exp(-im_2\gamma_2) \qquad (8\text{-}18)$$

However, for the coupled problem both angles must be the same. Thus,

Eq. 8-18, for $\gamma_1 = \gamma_2 = \gamma$, is

$$\chi^{j1}(\gamma)\chi^{j2}(\gamma) = \sum_{m_1,m_2} \exp[-i(m_1 + m_2)\gamma] \tag{8-19a}$$

$$= \sum_{J=0}^{\infty} a(J) \sum_{M=-J}^{J} \exp[-iM\gamma] \tag{8-19b}$$

where $m_1$ and $m_2$ take the expected values as shown in Eq. 8-18 and we would like to determine $a_J$ for all J. Clearly a(J) is zero for $J > j_1 + j_2$ because the largest value of M is $j_1 + j_2$. There is only one term in Eq. 8-19a with $M = j_1 + j_2$, so $a(j_1 + j_2) = 1$. There are two terms in Eq. 8-19 that give $M = j_1 + j_2 - 1$ which arise when $m_1 = j_1 - 1$, $m_2 = j_2$, and when $m_1 = j_1$, $m_2 = j_2 - 1$. However, one of these terms is accounted for in the sum of M for $J = j_1 + j_2$. Thus, $a(j_1 + j_2 - 1) = 1$. The same result is obtained until the smallest positive value of the exponent in Eq. 8-19 which is $|j_1 - j_2|$ is obtained. So $a(|j_1 - j_2|) = 1$ and for smaller values of J, a(J) = 0. Thus, Eq. 8-19 can be written in terms of characters or representations

$$\chi^{j1}(\gamma)\chi^{j2}(\gamma) = \sum_{J=|j_1-j_2|}^{j_1+j_2} \sum_{M=-J}^{J} \exp(-iM\gamma) = \sum_{J=|j_1-j_2|}^{j_1+j_2} \chi^J(\gamma) \tag{8-20a}$$

$$\mathbf{D}^{j1} \times \mathbf{D}^{j2} = \sum_{J=|j_1-j_2|}^{j_1+j_2} \mathbf{D}^J = \mathbf{D}^{j1+j2} + \mathbf{D}^{j1+j2-1} + \ldots + \mathbf{D}^{|j1-j2|} \tag{8-20b}$$

This is just the statement of the usual addition of angular momentum of the vector model. The expression in Eq. 8-20b is not the normal addition of matrices (the dimensions of the matrix are different) but the addition of the various irreducible blocks along the diagonal of the large $(2j_1 + 1)(2j_2 + 1)$ by $(2j_1 + 1)(2j_2 + 1)$ matrix as originally discussed in Section 8-4a. As an example, consider two spinless particles in a $p(j_1 = 1)$ and $d(j_2 = 2)$ state. If the particles are coupled, then the resulting eigenstates that can be formed transform as the J = 1, 2, and 3 irreducible representations of the full rotation group having degeneracies (2J + 1) = 3, 5, and 7, respectively. Coupled states, designated by capital letters, would be P, D, and F states.

The above discussion will be all that is required for the remainder of this chapter. However, we can inquire into the question of the basis functions. If the $j_1$ and $j_2$ basis functions $u_{m1}^{j1}$ and $v_{m2}^{j2}$ are known, we would like to determine the basis functions $\psi_m^J$ that transform as $\mathbf{D}^J$ in the reduction Eq. 8-20b. In Chapter 13 we determine the $S_{JM;m_1m_2}$ coefficients

$$\psi_M{}^J = \Sigma_{m_1,m_2} S^*{}_{JM;m_1m_2}\, u_{m_1} v_{m_2}$$
$$u_{m_1} v_{m_2} = \Sigma_{J_1M} S_{JM;m_1m_2}\, \psi_M{}^J \qquad (8\text{-}21)$$

where the superscripts $j_1$ and $j_2$ are dropped for simplicity. These coefficients, known as the Wigner coefficients, Clebsch–Gordan coefficients, or vector coupling coefficients can now be determined in general. However, we will discuss this in Chapter 13, and instead show by a simple example how Eq. 8-21 can, and often is, obtained without calculating the coefficients in general. Consider the three basis functions $u_{m_1}$ that transform as $j_1 = 1$ (we continue to suppress the superscript $j_1$ and $j_2$ for simplicity), and the three $v_{m_2}$ for $j_2 = 1$. This can be the problem of two p-electrons on an atom. Then if the electrons are considered to be coupled, the resulting system transforms as $\mathbf{D}^1 \times \mathbf{D}^1 = \mathbf{D}^0 + \mathbf{D}^1 + \mathbf{D}^2$ from Eq. 8-20 (S, P, and D states). To find the basis functions, we start with the representation with the largest J, which is $\mathbf{D}^2$ in this example. As noted in the proof of Eq. 8-20, for the largest J there is only one way to from $\psi^J{}_{M=J}$ and that is from $u_{m_1=j_1} v_{m_2=j_2}$ which in this example is $(u_1)(v_1)$. With N as a normalizing factor, one has $\psi_2{}^2 = Nu_1v_1$. $\psi_1{}^2$ is obtained by operation on both sides of this equation with a lowering operator $J_-\psi_2{}^2 = NJ_-(u_1v_1)$ and using Eq. 7-26. Remember that $J_-$ is a linear operator so the right side is $(J_-u_1)v_1 + u_1(J_-v_1)$. Continuing this process, all 2J + 1 states $\psi_m{}^J$ are obtained for $J = j_1 + j_2$. These are listed in Eq. 8-22a.

$$\begin{aligned}
\psi_2{}^2 &= Nu_1v_1\\
\psi_1{}^2 &= (N/\sqrt{2})(u_0v_1 + u_1v_0)\\
\psi_0{}^2 &= (N/\sqrt{6}(u_{-1}v_1 + 2u_0v_0 + u_1v_{-1})\\
\psi_{-1}{}^2 &= (N/\sqrt{2})(u_{-1}v_0 + u_0v_{-1})\\
\psi_{-2}{}^2 &= Nu_{-1}v_{-1} \qquad (8\text{-}22a)
\end{aligned}$$

$$\begin{aligned}
\psi_1{}^1 &= (M/\sqrt{2})(u_0v_1 - u_1v_0)\\
\psi_0{}^1 &= (M/\sqrt{2})(u_{-1}v_1 - u_1v_{-1})\\
\psi_{-1}{}^1 &= (M/\sqrt{2})(u_{-1}v_0 - u_0v_{-1}) \qquad (8\text{-}22b)
\end{aligned}$$

$$\psi_0{}^0 = (P/\sqrt{3})(u_{-1}v_1 - u_0v_0 + u_1v_{-1}). \qquad (8\text{-}22c)$$

Then the state $\psi_1{}^1$ can only contain terms $u_0v_1$ and $u_1v_0$ and as noted in the proof of Eq. 8-20, there are two terms like this but one is contained in the $\psi_1{}^2$ function. Thus $\psi_1{}^1 = au_0v_1 + bu_1v_0$ and the a and b coefficients are determined by $\langle\psi_1{}^1|\psi_1{}^2\rangle = 0$, and $\langle\psi_1{}^1|\psi_1{}^1\rangle = 1$. The result appears in Eq. 8-22b. The other functions are obtained by using the lowering operator as before. The function $\psi_0{}^0$ can only contain terms like $u_{-1}v_1$, $u_0v_0$, and $u_1v_{-1}$. The function is obtained by making it orthogonal to $\psi_0{}^2$ and $\psi_0{}^1$, and $\langle\psi_0{}^0|\psi_0{}^0\rangle = 1$. The approach outlined here of obtaining basis functions for the coupling of two angular momentum states is general. Heine (Appendix I) gives the coefficients for a number of cases.

### c. Equivalent electrons — method of descent in symmetry

We determine the correlation diagram for the case of two d-electrons, as in Section 8-4a, but now consider these electrons to have the same principal quantum numbers. For the strong crystal field approximation where several electrons have the same principal and azimuthal quantum numbers, care must be taken so that only terms allowed by the Pauli principle are considered. The straightforward method to do this, used by Bethe, is called the method of descent in symmetry. The idea is to remove the orbital degeneracy found for the O point group for the E- and T-irreducible representation by a gedanken experiment. Lowering the symmetry to tetrahedral $D_4$ or orthorhombic $D_2$ will remove all of the orbital degeneracy and the interaction of the spins can be treated trivially in terms of the Pauli principle. Then these nondegenerate levels in low symmetry are recombined to yield the results for the O-point symmetry. While the symmetry is being changed, one keeps count of the degeneracy to help make sure a mistake is not introduced. (See the Appendix to this chapter for a statement of the Pauli principle.)

Consider the two equivalent d-electron problem ($d^2$). In a d-state, there are five orbital levels and each orbitally degenerate state can have spin up or down resulting in 10 states. The second d-electron has nine available states since the two electrons can not have the same quantum numbers. However, because the two electrons are indistinguishable the total degeneracy for $d^2$ is $(10)(9)/2 = 45$. At all times this number must remain invariant.

Now consider an infinite crystal field of O-symmetry acting on $d^2$. Each d-level is split into a 2- and 3-fold orbital degeneracy that transforms as the e and $t_2$ irreducible representation. These are shown in Fig. 8-4 which is similar to Fig. 8-3. The separation between the $t_2t_2$ and $t_2e$ is 10Dq (Eq. 8-16) which also is equal to the separation between the $t_2$ and ee states. The degeneracy is calculated as in the previous paragraph. For the ee level, the electron in the first e-orbital has a 2-fold orbital degeneracy and spin can be up or down so the degeneracy is four. Then the total degeneracy for ee is $(4)(3)/2 = 6$. Similarly for the $t_2t_2$ level, the total degeneracy is $(6)(5)/2 = 15$. However, for the $et_2$ level, one electron has an e-orbital wave function and the other a $t_2$ orbital wave function, so the electrons are distinguishable and the second electron can

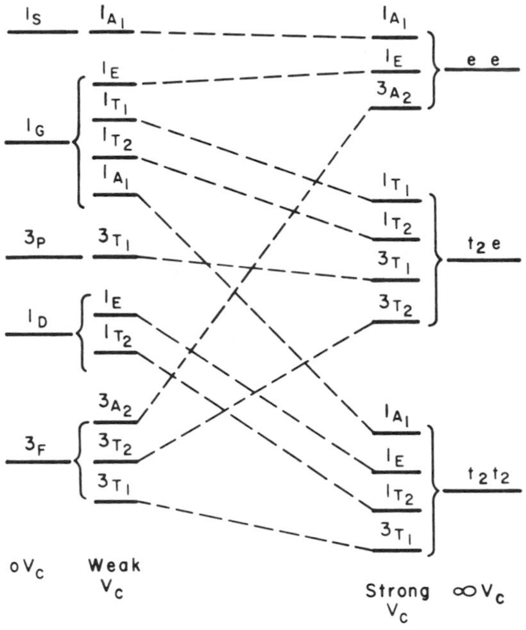


**Fig. 8-4** The same as Fig. 8-3 but for the two-electron $nd^2$ case.

be placed in any of its quantum states independent of the first. Thus, the degeneracy is (4)(6) = 24. The total degeneracy 6 + 15 + 24 = 45 is maintained.

We now want a strong crystal field but the spins are allowed to couple. Since there are only two spins, they can couple parallel with S = 1 and a degeneracy 2S + 1 = 3, or antiparallel S = 0 and 2S + 1 = 1. The spin degeneracy is traditionally written as a left superscript to the orbital level signature.

Consider the $et_2$ level first since it is easiest to assign the spin multiplicity to this level. In a cubic crystal field the combined orbital levels transform as the direct product of functions that separately transform as the e and $t_2$ irreducible representations of the O-point group. This occurs for the same reason as it does for the full rotation group as already discussed. For the coupled problem the angle between the two electrons must be the same under all symmetry operations so that the Hamiltonian will transform into itself. Thus, the product orbital eigenfunctions transform as $e \times t_2 = T_1 + T_2$. As noted before, since the e and $t_2$ orbital states are distinguishable, the spin of this two-electron system can be paired (S = 1) or unpaired (S = 0). So the Pauli exclusion principle does not reduce the number of possible states and the allowed states are $^1T_1$, $^3T_1$, $^1T_2$, and $^3T_2$. These states have degeneracies of 3, 9, 3, and 9 to give the total of 24 as required. We will ignore spin–orbit coupling until Section 8-6. The reason is 2-fold. First, the d-electron states have large crystal field and much smaller spin–orbit coupling, so most of the features of the 3d-levels can be underscored without it. Second, it complicates the problem. As will be seen in Section 8-6, the direct product of the spin and orbit irreducible representation will be required when a Hamiltonian with spin–orbit coupling is considered because the spins and orbits must be rotated through the same angle to keep the Hamiltonian invariant under all the operations of the full rotation group. This is the vector addition concept of total angular momentum $\mathbf{J} = \mathbf{L} + \mathbf{S}$.

Next consider the ee level with total degeneracy of six. The combined orbital levels transform as $e \times e = A_1 + A_2 + E$, but it is not clear which levels are singlets and which are triplets. Two possibilities $^1A_1 + {}^3A_2 + {}^1E$ or $^3A_1 + {}^1A_2 + {}^1E$ give the correct total degeneracy. To determine the correct states we use the **method of descent in symmetry**. That is, we consider the O-symmetry lowered to $D_4$ (by a strain along one of the 4-fold axis). The correlations taken from Appendix 7 for O → $D_4$ → $D_2$ is shown in Table 8-4 for convenience. As can be seen E in O-symmetry splits into $A_1 + B_1$ in $D_4$. (It will be most convenient to "lower" the symmetry until only one-dimensional irreducible representation is obtained.) Thus the problem in $D_4$ is $(A_1 + B_1)(A_1 + B_1)$. Table 8-5a shows the solution. However, $a_1 \times b_1$ can have S = 0 or 1 because the oribital states are different; hence $^1B_1 + {}^3B_1$. When one correlates back to O-symmetry, the crystal field reduction does not affect the spin states. Using Table 8-4 there is only one possible result as shown in Table 8-5a.

For $t_2t_2$, we have in O-symmetry $t_2 \times t_2 = A_1 + E + T_1 + T_2$. Since the total degeneracy is 15, simple algebra shows that there are three possibilities: $^1A_1 + {}^1E + {}^1T_1 + {}^3T_2$; $^1A_1 + {}^1E + {}^3T_1 + {}^1T_2$; $^3A_1 + {}^3E + {}^1T_1 + {}^1T_2$. To see which is correct, we must lower the symmetry so that the $t_2$ level is entirely split. Orthorhombic $D_2$ symmetry naturally follows from $D_4$ symmetry. Table 8-4 is the correlation table and Table 8-5b presents the results for $t_2 \times t_2$ is O becoming $(b_1 + b_2 + b_3) \times (b_1 + b_2 + b_3)$ in $D_2$. As in Table 8-5a, when the two orbitals are the same, only an antiparallel spin arrangement is allowed by the exclusion principle. If the orbitals are different, then S = 0 and 1 is allowed as indicated. Now we have three triplet states $^3B_1 + {}^3B_2 + {}^3B_3$ and from the correlation tables or Table 8-4, we can obtain either $^3T_1$ or $^3T_2$. We are still left with two out of three possibilities. (Note that the rest of the states give

**Table 8-4**  Correlation table for the irreducible representations of the point groups O, $D_4$, and $D_2$

| O | | $D_4$ | | $D_2$ |
|---|---|---|---|---|
| $A_1$ | | $A_1$ | | A |
| $A_2$ | → | $B_1$ | → | A |
| E | → | $A_1$ | → | A |
|   |   | $B_1$ | → | A |
| $T_1$ | | $A_2$ | → | $\begin{cases} B_1 \\ B_2 \\ B_3 \end{cases}$ |
|   |   | E | → |  |
| $T_2$ | | $B_2$ | → | $\begin{cases} B_1 \\ B_2 \\ B_3 \end{cases}$ |
|   |   | E | → |  |

**Table 8-5**  (a)  Recombining orbital singlet states in $D_4$ to states in O.  (b)  Direct product of orbital single states in $D_2$ showing the singlet, triplet possibilities

(a)

$D_4$ ⎸ 0

$a_1 \times a_1 = {}^1A_1 \longrightarrow {}^1A_1$

$a_1 \times b_1 = ({}^1B_1) + {}^3B_1 \longrightarrow {}^3A_2$

$b_1 \times b_1 = {}^1A_1 \longrightarrow {}^1E$

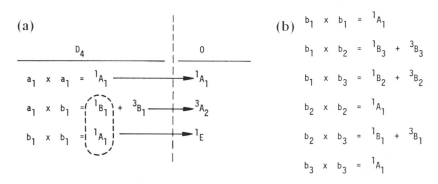

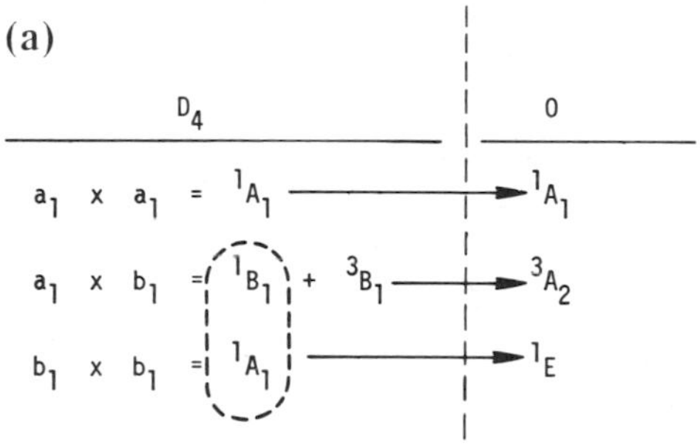

(b)  $b_1 \times b_1 = {}^1A_1$

$b_1 \times b_2 = {}^1B_3 + {}^3B_3$

$b_1 \times b_3 = {}^1B_2 + {}^3B_2$

$b_2 \times b_2 = {}^1A_1$

$b_2 \times b_3 = {}^1B_1 + {}^3B_1$

$b_3 \times b_3 = {}^1A_1$

${}^1A_1 + {}^1E + {}^1T_2$ as expected.) This is indeed true and obviously stems from the fact that $t_1$ and $t_2$ in O-symmetry both correlate with the same irreducible representation in $D_2$. So the choice was straightforward, but bad. We must descend to a symmetry where $t_1$ and $t_2$ correlate with different irreducible representations. The correlation tables show that $C_{2v}$ or $C_{2h}$ are subgroups of $O_h$ and a table similar to Table 8-5b will immediately yield for O-symmetry ${}^1A_1 + {}^1E + {}^3T_1 + {}^1T_2$. The degeneracy $1 + 2 + 9 + 3 = 15$ is as expected.  (If in working out this problem you descended to $C_{2h}$, then $t_{2g} \to A_g + A_g + B_g$; but this is better written as $A_g + \overline{A}_g + B_g$ to remind one that $A_g$ and $\overline{A}_g$ represent different orbital functions.) We have now completed the right side of Fig. 8-4.

The left side of Fig. 8-4 starts in the same manner as Fig. 8-3. Namely, each d-electron transforms as $\mathbf{D}^2$ in the free atom. Thus the coupled $d^2$ state transforms as $\mathbf{D}^2 \times \mathbf{D}^2 = \mathbf{D}^4 + \mathbf{D}^3 + \mathbf{D}^2 + \mathbf{D}^1 + \mathbf{D}^0$ from Eq. 8-20 or as G + F + D + P + S terms. For the nonequivalent electrons ndmd each of these terms can be a spin singlet and triplet since the principal quantum numbers are different. However, for the case of equivalent electrons the Pauli exclusion principle allows only certain terms. This free ion problem was worked out in atomic physics texts. See also the Appendix to this chapter where the results are listed for all $d^m$ and $p^m$. For example, it is clear that only $^1G$ will be allowed because to get a G term both electrons must be in the same orbital that has the maximum orbital angular momentum. Thus, only an antiparallel spin configuration is allowed. The orbital states with the allowed spin multiplets are listed on the left in Fig. 8-4. The order is arbitrary as far as symmetry is concerned although it is listed to correspond to that observed in doubly ionized vanadium ($V^{2+}$).

When the free ion states of the coupled $d^2$ configuration are subject to a weak crystal field, the orbital levels split as already discussed in connection with Fig. 8-3. Remember we are still ignoring spin–orbit coupling, so there is no effect on the spin multiplicity. The results are given in Fig. 8-4.

There is a great deal of interest in the energy levels of $3d^n$ configurations because spectra can be easily measured and the materials containing these ions are useful. These energy level diagrams have been worked out for a wide range of cubic crystal fields. They are reproduced in Appendix 9. In the figure for the $d^2$ case, in the appendix, the free ion terms on the left are for zero crystal field. For nonzero crystal field the states are labeled as in Fig. 8-4. As can be seen some of the free ion configurations vary linearly with crystal field while others have more complicated behavior.

<u>A remark on the use of the symmetric and antisymmetric direct products to determine singlet and triplet spin states.</u> Consider a two-electron case such as $nd^2$. According to the Pauli exclusion principle the total wave function (space and spin parts) must change sign if the coordinates of any two electrons are interchanged. Consider the two orbital states $\phi_a$ and $\phi_b$. When occupied with electron (1) they are written $\phi_a(1)$ and $\phi_b(1)$, and similarly for electron (2). The spin up and spin down functions are $\alpha$ and $\beta$ which if occupied by electron (1) are $\alpha(1)$ and $\beta(1)$

etc. The complete nonnormalized wave functions for this two-electron problem are

$$[\phi_a(1)\phi_b(2) + \phi_a(2)\phi_b(1)][\alpha(1)\beta(2) - \alpha(2)\beta(1)] \quad \text{(8-23a)}$$

$$[\phi_a(1)\phi_b(2) - \phi_a(2)\phi_b(1)] \begin{cases} [\alpha(1)\alpha(2)] & \text{(8-23b)} \\ [\alpha(1)\beta(2) + \beta(1)\alpha(2)] & \text{(8-23c)} \\ [\beta(1)\beta(2)] & \text{(8-23d)} \end{cases}$$

Equation 8-23a is antisymmetric with respect to the spin part of the wave function when the two electrons are interchanged, (1) → (2) and (2) → (1), and symmetric with respect to the space part, i.e., $[\phi_1(1)\phi_2(2) + \phi_1(2)\phi_2(1)] \rightarrow [\phi_1(2)\phi_2(1) + \phi_1(1)\phi_2(2)]$, which are identical. The three functions Eq. 8-23b–d are symmetric in the spin part but antisymmetric in the space part. Thus, for the two equivalent electron case the antisymmetrical direct product of the orbital states is a triplet spin state while the symmetric direct product of the orbital states can only have singlet spin states. This realization makes strong field states determination in Fig. 8-4 very easy since the antisymmetric and symmetric direct products are listed in Appendix 6 and discussed in Section 6-1b. The antisymmetric product of e × e is $A_2$ therefore $^3A_2$, and for $t_2 \times t_2$ it is $T_1$, therefore $^3T_1$. The symmetric direct product results are listed in Fig. 8-4. Although this approach is easy and fast, we hasten to add that it is not applicable to situations where there are more than two equivalent electrons. This results from the fact that for more than two equivalent electrons, it is not possible to write a wave function that is separately factored into space and spin parts as in Eq. 8-23.

The appendix to this chapter has a more complete discussion of permutation symmetry and the Pauli principle.

## 8-5 Crystal Double Groups

The discussion in this chapter has been concerned with the removal of the degeneracy of irreducible representations of the full rotation group O(3) when the symmetry is lowered to that of the point groups. We showed that, by using the spherical harmonics as basis functions of the $l$th irreducible representation the irreducible representations for the point groups could be obtained. We also calculated the characters of each irreducible representation Eq. 8-14 and showed how the direct product of irreducible representation can be reduced (Eq. 8-20). In this discussion we have been careful to work with integer $l$-values. It is only for integer values that the spherical harmonics are defined. However, we know that the electron itself has a half-integer spin and that total angular momentum can also be half-integer (should really be called half-odd-integer). Thus, we investigate the half-integer angular momentum case. We do this by the method developed by Bethe which is very simple but not rigorous. The simplicity is maintained by dealing only with the characters of the new representations that are obtained. From experience we know that characters are always relatively easy quantieis to handle. In Chapter 13 the basis function and represetations for $l$ half-integer will be obtained. The basis functions are spin functions which, from quantum mechanics, we know transform under rotation in a more complicated way then functions of x, y, z.

### a. Character tables for crystal double groups

Consider the character of the $l$th irreducible representation of the full rotation group Eq. 8-14 under a rotation of $\alpha + 2\pi$.

$$\chi^l(\alpha+2\pi) = \frac{\sin[(l+\frac{1}{2})(\alpha+2\pi)]}{\sin[(\alpha+2\pi)/2]} = (-1)^{2l}\,\chi^l(\alpha) \qquad (8\text{-}24)$$

For integer $l$ the expected result is obtained, $\chi^l(\alpha + 2\pi) = \chi^l(\alpha)$. However, for half-integer $l$, $-1$ is obtained on the right-side which is at first sight very surprising. Thus, an atom with an even multiplicity ($2l + 1$ is even if $l$ is half-odd-integer) has a double-valued representation of the full rotation group. That is, the character changes sign under a rotation of $2\pi$. The character of the identity operation is then

$$\chi^l(0) = 2l + 1 \qquad \chi^l(2\pi) = -(2l + 1) \qquad (8\text{-}25)$$

However, we notice that a rotation of $4\pi$ behaves as we expect

$$\chi^l(\alpha \pm 4\pi) = \chi^l(\alpha) \qquad (8\text{-}26)$$

for integer or half-integer $l$. In order to obtain the characters and character tables of these double-valued representation, we introduce the fiction that the identity operation is a rotation by $4\pi$ = E and we define the symmetry operation $2\pi$ = R. Thus, $R^2$ = E. The point groups now have the original symmetry operations (times E if you like) plus all the original symmetry operations times R. The resultant 32 point groups are called the **crystal double groups** or often just **double groups**.

The character tables of the double groups are obtained in a manner completely similar to the ordinary groups. The double groups contain more classes than the ordinary groups but not twice as many. For double groups a rotation by $\alpha$ and $\alpha + 2\pi$ do not have the same character in general so they cannot be in the same class. However, for $\alpha = \pi$ (a $C_2$ rotation)

$$\chi^l(C_2) = \chi^l(C_2 + 2\pi) = \chi^l(RC_2) = 0 \tag{8-27}$$

So, at least it is possible for the symmetry operations $C_2$ and $RC_2$ to be in the same class because they have the same character. In fact Opechowski (see the Notes) has shown that $C_2$ and $RC_2$ will be in the same class if and only if there is another 2-fold axis perpendicular to the $C_2$ axis.

Consider, as an example, the ordinary point group $D_4$ with eight symmetry operations separated into five classes (E), ($C_4 + C_4{}^3$), ($C_2$), ($2C_2'$), ($2C_2''$). By **convention** we will star the double group symbol, i.e., $D_4{}^*$. For $D_4{}^*$ there are 16 symmetry operations. When grouped according to class, by the standard techniques of Section 2-4c, the result is seven classes (E), (R), ($C_4 + RC_4{}^3$), $C_4{}^3 + RC_4$), ($C_2 + RC_2$), ($2C_2' + 2RC_2'$), ($2C_2'' + RC_2''$). Note that $C_2$ and $RC_2$ are in the same class for each 2-fold rotation; $C_4$ and $C_4{}^3$ are not in the same class; and note the physical result that $C_4$ and $RC_4{}^3$, which are in the same class, are rotations by $\pi/2$ and $4\pi - \pi/2$. For $D_4{}^*$ these symmetry operations have a similar physical effect as do $C_4$ and $C_4{}^3$ in $D_4$ and indeed $C_4$ and $C_4{}^3$ are in the same class for the ordinary group, $D_4$.

The character tables for the double groups are obtained in the same way as tables for the ordinary groups. We will discuss only some of the results from a physical point of view, and use the point group O and O* as an example. Basis functions or cigenfunctions that transform properly under ordinary group operations must continue to do so. That is, functions with $l$ = integer have $\chi^l(\alpha + 2\pi) = \chi^l(\alpha)$ or $\chi^l(C_n) = \chi^l(RC_n)$. Thus, for O* the first five irreducible representation must be the same as those for O. For example, for $\Gamma_5$

$$E = \begin{bmatrix} 1 & 0 & 0 \\ 0 & 1 & 0 \\ 0 & 0 & 1 \end{bmatrix}, \quad R = \begin{bmatrix} 1 & 0 & 0 \\ 0 & 1 & 0 \\ 0 & 0 & 1 \end{bmatrix} \quad \Gamma(C_n) = \Gamma(RC_n) \tag{8-28}$$

Similar results are obtained for $\Gamma_1$ through $\Gamma_4$. Thus, for $\Gamma_1$ through $\Gamma_5$ in O* the characters of the irreducible representation of $C_n$ and $RC_n$ are the same. This is shown in Table 8-6 for O*. Now consider the double group representations. For O* there are eight classes, so there are eight irreducible representations. The order of the group O* is $2 \times 24 = 48$. Thus, for the double-valued representations $\Gamma_6$, $\Gamma_7$, $\Gamma_8$, one has $48 = 24 + l_6{}^2 + l_7{}^2 + l_8{}^2$. The only integer solution is $l_6 = 2 = l_7$ and $l_8 = 4$. Also

**Table 8-6** Character table of the double group O*

| O* | E | R | $4C_3$<br>$4C_3^2R$ | $4C_3^2$<br>$4C_3R$ | $3C_2$<br>$3C_2R$ | $3C_4$<br>$3C_4^3R$ | $3C_4^3$<br>$3C_4R$ | $6C_2'$<br>$6C_2'R$ |
|---|---|---|---|---|---|---|---|---|
| $(\Gamma_1)$ $A_1$ | +1 | +1 | +1 | +1 | +1 | +1 | +1 | +1 |
| $(\Gamma_2)$ $A_2$ | +1 | +1 | +1 | +1 | +1 | -1 | -1 | -1 |
| $(\Gamma_3)$ E | +2 | +2 | -1 | -1 | +2 | 0 | 0 | 0 |
| $(\Gamma_4)$ $T_1$ | +3 | +3 | 0 | 0 | -1 | +1 | +1 | -1 |
| $(\Gamma_5)$ $T_2$ | +3 | +3 | 0 | 0 | -1 | -1 | -1 | +1 |
| (E') $(E_{1/2})$ $(\Gamma_6)$ $E_{1/2}$ | +2 | -2 | +1 | -1 | 0 | $\sqrt{2}$ | $-\sqrt{2}$ | 0 |
| (E") $(E_{5/2})$ $(\Gamma_7)$ $E_{5/2}$ | +2 | -2 | +1 | -1 | 0 | $-\sqrt{2}$ | $\sqrt{2}$ | 0 |
| (U') $(U_{3/2})$ $(\Gamma_8)$ G | +4 | -4 | -1 | +1 | 0 | 0 | 0 | 0 |

for these double-valued representations, $\Gamma(E) = -\Gamma(R)$, $\Gamma(C_m) = -\Gamma(RC_n)$, and $\Gamma(C_2) = 0 = \Gamma(RC_2)$ from Eq. 8-24–27. Using these results and the orthogenerality theorems (Section 4-1), the characters of the double-valued representations can be obtained. These representations must be joined to the ordinary group O to form the character table of the double group O* as in Table 8-6.

With the experience gained, the character table for a small group like $D_2$* can be determined very quickly. $D_2$ has four symmetry operations (E, $C_2^z$, $C_2^x$, $C_2^y$) each in a class by itself. Thus, there are four one-dimensional irreducible representations. For $D_2$* there are eight symmetry operations but, as already discussed, $RC_2^i$ and $C_2^i$ are in the same class and E and R are each in a class by themselves. Thus the five classes in $D_2$* are (E), (R), ($C_2^z + RC_2^z$), ($C_2^x + RC_2^x$), and ($C_2^y + RC_2^y$). Only the dimensionality of the double representation $\Gamma_5$ is unknown. However, it can immediately be found since the order of the group equals the sum of the squares of the dimensionals of the irreducible representations (Eq. 4-2) or $8 = 1 + 1 + 1 + 1 + a^2$ or $a = 2$. Thus, $\Gamma_5$

is 2-dimensional and the characters can be written down immediately as 2, −2, 0, 0, 0 corresponding to the order of the five classes of $D_2{}^*$ above. These values follow immediately from Eqs. 8-25 and 8-27.

In Appendix 10 the character tables of the double-valued representations are listed. In Appendix 6 the reduction of direct product tables, and symmetrized and antisymmetrized direct product tables are given for double-valued representations. As can be seen, all the usual properties of irreducible representations can be found for the double-valued representations.

A word on notation of the double-valued irreducible representations; it is bad. Three notations are shown in Table 8-6. The Bethe notation ($\Gamma_i$) has an advantage of arbitrariness. Most of the other notations employ E with some sort of superscripts and/or subscripts for 2-dimensional double-valued representations, and U or G for 4-dimensional double-valued representations. As will be seen, a spin 1/2 function in O* transforms as the $\Gamma_6$ (E′ or $E_{1/2}$) double-valued representations (which explains the $E_{1/2}$ notation). In most doubled-valued character tables the spin 1/2 function transforms as the first double-valued representation that is listed.

### b. Reduction of symmetry of double representations

In Section 8-3 we considered how the odd dimension ($2l + 1 =$ odd, for integer $l$), irreducible representations of the full rotational split when the symmetry is lowered to one of the 32 point groups. Here we would like to do the same thing for the double representations which are of even dimensions ($2l + 1 =$ even, for half-odd-integer $l$).

Using the expression for the character for any angle $\alpha$ and $l$ value in Eq. 8-14, the character for the even dimension irreducible representations of the full rotation group can be determined. The results for angles of interest in the 32 point groups are listed at the left side of Table 8-7. The result for a $C_2$ rotation is not listed because the character is zero (Eq. 8-27). This table is similar to Table 8-1 which is appropriate for integer $l$. Reducing these representations for O* is accomplished by use of Eq. 4-7 or by inspection, as was done for the ordinary groups for integer $l$ (Table 8-2). The result is shown at the right side of Table 8-7 for the crystal double groups O* and $D_6{}^*$. As we expect, the odd dimension irreducible representations of the full rotation group reduce to only the double group irreducible representations. As the symmetry is reduced from O* to lower symmetry, this effect will always remain because the characters of the single and double groups are orthogonal as are the basis functions. As

**Table 8-7** Characters of the lowest order double-valued irreducible representations for certain angles, and the reduction to the irreducible representations of $O^*$ and $D_6{}^*$

| $\ell$ | E | R | $C_3$ | $C_4$ | $C_6$ | Reduction of $\underline{D}^2$ to Irreducible Representations in O* | Reduction of $\underline{D}^2$ to Irreducible Representation in $D_6$* |
|---|---|---|---|---|---|---|---|
| 1/2 | 2 | -2 | 1 | $\sqrt{2}$ | $\sqrt{3}$ | $E_{1/2}$ | $E_{1/2}$ |
| 3/2 | 4 | -4 | -1 | 0 | $\sqrt{3}$ | G | $E_{1/2} + E_{5/2}$ |
| 5/2 | 6 | -6 | 0 | $-\sqrt{2}$ | 0 | $E_{5/2} + G$ | $E_{1/2} + E_{3/2} + E_{5/2}$ |
| 7/2 | 8 | -8 | 1 | 0 | $-\sqrt{3}$ | $E_{1/2} + E_{5/2} + G$ | $E_{1/2} + 2E_{3/2} + E_{5/2}$ |
| 9/2 | 10 | -10 | -1 | $\sqrt{2}$ | $-\sqrt{3}$ | $E_{1/2} + 2G$ | $E_{1/2} + 2E_{3/2} + 2E_{5/2}$ |
| 11/2 | 12 | -12 | 0 | 0 | 0 | $E_{1/2} + E_{5/2} + 2G$ | $2E_{1/2} + 2E_{3/2} + 2E_{5/2}$ |

can be seen, an angular momentum state of 1/2 will not split as the symmetry is lowered from the full rotation group to that of O*. In fact the $l = 1/2$ level will not split in any crystal field as can be seen in the double groups character tables in Appendix 10. An angular momentum state of $l = 3/2$ will not split in cubic symmetry but will split in any lower symmetry because only cubic point groups have double representation of dimension four. All the other 32 point groups have only 2-dimensional double representations. Thus, eigenfunctions that transform as $l > 3/2$ irreducible representations in O(3) will always split when the symmetry is lowered to one of the 32-point groups. However, there will always be at least a 2-fold degeneracy for the double representations. In fact, **Kramer's theorem** states that for half-odd-integer angular momentum, which must be the case for an odd number of electrons, a (crystal) electric field must leave at least a 2-fold degeneracy in all the levels. This result has been discussed and can also be proved by time reversal symmetry.

## 8-6 Correlation Diagrams including Double Groups

So far we have assumed that the spin–orbit coupling is zero. Thus the Hamiltonian is invariant separately to rotations of the orbits and of the spins. Now we want to allow a nonzero spin–orbit coupling term in the Hamiltonian. This term will be invariant under a simultaneous rotation of the orbits and spin. The resulting splitting is very important in free atoms and rare earths ($4f^n$) in crystals. In atoms the observed spectra are very sharp so any splitting is important. For crystalline rare earth spectra the spin–orbit effects are larger than the crystal field effects. However, for transition series ions ($3d^n$) the crystal field effects are much larger than the spin–orbit coupling and these considerations can usually be ignored (see the Notes).

Consider a single electron with spin angular momentum s and orbital angular momentum $l$. The electron moving about a nucleus will see a magnetic field due to the charged nucleus. The resulting term in the Hamiltonian is

$$\mathbf{H} = \tfrac{1}{2} \frac{Z e^2}{m^2 c^2 r^3} l \bullet \mathbf{s} \tag{8-29}$$

The factor 1/2 is not actually obtained in the usual simple derivation but in the Thomas factor that arises from relativistic effects, since the frame of the electron has a constant acceleration (angular) about the nucleus. For a many-electron atom the Hamiltonian for the spin–orbit coupling is somewhat more complicated. Normally before the inclusion of spin–orbit coupling, the interactions between only the orbits is considered and the eigenfunction that transform properly under simultaneous rotation of all orbits (i.e., $\mathbf{D}^{l1} \times \mathbf{D}^{l2} \ldots \mathbf{D}^{lm} = \Sigma\, \mathbf{D}^{L}$ for m-electrons) is obtained. A similar procedure is applied to the spin ($\mathbf{D}^{s1} \times \mathbf{D}^{s2} \times \ldots \times \mathbf{D}^{sm} = \Sigma\, \mathbf{D}^{S}$). Now, with a term in the Hamiltonian like Eq. 8-29 we must simultaneously rotate the spins and orbits. The resultant eigenfunctions transform as the J irreducible representations of the full rotation group

$$\mathbf{D}^{L} \times \mathbf{D}^{S} = \sum_{J=|L-S|}^{L+S} \mathbf{D}^{J} \tag{8-30}$$

from Eq. 8-20. Thus, the spin–orbit coupling term in the Hamiltonian will cause the previous eigenfunction to split into several levels each with a 2J + 1 degeneracy. Spectroscopic **notation** usually lists the J value as a right subscript. For example $^2F_{5/2}$ is a state with spin degeneracy of 2 or spin equals 1/2, orbital angular momentum of 3, and total angular momentum J = 5/2. Since S can be integer or half-integer, J can also be either. When J is half-integer, double-group character tables must be used.

As a simple, but experimentally useful, example consider $4f^1$ (one electron in the 4f shell). For the single f-electron, the total orbital angular momentum is L = 3 and spin angular momentum S = 1/2 or the term signature is $^2F$. When the spin–orbit coupling is considered, the eigenfunctions transform as $\mathbf{D}^{1/2} \times \mathbf{D}^{3} = \mathbf{D}^{5/2} + \mathbf{D}^{7/2}$ (term signature $^2F_{5/2}$ and $^2F_{7/2}$) irreducible representations of the full rotation group. These

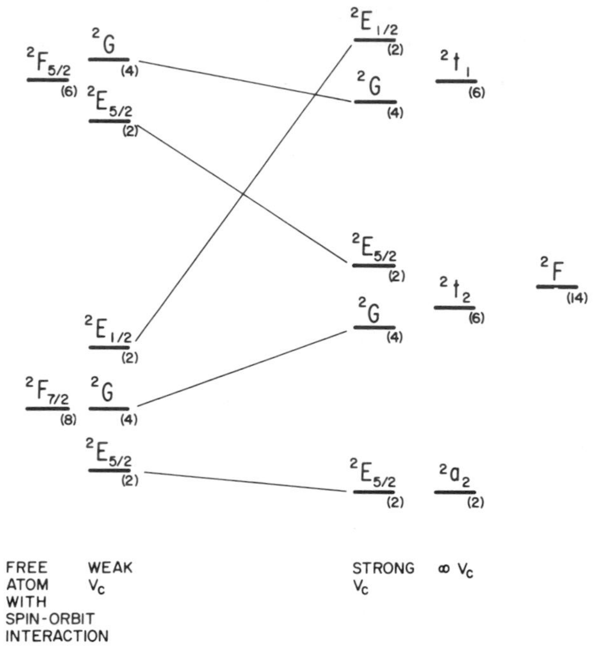


**Fig. 8-5** The correlation diagram for $4f^1$ with spin–orbit coupling taken into account.

are clearly double group irreducible representation. The left side of Fig. 8-5 shows eigenvalues of the free atom result. If a small (compared to the spin–orbit coupling) crystal field is added of $O_h$ symmetry, the degeneracy will be lifted. We only need consider O point symmetry since $O_h$ will just add u-subscripts to this f-electron problem. When the symmetry is reduced from the full rotation group to O-point symmetry, $\mathbf{D}^{5/2}$ and $\mathbf{D}^{7/2}$ reduce as was calculated in Table 8-7. The results are shown on the left in Fig. 8-5. The degeneracies of the various eigenvalues are also shown. Now consider the infinitely strong crystal field acting on an f-electron. As was calculated in Table 8-2 the f-orbital states will split $\mathbf{D}^3 = a_2 + t_1 + t_2$. These are listed on the right side of Fig. 8-5 with a left superscript to indicate the spin multiplicity. When the spin–orbit coupling is considered, "$\mathbf{V}_c$ strong" levels transform as: $a_2 \times \Gamma_6 = \Gamma_7$; $t_1 \times \Gamma_6 = \Gamma_6 + \Gamma_8$; $t_2 \times \Gamma_6 = \Gamma_7 + \Gamma_8$. These results are shown on the right of Fig. 8-5. To complete the correlation diagram we only need connect the same irreducible representation with the no-crossing rule.

The type of coupling that has been discussed so far in this section is known as **Russell–Saunders coupling**. The approach is summarized here.

From the difference between Eqs. 8-17a and 8-17b we have coupled the orbital angular momentum of the individual electrons $\mathbf{D}^{l1} \times \mathbf{D}^{l2} \times \ldots \times \mathrm{D}^{lm} = \Sigma\, \mathbf{D}^{\mathrm{L}}$ to determine **L**. From terms in the spin part of the Hamiltoanian that depend on the internuclear distance the same approach is used for the spins $\mathbf{D}^{s1} \times \mathbf{D}^{s2} \times \ldots \times \mathbf{D}^{sm} = \Sigma\, \mathbf{D}^{\mathrm{S}}$ to determine **S**. The spin–orbit term in the Hamiltonian was considered so **S** and **L** must be coupled to give a total angular momentum **J**, $\mathbf{D}^{\mathrm{S}} \times \mathbf{D}^{\mathrm{L}} = \Sigma\, \mathbf{D}^{\mathrm{J}}$. The interactions that are considered first are thought to be the largest. This results in the final coupling of **L** to **S** to give **J**. This is known as Russell–Saunders coupling. The resultant wave functions, $\psi_{\mathrm{M}}{}^{\mathrm{J}}$ as in Eq. 8-21, are made up from linear combinations of wave functions of **L** and **S**. Although the functions transform properly (as $\mathbf{D}^{\mathrm{J}}$), they are not eigenfunctions of the true Hamiltonian. Rather, what is implied or hoped is that they are a good approximation to the true eigenfunctions. As a better approximation, for the free atom for example, configuration interaction will be considered. That is, other states that also transform as $\mathbf{D}^{\mathrm{J}}$ but arise from different **L** and **S** values can be mixed into the state of interest. To the extent that this mixing is small, the Russel–Saunders approximation is good which is particularly true for light atoms. For electric dipole transitions, we obtain a **selection rule** $\Delta\mathrm{L} = 0, \pm 1$ except, $\mathrm{L} = 0$ to $\mathrm{L}' = 0$ is not allowed. This selection rule follows from the fact that the electric dipole operator transforms as $\mathbf{D}^1$, so $\mathbf{D}^{\mathrm{L}} \times \mathbf{D}^{1} \times \mathbf{D}^{\mathrm{L}'}$ contains $\mathbf{D}^0$ if $|\mathrm{L} - \mathrm{L}'| = 0, \pm 1$ except when $\mathrm{L} = 0 = \mathrm{L}'$. The same selection rule applies to J.

Another approximation, known as **jj coupling** is often more appropriate for heavy atoms. In this approximation the spin–orbit coupling is more important that the noncentrosymetric terms in Eq. 8-17a. Thus, each individual electron has its spin and orbit coupled, $\mathbf{D}^{l} \times \mathbf{D}^{1/2} = \mathbf{D}^{\mathrm{j}}$, to determine its j-states. Then, since the true Hamiltonian is only invariant to rotations if all the electrons are rotated through the same angle, the individual j-values are coupled together to give a total J, $\mathbf{D}^{\mathrm{j}1} \times \mathbf{D}^{\mathrm{j}2} \times \ldots \times \mathbf{D}^{\mathrm{jm}} = \Sigma \mathbf{D}^{\mathrm{J}}$. The wave functions that arise from this procedure will also have admixtures from other states that transform as the same $\mathbf{D}^{\mathrm{J}}$.

## 8-7 Other Crystal Field Effects

Crystal field theory is a very extensive subject and entire books have been devoted to it in general and on special areas such as the transition series ions, $3\mathrm{d}^{\mathrm{n}}$. Several references are mentioned in the Notes. In this section we cover several points that have only been briefly touched and several other points that follow in a straightforward way from the principles developed in the previous sections of this chapter.

### a. Cubic symmetry

Equation 8-15 shows the expansion of the potential about the origin due to charges at the six octahedral positions which result in $O_h$–point symmetry. However, a number of different arrangements of charges can result in $O_h$ symmetry as discussed below Eq. 8-15. We discuss this point and the ratio of the splittings that will result.

Taking the 4-fold axis of a cube as the z-axis, $b_4$ from Eq. 8-15 can be written in terms of the charge arrangment as

$$b_4 = -(e\langle r^4\rangle/264)\ \Sigma_i\ j_i(35\cos^4\theta_i - 30\cos^2\theta_i + 3)/a_i^5 \qquad (8\text{-}31)$$

where $e = +4.8 \times 10^{-10}$ esu, $\langle r^4\rangle = \langle\psi|r^4|\psi\rangle$, $\psi$ is the 3d or 4f wave function of the ion at the origin, the sum over the external ions with charge j, located at distance $a_i$ and angle $\theta_i$ ($\theta$ is the angle between the z-axis and the vector to the charge). $b_4$ in Eq. 8-15 gives the potential energy of the interaction of a set of external charges $j_i$ with the electrons on a central ion. For six charges arranged octrahedrally as described in the text just above Eq. 8-15 the rsult of the sum in Eq. 8-31 is $28j/a^5$. If the charges are placed at eight corners of a cube then the sum is $(-8/9)28j/a^5$. The point symmetry is still $O_h$ for these eight neighbors. If four charges are arranged tetrahedrally at half of the corners of a cube, then the sum is $(-4/9)28j/a^5$ and the point symmetry is $T_d$. If the charges are placed at the 12 midpoints of the sides of a cube, as shown in the figure in the Problems in Chapter 10, then the sum is $(-1/2)28j/a^5$ and the point symmetry is $O_h$. Thus, different cubic arrangements of charges have different magnitudes and signs of the crystal field.

The ratio of tetrahedral to octrahedral crystal field of $-4/9$ is the most widely known difference. The opposite sign will cause the ordering of the crystal field split levels to reverse for these two arrangements of charge. This is shown schematically in Fig. 8-2 and would apply to the results in Figs. 8-3–5 and anywhere this comparison arises.

### b. Hole formulation

The interaction energy of $3d^1$ on a central ion with a given set of external charges is the same but opposite in sign as the interaction energy of $3d^{10-1}$, or one hole in the 3d shell. Similarly, for $3d^n$ and $3d^{10-n}$.

Thus, the crystal field splittings between these two configurations are just the reverse of each other for the same arrangement of external charge. This fact saves a great deal of work in calculations.

### c. Actual $d^n$ splittings

In order to actually determine the energies of the various terms, the matrix elements of the interelectron repulsion term $1/r_{ij}$ in Eq. 8-17a must be calculated. This is done by straightforward but involved techniques. The $1/r_{ij}$ term is expanded in terms of the sum of products of spherical harmonics $Y_m^l(\theta_i\phi_i)Y_m^l(\theta_j\phi_j)$ where only terms up to $l = 4$ are needed for d-electrons, since the electron orbitals contain only $Y^2$ terms. The results contain the usual Coulomb and exchange integrals which can be simplified since all the electrons have the same orbital angular momentum. The energy splittings are usually expressed in terms of linear combinations of these integrals. A particular linear combination is called the Racah parameters A, B, and C. The differences in energy between the various terms is determined by the interelectron repulsion parameters B and C, and the crystal field. Thus, for a cubic crystal field there are three parameters that determine the splittings of $d^n$ terms 10Dq, B, and C. In crystals we find that the interelectron–electron integrals are slightly smaller than in free atoms possibly due to an expansion of the orbitals, but B/C is 4 to 5.

In order to compare observed experimental splittings to theory, Tanabe and Sugano calculated the cubic crystal field splittings in terms of B and C and expressed the results in a set of diagrams that are shown in Appendix 9. The ratio of the energy splitting to B is plotted versus 10Dq/B for a B/C ratio that is shown ($\approx$4.5). On the left side, the free electron terms can be seen and on the right are the strong field terms. In the Tanabe–Sugano diagrams the actual ground state is taken as the zero energy, thus a discontinuity in slope appears in $d^4$ to $d^7$ when the cubic crystal becomes large enough to cause the ground state to change from the weak field "high spin state" to the strong field "low spin state." Physically the energies of the various terms are smooth functions of 10Dq. This smooth change in energy is better seen in the so-called Orgel diagrams which take the zero of energy as the ground state of the free atom. The Notes should be consulted for several references to extensive literature to this field.

We describe the use of these diagrams for the electronic transitions of $Cr^{3+}$, in $Cr(H_2O)_6^{3+}$ for example. For this ion B $\approx$ 1000 $cm^{-1}$. For a cubic crystal field 10Dq = 20,000 $cm^{-1}$; the $^4T_2$, $^4T_1$, and higher lying $^4T_1$ states should occur at 20, 27, and 45 × $10^3$ $cm^{-1}$, respectively. Transitions from the $^4A_2$ ground state to these energy levels should be fairly intense since no spin charge is required, i.e., they are spin-allowed. Transitions to the doublet $^2E$ and $^2T$ states are spin-forbidden and expected to be much weaker. Their observed positions are in good agreement with the positions predicted from the diagrams.

### d. Lower symmetry

Cubic symmetry was considered in Section 8-7c. Thus, only one crystal field parameter is required for $d^n$-electrons as shown in Eq. 8-15 and in Section 8-7a. (Two terms are required for $f^n$-electrons since the f-electron orbitals have $Y^3$ spherical harmonics so the crystal field must be expanded to $Y^6$ terms. Equation 8-15 shows the expansion only to $Y^4$ terms.) If the symmetry is lower, more crystal field terms are required as mentioned at the end of Section 8-3. For example, if the symmetry is $D_{4h}$ there are axial crystal field terms $\sim Y^2$ and $Y^4$. The usual approach is to try to treat these terms as small compared to the cubic crystal field and attempt to fit the overall features with the cubic terms and splittings with the lower symmetry term. The references in the Notes should be consulted.

### e. Vibronic assisted absorption

These ideas, used to explain absorption lines that were thought to be not allowed, have already been discussed in Section 7-6f.

## Appendix to Chapter 8

### Permutation symmetry and the Pauli principle

In this chapter functions that are the proper linear combinations of the one-electron functions were found when the electron–electron repulsion term in the Hamiltonian was considered. A great deal of scientific understanding can result from using these functions. However, the Hamiltonian also has permutation symmetry. That is, it is unchanged if the electrons (or any identical particles) are interchanged or permuted among themselves.

The simple product wave function considered in this chapter do not transform as irreducible representations of the permutation group which was breifly discussed in the Appendix to Chapter 2. There it was shown that the permutation group of n objects, $\mathbf{P}_n$, always contains a symmetric $A_1$ and antisymmetric $A_2$ irreducible representation with respect to the interchange of two particles. The **Pauli exclusion principle** states that for a system of identical particles with half-integer spin (obey Fermi statistics) the eigenfunctions reverse sign if the coordinates of any two particles are interchanged. That is, the eigenfunction transforms as the $A_2$ irreducible representation of the permutation group. For particles with integral spin (obey Bose statistics) the eigenfunctions transform as the $A_1$ irreducible representation of the permutation group. This is a principle, i.e., a postulate found to agree with experiment.

We will consider only electrons (Fermions) here. Define $P_{ij}$ as the operator that interchanges the electron, that is described by the ith coordinates (space and spin), with the electron that is described by the jth coordinates. For a many-electron system with eigenfunction $\psi$ the Pauli principle states

$$P_{ij}\psi = -\psi \tag{8A-1}$$

Note that this equation must be true, according to the Pauli principle, for the many-electron wave function no matter how this function is written. In this book we always consider simple types of wave functions that are products of one-electron functions u. By $u_k(l)$ we mean the $l$th electron has coordinates labeled by k, so

$$P_{ij}\{u_a(1)u_b(2)...u_i(i)u_j(j)...u_n(n)\} = -u_a(1)...u_i(j)u_j(i)...u_n(n) \tag{8A-2}$$

Of course giving the electrons a label seems to contradict the idea behind the indistinguishability of the electrons, but this is overcome because the eigenfunction is taken as a linear combination of all distinct permutations of the electrons

$$\Psi = 1/\sqrt{n}!\, \Sigma_P\, (-1)^p\, P\{u_1(1)\, u_2(2)...u_n(n)\} \tag{8A-3}$$

The sum is over all possible distinct permutations P. Since there are n! permutations in the sum, $1/\sqrt{n}!$ is the normalizing factor since we take the individual $u_i$ to be orthonormal one-electron function. $(-1)^p$ is positive (negative) when there are an even (odd) number of two-electron interchanges required to bring the term in the sum back to some standard order which is usually the form as on the left side of Eq. 8-A2. The operation in Eq. 8-A3 is just the projection operator for the $A_2$ irreducible representation of permutation groups. The projection operator for the

$A_1$ irreducible representation of the permutation group is the same expression without the $(-1)^p$ term.

As an example consider the three-electron system. The eigenfunction that has permutation symmetry and satisfies the Pauli principle is

$$\Psi = u_a(1)u_b(2)u_c(3) - u_a(2)u_b(1)u_c(3) - u_a(1)u_b(3)u_c(2) - u_a(3)u_b(2)u_c(1) + u_a(2)u_b(3)u_c(1) + u_a(3)u_b(1)u_c(2) \quad (8\text{-A4})$$

The first term is the "standard" order. The terms with a $(-1)$ are obtained from the standard order by an odd number of permutations namely the interchange of 1 and 2, 2 and 3, 1 and 3, respectively. The last two terms require an even number of permutations, for example the last term requires the interchange of 1 and 2 followed by 2 and 3 while the next to the last term requires the interchange of 1 and 2 followed by 1 and 3. Another way to write the many-electron wave function in Eq. 8-A3 is

$$\Psi = 1/\sqrt{n!} \begin{vmatrix} u_1(1) & u_2(1)\ldots & u_n(1) \\ u_1(2) & u_2(2) & \\ \cdot & \cdot & \cdot \\ \cdot & \cdot & \cdot \\ \cdot & \cdot & \cdot \\ u_1(n) & u_2(n)\ldots & u_n(n) \end{vmatrix} \quad (8A\text{-}5)$$

This determinant, usually called a Slater determinant, has the same terms as Eq. 8-A3. One can check, for the three-electron case written in Eq. 8-A4.

Using the Slater determinant, the older statement of the **Pauli exclusion principle** can immediately be seen to be true. The statement is: No two electrons can have the same coordinates (space and spin) or alternately the same quantum numbers. Clearly from Eq. 8-A5, if two columns are the same, the wave function vanishes. This can be seen in Eq. 8-A3 as well, or in the three-electron example where the terms cancel in pairs if $u_1 = u_2$.

The Pauli principle can have important effects even if the Hamiltonian does not contain spin dependent terms. We investigate this problem for the two-electron case where the antisymmetrized eigenfunction is

$$\Psi = (1/\sqrt{2})\,[u_a(1)u_b(2) - u_a(2)u_b(1)] \equiv (1/\sqrt{2})\,[u_1u_2 - u_2u_1] \quad (8\text{-A6})$$

The confusing **convention** is also defined in this equation. The spatial parts are assumed always to be in standard order so we can write $u_a(2)u_b(1) = u(2)u(1)$, but the electron label is taken out of the brackets and made a subscript, as in Eq. 8-A6. The one-electron orbital function u is a product of a space part $\phi$ and a spin part. Since the spin of an electron can only be 1/2, the projection quantum number can only be $\pm 1/2$ which is conventionally taken as $\alpha$ and $\beta$ for spin up (+1/2) and spin down(−1/2), respectively. For two electrons there are only four possible spin combinations: $\alpha(1)\alpha(2)$, $\alpha(1)\beta(2)$, $\beta(1)\alpha(2)$, and $\beta(1)\beta(2)$, where no subscripts are required. Again the electron label **conventionally** is taken out of the brackets and made into a subscript: $\alpha_1\alpha_2$, $\alpha_1\beta_2$, etc. Thus, for the two-electron problem four different antisymmetric wave functions of the type Eq. 8-A6 can be written

$$\begin{aligned}
\Psi_1 &= (1/\sqrt{2})\,(\phi_1\phi_2 - \phi_2\phi_1)\alpha_1\alpha_2 \\
\Psi_2 &= (1/\sqrt{2})\,(\phi_1\phi_2\alpha_1\beta_2 - \phi_2\phi_1\beta_1\alpha_2) \\
\Psi_3 &= (1/\sqrt{2})\,(\phi_1\phi_2\beta_1\alpha_2 - \phi_1\phi_2\alpha_1\beta_2) \\
\Psi_4 &= (1/\sqrt{2})\,(\phi_1\phi_2 - \phi_2\phi_1)\beta_1\beta_2
\end{aligned} \tag{8-A7}$$

These four functions are eigenfunctions of $\mathbf{S}_z = \mathbf{s}_{z1} + \mathbf{s}_{z2}$ with eigenvalues h, 0, 0, −h as can readily be seen. However only $\Psi_1$ and $\Psi_4$ are eigenfunctions of $\mathbf{S}^2 = \mathbf{S}_z^{\ 2} + 1/2[\mathbf{S}_+^{\ 2} + \mathbf{S}_-^{\ 2}]$ with S = 1 and $m_s = 0$. This function can be found by applying a lowering operator to $\Psi_1$.

$$\begin{aligned}
\mathbf{S}_-\,\Psi_1 &= \mathbf{S}_-\,|S{=}1, m_s{=}1\rangle = \sqrt{2}\,|S{=}1, m_s = 0\rangle \\
&= \mathbf{s}_{1-}\Psi_1 + \mathbf{s}_{2-}\Psi_1 = (\phi_1\phi_2 - \phi_2\phi_1)\beta_1\alpha_2 + (\phi_1\phi_2 - \phi_2\phi_1)\alpha_1\beta_2 \\
&= \sqrt{2}\,(\Psi_2 + \Psi_3)
\end{aligned} \tag{8-A8}$$

The usual relations for raising and lowering operators (Eq. 7-26) are used. Since $|S{=}0, m_s{=}0\rangle$ is orthogonal to $|S{=}1, m_s{=}0\rangle$, it must equal $\sqrt{2}(\Psi_2 - \Psi_3)$. So the eigenfunctions can be grouped

$$\begin{aligned}
|1, 1\rangle &= (1/\sqrt{2})\,(\phi_1\phi_2 - \phi_2\phi_1)\alpha_1\alpha_2 \\
|1, 0\rangle &= (1/\sqrt{2})\,(\phi_1\phi_2 - \phi_2\phi_1)\,(\alpha_1\beta_2 + \beta_1\alpha_2) \\
|1, -1\rangle &= (1/\sqrt{2})\,(\phi_1\phi_2 - \phi_2\phi_1)\beta_1\beta_2 \\
|0, 0\rangle &= (1/\sqrt{2})\,(\phi_1\phi_2 + \phi_2\phi_1)(\alpha_1\beta_2 - \beta_1\alpha_2)
\end{aligned} \tag{8-A9}$$

Note that the three eigenfunctions with S=1 have the same spatial wave function and that this is different from the spatial part of the eigenfunction for S=0. In fact the spatial part of the eigenfunctions of S=1 is antisymmetric under the interchange of the two electrons, while for the S=0 the spatial part of the wave function is symmetric under the interchange of the two electrons. For the spin parts of the wave function, just the opposite occurs. The Pauli principle demands that the total wave function $\Psi$ is antisymmetric under the permutation of any two electrons. Equation A8-9 shows that this occurs in different ways for the two-electron problem for the S=1 and S=0 eigenfunctions. $\Psi$ is the product of a space part $\Psi_{space}$ and a spin part $\Psi_{spin}$

$$\Psi_{S=1} = \Psi(A_2)_{space}\,\Psi(A_1)_{spin}, \qquad \Psi_{S=0} = \Psi(A_1)_{space}\,\Psi(A_2)_{spin} \tag{8-A10}$$

$A_1$ refers to symmetric irreducible representation under the interchange of two electrons and $A_2$, an antisymmetric irreducible representation. It is clear that the probability that both electrons can be at the same point in space is zero for $\Psi(A_2)_{space}$ and finite for $\Psi(A_1)_{space}$. Thus, due to Coulomb repulsion, $\Psi_{S=1}$ will have a lower energy than $\Psi_{S=0}$ even though there are no spin dependent terms in the Hamiltonian.

Thus, the importance of the Pauli principle on actual energies can be seen. There are several rules, **Hund's rules**, that describe the **S**, **L**, and **J** that will occur in an atom that contain several electrons. (a) The lowest energy state will have the maximum **S**. The reason is that parallel electron spins will favor a spatial electron distribution where the electrons are well separated which leads to a smaller Coulomb repulsion as in the two-electron case above. (b) For a given **S**, the term with the greatest **L** will have the lowest energy. This is associated with the fact that for the maximum **L** the electrons tend to move in the same direction in phase with one another. Thus they are in close proximity with each other less often, again leading to a smaller Coulomb interaction. (c) for a less then half-filled shell the lowest energy state has **L** and **S** antiparallel or $\mathbf{J} = \mathbf{L} - \mathbf{S}$ and $\mathbf{L} = \mathbf{L} + \mathbf{S}$ correspond to the highest energy state.

### Allowed configurations for np$^2$

As a last example of the use of the Pauli exclusion principle we consider the np$^2$ problem, i.e., two equivalent p-electrons. and show that only $^1$D, $^3$P, and $^1$S terms are allowed instead of $^3$D, $^1$D, $^3$P, $^1$P, $^3$S, and $^1$S which are allowed for npmp. Table 8-A1 shows a simple construction that helps to solve the problem. Each individual electron can have $m_l$ = +1, 0, or −1. This electron can also have spin up, represented by a diagonal line upper left to lower right (\) or spin down (/). The one condition on filling the boxes is that the two electrons which must be accounted for on each line cannot have all their quantum numbers the same according to the Pauli principle. Since the principal quantum number is the same, either $m_l$ or $m_s$ must differ for the two electrons. On the

**Table 8-A1**

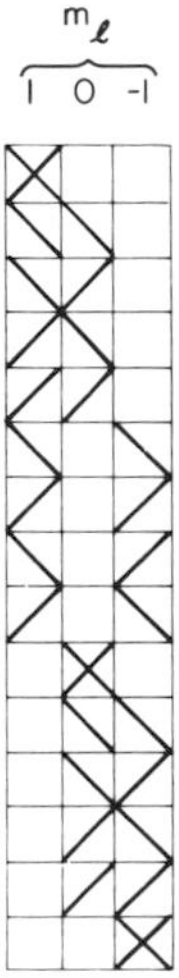


| | $M_L = \Sigma m_\ell$ | $M_S = \Sigma m_s$ | | TERM |
|---|---|---|---|---|
| 1 | 2 | 0 | | $^1D$ |
| 2 | 1 | 1 | | $^3P$ |
| 3 | 1 | 0 | $^1D$, (rows 3–4) | $^3P$ |
| 4 | 1 | 0 | | |
| 5 | 1 | -1 | | $^3P$ |
| 6 | 1 | 0 | | $^3P$ |
| 7 | 0 | -1 | | $^3P$ |
| 8 | 0 | 0 | | |
| 9 | 0 | 0 | $^1D$, (rows 8–10) | $^3P$, $^1S$ |
| 10 | 0 | 0 | | |
| 11 | -1 | 1 | | $^3P$ |
| 12 | -1 | 0 | $^1D$, (rows 12–13) | $^3P$ |
| 13 | -1 | 0 | | |
| 14 | -1 | -1 | | $^3P$ |
| 15 | -2 | 0 | | $^1D$ |

first line, the first electron is put in with $m_l = 1$ and spin up. For the next electron to have $m_l = 1$ the spin must be down. This procedure is continued until all 15 [= (6)(5)/2] possibilities are considered, as in the figure. Now we can determine the terms that have been found. The maximum value of $M_L = 2$ which is the maximum value of L. The value of L = 2 only occurs with $M_S = 0$, so the term must be a singlet or $^1D$. This is entered on the right. However, $^1D$ has $M_L = 2, 1, 0, -1, -2$ all with $M_S=0$. These are labeled on the right. The next highest value is $M_L = 1$, but this term has $M_S = 1$ so we must have a triplet P or $^3P$. There are nine partners for $^3P$ since $M_L = 1, 0, -1$ and $M_S = 1, 0, -1$. These are also listed on the right. Only one term is left and that is $M_L = 0$ and $M_S = 0$ or $^1S$. This term is listed also. Thus for the $np^2$ problem only the $^1D$, $^3P$, and $^1S$ terms are obtained.

In a similar manner the $nd^2$ problem will have $^1G$, $^3F$, $^1D$, $^3P$, and $^1S$ terms.

### Terms for equivalent $p^n$ and $d^n$ configurations

We list terms allowed by the Pauli principle for equivalent $p^n$ and $d^n$ electrons in Table 8-A2. Terms for the closed shells $s^2$, $p^6$, $d^{10}$ configurations are $^1S$ and are not listed. The number 2 or 3 in parentheses after some of the terms means that the term occurs 2 or 3 times. The ground term is written first but then the order is arbitrary.

**Table 8-A2**

| Configuration | Terms |
|---|---|
| $s^1$ | $^2S$ |
| $p^1, p^5$ | $^2P$ |
| $p^2, p^4$ | $^3P$, $^1S$, $^1D$ |
| $p^3$ | $^4S$, $^2P$, $^2D$ |
| $d^1$, $d^9$ | $^2D$ |
| $d^2$, $d^8$ | $^3F$, $^3P$, $^1S$, $^1D$, $^1G$ |
| $d^3$, $d^7$ | $^4F$, $^4P$, $^2P$, $^2D(2)$, $^2F$, $^2G$, $^2H$ |
| $d^4$, $d^6$ | $^5D$, $^3P(2)$, $^3D$, $^3F(2)$, $^3G$, $^3H$, $^1S(2)$, $^1D(2)$, $^1F$, $^1G(2)$, $^1I$ |
| $d^5$ | $^6S$, $^4P$, $^4D$, $^4F$, $^4G$, $^2S$, $^2P$, $^2D(3)$, $^2F(2)$, $^2G(2)$, $^2H$, $^2I$ |

**Notes**

Ballhausen ("Ligand Field Theory," Chapter 1, McGraw-Hill, New York, 1952) has a short history with references to the early work in crystal field theory. Van Vleck's classic book ("The Theory of Electric and Magnetic Susceptibilities," Oxford Univ. Press, London and New York, 1932) should also be consulted. In this book he explains the quenching of the orbital angular momentum.

Bethe's original paper is available in English and is worthwhile reading. The character tables are determined in Bethe's paper. They are also discussed by Hamermesh (Section 9-7). Tinkham (Sections 4-7 and 8), discusses the splittings of $\mathbf{D}^l$ (half-integer $l$) in crystal fields. Opechowski [Physica **7**, 552 (1940)] and Heine (p. 137) also discuss the double group character table.

There are many different conventions followed in the definitions of the Euler angles. Left-handed coordinate systems also have been used; we follow Rose, p. 48. For some of the differences see H. Goldstein, "Classical Mechanics," p. 107 (Addison-Wesley, Massachusetts, 1953).

There are a very large number of books covering the general field of "Ligand Field Theory." We mention only one of the earlier ones that is also good, Ballhausen, in which experimental results for the 3d, 4d, and 5d ions are compared with theory in Chapter 10. Reading this chapter will give one a good working knowledge of ligand field theory.

The Tanabe and Sugano diagrams are discussed in J. Phys. Soc. Japan **9**, 753 (1954) and by S. Sugano, H. Kamimura and Y. Tanabe,

"Multiplets of Transition-Metal Ions in Crystals." (Academic Press, New Yor, 1970.)

McWeeny's Appendix 1 has the appropriate linear combinations of spherical harmonics that transform as the s-, p-, d-, f-atomic orbitals for all of the 32 point groups. In Appendix 2 he shows how to write the appropriate linear combinations for the cubic point groups for alternative equivalent axes. For example, instead of the x, y, z axes of the cube, the [1,1,0], $[1,\bar{1},0]$, z axes might be used if a phase transition to an orthorhombic form were expected. Then, the linear combination of functions is immediately adapted for the appropriate reduction in the lower symmetry group and in the cubic group, both sets of axes are equivalent.

## Problems

**1.** Prove that Eqs. 8-1 and 8-2 give identical results.

**2.** Find the recursion relation for the coefficients $N_{ab}$ in Eq. 8-10 and determine all the polynomials for $l = 2$ and 3. Compare the results to those obtained in Appendix 8.

**3.** For atoms placed at the midpoints of the 12 edges of a cube (see the figure in Problem 4 of Chapter 10) find the point symmetry of the molecule.

**4.** The particular values of $m = \pm 4$ in Eq. 8-15 arise because the axis of quantization is the 4-fold axis. Express the crystal field in terms of spherical harmonics if the 3-fold axis is picked as the axis of quantization. This would be the sensible choice if the cubic symmetry is lowered by a strain along the 3-fold axis.

**5.** (a) Verify explicitly Eqs. 8-22. (b) Show that $\Sigma\ (2J+1) = (2j+1)(2j'+1)$ where the sum is from $|j-j'|$ to $j+j'$. Comment.

**6.** Using Eq. 8-21 write the $u_m v_n$ from Eq. 8-22 in terms of $\psi_M{}^J$.

**7.** Using the method of descent in symmetry show for the point group O that $t_2 \times t_2 = A_1{}^1 + {}^1E + {}^3T_1 + {}^1T_2$ is obtained to satisfy the Pauli principle. In the same manner, find $t_1 \times t_1$.

**8.** Prove the statements made about the hole formulation (Section 8-7b).

**9.** How can $Cr(H_2O)_6{}^{3+}$ have $O_h$ symmetry? What would you expect in an aqueous solution? Using the expected values for the extended dipole moment of $H_2O$, calculate the cubic crystal field at the $Cr^{3+}$ ion.

**10.** Consider a 3d electron on an A-atom in a planar $AB_3$ molecule. Find the splitting of the 3d electron in electron volts using the point charge model and assuming each B-atom to have one negative electronic charge.

**11.** It is found that in a solution of $CoCl_2 \bullet 6H_2O$ the complex is trans-$[CoCl_2(H_2O)_4]$, i.e., $D_{4h}$ symmetry. Using the accepted radii and dipole moment, calculate the cubic component of the potential. Calculate the axial component of the potential that transforms as $Y_0^2$.

**12.** Draw the energy levels of the d-electrons in $O_h$ symmetry as in Fig. 8-2. Fill up the orbital levels with electrons according to Hunds' rule (maximize spin) for small 10Dq and for very large 10Dq, showing that for $d^4$ to $d^7$ one can distinguish a high-spin and a low-spin configuration. For what values of n in $d^n$ can these two spin configurations be distinguished in $T_d$ symmetry?

**13.** For $s^2$ with a first excited state of $s'p'$, find the energy levels in the Russell–Saunders and the jj approximation. Show allowed electric dipole transitions and draw a correlation diagram. The two limits might apply to Be $2s^2$ and Hg $6s^2$.

**14.** For an $nd^2$ configuration find the allowed terms that are consistent with the Pauli exclusion principle.

## Chapter 9

# HYBRID FUNCTIONS

*...it is time to make rich and profitable campaigns and to gain the great reward for your labors, after having accomplished such a length of journey over so many mountains and rivers...*

*Hannibal, "To his soldiers"*

## 9-1 Introduction

Exact solution of the wave equation for all but the simplest molecules is impossible. Good approximate ground state eigenfunctions and eigenvalues of simple molecules can be calculated with the aid of a large digital computer. However, a relatively simple molecular orbital approach can be taken, as will be discussed in Chapter 10. In this chapter the more elementary **maximum overlap** method will be presented, since it is an excellent example of how group theory can be applied to chemistry. We will find functions, centered on a central atom, that have a large overlap with neighboring atoms. These functions are called **hybrid functions**. For example, we will find hybrid functions of the carbon atom in the molecule $CH_4$. These hybrid functions will be composed of linear combinations of the atomic eigenfunctions. The determination of these hybrid functions also can be thought of as a problem of finding **equivalent functions**. That is, functions that are equivalent to each other except for their spatial orientation. Thus the symmetry operations permute these functions among themselves. These equivalent functions transform among themselves as a reducible representation of the group. In this sense they are different from basis functions which transform among themselves as an irreducible representation.

The basic chemical idea is that in the calculation of the binding energy of a molecule, a large fraction of the energy comes from the exchange terms. The magnitude of these terms is determined by the amount of overlap of the charge clouds of the constituent atoms. Thus, to maximize the binding energy (larger negative value), the atoms that make up the molecule should overlap as much as possible. In Chapter 10 we will find linear combinations of atomic orbitals that can overlap and form molecular orbitals (LCAO-MO). These LCAO-MO's will transform as an irreducible representation of the point group of the molecule. Thus, they have the proper transformation properties of an eigenfunction even though they are only approximations or severely truncated expansions in the sense of Eq. 7-37. These LCAO-MO's will also give excited state eigenfunctions and eigenvalues as well as ground state values.

In this chapter we do something much simpler. We find a linear combination of atomic orbitals for a central atom that forms hybrid functions that can overlap to a large extent with other atoms. A number of points that will come up throughout this chapter should be kept in mind with respect to these hybrid functions. The hybrid functions in general will not transform as an irreducible representation for the point symmetry of the particular central atom involved. Rather they transform as a reducible representation of the group. Thus the hybrid functions do not have the proper symmetry of an eigenfunction of the atom except in the case of accidental degeneracy. This is no problem since hybrid functions are used in a much more empirical manner along with experimental data. They are most useful for ground state energies and charge distributions. What is hoped is that the hybrid functions are reasonable approximations to the ground state eigenfunctions of the molecule. Within the spirit of the hybridization approach, we assume that the energy difference between the atomic orbitals used to make the hybrid function is small compared to the exchange energy obtained upon the formation of the bonds of the molecule. The large qualitative advantage of the hybrid functions and the overlap formed from them is that, in a natural way, they emphasize the chemical bond. That is, the hybrid functions lead in a very simple way to the localized electron-pair bond. It will turn out that if the expectation values of LCAO-MO is calculated, the results are similar in that the electron density will be found between neighboring atoms (bonds). However, it is difficult for one to even qualitatively appreciate this by just looking at the linear combinations of atomic orbitals that make up the LCAO-MO. On the other hand the hybridization approach, discussed in this chapter, is used qualitatively or semiqualitatively in chemistry but is not used for detailed quantum mechanical calculations.

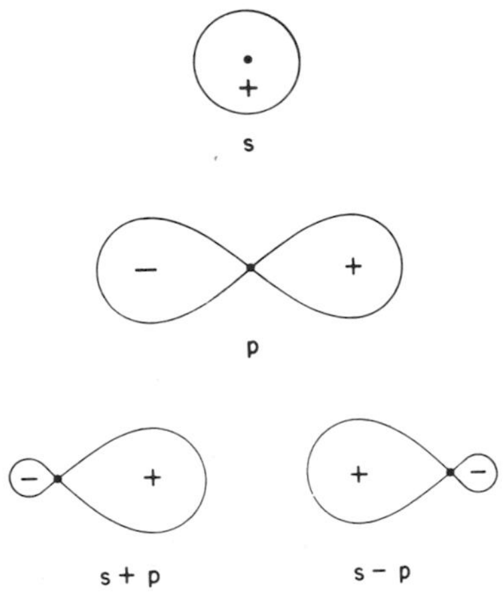


**Fig. 9-1** The hybridization of an s and p function

## 9-2 Simple Hybrid Functions and Bonding

As a very simple example of hybridization (forming hybrid functions), consider an atom with an electron in an s and p state. These eigenfunctions are normalized; then two other normalized orthogonal functions can be formed

$$f_1 = (1/\sqrt{2})\,(s + p), \qquad f_2 = (1/\sqrt{2})\,(s - p) \tag{9-1}$$

The resulting functions are pictured in Fig. 9-1. Note that $f_1$ has a large charge density to the right. Thus, $f_1$ can overlap a great deal with an atom to the right. Also note that $f_1$ is not in general an eigenfunction. Let H be the Hamiltonian of the atom, then $Hf_1 = (E_s s + E_p p)/\sqrt{2}$ so unless there is an accidental degeneracy, $E_s = E_p$, $f_1$ is not an eigenfunction. The function $f_2$ has similar properties except that it has a large charge density and can overlap a great deal with an atom to the left.

Consider two well-separated atoms, each with a closed shell and one electron in an s-orbital outside the closed shell. As these two atoms are brought together the s-orbitals overlap, the exchange terms lower the energy, and a molecule is formed. Each atom contributes an electron to the bond and the bonded electrons have opposite spin so they do not contribute to the paramagnetic susceptibility. This example would apply to $H + H \rightarrow H_2$. Now consider two well-separated atoms but each having two electrons in an s-orbital ($ns^2$) outside a closed shell. One cannot form a bond of these two atoms if the electrons remains in the s-orbital. However, if there is a p-orbital with an energy close to the energy of the s-orbital, the following process can be conceived. On one atom, one of the s-electrons will be promoted to the p-state, $ns^2 \rightarrow nsnp$. Then a hybrid function Eq. 9-1 is formed. The same process can take place on the other atom. Then a molecular bond can be formed where the exchange energy obtained by the formation of the molecule amounts to more than the energy required to promote the electrons from the s- to the p-orbital. In the molecular bond the spins of the p-electrons from the two atoms are oppositely directed as are the spins from the s-electrons. This bond will be formed if the promotional energy required is smaller than the exchange energy obtained.

Consider the $H_2O$ molecule. As separated atoms, each hydrogen has an electron in the 1s-orbit and oxygen has, outside a closed shell, a $2s^22p^4$ configuration with two unpaired p-electrons. To form a molecule, each hydrogen s-electron can overlap with one of the unpaired oxygen p-electrons and form a molecular bond. No promotion of electrons is needed. Since the p-orbitals are 90° apart, the H–O–H angle should be 90°. The angle is actually somewhat larger (105°) because of the interaction of the two hydrogen atoms. For $H_2S$ the angle is 90°. This is expected because the hydrogen atoms are "farther inside" the highly polarizable sulfur atoms, so they "see" less of each other.

For nitrogen the normal state is $2s^22p^3$ with all three p-electrons unpaired. For $NH_3$, reasoning along the same lines as above, one will predict the bonds in $NH_3$ to be 90° from one another with a slight increase due to the interaction of the hydrogen atoms. This is in agreement with experiment.

For carbon the normal state is $2s^22p^2$ with the two p-electrons unpaired. Thus, bonds similar to those found for oxygen will be expected which is not in agreement with experiment. In the next section we discuss how this rather fundamental problem is solved.

## 9-3 Tetrahedral Hybridization

Consider how carbon might form compounds. The free atom electronic configuration is $ls^22s^2p^2$. Thus, the valence of carbon might be 2 because the $2p^2$ electrons can be ionized or they can be used to form bonds with two other atoms to form compounds like $H_2O$. Actually the experimental evidence shows that carbon behaves as a tetravalent atom. Pauling was the first to show how this could be understood in terms of hybridization. If one of the 2s electrons were promoted to the 2p state, the result is $2sp^3$. Then there would be four unpaired electrons available

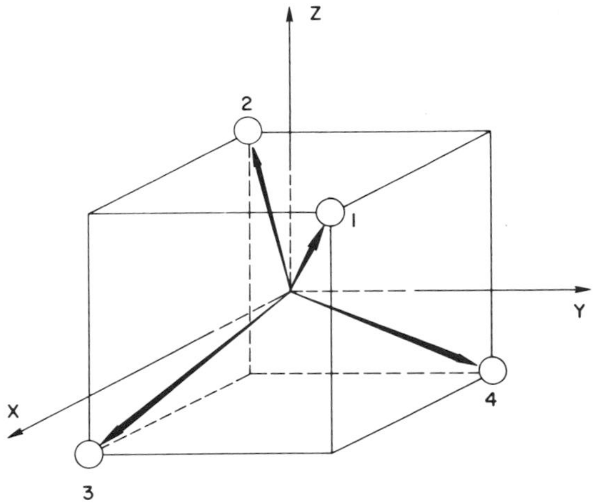


**Fig. 9-2** A carbon atom at the center of a tetrahedron formed by four atoms labeled 1 to 4.

for bonding. We will see how hybrid can be formed to maximize the overlap in, for example, $CH_4$.

Figure 9-2 shows a carbon atom at the center of a tetrahedron formed by four atoms, 1–4 as pictured. These four atoms could be hydrogen atoms for the $CH_4$ example. However, their only importance is to establish the point symmetry at the origin where the carbon atom is located. The point symmetry is $T_d$. The four dark arrows in the figure represent equivalent functions or charge lobes of the carbon atom that we would like to form to maximize the overlap with the four atoms 1–4. We call these functions 1, 2, 3, 4, respectively. Thus, we would like to find a suitable set of orbitals of the carbon atom that are properly directed. The key to the problem is to realize that the required orbitals must transform among themselves under all the symmetry operations of the group. Therefore, they form a basis of a representation. The reason we want four functions is that we have in mind the central atom bonding to four other atoms. Each orbital on the central atom will contribute an electron with spin up and each of the other atoms will contribute an electron with spin down. Thus each bond has two paired electrons. For $T_d$ symmetry consider the E symmetry operation. All four functions transform into themselves so the $4 \times 4$ matrix representing the tranformation has 1 along the diagonal and 0 everywhere else. The character of the matrix is four. For $C_3^{xyz}$ $1 \to 1, 2 \to 3, 3 \to 4, 4 \to 2$. The $4 \times 4$ matrix representation has four 1 and the rest 0 but only one of the 1 appears along the diagonal. Thus the character is 1. This result is shown in Table 9-1. We note that the character of the representation will always be given by the number of unshifted functions or charge lobes. The results for the other symmetry operations are given in Table 9-1a. The resulting representations are clearly reducible since there are no 4-dimensional irreducible representations in $T_d$ symmetry. Reducing it by inspection or by Eq. 4-7 yields

$$\Gamma_{tet} = A_1 + T_2 \tag{9-2}$$

This means that if we take one function that transforms as the $A_1$ irreducible representation and three functions that are partners of the $T_2$ irreducible represention we can form linear combinations, in the sense of Eq. 9-1, so that they will be directed in the four directions as in Fig. 9-2. Note that Eq. 9-2 makes no distinction as to which of the infinite number of functions that transform as $A_1$ (or $T_2$) should be used. In fact any linear combination of some or all of these functions will be appropriate. From the character table for $T_d$ we notice that s-wave functions transform as $A_1$ and the three p-, as well as d-wave functions transform as $T_2$. For carbon some linear combination of $sp^3$ will result in hybrid functions that have lobes in the four tetrahedral directions. We ignore the d-functions for carbon since the 3d atomic orbital has a much higher energy than the $2sp^3$ atomic orbitals.

Now it remains to write out the four hybrid functions $\phi_1, \phi_2, \phi_3, \phi_4$ of the central ion which "point" to the respective atoms 1, 2, 3, 4 in terms of the normalized eigenfunctions s, $p_x$, $p_y$, $p_z$. Thus

$$\phi_i = a_i s + b_i p_x + c_i p_y + d_i p_z, \qquad i = 1, 2, 3, 4 \tag{9-3}$$

We want our four hybrid functions to be orthonormal

$$<\phi_i \,|\, \phi_j> = \delta_{ij} \tag{9-4}$$

This condition and the symmetry of the problem will enable us to determine systematically all the coefficients in Eq. 9-3. However, one can often determine the coefficients by inspection. For example, the s-part of the hybrid function transforms into plus itself (transforms as $\Gamma_1$) under all the operations of the group while the $\phi_i$'s transform among themselves, thus $a_1 = a_2 = a_3 = a_4$. Also the p-eigenfunctions transform as arrows or vectors with the arrowhead corresponding to the positive phase part of the function which is in the +x direction for $p_x$, etc. Thus, to make up a directed charge with positive phase in the 1-direction for $\phi_1$, we need $b_1 = c_1 = d_1$. Similarly to get a positive phase in 2-direction we need $-p_x - p_y + p_z$ and the same amounts of the absolute value of each. i.e., $-b_2 =$

**Table 9-1a** Characters of the representation formed by tetrahedral orbitals transforming among themselves

| $T_d$ | E | $8C_3$ | $3C_2$ | $6S_4$ | $6\sigma_d$ |
|---|---|---|---|---|---|
| $\chi(\Gamma_{tet})$ | 4 | 1 | 0 | 0 | 2 |

$-c_2 = d_2$. From $<\phi_1 | \phi_2> = 0 = a_1^2 - b_1b_2 - b_1b_2 + b_1b_2$ and the normalization $a_1^2 + 3b_1^2 = 1 = a_1^2 + 3b_2^2$, we obtain $a_1 = b_1 = 1/2$. We can also obtain $4a_1^2 = 1$ since we want to use all of the s-eigenfunction in the hybrid bond. The same applies to each atomic orbital. The problem can be approached more systematically by realizing we can obtain a relationship among the coefficients by using the transformation properties of the hybrid functions. For example,

$$C_2{}^z\phi_1 = \phi_2 = a_1C_2{}^zs + b_1C_2{}^zp_x + c_1C_2{}^zp_y + d_1C_2{}^zp_z$$
$$= a_1s - b_1p_x - c_1p_y + d_1p_z$$

$$a_1 = a_2, \quad -b_1 = b_2, \quad -c_1 = c_2, \quad d_1 = d_2 \tag{9-5}$$

This procedure can be repeated to get a relationship between all the $a_i$'s, i.e., $a_1 = a_2 = a_3 = a_4$ and the $b_i$, $c_i$, $d_i$ so that there are just two unknown coefficients a and b

$$\begin{bmatrix} \phi_1 \\ \phi_2 \\ \phi_3 \\ \phi_4 \end{bmatrix} = \begin{bmatrix} a & b & b & b \\ a & -b & -b & b \\ a & b & -b & -b \\ a & -b & b & -b \end{bmatrix} \begin{bmatrix} s \\ p_x \\ p_y \\ p_z \end{bmatrix} \tag{9-6}$$

It is obvious that a relationship between all the coefficients of the p-functions will always be found this way since they are partners of an irreducible representation and will of course transform among themselves. The same will hold true for any set of partners. The coefficients in Eq. 9-6 can be determined by the fact that any two hybrid functions are orthogonal and any one is normalized, i.e., $a^2 + 3b^2 = 1$ and $a^2 - b^2 = 0$. So the hybrid functions can be taken as $a = b = 1/2$ in Eq. 9-6.

In this discussion we ignored the three d-functions ($d_{xy}$, $d_{xz}$, $d_{yz}$) that transform as $T_2$ and only considered the p-functions. For carbon, that makes sense on energetic grounds although one must remember that there will always be some small amount of mixing of the p- with d-functions as discussed in the last chapter. For the ion $CrO_4^{2-}$, energy considerations tell us it makes sense to ignore the p-functions and use only three d-functions that are partners of the $T_2$ irreducible representation. The result is identical with Eq. 9-6 if in the column matrix on the right $d_{yz}$ replaces $p_x$, $d_{xz}$ replaces $p_y$, $d_{xy}$ replaces $p_z$.

We now describe yet another way to determine systematically the 4 × 4 matrix of coefficients in Eq. 9-3 or 9-6. Let $\phi$ and $\alpha$ be the 1 × 4 column matrices given by $\phi_1 \ldots \phi_4$ and s, $p_x$, $p_y$, $p_z$, respectively, as shown in Eq. 9-6. The Eq. 9-3 or 9-6 can be written as $\phi = A\alpha$, where A is the 4 × 4 matrix of coefficients to be determined. Projection operator techniques cannot be used on $\phi_1$ because it is not a basis function of an irreducible representation. However, consider inverse transformation $\alpha = B\phi$ where B is the inverse of the A matrix, $B = A^{-1}$. Each row (single atomic orbital) transforms as an irreducible representation so projection operator techniques can be used. For example, $\alpha_1$ = s atomic orbital which transforms as $\Gamma_1$. Applying the projection operator $V^{A1} = \phi_1 + \phi_2 + \phi_3 + \phi_4$. Normalizing the wave function to 1 gives 1/2 multiplying each $\phi_i$. All the coefficients in the 4 × 4 B matrix can be obtained in a similar manner. Determination of the A matrix is easy since both A and B are orthogonal matrices so the transpose of B is its inverse.

## 9-4 Other Hybrid Functions

Consider a plane $AB_4$ molecule, such as $XeF_4$ as in Fig. 9-3, with $D_{4h}$ symmetry. Table 9-1b shows the character of the representation of the four functions that point toward the four atoms. Both functions and atoms are labeled 1, 2, 3, 4. The reduction of $\Gamma_{sq} = A_{1g} + B_{1g} + E_u$. We notice that s- and $d_{z^2}$-functions transform as $A_{1g}$, $d_{x^2-y^2}$ transforms as $B_{1g}$, $p_x$ and $p_y$ are partners of the $E_u$-irreducible representation. Thus we might have $sdp^2$ or $d^2p^2$ hybridization depending on the energies involved.

Considering $sdp^2$ we have four hybrid functions $\phi_i = a_i s + b_i d_{x^2-y^2} + c_i p_x + d_i p_y$ for i = 1, 2, 3, 4. We can get the coefficients as described in the last section. However, since we want to use all the s-functions in the four functions, it is immediately obvious that $a_1 = a_2 = a_3 = a_4 = 1/2$. The same holds true for $d_{x^2-y^2}$ except that we want positive phases to be directed toward the four b-atoms so $b_1 = b_3 = -b_2 = -b_4 = 1/2$. Similarly, $c_2 = 0 = c_4$ and $c_1 = -c_3$; therefore, $c_1 = 1/\sqrt{2}$. Similarly, $d_1 = 0 = d_3$ and $d_2 = -d_4 = 1/\sqrt{2}$.

However, if the energies of the s and $d_{z^2}$-atomic eigenfunctions are not very different, they will both be involved in forming the hybrid function since symmetry treats the infinite number of functions that transform as $A_1$ (or any $\Gamma_n$) in the same manner. We see that in the

**Table 9-1b** Characters of the representation for the square ($D_{4h}$) shown in Fig. 9-3

| $D_{4h}$ | E | $2C_4$ | $C_2$ | $2C_2'$ | $2C_2''$ | i | $2S_4$ | $\sigma_h$ | $2\sigma_v$ | $2\sigma_d$ |
|---|---|---|---|---|---|---|---|---|---|---|
| $\chi(\Gamma_{sq})$ | 4 | 0 | 0 | 2 | 0 | 0 | 0 | 4 | 2 | 0 |

xy-plane the s and $d_{z^2}$ function have the same cylindrical symmetry. Thus, in place of the s-functions in the above hybrid orbitals we would like to have a linear combination of s and $d_{z^2}$-orbitals keeping the normalization. This is generally accomplished by using a parameter $\Theta$ with the property $\cos^2\Theta + \sin^2\Theta = 1$ which keeps the normalization. Thus, instead of only the above functions $sd_{x^2-y^2}\,p_xp_y$ being used to form the hybrid function, we can have an arbitrary amount of $d_{z^2}$.

$$\phi_1 = (1/2)\,(\cos\Theta\, s + \sin\Theta\, d_{z^2}) + (1/2)\, d_{x^2-y^2} + (1/\sqrt{2})p_x$$
$$\phi_2 = (1/2)\,(\cos\Theta\, s + \sin\Theta\, d_{z^2}) - (1/2)\, d_{x^2-y^2} + (1/\sqrt{2})p_y$$
$$\phi_3 = (1/2)\,(\cos\Theta\, s + \sin\Theta\, d_{z^2}) + (1/2)\, d_{x^2-y^2} - (1/\sqrt{2})p_x$$
$$\phi_4 = (1/2)\,(\cos\Theta\, s + \sin\Theta\, d_{z^2}) - (1/2)\, d_{x^2-y^2} - (1/\sqrt{2})p_y \qquad (9\text{-}7)$$

The value of $\Theta$ must be determined by energy considerations.

Consider octahedral $O_h$ symmetry, for example $SF_6$, where we can solve the problem of six bonds that have charge lobes directed toward the six fluorine atoms. The result is $\Gamma = A_{1g} + E_g + T_{1u}$. We notice that of the usual low lying atomic orbitals only an s-function transforms as $A_{1g}$, only ($d_{x^2-y^2}$, $d_{z^2}$) transform as E, only ($p_x$, $p_y$, $p_z$) transform as $T_{1u}$. For normal molecules these atomic orbitals are well separated in energy from higher lying f, g, etc. orbitals, so that there are no problems of mixing as in Eq. 9-7. Using the labels for the atoms and hybrid function of Fig. 9-3 with directions 5 and 6 along the +z and −z directions, respectively, it is obvious that

$$\begin{bmatrix} \phi_1 \\ \phi_2 \\ \phi_3 \\ \phi_4 \\ \phi_5 \\ \phi_6 \end{bmatrix} = \begin{bmatrix} a & b & 0 & 0 & c & -d \\ a & 0 & b & 0 & -c & -d \\ a & -b & 0 & 0 & c & -d \\ a & 0 & -b & 0 & -c & -d \\ a & 0 & 0 & b & 0 & 2d \\ a & 0 & 0 & -b & 0 & 2d \end{bmatrix} \begin{bmatrix} s \\ p_x \\ p_y \\ p_z \\ d_{x^2-y^2} \\ d_{z^2} \end{bmatrix} \qquad (9\text{-}8)$$

Thus, $a = 1/\sqrt{6}$, $b = 1/\sqrt{2}$, $c = 1/2$, $d = 1/\sqrt{12}$. The only function that requires a comment is $d_{z^2} \propto 3\cos^2\theta - 1$. In the xy-plane −1 is

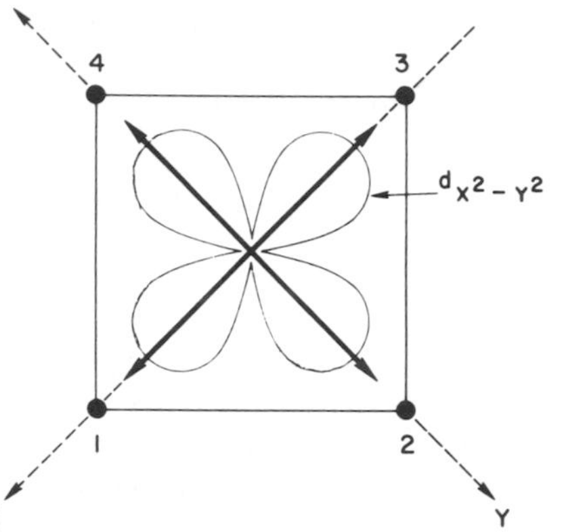


**Fig. 9-3** An $AB_4$ molecule having $D_{4h}$ symmetry showing four orbitals on the A-atom.

obtained while for $\theta = 0$, +2 is obtained. Thus $-d_{z^2}$ is needed to get a positive phase for bonding in the xy-plane so the bonding along the z-axis will be twice as effective as in the xy-plane for this function.

As a last example, consider a planar $AB_3$ ($D_{3h}$) molecule such as $CO_3^{2-}$ as in Fig. 9-4. $\Gamma_{trig} = A_1' + E'$, resulting in a number of possible combinations of atomic orbitals to make up the hybrid function. Some possible combinations are $sp^2$, $dp^2$, $sd^2$, $d^3$. Writing out the hybrid function for $sp^2$, we have

$$\begin{bmatrix} \phi_1 \\ \phi_2 \\ \phi_3 \end{bmatrix} = \begin{bmatrix} a & 2b & 0 \\ a & -b & c \\ a & -b & -c \end{bmatrix} \begin{bmatrix} s \\ p_x \\ p_y \end{bmatrix} \tag{9-9}$$

Hence, $a = 1/\sqrt{3}$, $b = 1/\sqrt{6}$, $c = 1/\sqrt{2}$. Note that the $p_z$-function is not involved with the hybrid function at all. The same will be true in all planar molecules. Thus, the $p_z$-function will be available to form bonds with other neighboring atoms or can be used for $\pi$-bonding, which we will now discuss.

## 9-5 $\pi$-Hybrid Functions

### a. General

So far we have been discussing hybrid functions that would cause the lobes of charge to be directed toward the neighboring atom. These are called $\sigma$-orbitals and give rise to $\sigma$-bonding. We now want to consider other types of orbitals.

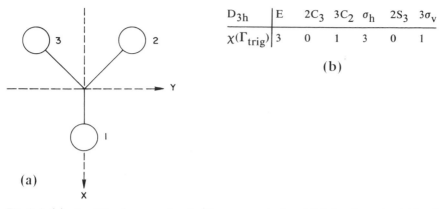

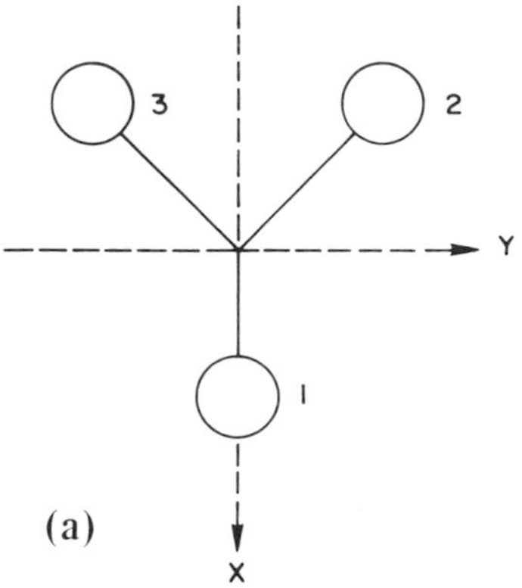

(a)

| $D_{3h}$ | E | $2C_3$ | $3C_2$ | $\sigma_h$ | $2S_3$ | $3\sigma_v$ |
|---|---|---|---|---|---|---|
| $\chi(\Gamma_{trig})$ | 3 | 0 | 1 | 3 | 0 | 1 |

(b)

**Fig. 9-4** (a) An $AB_3$ planar molecule ($D_{3h}$ symmetry) and (b) the character table of the reducible representation formed from the hybrid orbitals.

A **nodal surface** is the locus of points for which the wave function is zero. For $p_z$ the nodal surface is the xy-plane. The surface is formed because the wave function is changing sign. A $\sigma$-bond is defined as one for which the bond direction does not contain a nodal plane. A **$\pi$-bond is defined** as one for which the bond direction contains one nodal plane ($\delta$-two nodal planes, etc.).

Figure 9-5 shows what happends if two $p_z$-wave functions from different atoms are brought together along the x-axis. As pictured, overlap is obtained and there is one nodal surface, the xy-plane. For a given internuclear distance there is usually less overlap of the $\pi$-orbitals than the $\sigma$-orbitals ($p_x$) as can be seen in the figure. Thus, the $\pi$-bonds usually are not as strong as $\sigma$-bonds. It is just this feature that makes them so important in the study of chemical reactions. Since the electrons in the $\pi$-bond are not so tightly bound, the bond can be broken and these electrons can be used to bond with other complexes resulting in new molecules.

### b. Double and triple bonds

The molecule ethylene, $C_2H_4$ is schematically shown in Fig. 9-6. Each central carbon atom forms a trigonal $sp^2$ hybrid function, Eq. 9-9, and bonds to two hydrogens and one carbon as shown. Each of these bonds is filled with two electrons of opposite spin. This leaves an unpaired $p_z$-electron on each carbon atom. As shown in Fig. 9-5, these orbitals overlap and lower the energy of the entire molecule. This will also cause the carbon–carbon distance to become shorter in ethylene than

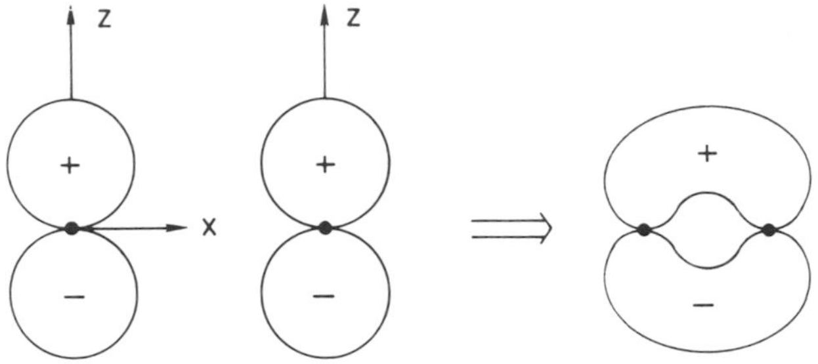


**Fig. 9-5** Two $p_z$ orbitals forming a $\pi$-bond.

if there was no $\pi$-bonding. Carbon atoms connected by one $\sigma$-bond and one $\pi$-bond are said to be "double bonded" and the formula is often written as $H_2C = CH_2$. Since energy is needed to break the $\pi$-bond, we can see why ethylene is a planar molecule and does not freely rotate about the double bond axis.

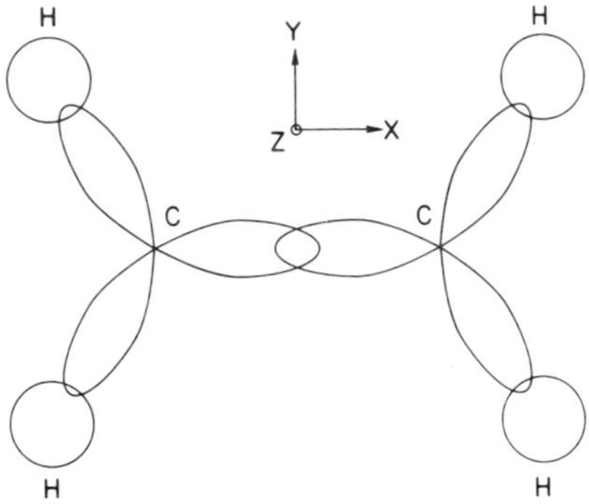


**Fig. 9-6** A "top" view of an ethylene molecule ($C_2H_4$) showing the hybrid $\sigma$-bonds. The $\pi$=bonds are not shown.

By extending these ideas, we can see how the linear molecule acetylene $C_2H_2$ has a triple bond between the carbon atoms. If the axis of the molecule is taken as the x-axis, then each carbon atom can have the two $s \pm p_x$ hybrid functions as in Eq. 9-1 overlapping with the atoms on its right and left. Thus, each carbon is $\sigma$-bonded to a hydrogen and a carbon. These leave each carbon with unpaired $p_y$ and $p_z$-electrons. Of course, these electrons have orbitals perpendicular to the axis of the molecule. As in the double bond case, we expect overlap of these electrons on the two carbon atoms to lower the energy of the molecule and pull the carbon atoms closer together (to increase the overlap). Then for a triple bond we write the chemical formula as HC≡CH as in acetylene. The carbon–carbon distance for a single bond (as in diamond) is 1.54Å; for a double bond, 1.33Å; for a triple bond, 1.20Å. Similarly, the frequency of vibration of the carbon–carbon bond will increase since the "spring" associated with the bond becomes stronger as one progresses from single to double to triple bonds.

### c. $\pi$-Hybrid orbitals

In Section 9-5b we discussed actual $\pi$-bonding for two very simple molecules. In this part of the section we go back to the group theoretical aspects of forming a $\pi$-orbital on a central atom and ignore the details of forming the actual bonds with these orbitals. However, we will consider these details if the central atom has enough electrons to half fill the $\pi$-orbitals in anticipation of an electron of opposite spin coming from another atom. To sec how the $\pi$-hybrid functions are obtained, we consider an example.

Consider a planar $AB_4$ molecule as in Section 9-4 and Fig. 9-3. Previously a set of $\sigma$-hybrid functions of the central atom were found that pointed toward the four atoms 1–N. We still want to find a set of hybrid functions of the central atom but this time they must be $\pi$-hybrid functions and we would like them to have a correct symmetry to be able to overlap with the four atoms 1–4. Thus they should look like the eight arrows shown in Fig. 9-7 that would be labeled $1_{\parallel},\ldots,4_{\parallel}$ and $1_{\perp},\ldots,4_{\perp}$. The result would be very similar to the acetylene example in the last section. Namely, the bond between the central A-atom and atom 1 would be a $\sigma$-bond as determined in Section 9-4 and two $\pi$-bonds, the hybrid functions for which we will now determine. Of course, for this problem there is a symmetry difference for $\pi$-orbitals between A and 1 in the xy-plane and a $\pi$-orbital parallel to the z-axis. This difference is due to the rest of the molecule, which does not exist in the acetylene case. The <u>four functions</u> centered on the A-atom that have lobes that can overlap the four arrows $1_{\perp},\ldots,4_{\perp}$ <u>transform among themselves under all the symmetry operations of the group</u> $D_{4h}$. Therefore they form a representation of the group the character of which is given in Fig. 9-7 as $\chi(\Gamma_{\perp})$. Also given is the character of the representation of the four functions that can overlap the four arrows $1_{\parallel},\ldots,4_{\parallel}$, $\chi(\Gamma_{\parallel})$. (The entries should be checked by the reader.) The reduction of the representations by inspection of Eq. 4-7 yields

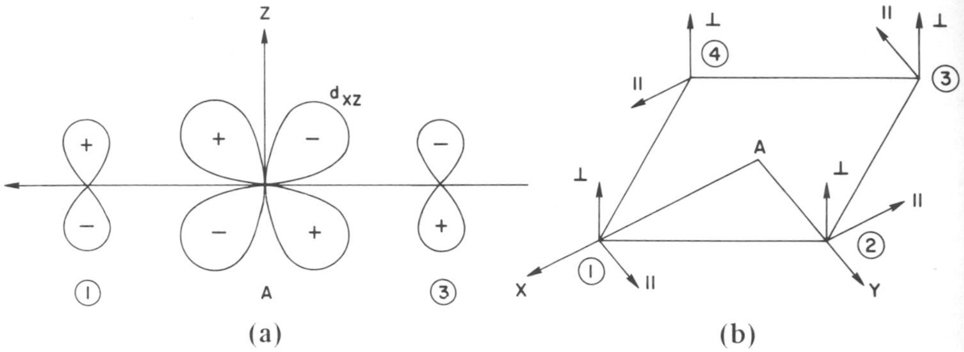


| (c) | E | $2C_4$ | $C_2$ | $2C_2'$ | $2C_2''$ | i | $2S_4$ | $\sigma_h$ | $2\sigma_v$ | $2\sigma_d$ |
|---|---|---|---|---|---|---|---|---|---|---|
| $\chi(\Gamma_\perp)$ | 4 | 0 | 0 | −2 | 0 | 0 | 0 | −4 | 2 | 0 |
| $\chi(\Gamma_\parallel)$ | 4 | 0 | 0 | −2 | 0 | 0 | 0 | 4 | −2 | 0 |

**Fig. 9-7** (a) A planar $AB_4$ molecule as in Fig. 9-3 but the coordinates used in the $\pi$-bonding discussion are shown. (b) A $d_{xz}$ orbital on the A-atom is shown. How this orbital $\pi$-bonds the atoms 1 and 3 can be seen. (c) The characters of the reducible representations for the $\pi$-orbitals.

$$\Gamma_\perp = A_{2u} + B_{2u} + E_g, \quad \Gamma_\parallel = A_{2g} + B_{2g} + E_u \tag{9-10}$$

First consider atomic orbitals that can be used for $\pi$-orbitals corresponding to $\Gamma_\parallel$. Of the low lying atomic orbitals (s, p, or d) that transform as the three irreducible representations in Eq. 9-10, we find from the character tables $A_{2u}$ – $p_z$; $B_{2u}$ – none; $E_g$ – ($d_{xz}$, $d_{yz}$). Finding $p_z$ is no surprise since we expected it to be used for $\pi$-hybrids from Section 9-5b. Figure 9-7 shows a $d_{xz}$ and how it is capable of overlap with, for example, $p_z$ atomic orbitals on atoms 1 and 3. $d_{yz}$ behaves similarly with respect to atoms 2 and 4. Notice that we only have three and not four possible low lying atomic orbitals for the perpendicular $\pi$-hybrid functions because there are no low lying atomic orbitals that transform as $B_{2u}$. Thus the $d_{xz}$ would contribute on the average half an electron to the A-1 bond and half to the A-3 bond. Similarly, $d_{yz}$ contributes half to the A-2 and A-4 bonds. The $p_z$ contributes 1/4 to each bond; thus, each $\pi$-bond on the average will only have 3/4 of an electron from the central atom.

Consider the atomic orbitals corresponding to $\pi$-orbitals for $\Gamma_\parallel$ in Eq. 9-10. We find $A_{2g}$ – none; $B_{2g}$ – $d_{xy}$; $E_u$ – ($p_x$, $p_y$). Notice that ($p_x$, $p_y$) appear in the $\pi$-case as well as the previously considered $\sigma$-orbital case. This is to be expected in this example. $p_x \pm s$ is obviously a good hybrid function for large overlap with atoms 1 and 3 in Fig. 9-7. At the same time $p_x$ can obviously form $\pi$-bonds with atoms 2 and 4. The chemical approach is to realize that $\sigma$-bonds are usually stronger (lower in energy) that $\pi$-bonds. So the approximation is made that the $(p_x, p_y)$ electrons are used up in $\sigma$-bonds leaving for this example only the $d_{xy}$ atomic orbital. Thus, if this orbital is low enough in energy one can use it to $\pi$-bond to the four atoms 1–4. On the average the central ion can contribute 1/4 electron to each bond. However, it must be remembered that using up the $(p_x, p_y)$ electrons in $\sigma$-bonding is only an approximation. All that the symmetry shows is that the $(p_x, p_y)$ orbitals can be used for both $\sigma$- and $\pi$-bonding, and the way they should be divided depends on the details of the problem. This can be similar to the problem of s and $d_{z^2}$ orbital mixing in Eq. 9-7.

## 9-6 Comment on Hybrid Orbitals (Slater Determinant)

This section can be skipped by most readers. The advantage of hybrid orbitals is that they are localized in space and so lead in a natural way to thinking in terms of a chemical bond. The question arises, do these hybrid orbitals have any relation to an antisymmetric eigenfunction of the atom that, for example, one might write as a Slater determinant? The answer is yes. The hybrid orbitals give functions that are in principle just as valid as the "original" nonhybridized orbitals. (However, the problem still arises, is the resultant Slater determinant an eigenfunction as discussed in Sections 9-1 and 9-2?)

To show the equivalence of the "original" orbitals and a hybridized set, consider four electrons (labeled 1–4) in two orbitals ($\psi_a$ and $\psi_b$) with spin up ($\alpha$) or spin down ($\beta$). The Slater determinant for the paired electrons with opposite spin is

$$\Psi = (1/\sqrt{4!}) \begin{vmatrix} \psi_a\alpha(1) & \psi_a\beta(1) & \psi_b\alpha(1) & \psi_b\beta(1) \\ \psi_a\alpha(2) & \cdot & \cdot & \cdot \\ \psi_a\alpha(3) & \cdot & \cdot & \cdot \\ \psi_a\alpha(4) & \cdot & \cdot & \psi_b\beta(4) \end{vmatrix} \qquad (9\text{-}11)$$

By adding column 3 to 1 as well as column 4 to 2, taking out of the determinant a factor 2 from the resulting column 1 (and 3), substracting from column 3 column 1 (and from 4 column 2), and putting the factor 4 back into the determinant, we arrive at Eq. 9-12 which is equal to that in Eq.

9-11. This expression in Eq. 9-12 should be multiplied by $(1/4!)^{1/2}$

$$\begin{vmatrix} \frac{1}{\sqrt{2}}(\psi_a + \psi_b)\alpha(1) & \frac{1}{\sqrt{2}}(\psi_a + \psi_b)\beta(1) & \frac{1}{\sqrt{2}}(\psi_b - \psi_a)\alpha(1) & \frac{1}{\sqrt{2}}(\psi_b - \psi_a)\beta(1) \\ \vdots & \vdots & \vdots & \vdots \\ \frac{1}{\sqrt{2}}(\psi_a + \psi_b)\alpha(4) & & & \frac{1}{\sqrt{2}}(\psi_b - \psi_a)\beta(4) \end{vmatrix} \tag{9-12}$$

This is just the Slater determinant if the two orbitals were $(\psi_a + \psi_b)\sqrt{2}$ and $(\psi_b - \psi_a)/\sqrt{2}$, and if two electrons with opposite spin were in each orbital. So in principle the description of the total wave function in terms of Eq. 9-11 or 9-12 is equally valid. The orbitals in Eq. 9-12 are just like the hybrid function in Eq. 9-1. This emphasizes the point made at the end of Section 9-1 that in principle the hybridization and LCAO–MO approaches are interchangeable. When the wave equation is really solved and eigenfunctions are obtained, it is found that hybridized orbitals are reasonable approximations in the sense that the true eigenfunctions show that charge is built up between the atoms in a manner similar to what is found by the hybridized orbitals.

## Notes

Pauling's original paper is clearly written and interesting to read from an informational as well as historical point of view [J. Amer. Chem. Soc. **53**, 1367 (1931)].

Other particularly useful or interesting references for this and the next chapter are F. A. Cotton (Chapters 7 and 8); C. A. Coulson "Valence" (Oxford Univ. Press, London and New York, 1952), particularly Chapter 8; M. J. S. Dewar "The Molecular Orbital Theory of Organic Chemistry" (McGraw-Hill, New York, 1969); R. B. Woodward and R. Hoffmann; C. J. Ballhausen "Ligand Field Theory" (McGraw-Hill, New York, 1962); McWeeny (Section 7-7 and 7-8). There are many other references that can be consulted.

Eyring, Walter, and Kimball, "Quantum Chemistry," p. 231 (Wiley, New York, 1944) have an extensive list of possible hybrid functions for many different geometric arrangements.

**Problems**

**1.** Instead of $AB_6$ in $O_h$ symmetry, consider $AB_4C_2$ with $D_{4h}$ symmetry. How will the $\sigma$-hybrid functions, Eq. 9-8, change?

**2.** The molecule $PF_5$ has $D_{3h}$ symmetry. What are the possible $\sigma$-hybridization schemes? Write out a set of $\sigma$-hybrid functions.

**3.** For the molecule $AB_8$ with $O_h$ symmetry, what are the hybrid functions of the central atom that could be used for $\sigma$-bonding? Use f-functions. For the molecule $AB_{12}$ with $O_h$ what are the hybrid functions of the central atom that can be used for $\sigma$-bonding? Use f-functions. [Hint: See G. Burns and J. D. Axe in "Optical Properties of Ions in Crystals," Wiley (Interscience), New York, 1967.]

**4.** Work out the $\pi$-hybrid functions that would be useful for $\pi$-bonding for an $AB_3$ molecule in $D_{3h}$ symmetry.

**5.** Repeat Problem 4 for the octahedral molecule $AF_6$.

Chapter 10

# MOLECULAR ORBITAL THEORY

*But we ought first to examine ourselves, next the business which we wish to undertake, next those for whose sake or with whom we have to act. Above all things it is necessary to have a proper estimate of oneself, because we usually think that we can do more than we really can.*

*Seneca, "On Tranquility"*

In this chapter the fundamentals of the molecular orbital (MO) theory are discussed. It is applied to two different kinds of problems. The first is the transition metal complex, for example $[FeF_6]^{3-}$. The MO approach enables the eigenfunction of this complex to be obtained, and shows how the eigenvalues can be calculated. This problem is of great interest for the iron series transition metals ions. The second application of MO theory is to $\pi$-electrons of conjugated hydrocarbons. In this problem the eigenvalues are usually estimated with the aid of the simple Hückel approximation although more sophisticated MO treatments are also used. This problem is of wide interest in organic chemistry. These two quite different problems are treated in this one chapter since they both illustrate straightforward uses of the MO theory.

## 10-1 Hydrogen Molecular Ion

The $H_2^+$ molecular ion is to molecules what the hydrogen atom is to atoms. Therefore we discuss it to form our ideas of solution of the secular equation before going on to more complex systems. Consider two hydrogen nuclei separated by a distance $r_{12}$ with only one electron shared by the nuclei. The Hamiltonian is

$$\mathbf{H} = -\frac{\hbar^2}{2m}\nabla^2 - \frac{e^2}{r_1} - \frac{e^2}{r_2} + \frac{e^2}{r_{12}} \tag{10-1}$$

with $r_1$ and $r_2$ the distances from each nucleus to the electron. Since we only want to determine the low lying energy levels of the molecule, we take for the atomic orbitals the normalized 1s wave function centered on each nucleus $\phi_1$ and $\phi_2$ to form the LCAO–MO as in Eq. 7-37

$$\psi = a\phi_1 + b\phi_2 \tag{10-2}$$

Taking $\beta = H_{12}$ $(\equiv\langle\phi_1 | \mathbf{H} | \phi_2\rangle)$, the secular determinant Eq. 7-41 is

$$\begin{vmatrix} H_{11} - E & \beta - SE \\ \beta - SE & H_{22} - E \end{vmatrix} = 0, \qquad (H_{11} - E)(H_{22} - E) = (\beta - SE)^2 \tag{10-3}$$

where $S = \langle\phi_1 | \phi_2\rangle$. For our problem, $H_{11} = H_{22}$ hence the two solutions obtained for the energy are

$$E_{\pm} = (H_{11} \pm \beta)/(1 \pm S) \tag{10-4}$$

Since $H_{11}$ and $\beta$ are negative, $E_+$ will be lower in energy than $E_-$. The normalized wave function Eq. 10-2 can be determined for $E_+$ by substituting $E = E_+$ into the secular equation, Eqs. 10-3 and 7-39 and normalizing $a^2 + b^2 + 2abS = 1$ to determine $\psi_+$. The wave functions corresponding to $E_+$ and $E_-$, respectively, are

$$\psi_+ = \frac{1}{\sqrt{2}}\frac{\phi_1 + \phi_2}{(1 + S)^{1/2}} \qquad \psi_- = \frac{1}{\sqrt{2}}\frac{\phi_1 - \phi_2}{(1 - S)^{1/2}} \tag{10-5}$$

Figure 10-1 shows the energy level diagram for the hydrogen molecular ion and the wave function along the internuclear axis. $\psi_+$ is called a **bonding state** because charge density is accumulated between the nuclei to form a bond, while $\psi_-$ is called an **antibonding state** because there is a "node of charge" between the two nuclei. Since the $\phi_1 = \phi_2 = 1s$ hydrogen wave function is known, $E_+$ can be evaluated as a function of $r_{12}$ to determine the $r_{12}$ corresponding to the minimum energy. Experimental results are $r_{12} = 1.06$Å and 2.791 eV. For the above simple theory, 1.32Å and 1.76 eV are obtained. (See Eyring, Walter, and Kimball, Section 11b.) As can be seen, this simple theory is only in modest agreement with experiment. The lack of agreement arises from the severely truncated expansion for the wave function, which in this case consists of only the 1s hydrogen atom wave function.

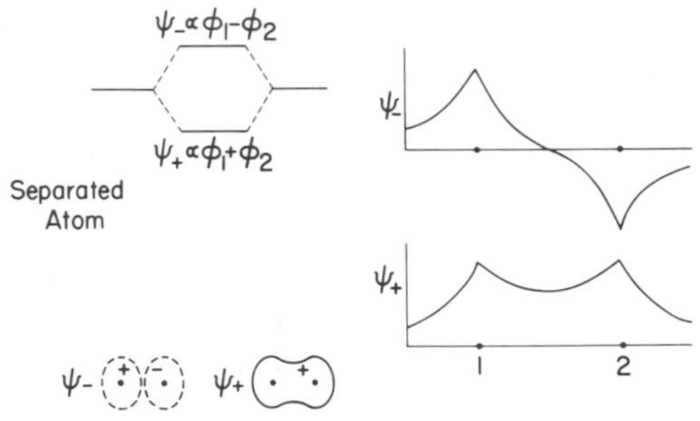


**Fig. 10-1** The energy level diagram for the hydrogen molecule ion. Diagrams of the bonding and antibonding wave functions are shown in two different formats.

To get better agreement with experiment, either the basis set can be expanded or other parameters can be varied using variation of parameter techniques. For example, if the nuclear charge is taken as a parameter and varied to minimize the energy, 1.06Å and 2.25 eV are obtained using a nuclear charge of 1.228 instead of 1. (The reason for allowing a variation in the nuclear charge is that the electron "feels" both nuclear charges.) If the basis set Eq. 10-2 included 2p orbitals as well as 1s, then 1.06Å and 2.71 eV are obtained. Thus, we see that as more atomic orbitals are used in the LCAO–MO better results are obtained.

The bonding orbital has a larger binding energy (larger negative value) than the antibonding orbital because it has a greater charge density between the two nuclei. Thus, an electron in the bonding orbital has a greater probability of being attracted to both nuclei than an electron in the antibonding orbital. (It is often said that the antibonding orbital has less binding energy because the extra node leads to a larger kinetic energy. However this is not true because the virial theorems require that the kinetic energy equal half the magnitude of the potential energy.)

The result from the above simple calculation, namely a bonding and an antibonding state, is a general one. This result will also be found when the two atoms are dissimilar. In fact, $\phi_1$ can be an atomic orbital of a central atom and $\phi_2$ can be a linear combination of atomic wave functions of many atoms that bond directly to atom 1. The only requirement on $\phi_2$ is that the off-diagonal elements in Eq. 10-3 are nonzero. For example, we can discuss the comples $[FeF_6]^{3-}$ where Fe is atom 1 and $\phi_2$ is formed from a linear combination of atomic orbitals of the six $F^{1-}$ ions. The off-diagonal elements will be nonzero if $\phi_1$ and $\phi_2$ transform as the same irreducible representation. If they transform as different irreducible representations, the off-diagonal terms will be zero and there will be no interaction. If there is an interaction, then a bonding and antibonding state will form as in Eqs. 10-4 and 10-5. This is similar to the results in Section 7-5b, particularly Eq. 7-30.

## 10-2 Simple MO Theory

In this section we return to the solution of Eq. 10-3 with $H_{11} \neq H_{22}$. This could be the problem of a central ion surrounded by different ions, $[FeF_6]^{3-}$ or else it could be a problem of dissimilar atoms such as HCl. We only require that $\phi_1$ and $\phi_2$ transform as the same irreducible representation so that the off-diagonal terms are nonzero. If $\phi_1$ is an s-wave function of $Fe^{3+}$, and $[FeF_6]^{3-}$ has octahedral symmetry, then $\phi_2 \propto (1 + 2 + 3 + 4 + 5 + 6)$ where each number represents an s-wave function of each of the six $F^-$ ions.

In Eq. 10-3 for convenience, assume $H_{11} > H_{22}$ ($H_{22}$ is more negative as in Fig. 10-2). This means that the $F^-$ states are more stable then the $Fe^{3+}$ states which is found experimentally. If the effect of bonding is small, a perturbation solution to second order in energy can be obtained

$$E_- = H_{11} - \frac{(\beta - H_{22}S)^2}{H_{22} - H_{11}} \equiv E^*, \qquad E_+ = H_{22} + \frac{(\beta - H_{22}S)^2}{H_{22} - H_{11}} \tag{10-6}$$

This is obtained by replacing the off-diagonal ES terms by the appropriate $H_{ii}S$. Note that $H_{22} - H_{11}$ is a negative number, so $E_+$ is more negative than $H_{22}$ alone. Similarly $E_-$ is less negative than $H_{11}$ alone. This is shown in Fig. 10-2 where stabilization due to bonding corresponds to a repulsion of levels, as was first encountered in Eq. 7-30.

The wave functions for the bonding and antibonding ($E_+$ and $E_-$) states are

$$\psi_- = N_-(\phi_1 - \lambda\phi_2), \qquad \psi_+ = N_+(\phi_2 + \gamma\phi_1) \tag{10-7}$$

where $N_\pm$ are normalization constants, and in our perturbation expansion we expect $\lambda$ and $\gamma$ to be small compared to one. From $\langle\psi_+ | \psi_-\rangle = 0$

$$\lambda = \frac{\gamma + S}{1 + \gamma S} \approx \gamma + S \tag{10-8}$$

By inserting $E_+$ from Eq. 10-6 into the secular equation Eq. 7-39, $\gamma$ can be determined; the coefficient $\lambda$ can be determined by inserting $E_-$. The

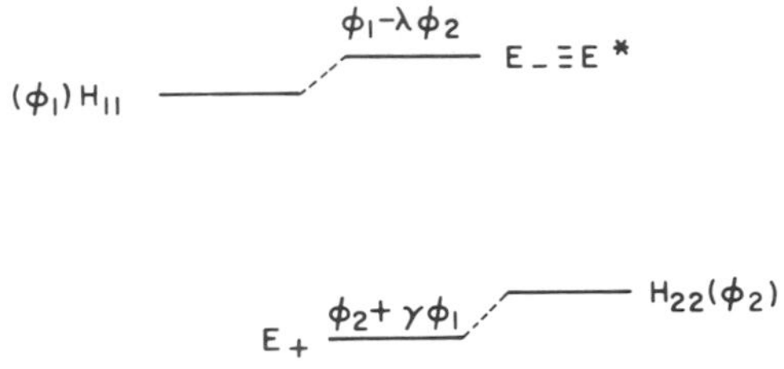


**Fig. 10-2** An energy diagram showing, in general, the formation of a bonding and antibonding level.

results are

$$\lambda = -\frac{\beta - H_{11}S}{H_{11} - H_{22}}, \quad \gamma = -\frac{\beta - H_{22}S}{H_{11} - H_{22}} \tag{10-9}$$

Two problems remain. The first is to find the linear combination of atomic orbitals of the bonding ions, the $F^-$ in $[FeF_6]^{3-}$, that has the correct symmetry for use in $\phi_2$. This is a straightforward group theory problem and will be considered in this chapter for various systems. The second problem is the evaluation of matrix elements and overlap S. The overlap can be calculated in a straightforward manner (see the Notes). The diagonal $H_{ii}$ and off-diagonal $\beta = H_{ij}$ matrix elements can be calculated in principle if the Hamiltonian is known. However, this is a very involved problem, and empirical approaches are often used. The Wolfsberg–Helmholtz approach approximates the off-diagonal terms by $H_{ij} = gS(H_{ii} + H_{jj})/2$ where g is usually taken as 2.0 or 1.75. The diagonal matrix elements can be taken as experimental ionization energies. For details of such calculations references in the notes should be consulted.

## 10-3 Transition Metal Complexes

### a. Orbitals

In this section we consider the general group theoretical problem of finding the LCAO for $\phi_2$. For example, in the complex $[FeF_6]^{3-}$ $\phi_1$ is an atomic orbital of the central $Fe^{3+}$ ion and $\phi_2$ is a LCAO of the ligand (neighboring) ions which are the six $F^-$ ions.

Figure 10-3 shows the six coordinate systems of the ligands used to describe the $\sigma$- and $\pi$-bonds of the six X ions to the M ion in the $MX_6$

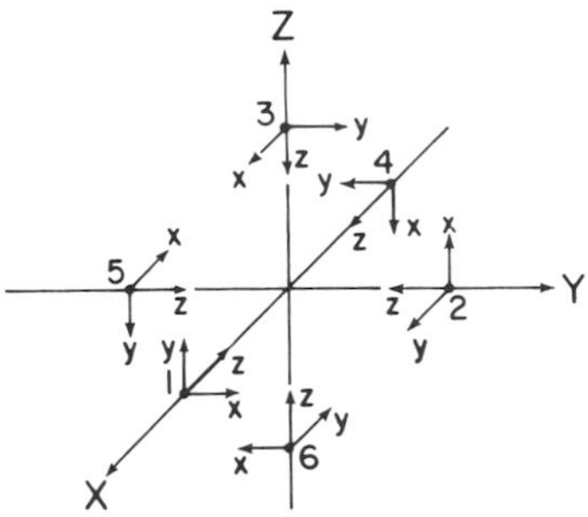

(a)

| $O_h$ | E | $8C_3$ | $6C_2$ | $6C_4$ | $3C_2$ | i | $6S_4$ | $8S_6$ | $3\sigma_h$ | $6\sigma_d$ |
|---|---|---|---|---|---|---|---|---|---|---|
| $\chi(\Gamma_\sigma)$ | 6 | 0 | 0 | 2 | 2 | 0 | 0 | 0 | 4 | 2 |
| $\chi(\Gamma_\pi)$ | 12 | 0 | 0 | 0 | −4 | 0 | 0 | 0 | 0 | 0 |

(b)

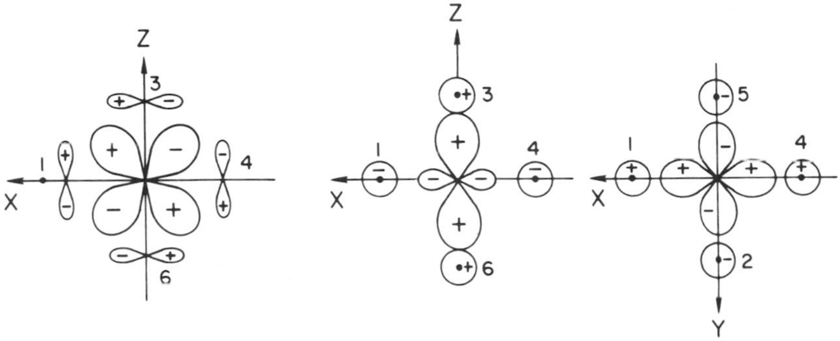

(c)

**Fig. 10-3** (a) Coordinates on the six ligands in the octahedral $MX_6$ problem. (b) The $\sigma$- and $\pi$-reducable representations of the linear combination of ligand atomic orbitals. (c) Some of the resulting combination of orbitals.

system with $O_h$ symmetry. The z-axes always point to the central M ion and each system is left-handed while that of the central ion is right-handed; this is the conventional way (see the Notes).

We shall first consider $\sigma$-bonding of the six X ions to the M ion. The $\sigma$-bonds of the X ions with M would be formed by s or $p_z$-atomic orbitals of the X ions, where $p_z$ refers to the particular X-ion coordinate system. Thus each of the six ligand $p_z$ wave functions point toward the central ion. To form a $\sigma$-bond to the central ion, a charge lobe of positive phase must point toward the central ion. These charge lobes (s or $p_z$ functions) can be represented by the six z-coordinates $z_1, z_2, \ldots, z_6$. These six functions must transform among themselves under all the symmetry operations of the group, $O_h$ symmetry in this case. The 6 × 6 matrix representation formed is obviously reducible. Figure 10-3 shows the character of the representation for each symmetry operation of $O_h$ for σ-bonding under $\chi(\Gamma_\sigma)$. Under the identity E the six functions $z_1, z_2, \ldots, z_6$ transform into themselves hence the character of the representation is 6. Under the $C_3[111]$ rotation, $z_1 \to z_2$, $z_2 \to z_3$, $z_3 \to z_1$, $z_4 \to z_5$, etc., so the character of the representation is zero. The next nonzero character comes from the $C_4$ operations. For $C_4[001]$, $z_1 \to z_2$, $z_2 \to z_4$, $z_3 \to z_3$, etc., and the character is 2. It is a general rule that the character of the representation will always be given by the number of unshifted functions. Reducing the representation by Eq. 4-7 or by inspection yields

$$\Gamma_\sigma = A_{1g} + E_g + T_{1u} \tag{10-10}$$

Of course, this is the same result as obtained in the hybridization problem in Section 9-4 for octahedral symmetry $SF_6$. That problem was concerned with charge lobes with a positive phase on the central ion directed to the six neighbors. That problem from a group theory point of view, is the same as the LCAO problem because in both cases the charge lobes transform among themselves in precisely the same way under all the symmetry operations of the group.

It now remains to obtain the actual LCAO that transform as each of the irreducible representations in Eq. 10-10. With a little experience this can be done by inspection. We shall use projection operator techniques (Section 5-7) to gain this experience. For the LCAO that transform as the one-dimensional $A_{1g}$, the character projection operator Eq. 5-26 will give the function uniquely. Remembering that the sum is over all the symmetry operations (48 for $O_h$) we obtain

$$\{V^{A_{1g}}\}z_1 = (1/48)\,(7)\,(z_1 + z_2 + z_3 + z_4 + z_5 + z_6) \tag{10-11}$$

Note that this is not a normalized function. If each of the atomic orbitals, $z_1$, etc., is normalized then instead of the factor 7/48 a factor $1/\sqrt{6}$ is needed. This will be used when the LCAO are tabulated. To obtain the two LCAO that are partners of the $E_g$ irreducible representation in a straightforward manner, either Eq. 5-22 or 5-25 can be used. However, the character projection operator will be used here, Eqs. 5-26 and 27, since it is easy to apply and it is worthwhile to gain some experience in manipulating the results to yield orthogonal partners.

$$\{V^{E_g}\}z_1 = (2/48)(4)(2z_1 + 2z_4 - z_2 - z_3 - z_5 - z_6) \equiv M/6$$
$$\{V^{E_g}\}z_2 = (2/48)(4)(2z_2 + 2z_5 - z_1 - z_3 - z_4 - z_6) \equiv N/6 \qquad (10\text{-}12)$$

These are not orthogonal; however, multiplying the sum of these equations by $-1$ and taking their difference, one has

$$-(M + N) = 2z_3 + 2z_6 - z_1 - z_2 - z_4 - z_5,$$
$$M - N = z_1 + z{+}_4 - z_2 - z_5 \qquad (10\text{-}13)$$

These functions are of the same form as the d-wave functions that are also partners of the $E_g$, namely $(2z^2 - x^2 - y^2, x^2 - y^2)$. Functions that transform as $T_{1u}$ are also obtained easily from the character projection operator.

$$\{V^{T_{1u}}\}z_1 = (3/48)(8)(z_1 - z_4)$$
$$\{V^{T_{1u}}\}z_2 = (3/48)(8)(z_2 - z_5)$$
$$\{V^{T_{1u}}\}z_3 = (3/48)(8)(z_3 - z_6) \qquad (10\text{-}14)$$

These functions look just like the functions (x, y, z) which are also partners of $T_{1u}$. Table 10-1 lists the LCAO for $\sigma$-bonding, normalized to one. The point is that we can now write a wave function for the complex having the proper symmetry for an eigenfunction (it transforms as an irreducible representation of the group) as well as being a LCAO-MO. Thus, as in Eq. 10-2, the wave function for the entire complex is $\psi = a\phi_1 + b\phi_2$, where $\phi_1$ consists of atomic orbitals of the central ion and $\phi_2$ is derived from atomic orbitals of the ligand atoms. For the $\sigma$-bonding MO the three types of functions allowable are

$$\psi(A_{1g}) = as + (b/\sqrt{6})(z_1 + z_2 + z_3 + z_4 + z_5 + z_6)$$
$$\psi(E_g) = a'd_{z^2} + (b'/\sqrt{12})(2z_3 + 2z_6 - z_1 - z_2 - z_4 - z_5)$$
$$\psi(E_g) = a'd_{x^2-y^2} + (b'/2)(z_1 + z_4 - z_2 - z_5)$$
$$\psi(T_{1u}) = a''p_x + (b''/\sqrt{2})(z_1 - z_4) \qquad (10\text{-}15)$$

where we have not bothered to include the last two parameters of $T_{1u}$. The coefficients a′ and b′ in the middle two equations are the same because both equations are for partners of the $E_g$ irreducible representation. Similarly for the three LCAO-MO that transform as $T_{1u}$ the coefficient of the p-functions a″ will be the same as well as b″. For each of the three

**Table 10-1** Symmetry functions for the $MX_6(O_h)$ problem

| Symmetry | Central atom | Ligand orbitals $\sigma$ orbitals | Ligand orbitals $\pi$ orbitals |
|---|---|---|---|
| $A_{1g}$ | s | $\frac{1}{\sqrt{6}}(z_1 + z_2 + z_3 + z_4 + z_5 + z_6)$ | |
| $E_g$ | $d_{x^2-y^2}$ | $(1/2)(z_1 - z_2 + z_4 - z_5)$ | |
| | $d_{z^2}$ | $\frac{1}{3\sqrt{2}}(2z_3 + 2z_6 - z_1 - z_2 - z_4 - z_5)$ | |
| $T_{1u}$ | $p_x$ | $\frac{1}{\sqrt{2}}(z_1 - z_4)$ | $(1/2)(x_3 + y_2 - x_5 - y_6)$ |
| | $p_y$ | $\frac{1}{\sqrt{2}}(z_2 - z_5)$ | $(1/2)(x_1 + y_3 - x_6 - y_4)$ |
| | $p_z$ | $\frac{1}{\sqrt{2}}(z_3 - z_6)$ | $(1/2)(x_2 + y_1 - x_4 - y_5)$ |
| $T_{2g}$ | $d_{xz}$ | | $(1/2)(x_3 + y_1 + x_4 + y_6)$ |
| | $d_{yz}$ | | $(1/2)(x_2 + y_3 + x_6 + y_5)$ |
| | $d_{xy}$ | | $(1/2)(x_1 + y_2 + x_5 + y_4)$ |
| $T_{2u}$ | $x(y^2-z^2)$ | | $(1/2)(x_3 + x_5 - y_2 - y_6)$ |
| | $y(z^2-x^2)$ | | $(1/2)(x_1 + x_6 - y_3 - y_4)$ |
| | $z(x^2-y^2)$ | | $(1/2)(y_1 + y_5 - x_2 - x_4)$ |

irreducible representations, a 2 × 2 secular determinant must be solved. Since the partners of an irreducible representation are picked to be orthogonal to each other, one 2 × 2 determinant is all that is required for each representation. Each secular determinant will yield two energies which will give two sets of parameters for a and b (or a′ and b′ or a″ and b″). One set corresponds to a bonding orbital and one corresponds to an antibonding orbital as described in Sections 10-1 and 10-2. The types of atomic orbitals of the central ion are clearly shown in Eq. 10-15. It is of course understood that in place of the s-atomic orbital in the first equation a linear combination of s with any other atomic orbital transforming as $A_{1g}$ is allowed by symmetry (see Eq. 9-7). However, we will only think in terms of low lying atomic orbitals to get the low lying MO's. The types of atomic orbitals of the ligands represented in Eq. 10-15 need further comment. Since we are considering $\sigma$-bonding, the z's refer to any type of orbital on the ligand that can form a $\sigma$-bond. Referring to the coordinate system with the origin on the ligand as in Fig. 10-3, the ligand atom orbitals could be s, $p_z$, $d_{z^2}$, etc., and linear combinations of these, depending on what is available or energetically possible.

Similar arguments hold for $\pi$-orbitals. All 12 functions $x_1,\ldots,x_6$, $y_1,\ldots,y_6$ transform among themselves under the symmetry operations of $O_h$. Figure 10-3 shows the character of the representation for each operation.

$$\Gamma_\pi = T_{1g} + T_{2g} + T_{1u} + T_{2u} \tag{10-16}$$

Again the actual orbitals that transform as the individual irreducible representations can be determined as in the $\sigma$-bonding case. The appropriate orbitals are listed in Table 10-1. Figure 10-3 also shows a $d_{xz}$ orbital of the central ion and corresponding orbitals on the ligands that are partners of the $T_{2g}$ irreducible representation. It is immediately obvious that $\pi$-bonding will occur. In fact, the LCAO often can be written by inspection when it is fully realized that the only requirement of the LCAO partners is that they are orthonormal and must overlap with the orbitals which correspond in symmetry to the functions of the central ion. For example, observe that in Fig. 10-3 it is immediately obvious that the function $1 + 4 - 2 - 5$ has the same symmetry as $d_{x^2-y^2}$. In the same way the ligand LCAO's shown in Table 10-1 can all be immediately written down. Try it!

Figure 10-4 shows the coordinate systems of the ligands used to describe $\sigma$- and $\pi$-bonds of the four X ions to the central M ion for the tetrahedral $MX_4$. Table 10-2 lists the appropriate LCAO which will form an MO in the same sense as Eq. 10-15. As can be seen $\Gamma_\sigma = A_1 + T_2$ and $\Gamma_\pi = E + T_1 + T_2$. Note how the description of the orbitals differs from the $sp^3$ hybridization of the central ion orbitals in the last chapter. In the molecular orbital theory an MO is formed by LCAO of the central ion s-orbital with the appropriate LCAO of the ligands. This MO transforms as an irreducible representation, and thus has the symmetry requirement of an eigenfunction of the molecule. The secular determinant for such an orbital can be solved to yield a bonding and antibonding state as in Eq. 10-2 to 10-9. Separately, a LCAO-MO made up of p-orbitals of the central ion and the appropriate LCAO of the ligands is considered. Again the secular determinant can be solved yielding bonding and antibonding states. It can be shown (Dewar, Section 4-10) that in the ground state the

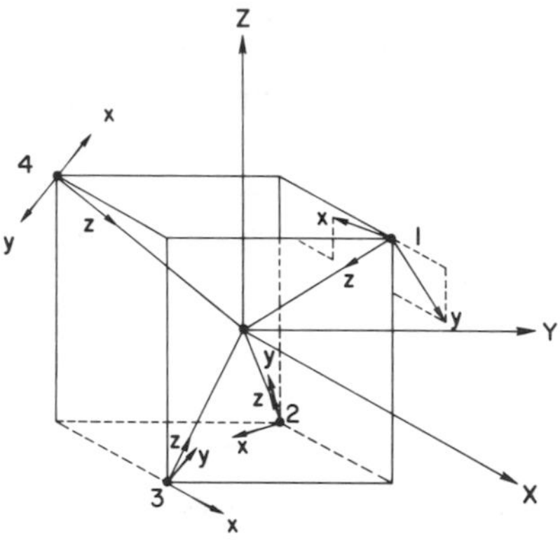


**Fig. 10-4** The coordinate system on the ligands used for the tetrahedral $MX_4$ problem.

**Table 10-2** Symmetry functions for the $MX_4$ ($T_d$) problem

| Symmetry | Central atom | Ligand orbitals σ orbitals | Ligand orbitals π orbitals |
|---|---|---|---|
| $A_1$ | s | $(1/2)[z_1 + z_2 + z_3 + z_4]$ | |
| E | $d_{z^2}$ | | $(1/4)[x_1 + x_2 + x_3 + x_4 - \sqrt{3}(y_1 + y_2 + y_3 + y_4)]$ |
| | $d_{x^2 - y^2}$ | | $(1/4)[y_1 + y_2 + y_3 + y_4 + \sqrt{3}(x_1 + x_2 + x_3 + x_4)]$ |
| $T_1$ | $x(5x^2 - 3r^2)$ | | $(1/4)[y_2 + y_4 - y_3 - y_1 + \sqrt{3}(x_1 + x_3 - x_2 - x_4)]$ |
| | $y(5y^2 - 3r^2)$ | | $(1/2)[y_1 + y_2 - y_3 - y_4]$ |
| | $z(5z^2 - 3r^2)$ | | $(1/4)[y_2 + y_3 - y_1 - y_4 + \sqrt{3}(x_2 + x_3 - x_1 - x_4)]$ |
| $T_2$ | $p_x$ | $(1/2)[z_1 + z_3 - z_2 - z_4]$ | $(1/4)[x_4 + x_2 - x_1 - x_3 + \sqrt{3}(y_4 + y_2 - y_1 - y_3)]$ |
| | $p_y$ | $(1/2)[z_1 + z_2 - z_3 - z_4]$ | $(1/2)[x_1 + x_2 - x_3 - x_4]$ |
| | $p_z$ | $(1/2)[z_1 + z_4 - z_2 - z_3]$ | $(1/4)[x_3 + x_2 - x_1 - x_4 + \sqrt{3}(y_4 + y_1 - y_2 - y_3)]$ |

total energy and total electron distribution of a simple molecule like $CH_4$ is the same if the calculation is a simple LCAO-MO as mentioned or a $sp^3$ hybridization (four localized bonds) approach. However, the hybridization approach should not be extended, for example, to localized CH bonds and properties of excited states. Although the MO theory uses a limited basis set for ease of calculation, it does offer a proper and straightforward approach to compute the eigenfunctions and eigenvalues. This approach also gives excited state wave functions and can easily be improved upon by expanding the basis set of atomic orbitals used to describe the MO. If one-electron properties such as absorption of light or ionization potentials, are to be calculated, MO eigenfunctions such as Eq. 10-15 should be used.

### b. Bonding

We now have LCAO-MO that transform as irreducible representations of the appropriate point group. The next step is to calculate the energies and eigenfunctions by solving the secular equation. The approach is shown in Eqs. 10-2 to 10-9. As discussed in Section 7-6, group theory enables one to factor the secular determinant. The solution of the secular determinant will yield bonding and antibonding states which in general repel each other as shown in Fig. 10-2. The states are then filled with the required number of electrons. Each orbital state can hold two electrons with opposite spin; thus a triply degenerate $T_{1u}$ MO can accomodate six electrons.

We show schematically the results for the $[FeF_6]^{3-}$ case which has $O_h$ symmetry. Consider the 3d-electrons of the $Fe^{3+}$ ion which comprises a semifilled shell outside the closed argon core of the ion, and the highest lying closed (2s and 2p) shells of the six $F^-$ ions. Figure 10-5 shows that the energies of the MO's shift with respect to the atomic orbital energies. In this diagram it has been assumed that the $F^-$ ion s-orbitals are lower in energy than the $F^-$ ion p-orbitals which in turn are lower in energy than the $Fe^{3+}$ ion 3d-orbitals. These assumptions are quite reasonable on the basis of the known free ionization energies. There is a LCAO of the $F^-$ s-orbitals that transform as the irreducible representation of the $O_h$ point group which will then interact with the $Fe^{3+}$ d- orbitals that transform as $E_g$. As expected from Eq. 10-2 to 10-9 and Fig. 10-2, the 3d-orbital will be raised in energy (antibonding) and the LCAO 2s-orbitals will be lowered in energy (bonding). This is shown in Fig. 10-5 where the bonding $E_g$ s-orbitals are labeled $E_g$ ($\sigma$) as a reminder that they form a $\sigma$-bond. The antibonding energy is a dashed line and labeled $E_g^*$(s) as a reminder that is is forming $\sigma$-bonds with ligand s-orbitals. However, the d-orbitals that transform as the $E_g$ irreducible representation also interact with the $\sigma$-LCAO of $p_z$-orbitals of the ligands. Thus, the ligand $p_z$-orbital energy is lowered and is labeled by $E_g(\sigma)$ as a reminder that they are forming a

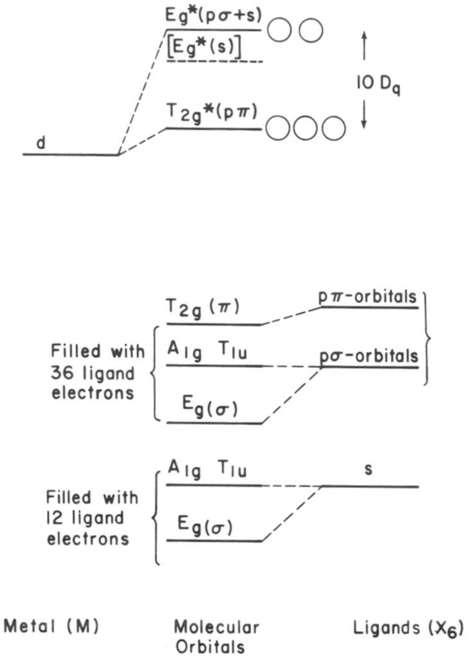


**Fig. 10-5** The energy level diagram showing the interaction of a high lying d-orbital on a metal atom with the LCAO on the ligand atoms.

σ-bond. The antibonding $E_g^*$ orbital is raised in energy and is labeled $E_g^*(p\sigma + s)$ as a reminder that it is interacting with a LCAO of s- and pσ-orbitals of the ligands. Thus the dashed line $E_g^*(s)$ has no real meaning and is there only for tutorial purposes and should now be erased. The d-orbitals interact with neither s- or pσ-orbitals "first," they just interact with both to yield $E_g^*(p\sigma + s)$. The other ligand LCAO σ-orbitals, $A_{1g}$ and $T_{1u}$ do not have the proper symmetry to interact with the central ion d-orbitals. Thus, their energies are not shifted in Fig. 10-5 and they are called nonbonding. (More will be said about these orbitals after π-bonding is discussed.)

The d-orbitals on the central ion that transform as the $T_{2g}$ irreducible representation do not form σ-bonds with the ligands, at least not for $MX_6$ in $O_h$ symmetry. (They do form σ-bonds for $MX_4$ in $T_d$ symmetry.) However, they interact and form π-bonds with the LCAO of ligand pπ-electrons that transform as $T_{2g}$. Thus, the ligand orbitals are lowered in energy and labeled $T_{2g}(\pi)$. The antibonding d-orbitals are raised in energy and labeled $T_{2g}^*(p\pi)$. In general, σ-bonding is stronger than

$\pi$-bonding so the antibonding $\sigma$-orbitals are raised a greater amount than the antibonding $\pi$-orbitals, as shown in Fig. 10-5. (Naturally s-orbitals can not form $\pi$-bonds.)

In general, a central ion orbital can have the appropriate symmetry to interact with ligand, s, p$\sigma$, as well as p$\pi$ LCAO. The secular determinant Eq. 10-3 will be larger, but making the approximations leading to Eq. 10-6, the antibonding metal ion (M) will be

$$E_- \equiv E^* = H_{MM} - \Sigma_i \frac{(H_{Mi} - H_{MM} S_{Mi})^2}{H_{ii} - H_{MM}} \qquad (10\text{-}17)$$

The sum is over s, p$\sigma$, p$\pi$, etc., i.e., over the LCAO of ligands that have the appropriate symmetry to interact with the central ion orbitals. $S_{Mi}$ is the overlap of the central ion orbital with the s, p$\sigma$, p$\pi$, etc., LCAO which is different for each type of ligand orbital. We see that formally there is no more difficulty if the metal ion orbitals interact with more than one ligand LCAO than in the simple case where the metal interacts with just one type of ligand as in Eq. 10-6. The equation for the ligand orbitals is obvious and similar to $E_+$ in Eq. 10-6.

For the sake of simplicity we have left out of Fig. 10-5 the fact that the ligand $A_{1g}$ and $T_{1u}$ LCAO can interact with the higher lying central ion s- and p-orbitals. For $[FeF_6]^{3-}$ in $O_h$ symmetry, the ligand $A_{1g}$ and $T_{1u}$ LCAO are nonbonding with respect to the central ion d-orbitals. However, at higher energies the central ion has s- and p-eigenfunctions which can form $\sigma$-bonds with the ligand $A_{1g}$ and $T_{1u}$ LCAO, respectively. Again the ligand energies will be lowered and thc metal energy will be raised. Usually the effects will be smaller than for the d-orbitals because the energy denominators in Eq. 10-17 are larger. However, care must be exercised because the overlap and off-diagonal matrix elements are usually larger. This interaction with higher lying central ion s- and p-orbitals becomes more important for the heavier transition metals, and it is taken into account in quantitative calculations.

Now we turn to the problem of putting electrons in the orbitals. The level scheme in Fig. 10-5 is a fairly general one for transition metal complexes $MX_6$ in $O_h$ symmetry where the d-electron energies are usually higher than those of the ligand. Thus we could apply it to any number of d-electrons. The ligand atoms will have just exactly enough electrons to fill all the bonding orbitals as shown in Fig. 10-5. If the central atom is $Ti^{3+}$ with one 3d-electron, the electron will go into the lowest available orbit which is $T_{2g}{}^*$. The next two electrons will also enter with parallel spins the $T_{2g}{}^*$ orbitals. When the fourth electron is added to the system,

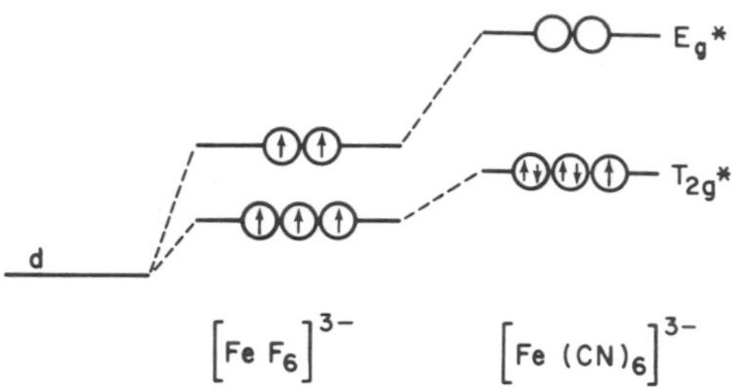


**Fig. 10-6** The splitting of the d-orbitals, as in Fig. 10-5, for different ligands showing that different electron spins can result.

a decision must be made based on energy considerations. Will the fourth electron go into a $T_{2g}$* orbital with antiparallel spin, thus increasing the orbital energy? Hund's rule states that the condition with the maximum spin multiplicity will have lower energy. However, Hund's rule does not take orbital level differences into account so the answer depends on the separation of the orbital states as gauged by the splitting 10Dq in Fig. 10-5 compared to the electron–electron repulsion energy. Figure 10-6 shows the same $Fe^{3+}$ $3d^5$ system surrounded in one case by six $F^-$ ions and in the other case by six $CN^-$ ions. The $CN^-$ ions form a stronger bond with $Fe^{3+}$ than the $F^-$ ions, thus lowering the bonding levels and raising the antibonding levels to a greater extent. For the weaker bonded case all the electron spins are parallel and the high spin form of $Fe^{3+}$ is found. For the stronger bond, the electrons pair with one another in the $T_{2g}$* level since the $E_g$* level is too high in energy thus resulting in the low spin form for $Fe^{3+}$ as pictured.

Most of the MO diagrams encountered in the literature appear somewhat more complicated than Fig. 10-5. In Fig. 10-5 the influence of the higher lying metals and p-states has been neglected as discussed. Often a more important complication will appear. The position before bonding of the metal states with respect to those of the ligand states can vary. If the two have very similar energies, a diagram similar to Fig. 10-5 will result but the wave functions will have a completely mixed character as described in Eq. 7-31 and the discussion with respect to level repulsion. Figure 10-5 implies that the antibonding orbitals are mostly d-like with a small admixture of the ligands but this is not always the case. When the states are more completely mixed, the electrons are still in the lowest energy orbitals, and transitions to empty orbitals are permitted. In this case it is more difficult to obtain a simple physical picture of what is happening. Another complication occasionally arises. As mentioned earlier, the wave function $\phi_1 - \lambda\phi_2$, Eq. 10-7, is the antibonding orbital

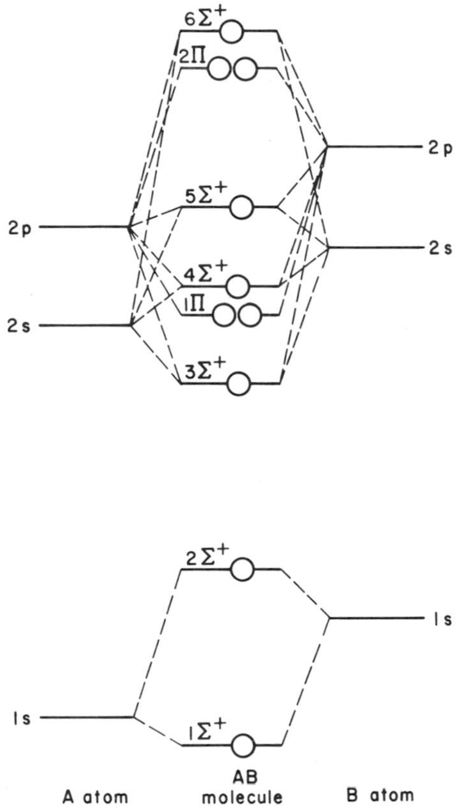


**Fig. 10-7** The energy levels for a diatomic AB molecule.

for the simple systems discussed here. Usually (for example in Eq. 10-9) S is positive and $\beta$ and $H_{ii}$ are negative and $|\beta| > |H_{22}S|$. However, in certain complicated systems $\beta$ and S can have either sign and $\beta$ can be smaller or larger than $H_{22}S$. Thus, the sign of $\lambda$ in Eq. 10-7 will occasionally turn out to be negative instead of positive as it is for most systems discussed here. This occasional reversal of sign of $\lambda$ will reverse the meaning of bonding and antibonding orbitals.

Figure 10-7 shows a conceptual example of an MO treatment of a diatomic molecule AB, $C_{\infty v}$ symmetry. The 1s-orbitals overalp and repel each other to form the $1\Sigma^+$ and $2\Sigma^+$ levels. Similarly, the 2s- and 2p-levels of the two atoms interact. Actually, the 2s- and 2p$\sigma$-orbitals of atom A can interact with the 1s of atom B. There should be dotted lines to indicate this interaction; however, for clarity this interaction has been omitted. The 2s-orbital is interacting with four MO levels. Loosely speaking, these are the bonding and antibonding levels of the 2s- and 2p$\sigma$-orbitals. The 2p$\sigma$-orbitals interact with these same four MO levels. The 2p$\pi$-orbitals of each atom interact only with each other to form the $1\Pi$ and $2\Pi$ bonding and antibonding MO's. The placement of the $3\Sigma^+$ to $6\Sigma^+$ levels relative to each other and to the 2s- and 2p-orbitals depends on the various overlap integrals and matrix elements. The position of the $\Pi$-levels with respect to $\Sigma$-levels also depends on the various terms in the secular equations.

To use a diagram like Fig. 10-7 for LiH requires some small modification in level order (Li = A and H = B). For the free atom, lithium has $1s^22s$ and hydrogen has 1s. The 1s-orbital of the hydrogen will overlap and interact strongly with the 2s-orbital of the lithium but very weakly with the 1s-orbital of the lithium. Thus, $2\Sigma^+$ will be lower in energy than the 1s-energy of the lithium atom. The $1\Sigma^+$ level will be filled with two electrons of opposite spin whose wave function is mostly like the two 1s-electron orbital of the lithium atom. The $2\Sigma^+$ state also will be filled with two electrons of opposite spin whose move function is most like the 1s-state of hydrogen, but one electron is contributed by each atom. Thus, the molecule has a considerably lower energy than the separate atoms.

## 10-4 LCAO–MO of $\pi$-Electrons in Conjugated Hydrocarbons

In the application of the LCAO-MO to $\pi$-bonding, a great deal of chemistry can be learned by using very little more than simple group theory. At the same time a wider appreciation of the use of symmetry operations and projection operators to obtain eigenfunctions from atomic orbitals is also obtained. In this section only the simplest of organic systems will be discussed. These will be mostly conjugated planar hydrocarbons. A conjugated molecule is one with a chain of unsaturated atoms linked by both $\sigma$ and $\pi$-bonds.

### a. Hückel approximations

To appreciate the Hückel approximations it would be useful to keep in mind the types of molecules that are to be treated in the remainder of this chapter. Figure 10-8 shows a benzene molecule $C_6H_6$. It is planar with $sp^2$ hybrid orbitals on each carbon atom that $\sigma$-bond to two other carbon atoms and one hydrogen, as discussed at the end of Section 9-4. The remaining $p_z$-function, perpendicular to the molecular plane, from each of the six carbon atoms is available for $\pi$-bonding. These

$p_z$-orbitals will bond and antibond in ways that will become clearer in the next section. (The measured distance between each carbon atoms in benzene is 1.40Å, which is in between the single and double bond distance.)

Hückel approached this $\pi$-bonding problem by treating the $\pi$- and $\sigma$-bonding systems separately. The in-plane $\sigma$-orbital system is considered localized and reasonably lower in energy than the $\pi$-orbitals. The assumption is that the $\pi$-electrons move in a framework of positively charged carbon atoms. For the Coulomb integrals in the secular equation, $H_{11} \equiv \alpha$ for all $\pi$-orbitals. The off-diagonal matrix elements are taken as $H_{ij} \equiv \beta$ if atom i and j are adjacent, and zero if the atoms are not nearest neighbors. Hückel assumed the value of $\beta$ to be the same for all the cyclic $C_nH_n$ compounds. Obviously $\beta$ depends upon other factors and on n but apparently the effects on $\beta$ are small. The last Hückel approximation is perhaps the most surprising: the off-diagonal $ES_{ij}$ terms in the secular determinant, $|H_{ij} - ES_{ij}| = 0$ (Eq. 10-3 or 7-41) are dropped. If the off-diagonal terms are evaluated as in Section 10-2, one finds that $H_{ij}$ is only somewhat more negative than $ES_{ij}$. Thus, the result is the difference of two terms of comparable size. However, this approximation does take into account symmetry in the sense that both $H_{ij}$ and $S_{ij}$ will be zero or nonzero at the same time. As discussed in Section 10-2, a good approximation is $H_{ij} \propto S_{ij}$. More recent calculation procedures do not drop the ES terms. Such calculations are termed extended Hückel methods but will not be discussed here.

### b. $\pi$-Orbitals in cyclic hydrocarbon systems ($C_nH_n$)

Equation 10-18 is the secular determinant for benzene $C_6H_6$ pictured in Fig. 10-8. As discussed, the Hückel approximations are used.

$$\begin{vmatrix} \alpha\text{-E} & \beta & 0 & 0 & 0 & \beta \\ \beta & \alpha\text{-E} & \beta & 0 & 0 & 0 \\ 0 & \beta & \alpha\text{-E} & \beta & 0 & 0 \\ 0 & 0 & \beta & \alpha\text{-E} & \beta & 0 \\ 0 & 0 & 0 & \beta & \alpha\text{-E} & \beta \\ \beta & 0 & 0 & 0 & \beta & \alpha\text{-E} \end{vmatrix} = 0$$

$$\alpha \equiv H_{ii}, \qquad \beta \equiv H_{ij} \text{ (neighbors)} \tag{10-18}$$

Thus all the off-diagonal overlap terms are dropped. The off-diagonal matrix elements are approximately by $\beta$ if the two $p\pi$-orbitals are neighbors, or dropped if they are next neighbors. Equation 10-18 is as difficult to solve as Eq. 7-41, because each $p\pi$-orbital is linked to each of its

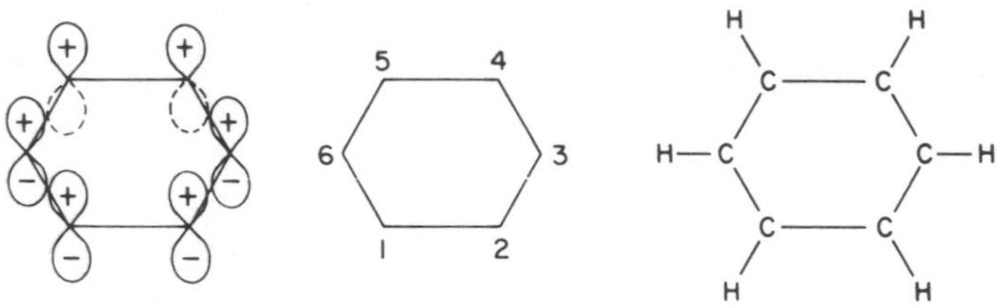


**Fig. 10-8** A diagram for benzene $C_6H_6$ showing the $p_z$-electrons.

neighbors by a matrix element. Such a secular determinant is easier to solve when factored by the use of symmetry coordinates as discussed in Section 7-7 (also in Chapters 6 and 9). Moreover, factoring the secular equations gives a physical understanding of the eigenfunctions. To find the symmetry coordinates we realize that all the $p_z$-orbitals transform among themselves under all the symmetry operations of the molecule. The point group of benzene is $D_{6h}$ but all the operations need not be used. Since we are dealing with $p_z$-orbitals, the point group $C_6$ can be used. The differences between the point groups are operations like $\sigma_h$ and $C_2{}^x$, etc., which merely describe the eveness or oddness of the $p_z$-orbitals which are already known. The operations of the point group $C_6$ (E, $C_6$, $C_3$, $C_2$, $C_3{}^2$, $C_6{}^5$) describe the interchange of the $p_z$-orbitals which is the crux of the problem. Naturally $C_n$ is a subgroup of $D_{nh}$. We shall also see that the use of the cyclic point groups $C_n$ enables one to treat the cyclic hydrocarbon systems $C_nH_n$ for any n with ease.

Table 10-3 shows the character table for the point group $C_6$ with the character of the reducible representation $\chi(\Gamma_b)$ which describes how the six $p_z$-wave functions transform among themselves under all the symmetry operations of the group. Obviously, all the operations of this cyclic group except E transform each atom into a different one resulting in a zero along the diagonal of the representation. For the E-operation one obtains just the number of atoms in the cyclic hydrocarbon system, six in this case. The reduction of $\Gamma_b$ is obvious ($= \Gamma_1 + \Gamma_2 + \Gamma_3 + \Gamma_4 + \Gamma_5 + \Gamma_6 = A + B + E_1 + E_2$). Each of the six irreducible representations must be contained once and only once. (Note that the cyclic $C_n$ groups have n-classes and thus n-irreducible representations, each being one-dimensional. Representations that are complex conjugates of each other are usually bracketed and labeled as an E-representation. This is because wave functions that are complex conjugate to each other and degenerate if no magnetic field is applied. However, this should not confuse the reader. The fact is that from a mathematical point of view there are n-one-dimensional irreducible representations.) Since the six

**Table 10-3** Character table for the cyclic group $C_6$

| $C_6$ | $E$ | $C_6$ | $C_3$ | $C_2$ | $C_3^2$ | $C_6^5$ | $[\varepsilon \equiv e^{2\pi i/6}]$ |
|---|---|---|---|---|---|---|---|
| A | 1 | 1 | 1 | 1 | 1 | 1 | $\rightarrow \phi_1 \rightarrow \psi_1$ |
| B | 1 | −1 | 1 | −1 | 1 | −1 | $\rightarrow \phi_2 \rightarrow \psi_2$ |
| $E_1$ | 1 | $\varepsilon$ | $-\varepsilon^*$ | −1 | $-\varepsilon$ | $\varepsilon^*$ | $\rightarrow \phi_3$ ; $\psi_3 \leftarrow \phi_3 + \phi_4$ |
| | 1 | $\varepsilon^*$ | $-\varepsilon$ | −1 | $-\varepsilon^*$ | $\varepsilon$ | $\rightarrow \phi_4$ ; $\psi_4 \leftarrow (\phi_3 - \phi_4)/i$ |
| $E_2$ | 1 | $-\varepsilon^*$ | $-\varepsilon$ | 1 | $-\varepsilon^*$ | $-\varepsilon$ | $\rightarrow \phi_5$ ; $\psi_5 \leftarrow \phi_5 + \phi_6$ |
| | 1 | $-\varepsilon$ | $-\varepsilon^*$ | 1 | $-\varepsilon$ | $-\varepsilon^*$ | $\rightarrow \phi_6$ ; $\psi_6 \leftarrow (\phi_5 - \phi_6)/i$ |
| $\chi(\Gamma_b)$ | +6 | +0 | +0 | +0 | +0 | +0 | |

irreducible representations are different and one-dimensional, the character projection operator, Eq. 5-26, can be used to write the symmetry-adapted functions. Table 10-3 shows the resulting six functions $\phi_1$ to $\phi_6$. To eliminate the complex coefficients from $\phi_3$ and $\phi_4$, two other orthogonal functions are formed by taking $\phi_3 + \phi_4$ and $(\phi_3 - \phi_4)/i$. Similarly for the functions that transform as the $E_2$-representations. Since $\varepsilon + \varepsilon^* = 1$ and $(\varepsilon - \varepsilon^*)/i = \sqrt{3}$, the normalized eigenfunction can be written immediately, where it is assumed that each of the $p_z$-atomic orbitals $z_1, z_2, \ldots, z_6$ each individually are normalized to one.

$$
\begin{aligned}
\psi_1(A) &= (1/\sqrt{6})\,(z_1 + z_2 + z_3 + z_4 + z_5 + z_6) \\
\psi_2(B) &= (1/\sqrt{6})\,(z_1 - z_2 + z_3 - z_4 + z_5 - z_6) \\
\psi_3(E_1) &= (1/\sqrt{12})\,(2z_1 + z_2 - z_3 - 2z_4 - z_5 + z_6) \\
\psi_4(E_1) &= (1/2)\,(z_2 + z_3 - z_5 - z_6) \\
\psi_5(E_2) &= (1/\sqrt{12})\,(2z_1 - z_2 - z_3 + 2z_4 - z_5 - z_6) \\
\psi_6(E_2) &= (1/2)\,(z_2 - z_3 + z_5 - z_6)
\end{aligned}
\tag{10-19}
$$

The eigenvalues of these eigenfunctions can be immediately calculated. We have j atoms and i eigenfunctions.

$$\psi_i = N_i \, \Sigma_j \, a_j z_j \tag{10-20}$$

For each eigenfunction the eigenvalue is

$$E = \langle \psi_i | H | \psi_i \rangle = N_i^2 \sum_j a_j^2 \langle z_j | H | z_j \rangle + \sum_{\substack{j,k \\ (j \neq k)}} a_j \, a_k \langle z_j | H | z_k \rangle \tag{10-21}$$

For example, for $\psi_6(E_2)$ in the Hückel approximation

$$E(E_2) = \alpha + (1/4)\,[-2\langle z_2 | H | z_3 \rangle - 2\,\langle z_5 | H | z_6 \rangle] = \alpha - \beta \tag{10-22}$$

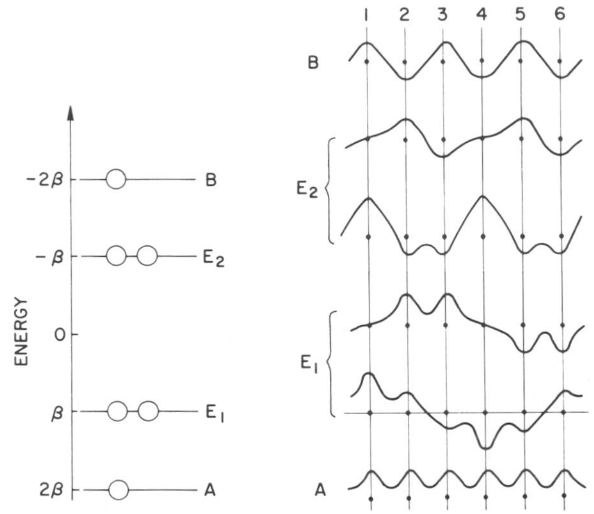


**Fig. 10-9** The energy levels and eigenfunctions for the $\pi$-electrons of benzene.

Figure 10-9 shows the energy level scheme for the four different eigenvalues. Also shown is a picture of the eigenfunctions of Eq. 10-19. The lowest energy eigenfunction has the electron charge spread out uniformly over all six carbon atoms, i.e., the electrons are completely **delocalized**. The higher energy eigenfunctions tend to have the charge more localized.

If the benzene molecule is neutral, there are six $\pi$-electrons that must be accommodated. Two, with antiparallel spin, can be placed in the lowest A-orbital and four, also with paired spins, can be accommodated in the $E_1$ level. Thus the energy of this delocalized $\pi$-bonding scheme is $6\alpha + 2(2\beta) + 4(\beta) = 6\alpha + 8\beta$. It is interesting to compare this energy to that obtained if benzene had three localized double bonds. Assume there is an ordinary double bond between atoms 1–2, as well as 3–4, and 5–6. Then the secular determinant is

$$\begin{vmatrix} \alpha\text{-E} & \beta & 0 & 0 & 0 & 0 \\ \beta & \alpha\text{-E} & 0 & 0 & 0 & 0 \\ 0 & 0 & \alpha\text{-E} & \beta & 0 & 0 \\ 0 & 0 & \beta & \alpha\text{-E} & 0 & 0 \\ 0 & 0 & 0 & 0 & \alpha\text{-E} & \beta \\ 0 & 0 & 0 & 0 & \beta & \alpha\text{-E} \end{vmatrix} = 0 \qquad (10\text{-}23)$$

Obviously each of the three double-bonded systems has a bonding and antibonding eigenstate in the usual way with eigenvalues $\alpha \pm \beta$. Filling

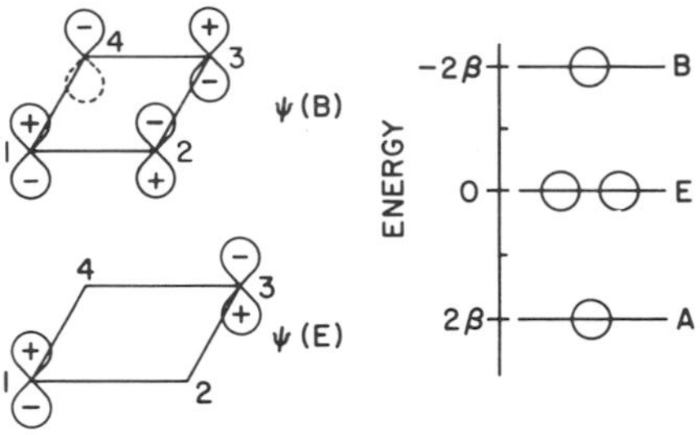


**Fig. 10-10** The energy levels and some eigenfunctions for cyclobutadiene.

each of the three bonding states with antiparallel spin electrons will result in an energy of the localized $\pi$-bonding scheme of $6\alpha + 6\beta$. The **delocalization or resonance energy** is defined as the energy of the filled delocalized orbitals minus the energy of the filled localized orbitals. For benzene, as discussed here, the delocalization energy is $(6\alpha + 8\beta) - (6\alpha + 6\beta) = 2\beta$. We conclude that the $\pi$-electrons in benzene will indeed be delocalized.

We will calculate the LCAO-MO energy for cyclobutadiene $C_4H_4$ (Fig. 10-10) which is theoretically interesting because the delocalization energy calculated by this simple method is zero. Thus, more sophisticated theories must be used to predict whether or not the molecule will have delocalized $\pi$-bonds. Using the character table for the point group $C_4$, the normalized wave functions and energies can easily be written down. Equation 10-24 shows the results where again we have found convenient linear combinations for the E-eigenfunctions.

$$\begin{aligned} &\psi_1(A) = (1/2)(z_1 + z_2 + z_3 + z_4) && E(A) = \alpha + 2\beta \\ &\psi_2(B) = (1/2)(z_1 - z_2 + z_3 - z_4) && E(B) = \alpha - 2\beta \\ &\left.\begin{aligned}\psi_3(E) &= (1/\sqrt{2})(z_1 - z_3) \\ \psi_4(E) &= (1/\sqrt{2})(z_2 - z_4)\end{aligned}\right\} && E(E) = \alpha \end{aligned} \qquad (10\text{-}24)$$

Figure 10-10 shows some of the wave functions involved and the energy level scheme. Note that the $\psi(E)$ levels are nonbonding. Thus, the $\pi$-electrons in a doubly ionized molecule will give an energy $2\alpha + 4\beta$ (antiparallel spin in the A-orbital). The addition of two more electrons to make the molecule neutral will result in an energy $4\alpha + 4\beta$, i.e., no increse in the overlap energy. If a localized electron picture is assumed, there would be a localized $\pi$-bond between atoms 1–2 as well as 3–4. The secular determinant will factor into two $2 \times 2$'s as in Eq. 10-23 where it factored into three $2 \times 2$'s. The result will be $E = \alpha \pm \beta$ for each $2 \times 2$ determinant. Four electrons will go into the two bonding orbitals to give an energy $4(\alpha + \beta)$ which is the same energy as calculated. Thus, the delocalization energy is zero.

### c. Open chain $\pi$-systems

In the simple cyclic systems already discussed, each irreducible representation occurred once when the cyclic point groups were used. This result made the group theory aspects of the problem very easy to handle. Some of the ireducible representations occur more than once for slightly more complicated systems. Projection operator techniques give symmetry-adapted functions which must then be used to diagonalize the secular determinant. The resulting energies can be used to determine the eigenfunctions.

Consider the open chain $\pi$-system $C_4H_6$(butadiene) as shown in Fig. 10-11 in its trans and cis forms. We will consider the cis form although it makes no difference as far as the energy is concerned for the approximations used here. Under all the symmetry operations of $C_{2v}$, the $p_z$ orbitals of carbon 1 and 4 transform between themselves forming a reducible representation. The characters of this representation are listed in Fig. 10-11 as are the characters of the representation formed from the transformation of the carbon 2 and 3 $p_z$ orbitals. Both representations are identical and reduce to $A_2 + B_1$. For each set of functions that transform between themselves projection operators can be used to write the symmetry-adapted functions. Since the irreducible representations are one-dimensional, the projection operators (Section 5-7) all give the same result. Consider first the $A_2$ eigenvalue problem. The normalized symmetry-adapted functions for atoms 1 and 4 as well as 2 and 3 are

$$\phi^a(A_2) = (1/\sqrt{2})\,(z_1 - z_4), \qquad \phi^b(A_2) = (1/\sqrt{2})\,(z_2 - z_3) \tag{10-25}$$

The secular determinant in the usual Hückel approximation is

$$\begin{vmatrix} H_{aa}\text{-}E & H_{ab} \\ H_{ab} & H_{bb}\text{-}E \end{vmatrix} = \begin{vmatrix} \alpha\text{-}E & \beta \\ \beta & \alpha\text{-}\beta\text{-}E \end{vmatrix} = 0$$

$$E = \alpha + \frac{1 \pm \sqrt{5}}{2}\beta = \alpha + \left\{ \begin{matrix} -1.62\beta \\ +0.62\beta \end{matrix} \right. \tag{10-26}$$

Similarly, for functions that transform as the $B_1$-irreducible representation

$$\phi^a(B_1 = (1/\sqrt{2})\,(z_1 + z_4), \qquad \phi^b(B_1) = (1/\sqrt{2})\,(z_2 + z_3)$$

$$E = \alpha + \frac{1 + \sqrt{5}}{2}\beta = \alpha + \left\{ \begin{matrix} -0.62\beta \\ +1.62\beta \end{matrix} \right. \tag{10-27}$$

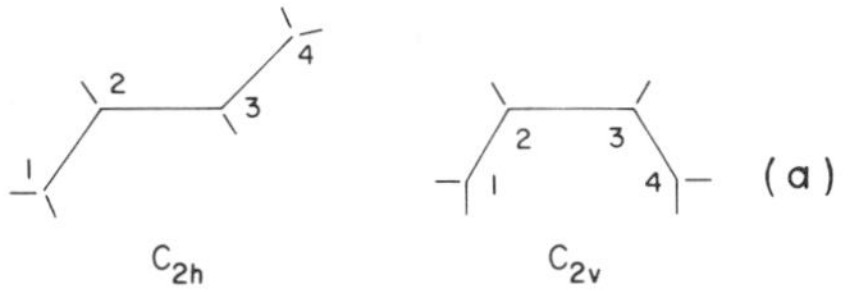


| $C_{2v}$ | E | $C_2$ | $\sigma_v(xz)$ | $\sigma_v(yz)$ | | |
|---|---|---|---|---|---|---|
| $\chi(\Gamma_{14})$ | 2 | 0 | 0 | -2 | → $A_2 + B_1$ | (b) |
| $\chi(\Gamma_{23})$ | 2 | 0 | 0 | -2 | → $A_2 + B_1$ | |

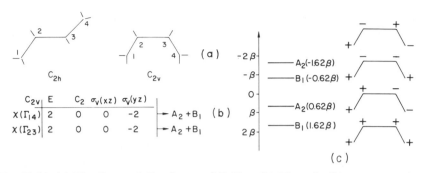

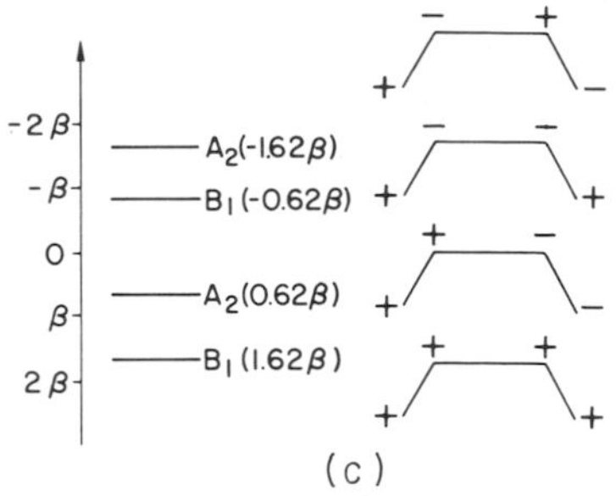


**Fig. 10-11** (a) The $C_{2h}$ and $C_{2v}$ forms of $C_4H_6$. (b) The reducible representations and (c) energy levels of the $C_{2v}$ form. (The molecule can go from one form to the other by a rotation about the single bond.)

Figure 10-11 shows the energy levels. Of the four $\pi$-electrons, two enter with spins antiparallel in the lowest $B_1$-state and the next two, in the lowest $A_2$-state. Thus the energy is $4\alpha + 4.48\beta$. If the $\pi$-bonds are localized between atoms 1 and 2 as well as atoms 3 and 4, the energy would be $4\alpha + 4\beta$. Thus, the delocalization energy os $0.48\beta$, indicating that the $\pi$-electrons will be delocalized.

As was shown in this section, open chain $\pi$-systems can be treated via the Hückel approximation in a manner similar to the closed chain. Only a slight bit of extra computational difficulty is involved. Therefore we can see, using the Hückel approximation and the simple group theory as shown in this section, that MO theory can be applied to many organic molecules. Applications to many molecules are covered in the problems. In the next section we apply these ideas to reactions.

## 10-5 Woodward–Hoffmann Rules

In this section the rules of Woodward and Hoffmann (1965) are discussed by means of several examples. It is exciting to realize that group theory can be used to predict the occurence of one reaction as opposed to a competing reaction. The basis of the idea is that the more bonding that is maintained throughout the reaction, the more readily the reaction will proceed. The ideas discussed here apply to a **concerted reaction**, that is, a reaction in which there is no intermediate species, no matter how short lived. The wave functions of the initial molecules transform in a continuous manner into those of the product. If an energy verses reaction coordinate diagram is drawn, the initial molecule has a minimum in energy with higher vibrational levels as does the product. As one proceeds along the reaction coordinate there are no other local minimum corresponding to intermediate products. Thus, the very high vibrational levels of the initial molecule are in "concert" with the very high vibrational levels of the product. Besides a concerted reaction, the Woodward–Hoffmann rules depend on at least one symmetry element being maintained continuously throughout the transformation of the initial molecules to the product. It is with respect to this symmetry operation that we label the orbitals and wave functions throughout the transformation. As always, attention is focused on the orbitals that are directly involved in the bonding or reaction in this case. We ignore the rest of the electrons which do not undergo significant changes.

### a. Cyclobutene–butadiene interconversion

As a first example consider the conversion of cyclobutene to butadient. Figure 10-12 shows that a cyclobutene molecule thermally converts to the cis,trans form of butadiene (Y) and not to the trans,trans form (N) (or to the cis,cis form which is not shown). The problem is to understand why one conversion occurs and not the other.

Figure 10-13a shows how the $\sigma$-bond of cyclobutene can be broken by a conrotatory mode to form cis,trans-butadiene. Figure 10-13b shows that trans,trans-butadiene could be formed by a disrotatory mode. (If the sense of both rotations of the disrotatory ring opening were reversed, the cis,cis form would be obtained. Try it.) For the disrotatory mode, a plane of symmetry that bisects the cyclobutene double bond persists as a symmetry element throughout the reaction. Thus, the orbital states, for the reaction of cyclobutene to butadiene by a disrotatory mode, can be classified according to irreducible representations of the point group $C_{1h}$ which contains ($E$, $\sigma_h$). For the conrotatory mode, a $C_2$ axis of symmetry that bisects the cyclobutene double bond persists as a symmetry element throughout the entire reaction, provided one assumes $S \equiv H$. Thus, the orbital states, for the reaction via a conrotatory mode, can be classified according to the point group $C_2$ ($E$, $C_2$). Since this reaction is remarkably stereospecific for $S = CH_3$, or deuterium, or several other possibilities, the results seem independent of the fact that the $C_2$-axis may only be an approximation.

Figure 10-13c shows the four MO's of interest of cyclobutene. We expect the lowest energy orbital to be the $\sigma$-orbital, then $\pi$, then $\pi^*$, with $\sigma^*$ being the highest energy orbital. In the orbital correlation diagram in Fig. 10-14 these four orbitals are shown in the center and they are labeled

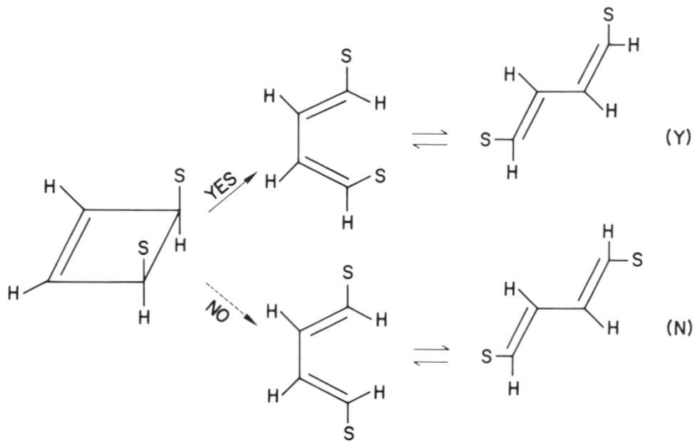


**Fig. 10-12** Cyclobutene can thermally convert to cis,trans-butadiene (top) but not to trans,trans-butadiene (bottom) or cis,cis-butadiene (not shown).

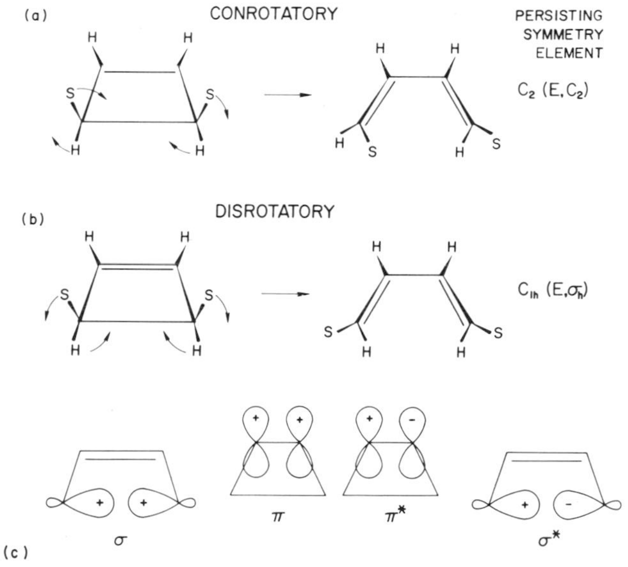


**Fig. 10-13** Conrotatory and disrotatory ring openings of cyclobutene showing show (a) cis,trans-butadiene and (b) trans,trans-butadiene can be formed. (c) The four MO orbitals of interest of cyclobutene. The energy increases from left to right as indicated.

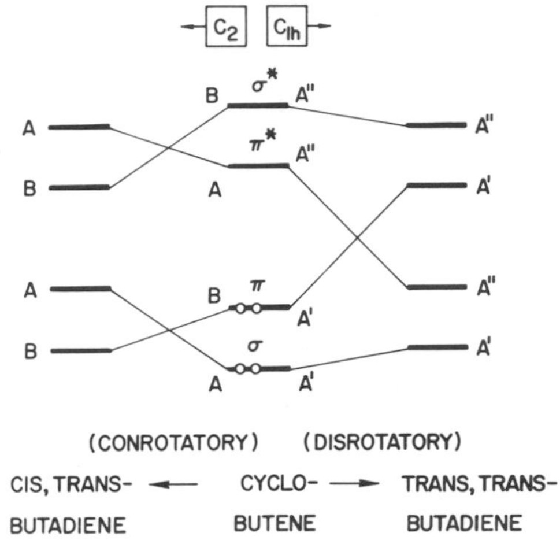


**Fig. 10-14** An orbital correlation diagram of cyclobutene for the two modes of ring openings. (The energies in this and the following figures are not to scale.)

according to the irreducible representations of the point group $C_2$ (A is totally symmetric and B antisymmetric under the 2-fold rotation) and the point group $C_{1h}$(A′ is totally symmetric and A″ antisymmetric under $\sigma_h$). The left and right sides of Fig. 10-14 show the orbitals of butadiene, taken from Fig. 10-11, labeled according to the irreducible representations of the point groups $C_{1h}$ and $C_2$. The appropriate orbital states of cyclobutene are connected to those of butadiene according to the no-crossing rule. The four electrons of cyclobutene fill the two lowest orbitals as shown. If the disrotatory mode were to occur, forming trans,trans-butadiene, then the continuous orbital correlation for this concerted reaction would leave the product in a high energy, antibonding state. In order to get to this high lying state, energy would have to be supplied. On the other hand, the conrotatory model will result in cis,trans-butadiene where the four electrons are in the ground state orbitals. Thus, a large activation energy is required to form trans,trans-butadiene while cis,trans-butadiene formation is expected to be thermally activated, in agreement with experiment.

We carry the analysis one step further by actually forming the state functions from the one-electron orbital functions that we have been considering so far. Since the state functions are solutions of the wave equation at all points in the correlation diagram, these functions and their symmetries really should be considered. For the nondegenerate orbitals found, the state function can be formed by taking the direct product of

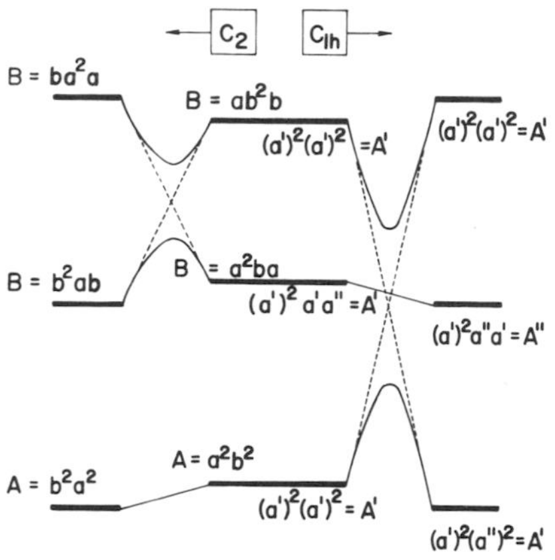


**Fig. 10-15** The various electronic states, labeled by the appropriate irreducible representation, for the two modes of ring openings of cyclobutene.

the orbitals concerned. This is shown in Fig. 10-15 where we have changed the notation to small letters for one-electron orbital, and capital letter for the direct product state function. Thus, the ground state of cyclobutene for the point group $C_{1h}$ is $(a')^2(a')^2 = A'$. The first excited state would be one-$\pi$-electron promoted to the $\pi^*$ orbital, so the state would be $(a')^2a'\,a'' = A''$. Figure 10-15 shows the various electronic states for the two modes of ring openings. The various total state functions are correlated with each other by dotted lines but the no-crossing rule is applied. Thus, we can see that cis,trans-butadiene can be formed thermally by a conrotatory ring opening of cyclobutene where a large activation energy must be overcome to form trans,trans-butadiene. In the first excited state of cyclobutene, the correlation is with the first excited state trans,trans-butadiene by a disrotatory mode. Thus, Fig. 10-15 shows that the excited state of cyclobutene correlates with a different end product than the ground state.

The results of more general electrocyclic ring opening reactions can be predicted. See the Notes for references.

### b. Ethylene dimerization

As a second example consider the formation of cyclobutane from two molecules of ethylene. Figure 10-16 shows the maximum-symmetry-possible approach of two ethylene molecules. There are a number of symmetry elements that are maintained throughout the reaction but we

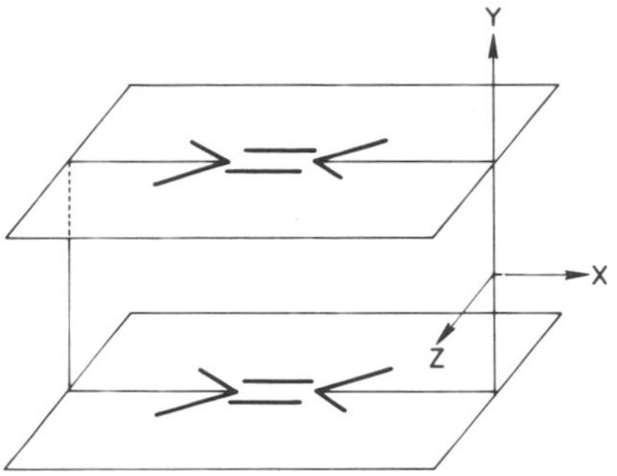


**Fig. 10-16** The coordinate system used to describe the "maximum-symmetry approach" of two ethylene molecules.

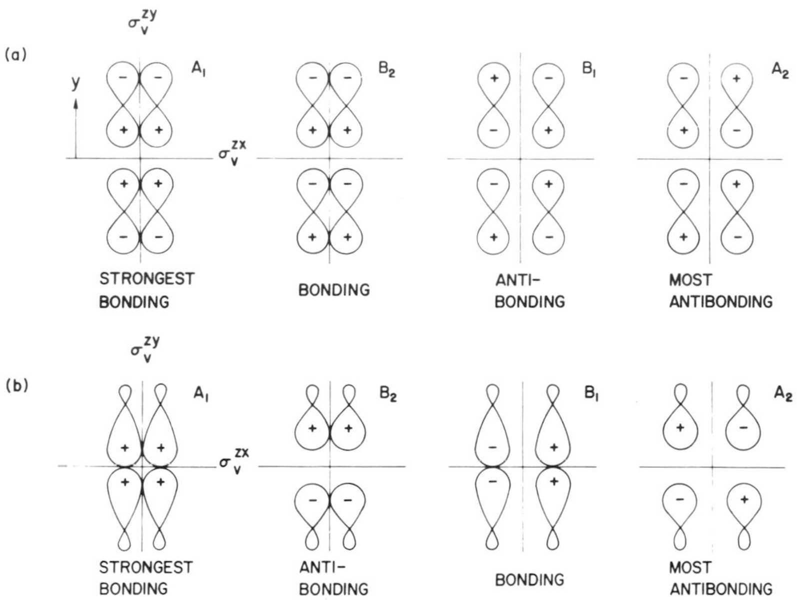


**Fig. 10-17** The possible arrangements of the MO's of two ethylene molecules (a) The molecules are far apart. (b) The molecules are much closer.

can ignore the $\sigma_v{}^{xy}$ plane since the $\pi$- and $\sigma$-orbitals are symmetric with respect to this plane at all stages of the reaction. Thus, only the $\sigma_v{}^{zx}$ and $\sigma_v{}^{zy}$ planes need be considered so the orbitals can be classified according to the irreducible representations of the point group $C_{2v}$.

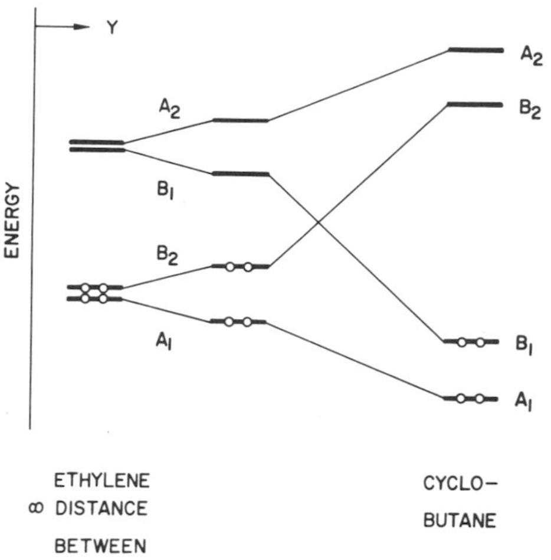


**Fig. 10-18** An orbital correlation diagram versus distance as two ethylene molecules approach each other to form cyclobutane.

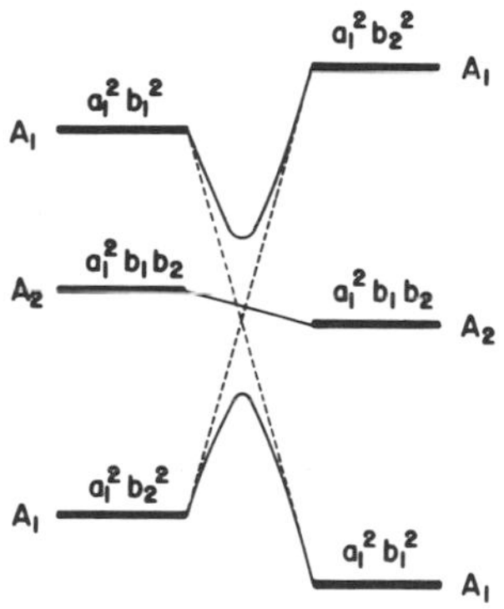


**Fig. 10-19** The various electronic states, labeled by the appropraite irreducible representation, for the ethylene dimerization to cyclobutane reaction.

Figure 10-17a shows the possible arrangement and ordering on an energy scale for two ethylene molecules that are still separated but are close enough so that some small amount of bonding and antibonding can occur between the two separated molecules. The states labeled $A_1$ and $B_2$ (irreducible representations of the point group $C_{2v}$) have the same energy when the molecules are separated by a very large distance. However, as the separation decreases, the state labeled $B_2$ becomes less strongly bonded then $A_1$ since the two different molecules are antibonding to each other. Figure 10-18 shows this result on a correlation diagram of energy versus distance. The right side of the diagram shows the orbital states for the product cyclobutane. The orbitals can be seen in Fig. 10-17b where energy ordering is clear.

From Fig. 10-18 we see that the thermally activated dimerization of two ethylene molecules to form cyclobutane is unlikely, since cyclobutane would have to be formed in an excited state and the energy is not available at normal temperatures. We proceed to draw a correlation diagram for the state functions, Fig. 10-19. Again, it can be seen that a potential barrier hinders the ehtylene dimerization. However, if one of the ethylene molecules is photoexcited, the state function transforms as the $A_2$ irreducible representation, and an excited state of cyclobutane will be formed. These predictions are in agreement with experiment.

The ideas of conservation of orbital symmetry presented here can be applied to many reactions. Consult the Notes for references.

## Notes

Molecular orbital theory is a very large field and there are many books and articles that can be referred to. We mention only a few of these as a general sampling: C. J. Ballhausen, "Ligand Field Theory" (McGraw-Hill, New York, 1962); Bishop, Cotton, Hochstrasser, and P. O'D. Offenhartz, "Atomic and Molecular Orbital Theory" (McGraw-Hill, New York, 1970); M. Orchin and H. H. Jaffe, "Symmetry Orbitals and Spectra" [Wiley (Interscience), New York, 1971]; D. S. Urch, "Orbitals and Symmetry" (Penguin, 1970).

The Wolfsberg–Holmholz method [J. Chem. Phys. **20**, 837 (1952)] is used extensively in many MO calculations. See Ballhausen, Chapter 7, Appendix I, and R. S. Mulliken, C. A. Rieke, D. Orloff, and H. Orloff, J. Chem. Phys. **17**, 1248 (1949) for the method of evaluation of group overlaps. As an example of the use of this method and of calculations of overlap, see the reference in Chapter 9, Problem 3, and J. D. Axe and G. Burns, Phys. Rev. **152**, 331 (1966).

There are a great many books and references to improved MO calculations including these mentioned. Some simple improvements of the Hückel modes can be found in T. E. Peacock, "The Electronic Structure of Organic Molecules," Chapter 5 (Pergamon Press, Oxford, 1972).

An excellent review of the Woodward–Hoffmann rules can be found in a book by the same authors "Conservation of Orbital Symmetry" (Academic Press, New York 1971). Discussion of these ideas also can be found in many other places including Orchin and Jaffe, Section 11-5 and Cotton, Section 7-8.

## Problems

**1.** For the planar $AB_4$ molecule, find the linear combination of B-ion wave functions that can form σ-bonds to the central A atom. For the complex $CoCl_4^{2-}$, sketch the MO-energy level diagram filling up the orbitals with the appropriate number of electrons. Do this first assuming the chlorine levels are below those of the cobalt, then assuming the reverse.

**2.** Find the irreducible representations to which the LCAO–MO of the ligand π-bonds of $MX_6$ with $O_h$ symmetry belong. Determine the appropriate LCAO's.

**3.** Solve the $BF_3$ MO problem so that an orbital energy level diagram, such as Fig. 10-5, can be drawn.

**4.** The figure shows an $MX_{12}$ system with $O_h$ symmetry. This would be needed to describe the bonding of the Eu-ion to 12 O-ions in $EuTiO_3$. Do the following for σ-bonding: Find all the irreducible representations that the LCAO of the oxygen ions transforms as. By inspection write out the LCAO in terms of normalized $z_1, z_2, \ldots, z_{12}$ that will form a σ-bond to the central ion f-orbitals that transform as $T_{2u}$, $T_{1u}$, $A_{2u}$.

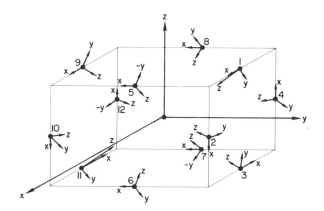

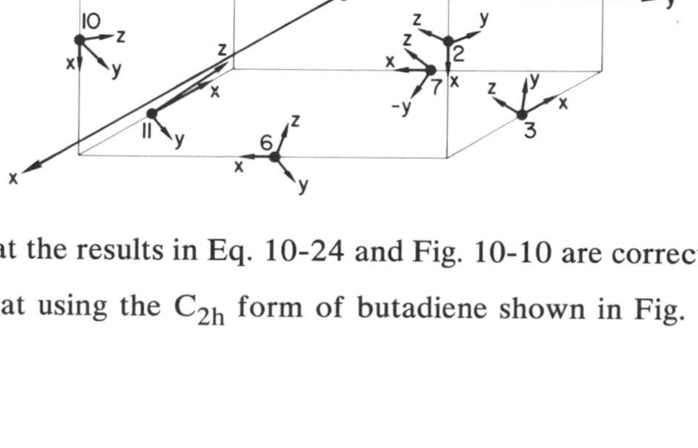

**5.** Show that the results in Eq. 10-24 and Fig. 10-10 are correct.

**6.** Show that using the $C_{2h}$ form of butadiene shown in Fig. 10-11, the same orbitals and energies are obtained as for the $C_{2v}$ form which is worked out in Eqs. 10-26 and 10-27.

**7.** Using the Hückel approximations and the LCAO-MO method, calculate the energies for a triangular arrangement of hydrogen atoms and compare the results to those for a linear arrangement. Do the same for $H_3^+$ and $H_3^-$. What is probably the worst aspect of the results as one changes the ionization state of the molecule?

**8.** (a) For cyclopropyl $C_3H_3$ determine the eigenvalues and eigenfunction of the $\pi$-bonding electrons in the Hückel approximation. What is the delocalization energy for a positively charged, neutral, and negatively charged molecule? (b) For the allyl compound which is an open chain $C_3H_3$ molecule, do the same calculations as (a).

**9.** (a) Consider a cyclic planar molecule $C_{10}H_{10}$. Using the Hückel approximation what is the eigenfunction and eigenvalue of the highest and lowest states? (b) Do the same for $C_{100}H_{100}$. (c) What is the connection between these problems and the band theory of solids?

**10.** Using the Hückel approximation determine the LCAO-MO for tetramethylene-cyclobutane. Determine the eigenvalues and eigenfunctions of the orbitals. (Hint: See Cotton, p. 149.)

**11.** Find the eigenvalues and eigenfunctions of the p$\pi$-orbitals of naphthalene using the Hückel approximation. (Use the coordinate and numbering system of Cotton, Fig. 7-1.)

**12.** To discuss the selection rules for electronic transitions, the state functions that are made up from the one-electron orbitals discussed in Section 10-4 must be considered. (a) Describe the method to form the eigenfunction of the total Hamiltonian ignoring spin (for example Cotton, Section 7-6) and including spin (for example Schonland Chapter 10). (b) Find the terms of the lowest excited configuration for neutral benzene, ignoring spin. Compare the results to Schonland Fig. 10-1 and Hochstrasser, Section 7-9. Which transitions are electron dipole allowed? (Don't forget that the point symmetry is centrosymmetric.) (c) Ignore spin and find the three lowest electronic-excited configurations in naphthalene. What are the electric dipole selection rules for these transitions? Since two of these configurations transform as the same irreducible representation, discuss the effects of configuration interaction. How will this affect the selection rules?

**13.** (a) Using the Wolfsberg–Helmholtz approximation for $H_{ij}$, in the simplest way, calculate the energy levels for a two-equal-atomic-orbit interaction including the off-diagonal ES term. (b) What will the inclusion of the ES terms do to the delocalization energy?

**14.** The concept of **bond order** is useful. (a) Define the bond order between two carbon atoms. (b) Determine the bond orders between all the carbon atoms in the ground state of naphthalene. (c) Why is the bond order concept useful? (See Orchin and Jaffe, Section 4-3; Cotton, p. 154; as well as other references.)

**15.** Very briefly discuss the extended Hückel approximation. See Orchin and Jaffe, Section 12-5; Cotton, Appendix 6; and R. Hoffmann, J. Chem. Phys. **39**, 1397 (1963) and **40**, 2480 (1964).

**16.** A $(MoO_4)^{2-}$ tetrahedral ion has a totally symmetric breathing vibrational mode at approximately 900 $cm^{-1}$ that is usually fairly sharp and well separated from the other $(MoO_4)^{2-}$ vibrations. A complicated molecule is formed that has four of these ions transforming among themselves under the $C_{2v}$ point symmetry operations. What are infrared and Raman selection rules for the coupled modes? What are the eigenfunctions and eigenvalues? Which eigenfunction will probably have the highest energy?

**17.** Discuss the symmetry groups of nonrigid molecules. See H. C. Longuet-Higgins, Mol. Phys. **6**, 445 (1963) and "Molecular Orbitals in Chemistry, Physics, and Biology," p. 113 (Academic Press, New York, 1964).

## Chapter 11

# SYMMETRY OF CRYSTAL LATTICES

*Two roads diverged in a wood, and I –*
*I took the one less traveled by,*
*And that has made all the difference.*

*Robert Frost, "The Road Not Taken"*

Various topics concerned with the symmetry of crystals are discussed in this chapter. In addition to some of these fundamental concerns, the international notations for symmetry operations, point groups, and space groups is introduced and used.

## 11-1 The Real Affine Group

The real affine group is the group that consists of orthogonal transformations (proper rotations as well as improper rotations) plus translations. Thus, an operation of this group on a position vector $\mathbf{r}$ produces a new vector $\mathbf{r}'$

$$\mathbf{r}' = \mathrm{R}\mathbf{r} + \mathbf{t} \equiv \{\mathrm{R} \mid \mathbf{t}\}\mathbf{r} \tag{11-1}$$

where R is an orthogonal transformation by an arbitrary amount, and $\mathbf{t}$ is a translation by an arbitrary amount. Thus, $\{\mathrm{R} \mid \mathbf{t}\}$ means operate first with R and then add an amount $\mathbf{t}$. The **notation** for the operator $\{\mathrm{R} \mid \mathbf{t}\}$ is a convenient way to describe the transformation and is sometimes called the Seitz operator. When the symmetry operation $\{\mathrm{R} \mid \mathbf{t}\}$ operates on a function $\psi(\mathbf{r})$, we have $\{\mathrm{R} \mid \mathbf{t}\}\psi(\mathbf{r}) = \psi(\{\mathrm{R} \mid \mathbf{t}\}^{-1}\mathbf{r})$ as always, e.g. Section 7-2.

Before making use of this relation we must make sure that the set of operations described by Eq. 11-1 forms a group. First, see that the multiplication of $\{\mathrm{R} \mid \mathbf{t}\}$ and $\{\mathrm{S} \mid \mathbf{u}\}$ are also members of the set.

$$\{\mathrm{R} \mid \mathbf{t}\}\{\mathrm{S} \mid \mathbf{u}\}\mathbf{r} = \{\mathrm{R} \mid \mathbf{t}\}\mathrm{S}\mathbf{r}+\mathbf{u} = \mathrm{RS}\mathbf{r}+\mathrm{R}\mathbf{u}+\mathbf{t} = \{\mathrm{RS} \mid \mathrm{R}\mathbf{u}+\mathbf{t}\}\mathbf{r} \tag{11-2}$$

Since **Ru** is a translation, it is evident that the set has closure and we write for the multiplication of two elements

$$\{R\,|\,\mathbf{t}\}\{S\,|\,\mathbf{u}\} = \{RS\,|\,R\mathbf{u}+\mathbf{t}\} \tag{11-3}$$

which immediately follows from Eq. 11-2. It is obvious that the identity element of the group is $\{E\,|\,0\}$. The inverse of $\{R\,|\,\mathbf{t}\}$, labeled $\{R\,|\,\mathbf{t}\}^{-1}$, is $\{R^{-1}\,|\,-R^{-1}\mathbf{t}\}$ since

$$\{R\,|\,t\}\{R^{-1}\,|\,-R^{-1}\mathbf{t}\} = \{E\,|\,0\} \tag{11-4}$$

as can be seen using Eq. 11-3. Also, from the law of combination in Eq. 11-3, associativity is obeyed. Thus, indeed the operations described in Eq. 11-1 form a group which is called the **real affine group**.

A subgroup of the real affine group is the set of pure translations $\{E\,|\,\mathbf{t}\}$. We can see that this subgroup is an invariant subgroup (consists of complete classes) since

$$\{R\,|\,\mathbf{u}\}^{-1}\{E\,|\,\mathbf{t}\}\{R\,|\,\mathbf{u}\} = \{R^{-1}\,|\,-R^{-1}\mathbf{u}\}\{R\,|\,\mathbf{u}+\mathbf{t}\} =$$

$$= \{E\,|\,R^{-1}\mathbf{u} + R^{-1}t - R^{-1}\mathbf{u}\} = \{E\,|\,R^{-1}\mathbf{t}\} \tag{11-5}$$

which is a pure translation. The fact that the subgroup of pure translations is an invariant subgroup will prove very useful in space group considerations.

## 11-2 Space Group

The **space group** of a crystal is defined as the group of symmetry operations (translation, point symmetry operations, and combinations of the two) which leaves the crystal invariant.

### a. Translational symmetry

We can define an infinite lattice in terms of translations of a **primitive cell** shown in Fig. 11-1. The primitive cell has edges $\mathbf{a}_1$, $\mathbf{a}_2$, $\mathbf{a}_3$ which define a lattice of points. Every point in this lattice can be obtained by a translation

$$\mathbf{t}_n = n_1\mathbf{a}_1 + n_2\mathbf{a}_2 + n_3\mathbf{a}_3, \quad n_i = \text{any integer} \tag{11-6}$$

Thus, in the notation of the last section, a **primitive lattice translation** is given by $\{E\,|\,\mathbf{t}_n\}$. We can define a space group as a group of symmetry operations $\{R\,|\,\mathbf{t}\}$ which contain the group $\{E\,|\,\mathbf{t}_n\}$ as a subgroup and contains no other pure translations. In terms of the Hamiltonian of the crystal we would say that at a point in the primitive cell r, the Hamiltonian has the exact same form as at all points $\mathbf{r} + \mathbf{t}_n$. Or we would say the Hamiltonian is invariant when expressed as $\mathbf{r} + \mathbf{t}_n$ for all $\mathbf{t}_n$.

We can see that a space group is a subgroup of the real affine group discussed in Section 11-1. The primitive lattice translations obviously form a group, as shown in Section 11-1. The inverse of any primitive translation $\mathbf{t}_n$ is $\mathbf{t}_{-n}$ where -n means replace $n_1$, $n_2$, and $n_3$ by minus themselves and we can use a short-hand notation of $\mathbf{t}_n$ instead of $\{E \mid \mathbf{t}_n\}$ for simplicity. $(\mathbf{t}_n)(\mathbf{t}_{-n}) = \mathbf{t}_o$ and we see that the law of combination of the translations is just additions of the $n_1$'s etc. as is clear from the meaning of $\mathbf{t}_n$ in Eq. 11-6. The primitive translation group is Abelian since $(\mathbf{t}_n)(\mathbf{t}_m) = (\mathbf{t}_m)(\mathbf{t}_n)$. Thus, each symmetry operation is in a class by itself, which also can be seen by $(\mathbf{t}_m)(\mathbf{t}_n)(\mathbf{t}_{-m}) = (\mathbf{t}_n)$. Therefore, there are as many one-dimensional irreducible representations of the pure translation group as there are symmetry operations. We shall see in the next chapter how the k-vector is used as a label for these irreducible representations.

### b. General comments on space groups

A general space group symmetry operation can be written in the form $\{R \mid \mathbf{t}\}$. R is a point operation taken with respect to a fixed point in the unit cell and $\mathbf{t}$ is a translation of the primitive unit cell but not necessarily by an amount $\mathbf{t}_n$. For example, in Section 1-5, a screw operation for $TiO_2$ was shown to be $\{C_4 \mid \mathbf{a}/2 + \mathbf{b}/2 + \mathbf{c}/2\}$. Clearly for this symmetry operation $\mathbf{t}$ is not one of the primitive cell translation symmetry operations.

The statement that $\{R \mid \mathbf{t}\}$ is a symmetry operation of the crystal means that after the application of $\{R \mid \mathbf{t}\}$ the crystal is identical in every way to what it was before. Another statement of this idea is to say that symmetry operation commutes with the Hamiltonian of the crystal. Or one can say that the potential energy has the symmetry of the space group, i.e., $\{R \mid \mathbf{t}\}V(\mathbf{r}) = V(\{R \mid \mathbf{t}\}^{-1}\mathbf{r}) = V(\mathbf{r})$. Note that the symmetry operation acts on the position vector in direct space (Section 7-2). The term direct space means ordinary x, y, z space as opposed to spin space or k-space as will be defined in Chapter 12. Since the space group is a subgroup of the real affine group we have already seen in Section 11-1, the law of combination of two or more symmetry operations, the inverse, etc., is already given.

The **point group of a space group** G is defined as the point group obtained when all $\mathbf{t} = 0$. We use the symbol $P_g$ for the point group of a space group. Thus, $P_g$ has the operations $\{R\,|\,0\}$. If, for a suitable choice of origin, all the operations of $P_g$ are operations of G, then the space group is a symmorphic space group. Or, expressing it another way if $P_g$ is a subgroup of G, the space group is **symmorphic**. This implies that there are no glide or screw operations. There are 230 different space groups. (See the Notes.) Of these, 73 are symmorphic and the other 157 are **nonsymmorphic**. From Eq. 11-5 we know that $\{E\,|\,R^{-1}\mathbf{t}_n\}$ is a pure translation. Then $R^{-1}\mathbf{t}_n$, or $R\mathbf{t}_n$, is a pure translation for all R. Thus, all the point operations R are symmetry operations that take the lattice into itself. (Remember the operations $\{E\,|\,\mathbf{t}_n\}$ define the lattice for each crystal system.) So the symmetry operations R in the space group operators are just those symmetry operations of the 32 crystal systems discussed throughout this book, since it is only those point operations that will take a space lattice into itself.

Define T as the subgroup of the space group G consisting of only primitive translations $\{E\,|\,\mathbf{t}_n\}$. In Section 11-1 it has been shown that T is an invariant subgroup. We know, from Section 3-4, that the irreducible representations of a factor group are also irreducible representations of the full group. In fact all the irreducible representations of a space group G can be found if the irreducible representations of T and $P_g$ are known. The proof of this is lengthy and references to it can be found in the notes. However, in the next chapter we will see how the irreducible representations are found.

Consider the factor group of the space group G with respect to T defined as G/T. A fundamental theorem of solid state physics is that G/T is isomorphic to $P_g$. Thus, the irreducible representations of $P_g$ are also the irreducible representations of G. Here we prove this theorem for symmorphic groups. For a general proof, see the Problems. If the point operations are R, S,...,W,..., then the factor group with respect to T is

$$\{E\,|\,\mathbf{t}_n\},\quad [\{E\,|\,\mathbf{t}_n\}\{R\,|\,0\}],\quad [\{E\,|\,\mathbf{t}_n\}\{S\,|\,0\}],\ldots\ = \{E\,|\,\mathbf{t}_n\},\quad C_R,\quad C_S,\ldots \tag{11-7}$$

$C_R$ etc. is the complex (Section 2-4) with $\{S\,|\,0\}$ etc. multiplying all the elements of $\{E\,|\,\mathbf{t}_n\}$. The square bracket is used as a reminder that the terms of the complex consist of as many entries as there are values of $\mathbf{t}_n$. To see that the multiplication table formed from this factor group is the same as the multiplication table formed by the point group $P_g$, form the multiplication of the two complexes

$$C_S C_R = [\{S\,|\,\mathbf{t}_{n'}\}]\,[\{R\,|\,\mathbf{t}_n\}] = [\{SR\,|\,S\mathbf{t}_n+\mathbf{t}_{n'}\}] = [\{SR\,|\,\mathbf{t}_{n''}\}] = C_W \tag{11-8}$$

Since $St_n$ is a translation and $\mathbf{t}_n$ ranges over all possible positions, the theorem is proved. (Note that $\{R\,|\,0\}\{E\,|\,\mathbf{t}_n\} \neq \{E\,|\,\mathbf{t}_n\}\{R\,|\,0\}$ from Eq. 11-3. Thus, the space group G is not the direct product of the group T and $P_g$. Direct product of groups is defined in Section 2-6c.) In Chapter 12 we will see how irreducible representations of space groups are found. In this chapter we discuss the different types of translational lattices and space group operations.

## 11-3 Translational Lattice

**A lattice** is defined as the set of all points obtained by starting at arbitrary origin and operating on the origin by a translation $\mathbf{t}_n$, as in Eq. 11-6 to obtain an infinite array of points. The environment of every point in the lattice is identical in arrangement and orientation with every other point. As we shall see there are 14 lattices in the 3-dimensional space. These are called the **14 space lattices** or the **14 Bravais lattices**.

A **crystal structure** is formed by associating with every point of a lattice an assembly of atoms, called a **basis** (or sometimes a **lattice complex**). Naturally the basis must be the same at every point in the lattice so that the translational symmetry remains. As will be discussed later, if the basis consists only of a single atom or a "spherical" arrangement of atoms there is no further restriction on the symmetry operations of the lattice, and this will lead to the crystal structure with the highest point symmetry (called the **holohedral point group**) compatible with the Bravais lattice. If this basis has more restrictions then the holohedral point group (fewer symmetry operations) then, depending on the details, some lower point symmetry of the crystal structure can be obtained.

In this section we sketch and try to clarify the ideas behind the development of the primitive cell, the 14 Bravais lattices, how they lead to the 32 point groups and then to the space groups. Some references to the extensive literature in this field can be found in the Notes.

### a. Primitive cell and Wigner–Seitz cell

We already have defined the infinite lattice in terms of translations of a primitive cell by an amount $\mathbf{t}_n$ in Eq. 11-6, implying that lattice points occur only at the corners of the primitive cell. Some comments on primitive cells are in order. There is some arbitrariness associated with the primitive cell. Figure 11-1 shows a 2-dimensional primitive cell with primitive lattice vectors $\mathbf{a}_1$ and $\mathbf{a}_2$. However, another cell is also shown with primitive lattice vectors $\mathbf{a}_1$ and $\mathbf{a}_2'$. Both of these sets of lattice

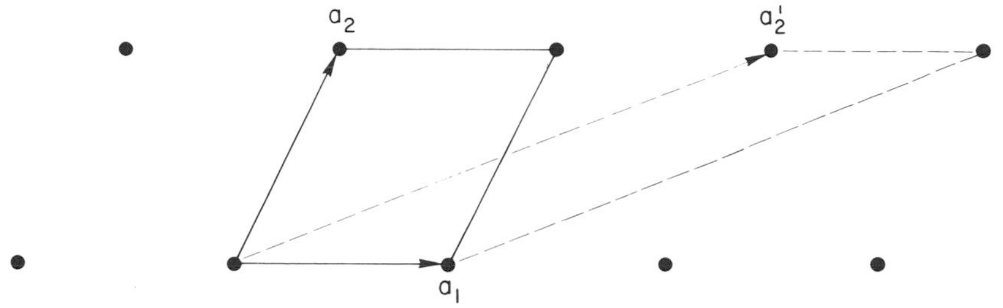


**Fig. 11-1** Two of the infinite number of primitive cells that can be used to define the lattice shown.

vectors could be used to generate the entire lattice. As can be seen there are infinite sets of primitive lattice vectors. Nevertheless all the primitive cells have the same volume and each contains one lattice point. For three dimensions the volume is $\mathbf{a}_1 \cdot (\mathbf{a}_2 \times \mathbf{a}_3)$.

One useful cell associated with the primitive cell is the **symmetrical unit cell** or **proximity cell** or **Wigner–Seitz cell.** (As we shall see in Chapter 12, this cell in reciprocal space is called the first Brillioun zone.) The Wigner–Seitz cell is obtained by starting at a lattice point which we call the origin and drawing vectors to all the nearby points of the lattice. Then perpendicular planes are constructed to pass through the midpoints of these vectors. The Wigner–Seitz cell is the cell which has the smallest volume about the origin bounded by these planes and all the points in this cell are closer to the origin then to any other lattice point. Figure 11-2 shows how a Wigner–Seitz cell is obtained for a body centered cubic lattice. As can be seen, the Wigner–Seitz cell displays the rotational symmetry of the lattice and this is one of the reasons why it is useful. These cells and how they are constructed will be discussed in more detail in Chapter 12 when Brillouin zones are discussed.

### b. The 14 Bravais lattices

In Section 4-5d we briefly discussed the seven crystal systems: triclinic, monoclinic, orthorhombic, tetragonal, trigonal, hexagonal, and cubic. The conditions on the lengths of the axes and the angles between the axes of the primitive cells can be found in Appendix 1. By associating a lattice point with the end point of the vectors that determine the primitive cell, seven lattices are obtained. We can ask the question: by placing lattice points within the primitive cells of any of these seven lattices, can another lattice be obtained that still satisfies the condition that the environment of every point is identical in arrangement and orientation with

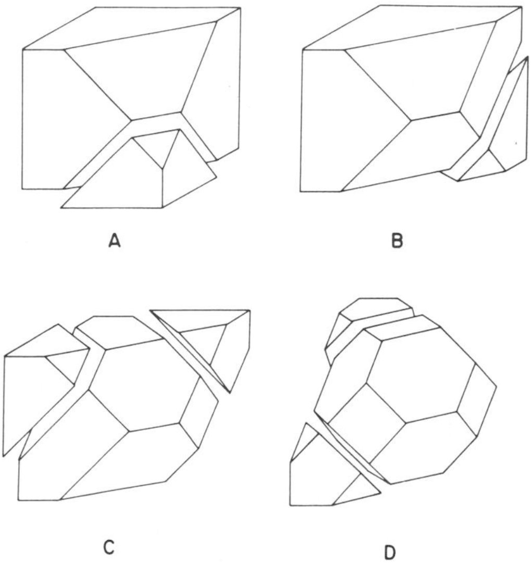


**Fig. 11-2** The construction of the Wigner–Seitz cell of a body-centered cubic lattice. Each of the cuts passes through the midpoints of the vectors from the center of the cell to the nearby lattice points.

every other point? The answer is yes. Fourteen space lattices are obtained and these are called the **14 Bravais lattices**. We sketch the approach that can be used to determine these lattices.

Consider a simple cubic lattice as shown in Fig. 11-3a. The primitive cell is outlined with dark lines. Now consider putting a lattice point in the body center position of each outlined primitive cell. The result is shown in Fig. 11-3b. Is this new arrangement, Fig. 11-3b, a lattice? The first objection one might have is that the cell, outlined with dark lines in Fig. 11-3b, is not a primitive cell because there are lattice points in places other than end points of the primitive lattice translation vector (Eq. 11-6). This is true but easy to correct, therefore let us consider the more important point: namely, is the environment of every point in Fig. 11-3b the same in arrangement and orientation? The answer is clearly yes. Each lattice point is surrounded by either other equivalent lattice points along the cube diagonal and all other neighbors have the same respective arrangement. Thus, Fig. 11-3b shows a lattice which is called a body-centered lattice and is usually referred to by the symbol I (Innenzentrierte) for an I-lattice or sometimes as bcc (body-centered cubic). We leave, for the moment, a description of the primitive cell until the third cubic space lattice is discussed.

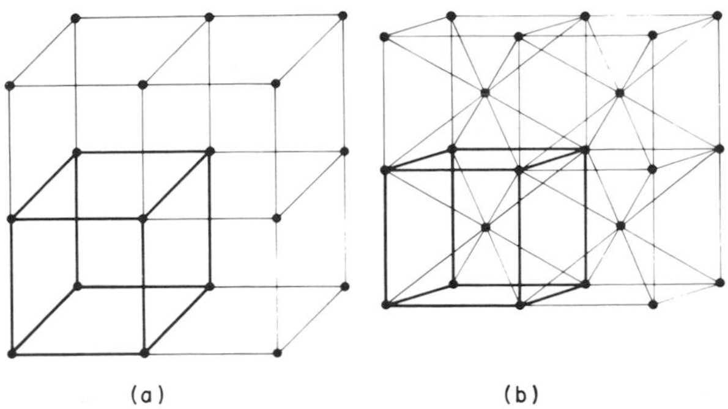

Fig. 11-3 (a) The primitive cell of a simple cubic lattice is outlined with dark lines. (b) A unit cell of a body-centered lattice is outlined with dark lines.

If, instead of putting a lattice point in the body center position of the simple cubic lattice in Fig. 11-3a, we place one on the face-centered position of every fact of the simple cubic lattice in Fig. 11-3a, Fig. 11-6 shows a small part of the lattice. It is clear that the environment of every point is identical, resulting in a space lattice. This lattice is called a face-centered cubic lattice and is usually referred to by the symbol F or sometimes fcc. (Convince yourself that a lattice is not formed if a point is placed at the center of every edge of a simple cubic lattice.)

We have obtained three space lattices within the cubic crystal system. The simple cubic lattice of Fig. 11-3a is clearly a primitive lattice and is referred to by the symbol P. The other two lattices are the I and F lattices. Note that each lattice point has six nearest neighbors in the P-lattice, eight in the I-lattice, and twelve in the F-lattice. These three possibilities were already alluded to in Section 8-7.

Now consider the true primitive cells (lattice points only at the end points of lattice vectors) for these three lattices. First define the term **unit cell** as a cell that can generate the lattice by the unit cell **translation vectors** in a manner as in Eq. 11-6, but this cell is not necessarily the smallest such cell. The primitive cell is the smallest such cell. Thus, unit cells that contain several primitive cells are sometimes called **multiply primitive cells.** We should emphasize that the choice of unit cell and primitive cell is arbitrary as can be seen in Fig. 11-1. The unit cell and primitive cell are usually picked in the conventional manner as is done in this chapter. However, other choices are occasionally made to highlight certain special aspects of a particular problem. The cubic P-lattice in Fig. 11-3a has the primitive cell outlined with dark lines, so the unit cell shown for this lattice is already primitive. For the cubic I-lattice the usual unit cell is outlined with dark lines in Fig. 11-3b. The three mutually orthogonal lattice vectors are called $\mathbf{a}_1$, $\mathbf{a}_2$, $\mathbf{a}_3$ so that the atom in the center is at $(\mathbf{a}_1 + \mathbf{a}_2 + \mathbf{a}_3)/2$. Similarly, the face-centered cubic lattice would have atoms located at the corners as well as at the face-centered positions, $(\mathbf{a}_1 + \mathbf{a}_2)/2$, etc. The primitive cell for the I- and F-lattices are easily obtained in terms of these lattice vectors $\mathbf{a}_i$. Defining $\mathbf{a}_i{}^p$ as the lattice vector of the primitive cell we have

| F | I |
|---|---|
| $\mathbf{a}_1{}^p = (\mathbf{a}_1 + \mathbf{a}_2)/2$ | $\mathbf{a}_1{}^p = (\mathbf{a}_1 + \mathbf{a}_2 + \mathbf{a}_3)/2$ |
| $\mathbf{a}_2{}^p = (\mathbf{a}_1 + \mathbf{a}_3)/2$ | $\mathbf{a}_2{}^p = (\mathbf{a}_1 - \mathbf{a}_2 + \mathbf{a}_3)/2$ |
| $\mathbf{a}_3{}^p = (\mathbf{a}_2 + \mathbf{a}_3)/2$ | $\mathbf{a}_3{}^p = (\mathbf{a}_1 + \mathbf{a}_2 - \mathbf{a}_3)/2$ |

for the lattice as listed. The primitive cells are shown in Fig. 11-4. We can see that the conventional I-cell shown in Fig. 11-3b has two atoms in the cell and twice the volume of its primitive cell shown in Fig. 11-4a. Similarly, the conventional F-cell has four times the number of atoms and volume of its primitive cell shown in Fig. 11-4b. This factor two and four are general results for all seven crystal systems. For completeness we mention that some crystal systems other then cubic can have a base-centered Bravais lattice. Usually such a lattice is called a C-lattice. This implies that the lattice is centered on the ab (or xy) plane (a plane that is perpendicular to the z-direction). Sometimes such a lattice is called an A- or B-lattice if, for convenience, it is described as centered on the bc or ac (yz or xz) plane. The C-lattice unit cell has two times as many atoms and twice the volume of its primitive cell, as will be seen later. For tutorial purposes Fig. 11-4c shows the Wigner–Seitz cell of an F-lattice which contains only one lattice point.

One might wonder why the C, F, and I cells are discussed and used since they are not primitive cells. The answer is fairly clear from Fig. 11-4. The primitive cells of the F and I-lattices are not convenient for visualization and by themselves do not display the full rotational symmetry of the lattice. The multiply primitive conventional unit cells display the rotational symmetry. Also it is more convenient, for many purposes, to describe all of the space lattices for one crystal system in terms of a similar set of axes. Thus, for the cubic crystal system we use the multiply primitive unit cells, F and I. However, it must be remembered that the primitive cell or any unit cell does have the full symmetry of the entire lattice when considered along with the infinite lattice translation operation. The primitive cell should be used in connection with problems where counting is concerned. For example, the number of $k \approx 0$ optic

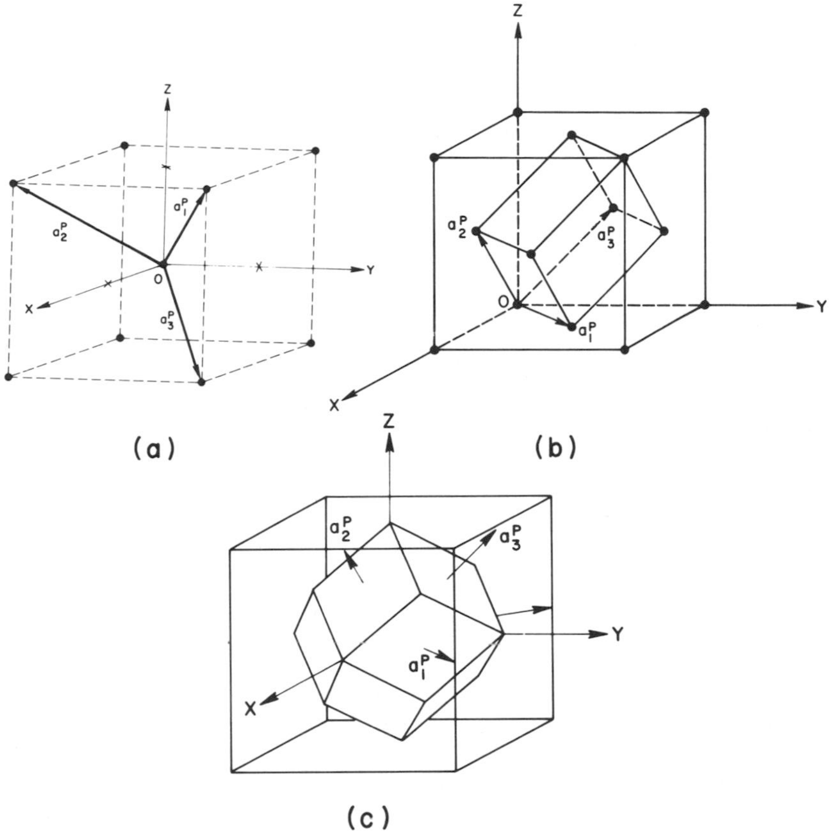


**Fig. 11-4** (a) The primitive translation vectors of a body-centered cubic lattice. (b) The primitive cell of a face-centered cubic lattice. (c) The Wigner–Seitz cell of a face-centered cubic lattice.

lattice vibrations in a crystal is 3n-3, where n is the number of atoms in the primitive cell, or 1/4 the number of atoms in the F-cell, or 1/2 the number of atoms in the I-cell (or 1/2 the number of atoms in the C-cell).

Now consider the possible lattices for the tetragonal crystal system. The primitive lattice and primitive unit cell is obvious from the tetragonal crystal system, so we have a P-space lattice. It is also easy to see that a body-centered lattice is a true lattice, so we have a tetragonal, I-space lattice. Some difficulties arise when various types of face-centered lattices are examined. Consider the (001), or base face, to be centered as in Fig. 11-5a. This will be called a C-lattice. This indeed is a lattice but it is not a new type since it is equivalent to a P-lattice with primitive lattice vectors $\mathbf{a}_1$ and $\mathbf{a}_2$ 45° to the original set as shown by the dotted lines in

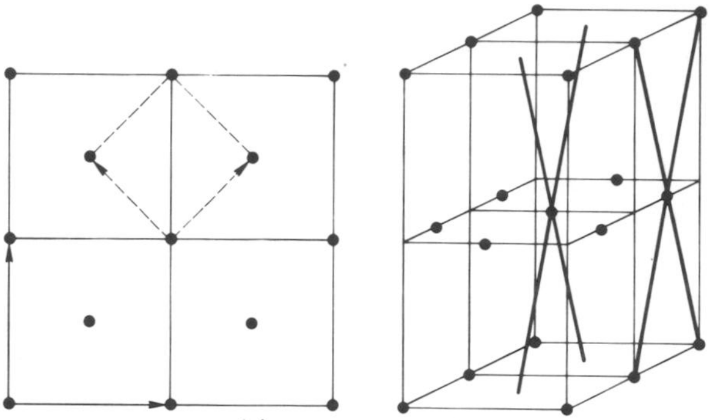


**Fig. 11-5** (a) A base-centered tetragonal lattice showing, with dotted lines, a smaller primitive cell. (b) A side-centered tetragonal arrangement showing that is is not a lattice.

Fig. 11-5a. Next consider the (100) and (010) faces to be centered as shown in Fig. 11-5b. It is clear that the environment of every point is not the same, so a lattice is not formed in this manner. The last type of centering to consider is a lattice point on all the faces. A true lattice is formed but is equivalent to an I-lattice in a different orientation. (See Problem 3.) Thus, for the tetragonal crystal system two space lattices are found, one P- and one I-lattice. (One could also say P ≡ C and I ≡ F lattices as discussed.)

In this way we have obtained three space lattices for the cubic crystal system and two for the tetragonal crystal system. In a similar manner the other five crystal systems can be investigated. Figure 11-6 shows the conventional unit cell (multiprimitive where appropriate) of the 14 space lattices that are obtained from the seven crystal systems (syngony). These space lattices are usually called the 14 Bravais lattices.

### c. Point groups and space groups

In the beginning of Section 11-3 we defined the terms lattice, basis, and crystal structure. In Section 11-3b we discussed the 14 Bravais lattices. Now we add a basis to a lattice to form a crystal structure. The translational symmetry Eq. 11-6 of the crystal structure is contained in the Bravais lattice. Now consider what happens as bases with different point symmetry are attached to lattice points of the 14 Bravais lattices. Then crystal structures with 32 different types of point symmetries are obtained. These are the 32 crystal classes corresponding to the 32 point groups. There are 73 crystal structures that can be obtained in this man-

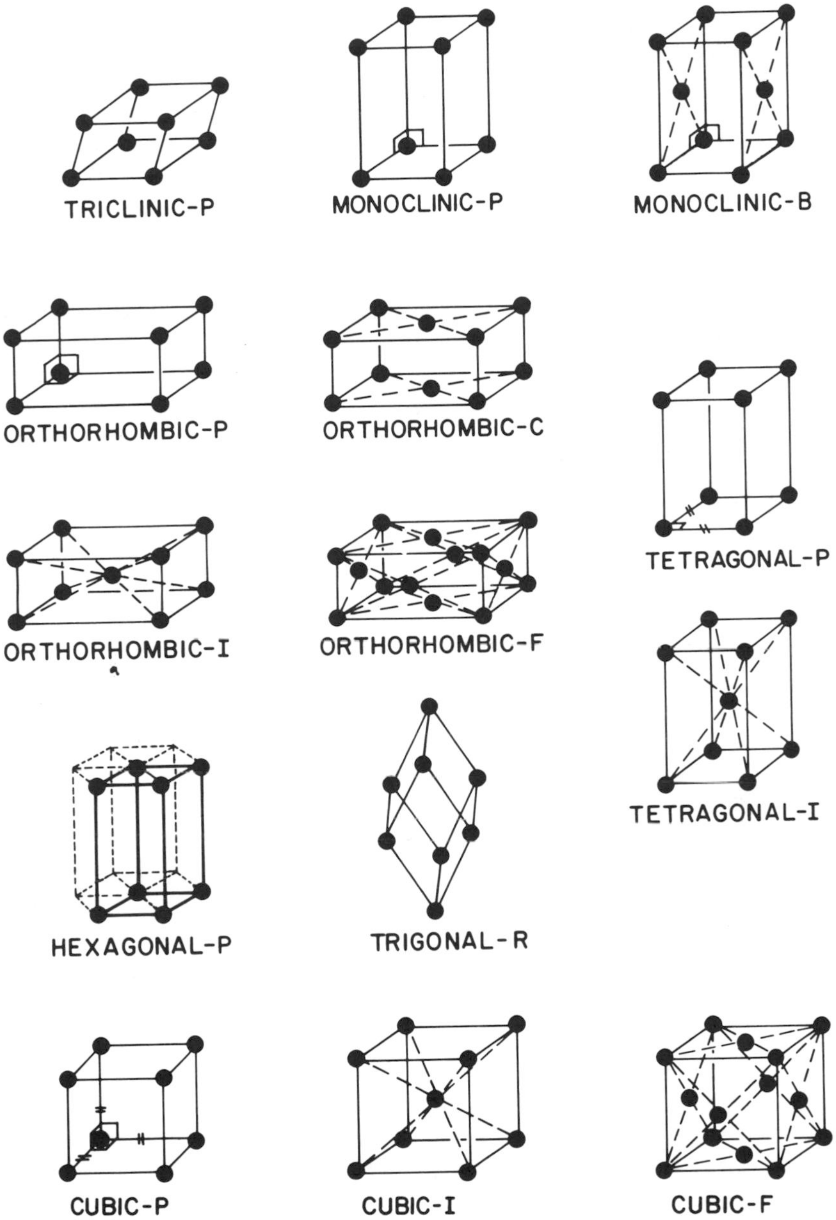


**Fig. 11-6** The conventional unit cell for the 14 Bravais lattices.

ner. These correspond to the 73 symmorphic space groups. We proceed by sketching how some of these results are obtained.

Consider the P, F, and I Bravais lattices for the cubic crystal system. Let the basis take the form of an octahedron and place it on each lattice point so that the 4-fold axes of the basis coincide with the 4-fold axes of the unit cell. Then, about an origin at the center of the multiply primitive cell, all 48 point symmetry operations corresponding to the point group $O_h$ will take the crystal structure into itself. Crystal structures with these three space lattices and the basis described are said to have $O_h$ point symmetry and are members of the $O_h$ crystal class. Note that all three cubic Bravais lattices can have the same point symmetry with the appropriate basis. Thus, three different space groups (combinations of a Bravais lattice and a basis) have been generated. (In international notation, discussed in the next section, these three space groups are labeled Pm3m, Fm3m, and Im3m; in the Schoenflies notation these are labeled $O_h^1$, $O_h^5$, $O_h^9$.) Instead of a basis in the shape of an octahedron, if a basis in the shape of a tetrahedron is used, many of the point symmetry operations of $O_h$ are not allowed. Instead the point symmetry $T_d$ is found for all three cubic Bravais lattices and three new space groups are formed. (In the international or Schoenflies notation these are labeled $P\bar{4}3m$, $F\bar{4}3m$, and $I\bar{4}3m$ or $T_d^1$, $T_d^2$, and $T_d^3$.) This procedure can be continued but a problem quickly arises. For example, can we put a basis with just $C_1$ point symmetry on a cubic Bravais lattice? This same type of problem will arise when other crystal systems are considered. Namely, can we put a basis with $O_h$ point symmetry on a triclinic Bravais lattice?

The answer to these two questions is no. The reason is that physically the forces due to a cubic Bravais lattice can be expanded about a lattice point, as in the crystal field problem in Chapter 8, and combinations of spherical harmonics result in cubic symmetry. There are no combinations that have lower symmetry, so on a cubic Bravais lattice, the basis must have one of the cubic point symmetries. Similarly for a triclinic Bravais lattice the forces expanded about a lattice point have triclinic symmetry, so these forces would physically distort a basis of higher symmetry to triclinic point symmetry.

We would now like to have a systematic way to relate the symmetry of the basis to that of the Bravais lattice, but first one must consider the hierarchy of the seven crystal systems as shown in Fig. 11-7. Every crystal system can be produced (from the one or two crystal systems connected by a line to the right) by an infinitesimal distortion. For example, if a cubic crystal system is pulled along one of the 3 fold axes, a [111] direction, a trigonal lattice results. If pulled along a 4-fold axis

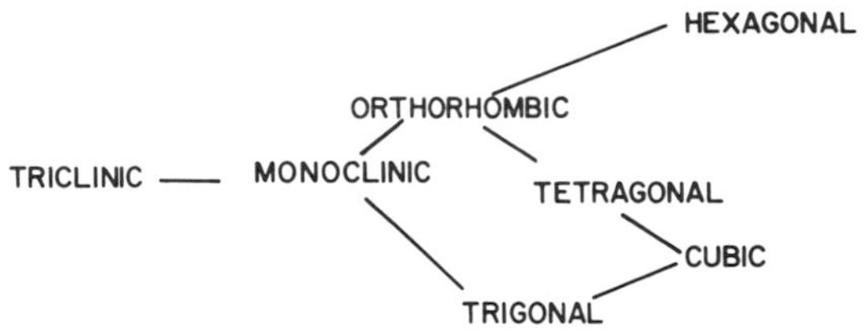


**Fig. 11-7** The hierarchy of the seven crystal systems.

[100], a tetragonal lattice results. Now we can state the rules that limit the symmetry of the basis, for a given Bravais lattice.

(1) All the symmetry operations possessed by the basis must also be possessed by the lattice. (The lattice can possess more symmetry elements.)

(2) The basis must possess at least one symmetry element that is not found in the lattice that is next lower on the hierarchy of the seven crystal systems.

The physical reason for these rules was discussed in the beginning of this paragraph. These two rules can be stated group theoretically as follows: In a crystal structure, the point group of the basis must be a subgroup of the point group of the Bravais lattice but not of a lattice that is lower in the hierarchy. Hence, the procedure to determine the various space lattices (crystal structures) is now straightforward. Consider for one crystal system one Bravais lattice. The point group of the lattice is the **holohedral point group** (highest point group compatible with the Bravais lattice). The symmetry of the basis can be that of any of the subgroups of the holohedral group that is still within the crystal system. For example, for any one of the three cubic Bravais lattices the basis can have point symmetry $O_h$, O, $T_d$, $T_h$, or T. Since there are three different Bravais lattices for the cubic crystal system, a total of 15 space groups are obtained in this manner. For the orthorhombic crystal system there are four different Bravais lattices (P, F, I, and C) and three point groups ($D_{2h}$, $D_2$, and $C_{2v}$) giving 12 space groups. In this manner the 14 Bravais lattices combined with the appropriate groups of the 32 point groups result in 66 space groups (see Problem 9). These are all symmorphic space groups. Actually, there are 73 symmorphic space groups; others are obtained because in a few cases the basis can be oriented in two ways in a given Bravais lattice (see Problem 10). After the international notation is discussed in the next section we return to the discussion of showing how nonsymmorphic space groups are obtained.

## 11-4 International Tables for X-Ray Crystallography, International Notation, etc.

In this section we discuss "other" notations, a topic that has been carefully avoided until now. The principal new notation that will be discussed is the so-called international notation or Hermann–Mauguin notation. This is used to describe symmetry operations, point groups, and space groups and is used in the standard reference "International Tables for X-Ray Crystallography" (see the Notes) and elsewhere.

### a. International notation

Table 11-1 shows the meaning of the symbols used for the international notation. Appendix 4 lists all 230 space groups in Schoenflies and international notation. For space groups the international notation is more descriptive. However, the Schoenflies notation immediately shows the point group that is isomorphic with the factor group of the space group, although this information can also be obtained from the international space group symbol.

The use of the international notation for the description of space groups (Appendix 4) and point groups (Appendix 2) is summarized in Table 11-1. It is worthwhile discussing a few examples for the point groups. A 2-fold axis perpendicular to a principal axis is X2. A vertical reflection plane ($\sigma_v$) containing the principal axis is Xm, while a horizontal reflection plane is X/m. Thus the point group $C_{4v}$ would be 4mm, while $C_{4h}$ is 4/m, and $D_{4h}$ is 4/mmm. However the symbol 4/mmm is the "short" symbol for the point group. There is also a more expressive and longer "full" symbol 4/m 2/m 2/m. The full symbol is not used very often except in tables so we will ignore it. Besides using the symmetry operations shown in the short symbol, the symmetry operations in the full symbol can be obtained. This can be seen by the use of stereograms (Appendix 1). The international symbols for the point groups $C_6$, $D_6$, $C_{6h}$, $C_{6v}$, and $D_{6h}$ are 6, 622, 6/m, 6mm, and 6/mmm, respectively. As will be seen, the international notation is extremely useful when describing the magnetic point groups.

We now discuss a few examples of the use of the notation for space groups. (For **notation** we use the international symbol and the Schoenflies symbol in parenthesis. This is not normally done but, hopefully, it will help the reader.) Consider space group number 75 which is the first listed tetragonal space group P4 ($C_4^1$). The symbol tells us that the Bravais lattice is a primitive lattice and the only symmetry operation is a 4-fold axis or 4. We also immediately see that the space group is symmorphic.

**Table 11-1** Summary of the meaning of the symbols used in international notation

| Symbol | |
|---|---|
| X | X-fold axis of rotation ($C_n$). X can be 1, 2, 3, 4, 6 |
| $\overline{X}$ | X-fold rotation followed by inversion ($iC_n$) |
| $X_p$ | Screw operations. X-fold rotation followed by a translation a fraction p/X along the c-axis. |
| m | Reflection plane ($\sigma$) |
| a,b, or c | Glide planes. A reflection followed by a translation a/2, b/2 or c/2 along the x, y or z axis, respectively (or an amount (a+b+c)/3 along [111] on rhombohedral axes) |
| n | Diagonal glide plane. A reflection followed by a glide on amount (a+b+c)/2 in tetragonal and cubic crystal systems or on amount (a+b)/2 or (b+c)/2 or (c+a)/2 in other systems depending on the position in the space group symbol. |
| d | Diamond glide plane. A reflection followed by a glide an amount (a+b+c)/4 in tetragonal and cubic systems or an amount (a±b)/4) or (b±c)/4 or (c±a)/4 in other systems |
| **Bravais lattice types** | |
| P | Primitive unit cell (In rhombohedral systems R is used.) |
| F | Face-centered unit cell |
| I | Body-centered unit cell |
| C | Base-centered unit cell (A or B if the a or b-face is centered rather than the c-face) |

The next space group number 76 $P4_1$ ($C_4^2$) is obviously not symmorphic, in fact the symmetry operation is just a 4 rotation followed by a translation by an amount 1/4 along the z-axis. The next two space groups are nonsymmorphic. Then number 79, I4 ($C_4^5$) is again symmorphic but with a body-centered Bravais lattice. In the next section, using the international notation, we show how some of the space groups are obtained in a systematic manner, and in the following section we discuss some of the information that can be obtained from the "International Tables for X-Ray Crystallography."

Recall, the point group is obtained by setting all translations to zero including the ones that are a fraction of a unit cell. So all a, b, c, n, or d planes are replaced with m. Similarly, all screw axes are replaced with a simple notation. So the space group Pmc2 or Cmc2 has a point group mm2 ($C_{2v}$); $F4_1/a$ has 4/m ($C_{4h}$). Appendix 2 shows all the point groups in both notations.

There is no such thing as an international notation for irreducible representations because crystallographers are mostly interested in the symmetry operations. However, there are other notations for irreducible representations besides the Mulliken and Bethe notation that we have been using. The Bouckaert–Smoluchowski–Wigner (BSW) notation is often used by solid state physicists. This will be mentioned in Chapter 12 and Appendix 3. There are various notations that one finds for the double groups (see Appendix 10 and Fig. 8-6) and care must always be exercised with respect to notation. Besides the Schoenflies and international notation for symmetry operations several other notations are occasionally used. We only mention one of these (Section 11-4d) and this is the notation used in Kovalev's book. The only other notation, besides the Schoenflies and international notation, that is sometimes used for space groups is that of Shubnikov (see the Notes), but we will not discuss it here.

### b. Determination of some space groups

In Section 11-2b the simple approach that enables one to obtain the 73 symmorphic space groups was outlined. Here we would like to extend these considerations to include glide planes and screw axes so that all 230 space groups can be obtained. We also show projections of some of the resultant crystal structures and symmetry operations.

Let us consider one of the monoclinic point groups and determine all (three) of the space groups that have this point group. The holohedral monoclinic point group is 2/m ($C_{2h}$), but let us consider a subgroup 2 ($C_2$) which is not a subgroup of a crystal system below monoclinic on the hierachy in Fig. 11-7. First investigate the P-monoclinic space lattice. Figure 11-8a shows a b–c plain view of the lattice. Attach a basis to a lattice point. **Conventionally** an open circle is used, as in Fig. 11-8a, where the + sign means a distance z above the plain of the paper. Operating with a 2 ($C_2$) point operation gives the arrangement as shown which still has the P-monoclinic space lattice translational symmetry. Thus, we have an arrangement that is a space group called $C_2^1$ in the Schoenflies notation or P2 in the international notation. Figure 11-8b shows the same space group but looking down the principle axis with only the symmetry operations shown; the presence of certain symmetry operations implies the existence of others which happens often for space groups and point groups (as discussed in this chapter as well as Chapter 1). In the

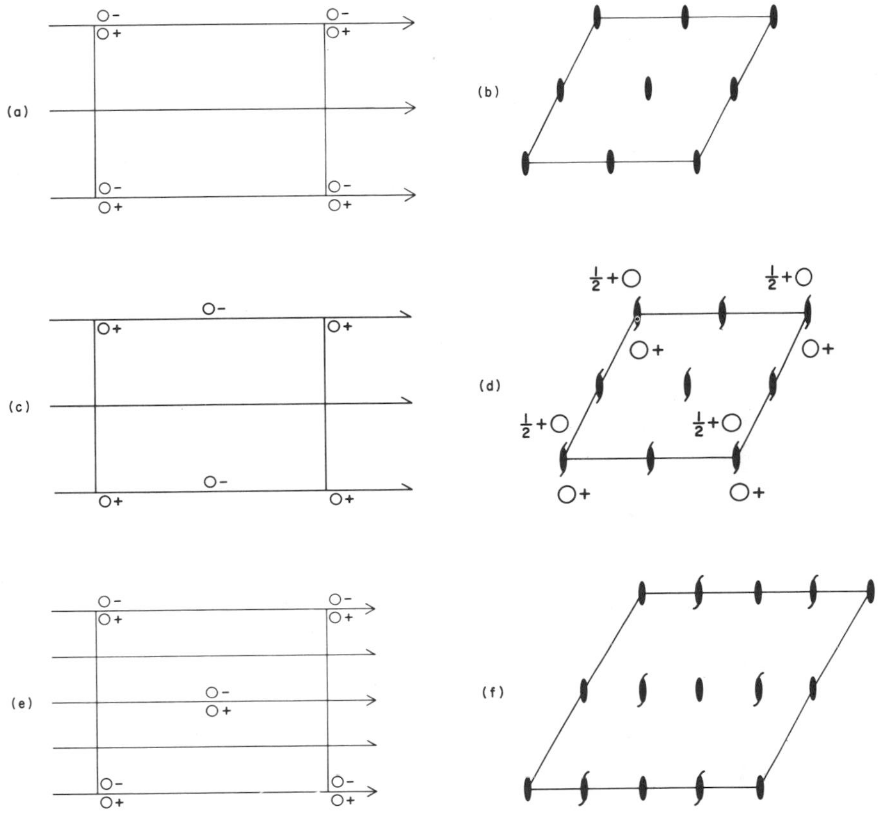


**Fig. 11-8** (a) The b–c plane view of a lattice with space group $P_2(C_2{}^1)$. (b) The same lattice, viewed down the principal axis, just showing the symmetry operations. (c) The b–c plane view of a lattice with space group $P2_1(C_2{}^2)$. (d) The same lattice as in (c), but viewed down the principal axis showing the symmetry operations and general positions. (e) The b–c plane view of a lattice with space group $C2(C_2{}^3)$. (f) principal axis view of the lattice with space group $C2(C_2{}^3)$.

"International Tables for X-Ray Crystallography," both types of figures for most of the space groups are given.

Figure 11-8c shows another arrangement that can be obtained from the P-monoclinic space lattice with a 2-fold screw axis along the unique c-axis. The resultant space group is called $C_2{}^2$ in the Schoenflies notation or $P2_1$ in the international notation. The **convention** of 1/2 + means the circle is located at 1/2 the unit cell height plus an amount z. Figure 11-8d shows the symmetry operation of the same space group with a principal axis view. Again note how other symmetry operations occur.

Figure 11-8e and 8f show the space group with a 2-fold rotation symmetry operation with a C-monoclinic Bravais lattice. The Schoenflies and international notation for this space group is $C_2{}^3$ and C2, respectively. In this example we can see how centering the Bravais lattice implies 2-fold screw axes, hence there is no new space group $C2_1$. This situation, various centering of the Bravais lattices reducing the number of centered space groups compared to P-lattice space groups, happens for many of the crystal systems. So for the monoclinic crystal system considered here there are two lattice (P and C) and two types of rotations (2 and $2_1$); one might expect four space groups but only three are obtained.

The **conventions** for the general points are in Appendix 1. The only other one needed here, besides the ones already mentioned ($\pm$ meaning a general point an amount z above or below the zero level of the unit cell, and 1/2 means a level 1/2 above the zero level to which one may add or subtract an amount z, i.e., 1/2 + or 1/2 −) is right-handed to left-handedness. These two positions are ○ and ⊙ and are related by a mirror reflection or center of inversion. Either of these operations will change the handedness of a molecule. The two positions ○ and ⊙ are **enantiomorphous**.

As another example of the determination of all possible space groups within some restricted point group, consider the orthorhombic crystal system with point group mm2 ($C_{2v}$). This is a simple system to consider since there is a unique axis namely, the 2-fold axis. Actually the two mirror planes automatically imply the 2-fold axis as seen with the aid of a stereogram. This carries over to the space groups. Consider now the P-lattice. First take the (100) plane (the x-axis is normal to the planes) to be an m plane, then the planes parallel to the (010) plane can be m, a, c, or n-planes (see Table 11-1). This leads to four space groups Pmm, Pma, Pmc, and Pmn where there are no symbols in the last place since the primitive lattice, and the first two planes determine the last symbols. (In fact the space groups are Pmm2, Pma2, $Pmc2_1$, and $Pmn2_1$.) If the (100) planes are c-glides, only Pca, Pcc, and Pcn are new (since Pcm is the same as Pmc with a new orientation). Again we leave the last 2 or $2_1$ out since it is determined from the other symmetry elements. Similarly, we may have Pba and Pbn and lastly, the (100) plane may be a diagonal glide plane n to give Pnn. (See the Problems for the determination of the third symbol.) Thus, ten space groups are obtained. Again we see how the number of space groups is limited.

In this manner the 230 space groups can be obtained and these are listed in the "International Tables for X-Ray Crystallography." We proceed by observing what other information about symmetry can be found from these tables.

**Table 11-2** Two pages from the book "International Tables for X-Ray Crystallography"

# *Pm*3*m* $O_h^1$

No. 221 $P\,4/m\,\bar{3}\,2/m$ *m* 3 *m* Cubic

Origin at centre (*m*3*m*)

| Number of positions, Wyckoff notation, and point symmetry | | | Co-ordinates of equivalent positions |
|---|---|---|---|
| 48 | *n* | 1 | $x,y,z;\ z,x,y;\ y,z,x;\ x,z,y;\ y,x,z;\ z,y,x;$ |
| | | | $x,\bar{y},\bar{z};\ z,\bar{x},\bar{y};\ y,\bar{z},\bar{x};\ x,\bar{z},\bar{y};\ y,\bar{x},\bar{z};\ z,\bar{y},\bar{x};$ |
| | | | $\bar{x},y,\bar{z};\ \bar{z},x,\bar{y};\ \bar{y},z,\bar{x};\ \bar{x},z,\bar{y};\ \bar{y},x,\bar{z};\ \bar{z},y,\bar{x};$ |
| | | | $\bar{x},\bar{y},z;\ \bar{z},\bar{x},y;\ \bar{y},\bar{z},x;\ \bar{x},\bar{z},y;\ \bar{y},\bar{x},z;\ \bar{z},\bar{y},x;$ |
| | | | $\bar{x},\bar{y},\bar{z};\ \bar{z},\bar{x},\bar{y};\ \bar{y},\bar{z},\bar{x};\ \bar{x},\bar{z},\bar{y};\ \bar{y},\bar{x},\bar{z};\ \bar{z},\bar{y},\bar{x};$ |
| | | | $\bar{x},y,z;\ \bar{z},x,y;\ \bar{y},z,x;\ \bar{x},z,y;\ \bar{y},x,z;\ \bar{z},y,x;$ |
| | | | $x,\bar{y},z;\ z,\bar{x},y;\ y,\bar{z},x;\ x,\bar{z},y;\ y,\bar{x},z;\ z,\bar{y},x;$ |
| | | | $x,y,\bar{z};\ z,x,\bar{y};\ y,z,\bar{x};\ x,z,\bar{y};\ y,x,\bar{z};\ z,y,\bar{x}.$ |
| 24 | *m* | *m* | $x,x,z;\ z,x,x;\ x,z,x;\ \bar{x},\bar{x},\bar{z};\ \bar{z},\bar{x},\bar{x};\ \bar{x},\bar{z},\bar{x};$ |
| | | | $x,\bar{x},\bar{z};\ z,\bar{x},\bar{x};\ x,\bar{z},\bar{x};\ \bar{x},x,z;\ \bar{z},x,x;\ \bar{x},z,x;$ |
| | | | $\bar{x},x,\bar{z};\ \bar{z},x,\bar{x};\ \bar{x},z,\bar{x};\ x,\bar{x},z;\ z,\bar{x},x;\ x,\bar{z},x;$ |
| | | | $\bar{x},\bar{x},z;\ \bar{z},\bar{x},x;\ \bar{x},\bar{z},x;\ x,x,\bar{z};\ z,x,\bar{x};\ x,z,\bar{x}.$ |
| 24 | *l* | *m* | $\tfrac{1}{2},y,z;\ z,\tfrac{1}{2},y;\ y,z,\tfrac{1}{2};\ \tfrac{1}{2},z,y;\ y,\tfrac{1}{2},z;\ z,y,\tfrac{1}{2};$ |
| | | | $\tfrac{1}{2},\bar{y},\bar{z};\ \bar{z},\tfrac{1}{2},\bar{y};\ y,\bar{z},\tfrac{1}{2};\ \tfrac{1}{2},\bar{z},\bar{y};\ \bar{y},\tfrac{1}{2},\bar{z};\ \bar{z},\bar{y},\tfrac{1}{2};$ |
| | | | $\tfrac{1}{2},y,\bar{z};\ \bar{z},\tfrac{1}{2},y;\ y,\bar{z},\tfrac{1}{2};\ \tfrac{1}{2},\bar{z},y;\ y,\tfrac{1}{2},\bar{z};\ \bar{z},y,\tfrac{1}{2};$ |
| | | | $\tfrac{1}{2},\bar{y},z;\ z,\tfrac{1}{2},\bar{y};\ \bar{y},z,\tfrac{1}{2};\ \tfrac{1}{2},z,\bar{y};\ \bar{y},\tfrac{1}{2},z;\ z,\bar{y},\tfrac{1}{2}.$ |
| 24 | *k* | *m* | $0,y,z;\ z,0,y;\ y,z,0;\ 0,z,y;\ y,0,z;\ z,y,0;$ |
| | | | $0,\bar{y},\bar{z};\ \bar{z},0,\bar{y};\ \bar{y},\bar{z},0;\ 0,\bar{z},\bar{y};\ \bar{y},0,\bar{z};\ \bar{z},\bar{y},0;$ |
| | | | $0,y,\bar{z};\ \bar{z},0,y;\ y,\bar{z},0;\ 0,\bar{z},y;\ y,0,\bar{z};\ \bar{z},y,0;$ |
| | | | $0,\bar{y},z;\ z,0,\bar{y};\ \bar{y},z,0;\ 0,z,\bar{y};\ \bar{y},0,z;\ z,\bar{y},0.$ |
| 12 | *j* | *mm* | $\tfrac{1}{2},x,x;\ x,\tfrac{1}{2},x;\ x,x,\tfrac{1}{2};\ \tfrac{1}{2},x,\bar{x};\ \bar{x},\tfrac{1}{2},x;\ x,\bar{x},\tfrac{1}{2};$ |
| | | | $\tfrac{1}{2},\bar{x},\bar{x};\ \bar{x},\tfrac{1}{2},\bar{x};\ \bar{x},\bar{x},\tfrac{1}{2};\ \tfrac{1}{2},\bar{x},x;\ x,\tfrac{1}{2},\bar{x};\ \bar{x},x,\tfrac{1}{2}.$ |
| 12 | *i* | *mm* | $0,x,x;\ x,0,x;\ x,x,0;\ 0,x,\bar{x};\ \bar{x},0,x;\ x,\bar{x},0;$ |
| | | | $0,\bar{x},\bar{x};\ \bar{x},0,\bar{x};\ \bar{x},\bar{x},0;\ 0,\bar{x},x;\ x,0,\bar{x};\ \bar{x},x,0.$ |
| 12 | *h* | *mm* | $x,\tfrac{1}{2},0;\ 0,x,\tfrac{1}{2};\ \tfrac{1}{2},0,x;\ x,0,\tfrac{1}{2};\ \tfrac{1}{2},x,0;\ 0,\tfrac{1}{2},x;$ |
| | | | $\bar{x},\tfrac{1}{2},0;\ 0,\bar{x},\tfrac{1}{2};\ \tfrac{1}{2},0,\bar{x};\ \bar{x},0,\tfrac{1}{2};\ \tfrac{1}{2},\bar{x},0;\ 0,\tfrac{1}{2},\bar{x}.$ |
| 8 | *g* | 3*m* | $x,x,x;\ x,\bar{x},\bar{x};\ \bar{x},x,\bar{x};\ \bar{x},\bar{x},x;$ |
| | | | $\bar{x},\bar{x},\bar{x};\ \bar{x},x,x;\ x,\bar{x},x;\ x,x,\bar{x}.$ |
| 6 | *f* | 4*mm* | $x,\tfrac{1}{2},\tfrac{1}{2};\ \tfrac{1}{2},x,\tfrac{1}{2};\ \tfrac{1}{2},\tfrac{1}{2},x;\ \bar{x},\tfrac{1}{2},\tfrac{1}{2};\ \tfrac{1}{2},\bar{x},\tfrac{1}{2};\ \tfrac{1}{2},\tfrac{1}{2},\bar{x}.$ |
| 6 | *e* | 4*mm* | $x,0,0;\ 0,x,0;\ 0,0,x;\ \bar{x},0,0;\ 0,\bar{x},0;\ 0,0,\bar{x}.$ |
| 3 | *d* | 4/*mmm* | $\tfrac{1}{2},0,0;\ 0,\tfrac{1}{2},0;\ 0,0,\tfrac{1}{2}.$ |
| 3 | *c* | 4/*mmm* | $0,\tfrac{1}{2},\tfrac{1}{2};\ \tfrac{1}{2},0,\tfrac{1}{2};\ \tfrac{1}{2},\tfrac{1}{2},0.$ |
| 1 | *b* | *m*3*m* | $\tfrac{1}{2},\tfrac{1}{2},\tfrac{1}{2}.$ |
| 1 | *a* | *m*3*m* | $0,0,0.$ |

Conditions limiting possible reflections

General:

*hkl*:
*hhl*: No conditions
0*kl*:

Special:

No conditions

**Table 11-2** (continued)

| | | | | |
|---|---|---|---|---|
| *$P4_2/mnm$* $D_{4h}^{14}$ | No. 136 | $P\,4_2/m\;2_1/n\;2/m$ | $4/m\;m\;m$ | Tetragonal |

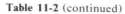

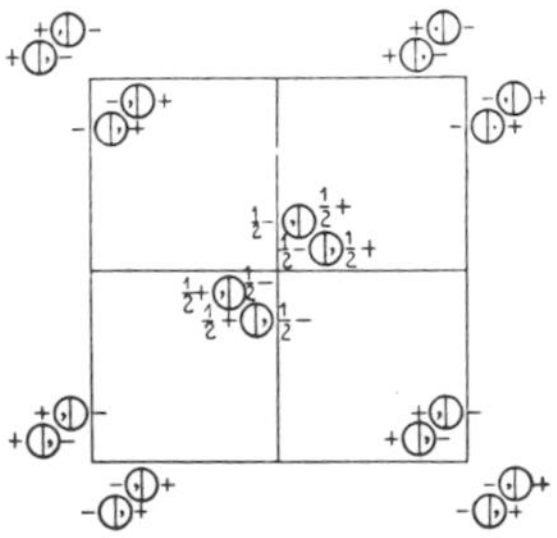

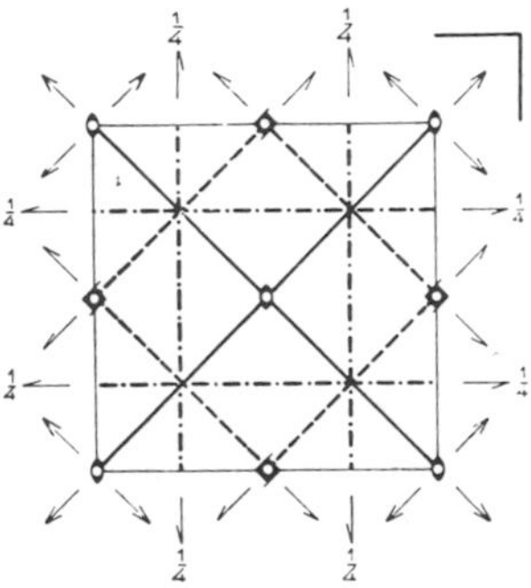


Origin at centre (*mmm*)

| Number of positions, Wyckoff notation, and point symmetry | | | Co-ordinates of equivalent positions | Conditions limiting possible reflections |
|---|---|---|---|---|
| | | | | General: |
| 16 | *k* | 1 | $x,y,z;\ \bar{x},\bar{y},z;\ \tfrac{1}{2}+x,\tfrac{1}{2}-y,\tfrac{1}{2}+z;\ \tfrac{1}{2}-x,\tfrac{1}{2}+y,\tfrac{1}{2}+z;$ | *hkl*: No conditions |
| | | | $x,y,\bar{z};\ \bar{x},\bar{y},\bar{z};\ \tfrac{1}{2}+x,\tfrac{1}{2}-y,\tfrac{1}{2}-z;\ \tfrac{1}{2}-x,\tfrac{1}{2}+y,\tfrac{1}{2}-z;$ | *hk*0: No conditions |
| | | | $y,x,z;\ \bar{y},\bar{x},z;\ \tfrac{1}{2}+y,\tfrac{1}{2}-x,\tfrac{1}{2}+z;\ \tfrac{1}{2}-y,\tfrac{1}{2}+x,\tfrac{1}{2}+z;$ | 0*kl*: $k+l=2n$ |
| | | | $y,x,\bar{z};\ \bar{y},\bar{x},\bar{z};\ \tfrac{1}{2}+y,\tfrac{1}{2}-x,\tfrac{1}{2}-z;\ \tfrac{1}{2}-y,\tfrac{1}{2}+x,\tfrac{1}{2}-z.$ | *hhl*: No conditions |
| | | | | Special: as above, plus |
| 8 | *j* | *m* | $x,x,z;\ \bar{x},\bar{x},z;\ \tfrac{1}{2}+x,\tfrac{1}{2}-x,\tfrac{1}{2}+z;\ \tfrac{1}{2}-x,\tfrac{1}{2}+x,\tfrac{1}{2}+z;$ | no extra conditions |
| | | | $x,x,\bar{z};\ \bar{x},\bar{x},\bar{z};\ \tfrac{1}{2}+x,\tfrac{1}{2}-x,\tfrac{1}{2}-z;\ \tfrac{1}{2}-x,\tfrac{1}{2}+x,\tfrac{1}{2}-z.$ | |
| 8 | *i* | *m* | $x,y,0;\ \bar{x},\bar{y},0;\ \tfrac{1}{2}+x,\tfrac{1}{2}-y,\tfrac{1}{2};\ \tfrac{1}{2}-x,\tfrac{1}{2}+y,\tfrac{1}{2};$ | |
| | | | $y,x,0;\ \bar{y},\bar{x},0;\ \tfrac{1}{2}+y,\tfrac{1}{2}-x,\tfrac{1}{2};\ \tfrac{1}{2}-y,\tfrac{1}{2}+x,\tfrac{1}{2}.$ | |
| 8 | *h* | 2 | $0,\tfrac{1}{2},z;\ 0,\tfrac{1}{2},\bar{z};\ 0,\tfrac{1}{2},\tfrac{1}{2}+z;\ 0,\tfrac{1}{2},\tfrac{1}{2}-z;$ | *hkl*: $h+k=2n;\ l=2n$ |
| | | | $\tfrac{1}{2},0,z;\ \tfrac{1}{2},0,\bar{z};\ \tfrac{1}{2},0,\tfrac{1}{2}+z;\ \tfrac{1}{2},0,\tfrac{1}{2}-z.$ | |
| 4 | *g* | *mm* | $x,\bar{x},0;\ \bar{x},x,0;\ \tfrac{1}{2}+x,\tfrac{1}{2}+x,\tfrac{1}{2};\ \tfrac{1}{2}-x,\tfrac{1}{2}-x,\tfrac{1}{2}.$ | no extra conditions |
| 4 | *f* | *mm* | $x,x,0;\ \bar{x},\bar{x},0;\ \tfrac{1}{2}+x,\tfrac{1}{2}-x,\tfrac{1}{2};\ \tfrac{1}{2}-x,\tfrac{1}{2}+x,\tfrac{1}{2}.$ | |
| 4 | *e* | *mm* | $0,0,z;\ 0,0,\bar{z};\ \tfrac{1}{2},\tfrac{1}{2},\tfrac{1}{2}+z;\ \tfrac{1}{2},\tfrac{1}{2},\tfrac{1}{2}-z.$ | *hkl*: $h+k+l=2n$ |
| 4 | *d* | $\bar{4}$ | $0,\tfrac{1}{2},\tfrac{1}{4};\ \tfrac{1}{2},0,\tfrac{1}{4};\ 0,\tfrac{1}{2},\tfrac{3}{4};\ \tfrac{1}{2},0,\tfrac{3}{4}.$ | *hkl*: $h+k=2n;\ l=2n$ |
| 4 | *c* | $2/m$ | $0,\tfrac{1}{2},0;\ \tfrac{1}{2},0,0;\ 0,\tfrac{1}{2},\tfrac{1}{2};\ \tfrac{1}{2},0,\tfrac{1}{2}.$ | |
| 2 | *b* | *mmm* | $0,0,\tfrac{1}{2};\ \tfrac{1}{2},\tfrac{1}{2},0.$ | *hkl*: $h+k+l=2n$ |
| 2 | *a* | *mmm* | $0,0,0;\ \tfrac{1}{2},\tfrac{1}{2},\tfrac{1}{2}.$ | |

## c. "Typical" International Tables, X-Ray Crystallography page

Since there are 230 space groups, it would be lengthy, to say the least, to discuss all the possibilities here. Instead we choose some "typical" space groups. We show several examples of the type of detailed information a noncrystallographer can obtain from the "International

Tables for X-Ray Crystallography." Table 11-2 shows two pages from these tables. Let us start across the top and work our way down for space group Pm3m. The top left lists the symbol for the space group in the international and Schoenflies notation. From the symbol Pm3m we see that the cell is a primitive one and the space group is symmorphic. We also see the types of symmetry operations, although not completely since there are many for this space group. Next, across the top is a number. The space groups are numbered from 1 to 230 starting with 1 for triclinic but these numbers should never be used to describe the space group; the symbol should be used. Next is the international space groups full symbol in which the symmetry operations are more fully described. Then the m3m ($O_h$) is the point group in the international notation and last is the crystal system. The line just below states that the origin of the coordinate is at a point with point symmetry (site symmetry) m3m ($O_h$). Since this is a symorphic group, the origin has the full-point symmetry. The next line below lists column headings. On the left is the number of positions for each type of site. For a general point there are 48 positions, since there are 48 symmetry operations for the point group $O_h$ or m3m. Then a symbol, starting at "a" and going through the alphabet, for each of the types of sites is given in the Wyckoff notation. This is important if you look up, in Wyckoff (see the Notes), the actual coordinates of the various atom positions. Then the point (site) symmetry in the international notation is given for the site. Next the coordinates of equivalent positions are listed. For the 48m sites (the general point), if there is an atom at x, y, z then there will be the same type of atom at z, y, x and at y, z, x, etc. These first three positions are obtained by a $C_3$ rotation about the [111]. If a crystal whose structure has been determined has this space group and if an atom is at the 48m site, Wyckoff will list a value for x, y, and z for the atom. For the 24m, 24*l*, and 24k there are only two unknowns that describe the atom positions. For other sites of higher point symmetry, the 3d and 3c with $D_{4h}$ site symmetry as well as the two with $O_h$ site symmetry, the positions are fixed entirely by the space group. Remember that if a crystal has this space group, atoms are not required to be at all these sites. In fact for most of the simple solids that are studied, many of the possible sites will not be filled. On the other hand there can be several, for example, 8g sites filled with different atoms. Each group of 8 will have a different value of x.

Table 11-2 also shows space group number 136 $P4_2/mnm$. We immediately see that we have a primitive lattice and that the space group is nonsymmorphic. This is seen again, because the "origin at center" has mmm ($D_{2h}$) which is not the point group of the space group. This is the same statement as the fact that the positions 2a and 2b have site symmetry mmm and not the symmetry of the point group 4/mmm. The general

**Table 11-3** Part of a page from the same book as in Table 11-2

Tetragonal $4/m\ m\ m$ $I\ 4/m\ 2/m\ 2/m$ No. 139 $I4/mmm$

$D_{4h}^{17}$

Origin at centre ($4/mmm$)

| Number of positions, Wyckoff notation, and point symmetry | | | Co-ordinates of equivalent positions $(0,0,0;\ \frac{1}{2},\frac{1}{2},\frac{1}{2})+$ | | | |
|---|---|---|---|---|---|---|
| 32 | $o$ | 1 | $x,y,z;$ | $\bar{x},\bar{y},z;$ | $x,y,\bar{z};$ | $\bar{x},\bar{y},\bar{z};$ |
| | | | $\bar{x},y,z;$ | $x,\bar{y},z;$ | $\bar{x},y,\bar{z};$ | $x,\bar{y},\bar{z};$ |
| | | | $\bar{y},x,z;$ | $y,\bar{x},z;$ | $\bar{y},x,\bar{z};$ | $y,\bar{x},\bar{z};$ |
| | | | $y,x,z;$ | $\bar{y},\bar{x},z;$ | $y,x,\bar{z};$ | $\bar{y},\bar{x},\bar{z}.$ |

point 16k has 16 positions as it should for this point group. The figures in this page of the table show the various symmetry operations of the space group. The **convention** is that the figures are always arranged so that the x-axis points to the bottom of the page, the y-axis to the right, and the z-axis up in a right-hand screw sense and the origin is in the upper left. The meanings of the various symbols associated with the figure are completely described in the international tables.

As a last example of the use of the International Tables, Table 11-3 shows only part of the page from these tables for space group number 139. From the symbol I4/mmm, we immediately see that the space group is symmorphic but that the unit cell is a body-centered multiply primitive cell. We also see that the general point is 32. There are 16 positions listed but the "coordinates of equivalent positions" tells one also to add (1/2, 1/2, 1/2) to each of the positions listed throughout the page to get twice as many positions as listed. This is the centering condition. However, the point group 4/mmm ($D_{4h}$) has only 16 symmetry operations, so how can there be 32-general points? The reason is that for many purposes it is much more convenient to use a nice symmetrical unit cell, as in Fig. 11-3b, rather than the primitive cells, as in Fig. 11-4a. The primitive cell of the 14 Bravais lattices are listed in Appendix 1. Within the primitive cells there are the correct number of atoms as discussed in Section 11-3.

Space group number 140, I4/mcm, is similar to the one shown in Table 11-3. However, it is nonsymmorphic so the point of highest symmetry does not have 4/mmm symmetry. In fact I4/mcm has four sites in the cell, all of equally high site symmetry 4d-mmm ($D_{2h}$), 4c-4/m ($C_{4h}$), 4b-$\bar{4}$2m ($D_{2d}$), 4a-42 ($D_4$). All of these point symmetries have eight

**Table 11-4** A small section of Kovalev's book which has the symmetry operations for all 230 space groups

CLASS $C_4$

| GROUP | TYPE | $h_{14}$ | $h_4$ | $h_{15}$ | $h_{37}$ | $h_{27}$ | $h_{40}$ | $h_{26}$ |
|---|---|---|---|---|---|---|---|---|
| $C^1_{4v}$ | $\Gamma_q$ | | | | | | | |
| $C^3_{4v}$ | $\Gamma_q$ | $0\ 0\ \tau_z$ | | $0\ 0\ \tau_z$ | | $0\ 0\ \tau_z$ | | $0\ 0\ \tau_z$ |
| $C^5_{4v}$ | $\Gamma_q$ | | | | $0\ 0\ \tau_z$ | $0\ 0\ \tau_z$ | $0\ 0\ \tau_z$ | $0\ 0\ \tau_z$ |
| $C^7_{4v}$ | $\Gamma_q$ | $0\ 0\ \tau_z$ | | $0\ 0\ \tau_z$ | $0\ 0\ \tau_z$ | | $0\ 0\ \tau_z$ | |

symmetry operations. Clearly the number of positions times the number of symmetry operations of the site symmetry divided by 1 for a P-lattice, 2 for a C or I-lattice, or 4 for an F-lattice is equal to the order of $P_g$.

### d. Determination of symmetry operations

It probably has been noticed that, while a great deal of information can be obtained about the space groups from the international notation and from the International Tables, the complete list of the actual symmetry operations has not been mentioned. Obviously the general points transform among themselves in a manner that is determined by the symmetry operations, including glide planes and screw axis, of the space group. From the set of general points the symmetry operations indeed can be determined. (See the Notes for a reference.) From the figures in the international tables, which are supplied for all of the seven crystal systems except the cubic system, the details of the transformation of the atoms among themselves is fairly clear.

However, there are more straightforward approaches. The book by Kovalev directly lists the symmetry operations for each of the 230 space groups as well as a great deal of other information. (References to several other similar books are listed in the Notes.) Unfortunately, the book is not easy to use without experience. Table 11-4 shows a small section of Kovalev's book for the first few space groups with the $C_{4v}$ point groups. As can be seen, the Schoenflies notation for the space groups is used but a totally arbitrary notation is used for the point symmmetry operations. The symbols $h_1$ to $h_{48}$ are used for point symmetry operations for the cubic and all their subgroups. These symbols are defined in the front of Kovalev's book. For the rhombohedral and hexagonal systems a different set of symmetry operations, also labeled $h_1$ to $h_{24}$, are defined and used. From Table 11-4 we see that $C_{4v}{}^1$ has only point symmetry operations. For the $C_{4v}{}^3$ space group the symmetry operations are, in the notation used in this book, $\{E\,|\,0\}$, $\{C_4\,|\,\tau\}$, $\{C_2\,|\,0\}$, $\{C_4^3\,|\,\tau\}$, $\{\sigma_d[\bar{1}10]\,|\,0\}$, $\{\sigma_v{}^y\,|\,\tau\}$, $\{\sigma_d[110]\,|\,0\}$, $\{\sigma_v{}^x\,|\,\tau\}$, where $\tau = 0\tau_x + 0\tau_y + (c/2)\tau_z$. These symmetry operations are taken with respect to a single origin and are very convenient to use for vibrational normal mode problems. Kovalev's book also has representation tables for the 32 point groups as well as various point within the Brillouin zone.

## 11-5 Magnetic Groups (Color Groups)

### a. Introduction

The derivation of the 230 space groups in the 1890's appeared to complete the symmetry aspect of the crystal problem. It remained for Shubnikov in 1951 to change this with the introduction of color symmetry which has led to a classification of the magnetic point and space groups.

The ordinary 230 space groups consider the symmetry of the time-averaged position of the charge density. For most crystals the time-averaged distribution of current density is zero, but for ferromagnetic and antiferromagnetic crystals this is not so. For these crystals, a certain spatial symmetry operation may bring the geometric crystal structure into coincidence with itself but possibly this resulting geometric structure will reverse the magnetic moments or spins on the atoms. Thus, the spatial operation by itself would not be a symmetry operation of the crystal. However, the spatial operations followed by an operation that reverses the magnetic moment of the atoms could be a symmetry operation of the crystal. This idea leads us to consider a new type of symmetry operator V which when operating on a crystal reverses the sign of the current density or magnetic moment at all points in the crystal. V is the antisymmetry or time-reversal operator. V does not act on the space coordinates and has the property $V^2 = E$. This operator commutes with all the space symmetry operations. We may think of V as physically changing + to −, or black to white, or changing the magnetic moment from up to down, or reversing the sign of time, although there are some subtle differences between time-reversal and color-reversal since time-reversal is an antiunitary operation. The distinction will become clearer in Section 12-5. For magnetic groups the international notation for point groups is often used. As an example of the effects of V, consider the symmetry operations in the point group 4mm ($C_{4v}$). The eight symmetry operations are E, $C_2$, $2C_4$, $2\sigma_v$, and $2\sigma_d$. Figure 11-9 shows two squares, both of which transform into themselves under all the symmetry operations of 4mm if the difference between the colors is ignored. However, if the difference is not ignored, then in the first square a rotation by $C_4$ ( or $C_4{}^3$) interchanges the black and white portions, thus it is not a symmetry operation; $C_4$ followed by V, $VC_4$, is a symmetry operation. Under each square in the figure the set of symmetry operations is listed along with the symbol for the magnetic point group $\underline{4}m\underline{m}$ and $4\underline{mm}$. The meaning of the magnetic point group symbol is straightforward in the international notation. An underlined symbol means the spatial symmetry operation is followed by V, the antisymmetry operation. For the two-color point groups in Fig. 11-9, we note that each has half of the symmetry operations without a V and half with a V. This will always be true for the new type of magnetic point groups.

### b. Magnetic point groups

There are three different types of magnetic point groups that should be considered.

(i) The first type is the ordinary 32 point groups with no V operation allowed. A ferromagnetic crystal clearly can have point-group symmetry $C_4$ or $C_{4v}$ since all the space symmetry operations will transform the "spin up" positions among themselves, and changing the sign of the magnetic moment will reverse the sign of the spins and will not be a symmetry operation.

(ii) The second type of group consists of the set of operations of any one of the 32 point groups $\{A_i\}$ plus the set of operations $\{VA_i\}$. Since E is in the set of operations $\{A_i\}$, we have V as one of the operations in the set $\{A_i\} + \{VA_i\}$. V by itself reverses the magnetic moment at all points, so only paramagnetic and diamagnetic crystals should have point groups of this type. These types of groups are sometimes called the **gray groups** since they mix black and white. (Although V by itself cannot be a symmetry operation for a ferromagnetic or antiferromagnetic crystal, V coupled with a translation $\tau$, $\{V \mid \tau\}$ can be a symmetry operation in some of the antiferromagnetic space groups. Here $\tau$ is a fraction of a magnetic primitive lattice vector.)

(iii) This type consists of new point groups. These new groups contain some operations found in the 32 point groups and some operations $VA_k$, where $A_k$ by itself is no longer a symmetry operation. These new types of groups were introduced in Section 11-5a. There are 58 of these new types of point groups and we discuss how these are obtained.

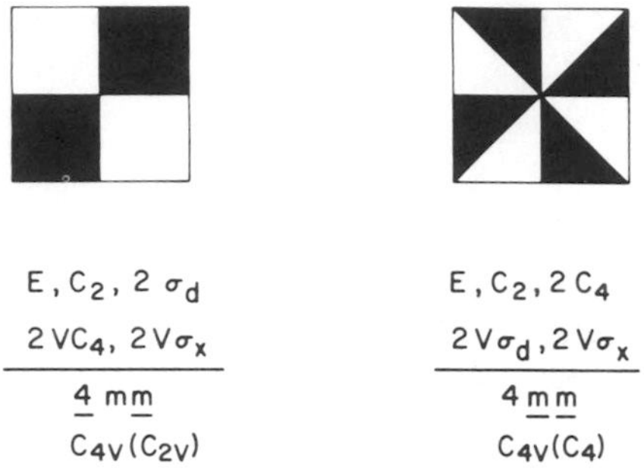


**Fig. 11-9** Squares that transform as 4mm if the colors are ignored but transform as the color groups $\underline{4}m\underline{m}$ and $4\underline{mm}$ if the colors are taken into account.

We want to find a systematic way to find all the point groups that have V together with a geometrical symmetry operation and not V by itself. These are the new point groups we are looking for. Let A be a geometric symmetry operation of the type we have been talking about so far in this book and let M = VA be a symmetry operation that interchanges the atoms and also interchanges the color. Note that, for a particular A, both A and M = VA cannot be members of the group. If they were, we would also have $MA^{-1} = V$ a member of the group, and this is not one of the new types of color groups we want. This leads us to consider a group G consisting of the h elements $\{A_j\}$. Now to get our new type of group, $\hat{G}$, we break up the h elements into two sets $\{A_i\}$ and $\{M_k\}$ as follows

$$A_i \ (i=1,2,\ldots,m) \qquad M_k = VA_k \ (k=m+1,m+2,\ldots,h) \qquad (11\text{-}9)$$

So we have $G = \{A_i\} + \{A_k\}$ and $\hat{G} = \{A_i\} + \{M_k\}$, where we have yet to prove that $\hat{G}$ is indeed a group. G is one of the 32 point groups that has been discussed so far in this book and this is true because if we let V = E we must be left with a group consisting of symmetry operations on the charge density. The set of elements $\{A_i\}$ must be an invariant subgroup of G, called S. It is clear the S is a group since any $A_i$ in the set $\{A_i\}$ operating on the crystal must be a symmetry operation, have an inverse, contain the identity, and the product of any two must also be a member of $\{A_i\}$. This is clear because any operation $A_iM_k$ must be contained within the set $\{M_k\}$ since it contains V. S is an invariant subgroup because the conjugate of $A_i$ with any element of G involves $V^2$ (=E) or no V at all, and $A_k^{-1}A_iA_k$ is contained in $\{A_i\}$ as can be seen by the rearrangement theorem. Thus, S is an invariant subgroup of one of the 32 point groups, which means it is also one of the 32 point groups. Therefore, we have a systematic way of obtaining the new type of point group. The procedure would be for a given point group G, find an invariant subgroup S containing the elements $A_i$. Take all the elements not in S, G − S, and multiply them by V, i.e., V(G − S), to obtain all $M_k$ as in Eq. 11-9. Then see if a group is formed by all the elements $A_i$ plus $M_k$; if a group is formed, it is the type that we seek. This approach can be simplified by realizing that the subgroup S must have exactly half the elements of G in order for the new set S + V(G − S) to be a group. This is proved in the next paragraph.

We want to prove that a necessary and sufficient condition for the set of elements $\hat{G} = \{A_i\} + \{M_k\}$ to form a group is that the order of the group S is one-half the order of group G. Remember $G = \{A_i\} + \{A_k\} = S + A_kS$, where $A_k$ are elements not in S as in Eq. 11-9. Consider the following set of operations

$$\hat{G} = \{A_i\}+\{M_k\} = S+M_kS = S+VA_kS = S+V(G-S) \qquad (11\text{-}10)$$

We can see that $\hat{G}$ indeed is a group. Remember that V commutes with spatial operations ($VA_k = A_kV$) and that $V^2 = E$, because

$$\begin{aligned} SS &= S & (VA_kS)(S) &= VA_kS \\ (S)(VA_kS) &= VA_kS & (VA_kS)(VA_kS) &= V^2A_kSA_kS = S \end{aligned} \qquad (11\text{-}11)$$

(Remember $A_k\{A_i\} = \{A_k\}$ by the rearrangement theorem.) Now consider this group $\hat{G}$. Multiplying the m elements of S by one element in the set $\{M_k\}$ results in m different elements of $\{M_k\}$. (Use the rearrangement theorem as in Eq. 11-11.) Also multiplying the (h − m) elements of $\{M_k\}$ by one of the elements of $\{M_k\}$ results in h − m different elements of S. So m = h − m or S has half the number of elements found in the group $\hat{G}$. Note, from Eq. 11-10, it is clear that there are as many elements in the group $\hat{G} = S + V(G - S)$ as in G. Thus, the theorem is proved and the procedure to obtain these new groups is straightforward. Consider any one of the 32 point groups, G. Find each of the subgroups S of order h/2. Then multiply each of the elements of S by V to obtain the set of elements S + V(G − S). This set of elements is the new type of group we are looking for.

Table 11-5 shows the 58 new types of groups that are obtained. The point groups $C_1$, $C_3$, and T do not have subgroup of order h/2 as shown. Also as can be seen, some groups have several different subgroups of order h/2. An example of this situation was shown in Fig. 11-9 for the point group 4mm ($C_{4v}$). The magnetic point group notation (the underlines) for the international symbol was mentioned in Section 11-5a. For the Schoenflies notation the invariant subgroup is put in parenthesis G(S) as shown in Table 11-5.

**Table 11-5** The 58 black and white point groups

| | Ordinary Point Groups | | | Black & White Point Groups | |
|---|---|---|---|---|---|
| | $C_1$ | 1 | 1 | None | |
| Monoclinic | $C_i$ | $\bar{1}$ | 2 | $C_i(C_1)$ | $\underline{\bar{1}}$ |
| | $C_2$ | 2 | 2 | $C_2(C_1)$ | $\underline{2}$ |
| | $C_{1h}$ | m | 2 | $C_{1h}(C_1)$ | $\underline{m}$ |
| | $C_{2h}$ | 2/m | 4 | $C_{2h}(C_2)$ | $2/\underline{m}$ |
| | | | | $C_{2h}(C_{1h})$ | $\underline{2}/m$ |
| | | | | $C_{2h}(C_i)$ | $\underline{2}/\underline{m}$ |
| Orthorhombic | $D_2$ | 222 | 4 | $D_2(C_2)$ | $2\underline{22}$ |
| | $C_{2v}$ | 2mm | 4 | $C_{2v}(C_2)$ | $2\underline{mm}$ |
| | | | | $C_{2v}(C_{1h})$ | $\underline{2m}m$ |
| | $D_{2h}$ | mmm | 8 | $D_{2h}(D_2)$ | $\underline{mmm}$ |
| | | | | $D_{2h}(C_{2v})$ | $mm\underline{m}$ |
| | | | | $D_{2h}(C_{2h})$ | $\underline{mm}m$ |
| Tetragonal | $C_4$ | 4 | 4 | $C_4(C_2)$ | $\underline{4}$ |
| | $S_4$ | $\bar{4}$ | 4 | $S_4(C_2)$ | $\underline{4}$ |
| | $D_4$ | 422 | 8 | $D_4(C_4)$ | $4\underline{2}$ |
| | | | | $D_4(D_2)$ | $\underline{4}2$ |
| | $C_{4h}$ | 4/m | 8 | $C_{4h}(C_4)$ | $4/\underline{m}$ |
| | | | | $C_{4h}(S_4)$ | $\underline{4}/\underline{m}$ |
| | | | | $C_{4h}(C_{2h})$ | $\underline{4}/m$ |
| | $C_{4v}$ | 4mm | 8 | $C_{4v}(C_4)$ | $4\underline{mm}$ |
| | | | | $C_{4v}(C_{2v})$ | $\underline{4}m\underline{m}$ |
| | $D_{2d}$ | $\bar{4}2m$ | 8 | $D_{2d}(S_4)$ | $\bar{4}\underline{2m}$ |
| | | | | $D_{2d}(D_2)$ | $\underline{\bar{4}}2\underline{m}$ |
| | | | | $D_{2d}(C_{2v})$ | $\underline{\bar{4}2}m$ |
| | $D_{4h}$ | 4/mmm | 16 | $D_{4h}(D_4)$ | $4/\underline{mmm}$ |
| | | | | $D_{4h}(C_{4v})$ | $4/\underline{m}mm$ |
| | | | | $D_{4h}(D_{2h})$ | $\underline{4}/mmm$ |
| | | | | $D_{4h}(D_{2d})$ | $\underline{4}/\underline{mm}m$ |
| | | | | $D_{4h}(C_{4h})$ | $4/m\underline{mm}$ |

| | Ordinary Point Groups | | | Black & White Point Groups | |
|---|---|---|---|---|---|
| Trigonal | $C_3$ | 3 | 3 | None | |
| | $D_3$ | 32 | 6 | $D_3(C_3)$ | $3\underline{2}$ |
| | $C_{3v}$ | 3m | 6 | $C_{3v}(C_3)$ | $3\underline{m}$ |
| | $S_6$ | $\bar{3}$ | 6 | $S_6(C_3)$ | $\underline{\bar{3}}$ |
| | $D_{3d}$ | $\bar{3}m$ | 12 | $D_{3d}(S_6)$ | $\bar{3}\underline{m}$ |
| | | | | $D_{3d}(C_{3v})$ | $\underline{\bar{3}}m$ |
| | | | | $D_{3d}(D_3)$ | $\underline{\bar{3}m}$ |
| Hexagonal | $C_6$ | 6 | 6 | $C_6(C_3)$ | $\underline{6}$ |
| | $C_{3h}$ | $\bar{6}$ | 6 | $C_{3h}(C_3)$ | $\underline{\bar{6}}$ |
| | $D_{3h}$ | $\bar{6}m2$ | 12 | $D_{3h}(C_{3h})$ | $\bar{6}\underline{m2}$ |
| | | | | $D_{3h}(C_{3v})$ | $\underline{\bar{6}}m\underline{2}$ |
| | | | | $D_{3h}(D_3)$ | $\underline{\bar{6}m}2$ |
| | $D_6$ | 622 | 12 | $D_6(C_6)$ | $6\underline{2}$ |
| | | | | $D_6(D_3)$ | $\underline{6}2$ |
| | $C_{6h}$ | 6/m | 12 | $C_{6h}(C_6)$ | $6/\underline{m}$ |
| | | | | $C_{6h}(S_6)$ | $\underline{6}/\underline{m}$ |
| | | | | $C_{6h}(C_{3h})$ | $\underline{6}/m$ |
| | $C_{6v}$ | 6mm | 12 | $C_{6v}(C_6)$ | $6\underline{mm}$ |
| | | | | $C_{6v}(C_{3v})$ | $\underline{6}m\underline{m}$ |
| | $D_{6h}$ | 6/mmm | 24 | $D_{6h}(D_{3h})$ | $\underline{6}/m\underline{m}m$ |
| | | | | $D_{6h}(D_{3d})$ | $\underline{6}/\underline{mm}m$ |
| | | | | $D_{6h}(D_6)$ | $6/\underline{mmm}$ |
| | | | | $D_{6h}(C_{6v})$ | $6/\underline{m}mm$ |
| | | | | $D_{6h}(C_{6h})$ | $6/m\underline{mm}$ |
| Cubic | T | 23 | 12 | None | |
| | $T_h$ | m3 | 24 | $T_h(T)$ | $\underline{m}3$ |
| | $T_d$ | $\bar{4}3m$ | 24 | $T_d(T)$ | $\underline{\bar{4}}3\underline{m}$ |
| | O | 432 | 24 | O(T) | $\underline{4}3$ |
| | $O_h$ | m3m | 48 | $O_h(O)$ | $\underline{m}3\underline{m}$ |
| | | | | $O_h(T_d)$ | $\underline{m}3m$ |
| | | | | $O_h(T_h)$ | $m3\underline{m}$ |

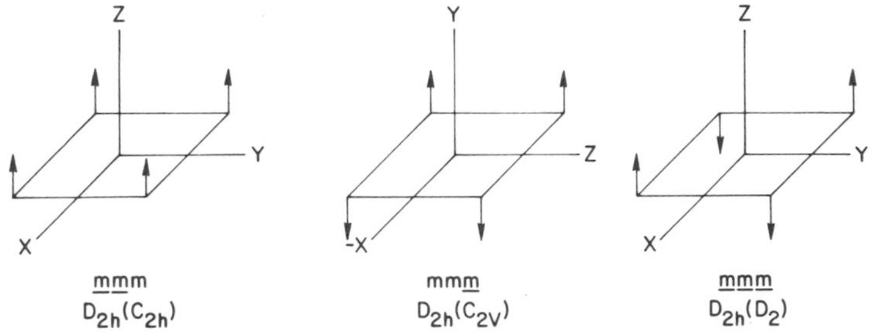


**Fig. 11-10** The types of magnetic ordering that can be derived from the point group mmm($D_{2h}$).

**Table 11-6** The number of colored point groups and space groups of the various types that can be obtained

| | Type of point group | Number of point groups | Number of space groups |
|---|---|---|---|
| (1) | Ordinary | 32 | 230 |
| (2) | Gray | 32 | 230 |
| (3) | Black and white | 58 | 1191 |
| | Total | 122 | 1651 |

### c. Magnetic structures

We very briefly ennumerate and discuss the types of magnetic crystal structures and the kind of magnetic order that can be found in magnetic crystals. Table 11-6 lists the number of point groups and space groups for each of the three types of point groups discussed in Section 11-5b. Each of the type 1 groups has the same number of symmetry operations as the crystallographic point groups discussed throughout this book, hence the name "ordinary." As already mentioned ferromagnetic crystal can have these point groups ($C_4$, $S_4$, $C_{4h}$,...) since the symmetry operations transform the charge densities as well as the current densities among themselves. Antiferromagnetic crystals can have some of the other point groups. (See the Problems.)

Type 2 or gray groups, as already mentioned refer to diamagnetic or paramagnetic materials. However, with the inclusion of a nonprimitive lattice translation, antiferromagnetic space groups are included. The order of the group for the gray groups is two times as large as the corresponding ordinary since each operation of the ordinary group is included plus the same operations multiplied by V, i.e., G + VG.

Type 3 or black and white groups $\hat{G}$ have as many symmetry operations as the ordinary group G that they are derived from. However, half of the symmetry operations involve an ordinary point operation multiplied by V as discussed. Figure 11-10 for example shows the types of magnetic ordering that can be obtained from the black and white groups derived from the point group $D_{2h}$(mmm). The black and white plus the gray groups are sometimes called magnetic groups or Shubnikov groups; there are 90 of these groups. Nevertheless, the "ordinary" groups can have magnetic structures giving a total of 122 groups to consider. Space groups derived from the black and white point groups are of two types. The first type comes from the black and white point groups combined with the 14 Bravais lattices in the same way as the "ordinary" space groups. The second type comes from black and white Bravais lattices, of which there are 22. References to the literature are given in the Notes to this specialized but important topic.

## Notes

A short history of crystal symmetry leading up to space groups can be found in Koster's article. The Notes to Chapter 1 also contain several references concerned with the history of the 230 space groups.

The "International Tables for X-Ray Crystallography, Volume I" is the reference for a description of the 230 space groups, the symbols used, and many related matters. There are many books that develop an appreciation for space groups from different points of view. We pick out several with different approaches somewhat arbitrarily: F. C. Phillips "An Introduction to Crystallography" (Wiley, New York, 1971); H. Megaw, "Crystal Structures: A Working Approach" (Saunders, Philadelphia, 1973), and G. Weinreich.

Koster's article is the best reference for most properties of irreducible representation of the space groups. However, the fundamental papers by Seitz [Ann. Math. **37**, 17 (1936)] and Bouckaert, Smoluchowski, and Wigner [Phys. Rev. **50**, 58 (1936)] can be read with profit. Both of these papers are reprinted in Cracknell and in Knox and Gold.

A simple prescription to determine the symmetry elements of a space group given the coordinates of the general point listed in the "International Tables of X-Ray Cyrstallography" is given by Wondratachek and Neubüser, Acta Cryst. **23**, 349 (1967). Kovalev and the book by Zak, Casher, Glück, and Gur list the symmetry operations for the 230 space groups.

Wyckoff's books list most of the crystal structure determinations and the space group and the position of the atoms in the crystal.

Color groups are covered in a book by Shubnikov which has reprints of some of the fundamental papers translated into English.

One can show that the set of irreducible representations of the space group are independent of the origin (Zak, p. 15). However, there can be origin effects. For example both gallium and phosphorus have the same site symmetry in GaP. However, the ordering of eigenfunctions of an impurity at a Ga- or P-site will be different because the potential energy is different at the two sites. This can lead to entirely different selection rules for the same impurity at the two different sites. See T. N. Morgan, Phys. Rev. Lett. **21**, 819 (1968).

## Problems

**1.** Do Problem 9 of Chapter 1.

**2.** Prove that G/T is isomorphic to $P_g$, where G/T is the factor group of the space group G with respect to the primitive translation group T. (Hint: See Cornwell, p. 164.)

**3.** Show that a face-centered tetragonal F-lattice is equivalent to an I-lattice in a different orientation.

**4.** Put a lattice point at the body center of a primitive rhombohedral cell. Draw another primitive trigonal cell showing an I-rhombohedral is not needed. Do the same for an F-rhombohedral cell.

**5.** (a) Draw a figure with $D_{2h}$, $C_{2h}$, $C_{2v}$, and $C_2$, and $D_2$ point symmetry. (b) Draw a figure with $O_h$, O, $T_h$, $T_d$, and T point symmetry.

**6.** (a) For the ten orthorhombic P-space groups in Section 11-4b draw a projection as in Fig. 11-8 (looking down the z-axis). Determine the symmetry operations along the z-axis and write out the full space group symbol. (To check, see the "International Tables for X-Ray Crystallography." Notice that in these tables, the origin need not be on a 2 or $2_1$ axis.) (b) Show why it makes no sense to have a Pam type of space group.

**7.** (a) Consider the space group Pban ($D_{2h}^4$). From these symmetry operations draw a view as in Fig. 11-8. Put in the symmetry operations that are implied to show that the full symbol is P2/b 2/a 2/n. Check the result in the tables. (b) Do the same for Pmma ($D_{2h}{}^5$). (c) Do the same for Pccm ($D_{2h}{}^3$).

**8.** For the monoclinic crystal system show that the three space groups (P2, $P2_1$, and C2) are the only ones that are isomorphic to the point group $C_2$.

**9.** Show that 66 symmorphic space groups are obtained by the simple considerations in Section 11-3c. (Hint: If you obtain only 61 see Weinreich, Chapter 1 and Nussbaum, "Applied Group Theory," Chapter 1 (Prentice-Hall, Englewood, New Jersey, 1971).

**10.** (a) Find the seven symmorphic space groups not found in Problem 9 by using Table 6.2.1 in the international tables. (b) Show a possible crystal structure for one of these seven space groups. (Hint: See Weinreich, Chapter 1.)

**11.** For the type 1 "ordinary" point groups in the triclinic, monoclinic, orthorhombic, and tetragonal crystal system determine if a ferromagnetic or antiferromagnetic crystal structure is allowed. (See Tinkham, Table 8-4.)

**12.** $CaCO_3$ occurs in two crystallographic forms, calcite and argonite. In solution the totally symmetric mode of the $CO_3{}^{2-}$ ion is not infrared active. Consider the observation or nonobservation of this mode in the two crystallographic forms. Use the site symmetry approximation. Use Wyckoff.

## Chapter 12

# BAND THEORY OF SOLIDS

*We will therefore turn to the less ambitious question of what men themselves show by their behaviour to be the purpose and intention of their lives.*

*Freud, "Civilization And Its Discontents"*

In this chapter we will discuss some of the elementary aspects of the band theory of solids. Group theory is extremely useful and necessary in this area and many qualitative and even semiquantitative aspects of the theory can be understood from just the symmetry properties of the solid. However, to be able to calculate which bands lie higher and which lie lower in energy, some model must be assumed. If the electrons are thought to be nearly free electrons, then the orthogonalized plane wave (OPW) might be used. If the electrons are thought to be nearly localized at one ion site, then the tight bonding model might be used. We will not discuss these details but restrict ourselves to the symmetry aspects of band theory.

Even with these restrictions the subject will be approached slowly. First only the translational symmetry of the crystal is considered assuming there are no point group operations, i.e., a triclinic crystal. This leads to the concept of the wave vector of the crystal **k** being a label of an irreducible representation. When cyclic boundary conditions are imposed on the crystal (crystals are finite in size so must be terminated in some manner) one is lead to the concept of a finite number of **k** values that can be restricted to a zone in k-space (reciprocal lattice space). This brings in, via reciprocal space, the concept of the Brillouin zones and reciprocal lattice vectors. Bloch functions are shown to be basis functions of the irreducible representations of the translational group. We then show how bands can be made up of 1s-, 2s-, 2p-like wave functions with the phase modulated over the unit cells as the **k** values are varied. These considerations apply to symmorphic as well as nonsymmorphic space groups.

## 12-1 Translational Symmetry

### a. Translational symmetry

We have an infinite crystal that can be made up of the translations of a primitive cell shown in Fig. 12-1. The primitive cell has edges $\mathbf{a}_1$, $\mathbf{a}_2$, $\mathbf{a}_3$ which define a lattice of points. Every point in this infinite lattice can be obtained by a translation

$$\mathbf{t}_n = n_1\mathbf{a}_1 + n_2\,\mathbf{a}_2 + n_3\,\mathbf{a}_3, \qquad n_i = \text{any interger} \tag{12-1}$$

We will sometimes use the notation for the translation operator $\{E\,|\,\mathbf{t}_n\}$ instead of just $\mathbf{t}_n$ to remind ourselves that there is no point group rotation involved.

Consider, for example, a hydrogen atom at every lattice point defined by Eq. 12-1. All the atoms transform among themselves under all the symmetry operations of the group. This group is the translation group, the operations are described by $\mathbf{t}_n$ for all n. If the hydrogen atoms are thought of as well separated, to a very good approximation each one will have an electron in a 1s-state. As the primitive lattice vectors get smaller the atoms begin to "feel" each other, i.e., the off-diagonal terms $H_{ij}$ and overlap terms $S_{ij}$ in the secular determinant must be taken into account. Just as in the problem of the LCAO-MO discussed for cyclic rings of carbon atoms, the eigenfunctions for the crystal are made up of linear combinations of the hydrogen 1s-functions. The eigenvalues of the crystal differ from the eigenvalue of the isolated 1s-hydrogen atom. We can see that the translational symmetry operations $\mathbf{t}_n$ will have much the same effect as the point symmetry operations in the cyclic carbon systems. This model of hydrogen atoms will be discussed later.

As discussed in Sections 11-1 and 11-2 the set of operations defined in Eq. 12-1 do indeed form a group and the group is Abelian since each operation is in a class by itself, i.e., the inverse of $\mathbf{t}_n$ is $\mathbf{t}_{-n}$ where $-n$ means $n_1$, $n_2$, $n_3$ are replaced by minus themselves

$$(\mathbf{t})_n(\mathbf{t}_{-n}) = \mathbf{t}_0 \tag{12-2a}$$

$$(\mathbf{t}_m)(\mathbf{t}_n)(\mathbf{t}_{-m}) = \mathbf{t}_n \tag{12-2b}$$

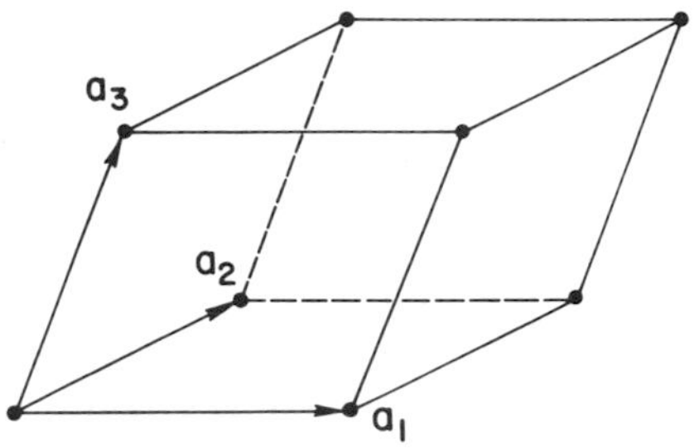


**Fig. 12-1** A primitive cell of a lattice. The three translation vectors are noncoplanar.

Thus there are a many irreducible representations as there are symmetry operations.

### b. Cyclic boundary conditions

If we are thinking about real crystals, rather than infinite lattices, some kind of boundary conditions should be used. For a one-cubic centimeter crystal there are many fewer atoms near the surface than there are in the bulk of the crystal. Hence, we would hope our arbitrary boundary conditions will not effect our understanding of the bulk of the crystal but care must be exercised if surface effects are considered. The boundary conditions most often used are the cyclic or periodic boundary conditions. Consider the finite crystal to have $N_1$, $N_2$, $N_3$ primitive cells in the $\mathbf{a}_1$, $\mathbf{a}_2$, $\mathbf{a}_3$ directions, respectively. We can restrict the $n_i$ intergers in Eq. 12-1 to $0 \leq n_i \leq N_i - 1$. Thus our boundary conditions are

$$t_{n_1, n_2, n_3} = t_{n_1+N_1, n_2, n_3} = t_{n_1, n_2+N_2, n_3} = \text{etc.} \qquad (12\text{-}3)$$

This is equivalent to imagining that the last primitive cell on the right-hand side of the crystals has on its right the first cell of the left-hand side of the crystal ($t_{0,n_2,n_3} = t_{N_1,n_2,n_3}$). Or Eq. 12-3 can be imagined to represent an ensemble of identical crystals adjoining the real crystal on each of its faces.

The cyclic boundary condition results in a cyclic group for the translational group. The generating element of the cyclic group is $\{E \mid \mathbf{a}_1\}$and the order of the group is $N_1N_2N_3$. We use $A \equiv \{E \mid a_i\}$ for one of the i's and then drop the i-subscript until Eq. 12-6 for convenience, thinking in terms of a 1-dimensional crystal. From the boundary conditions

$$t_n = A^n = \{E \mid a\}^n = \{E \mid a\}^{n+N} \qquad (12\text{-}4)$$

The elementary properties of cyclic groups were discussed in Section 4-1. A is a lattice translation of one primitive cell. Since A is the generating element, $A^N = E$, the identity operation. Thus a representation of A is the N roots of unity since these roots have the same multiplication table as the operations.

$$\Gamma\{E\,|\,a\} = \Gamma(A) = [E]^{1/N}$$
$$= [\exp(-2\pi\, i\, m)]^{1/N} = \exp(-2\pi\, i\, m/N) \qquad (12\text{-}5)$$

where m = any integer. The order of the group, N, is equal to the number of classes which is equal to the number of irreducible representations. Thus, while m can be any integer its values must be restricted. **By convention we pick $0 \leq m \leq N - 1$.** That is, there are N values of m.

It is very easy to write all the representations for the translation group. (See also Section 4-1b.) Since the representations are all 1-dimensional, the characters and the representations are identical. Table 12-1 shows the results. The first column under the symmetry operation A lists the N roots of unity as prescribed in Eq. 12-5 for the conventionally allowed values of m. These results should be thought of in the usual group theoretical way. For example, for the mth irreducible representation there are some basis functions which when operated on by the A-symmetry operation the result is $\varepsilon^m$ times the basis function. The rest of the columns, $A^2$, etc. are obtained by squaring, etc. the results in the first column. In the last column we obtain $A^N = E$ and all 1's for representations. This last column is usually listed as the first column so should be picked up and put before the "A" column. Thus, Table 12-1 shows the complete character table (and representation table) for the translational group. The restricted range of intergers m, label the mth irreducible representation

$$\Gamma_m\{E\,|\,na\} = \exp(-i2\pi mn/N) \qquad (12\text{-}6a)$$

$$\Gamma_{m_1,\, m_2,\, m_3}\{E\,|\,\mathbf{t}_n\} = \exp(-i2\pi m_1 n_1/N_1) \times \exp(-i2\pi m_2 n_2/N_2) \times \exp(-i2\pi m_3 n_3/N_3) \qquad (12\text{-}6b)$$

Equation 12-6b is the obvious extension of Eq. 12-6a to three dimensions since the translation operations along each of the three primitive axis directions are independent of each other. For example, it does not matter if the translation is along $\mathbf{a}_1$ or along $\mathbf{a}_2$ first and then $\mathbf{a}_1$. Thus, each

**Table 12-1** Character table of the cyclic translation group

| | $A$ | $A^2$ | ... | $A^{N-1}$ | $A^N = E \; [\varepsilon \equiv e^{\frac{-2\pi i}{N}}]$ |
|---|---|---|---|---|---|
| m = 0 | 1 | 1 | ... | 1 | 1 |
| 1 | $\varepsilon$ | $\varepsilon^2$ | ... | $\varepsilon^{N-1}$ | $\varepsilon^N = 1$ |
| 2 | $\varepsilon^2$ | $\varepsilon^4$ | ... | $\varepsilon^{2N-2}$ | $\varepsilon^{2N} = 1$ |
| ⋮ | ⋮ | ⋮ | | | ⋮ |
| N - 1 | $\varepsilon^{N-1}$ | $\varepsilon^{2N-2}$ | ... | | $\varepsilon^{N^2-N} = 1$ |

direction can be treated separately and the resulting group in three dimensions is Abelian with respect to the multiplication of the translation group along each separate primitive axis.

### c. Brillouin zones

By introducing the concept of reciprocal space and Brillouin zones we can give Eq. 12-6 an important and simple geometric interpretation. It should be remembered that besides finding irreducible representations, we will want to put electrons (or normal modes of vibration) in the lattice. The Brillouin zones will also be used to show contours of constant energy of the electrons.

To study the geometric interpretation of Eq. 12-6, define a reciprocal lattice vector $\mathbf{b}_j$ in terms of the direct primitive lattice vector $\mathbf{a}_i$ by

$$\mathbf{a}_i \cdot \mathbf{b}_j = 2\pi\,\delta_{ij} \qquad \mathbf{b}_1 = 2\pi\,\mathbf{a}_2 \times \mathbf{a}_3/[\mathbf{a}_1 \cdot (\mathbf{a}_2 \times \mathbf{a}_3)] \text{ etc.} \qquad (12\text{-}7)$$

where i and j = 1, 2, and 3. If the direct lattice is made up of orthogonal vectors (primitive orthorhombic, tetragonal, or cubic crystal system), then the reciprocal lattice vectors are parallel to the corresponding direct lattice vectors with magnitude given by

$$b_1 = 2\pi/a_1 \qquad b_2 = 2\pi/a_2 \qquad b_3 = 2\pi/a_3 \qquad (12\text{-}8)$$

Figure 12-2 shows a 2-dimensional example of a lattice of points made up from the reciprocal lattice vectors $\mathbf{b}_i$. The lattice of points is defined by

$$\mathbf{K}_n = n_1\mathbf{b}_1 + n_2\mathbf{b}_2 + n_3\mathbf{b}_3, \qquad n_i = \text{any interger} \qquad (12\text{-}9)$$

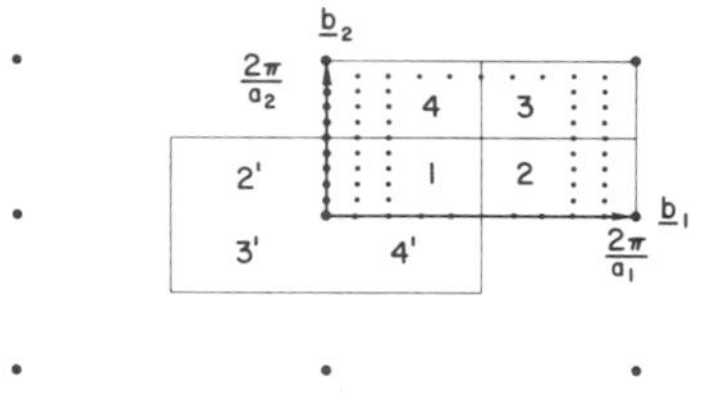


**Fig. 12-2** The large solid points form a lattice called the reciprocal lattice. The small circles represent allowed **k** values as discussed.

**The lattice defined in Eq. 12-9 is called the reciprocal lattice.** Since typical dimensions of a direct lattice unit cell are 6 Å, the magnitude of reciprocal lattice vectors is $b \sim 10^8\ \mathrm{cm}^{-1} = 1\ \text{Å}^{-1}$. **Reciprocal space** or **reciprocal lattice space** is the space defined by the lattice of points in Eq. 12-9.

In reciprocal space a set of vectors can be defined as

$$\mathbf{k} = p_1\,\mathbf{b}_1 + p_2\mathbf{b}_2 + p_3\mathbf{b}_3 \tag{12-10}$$

which looks like Eq. 12-9 but the $p_i$ will generally be restricted to values smaller than one. We will also see that the points formed by the end points of the **k** vector form a lattice of very high density within the reciprocal space. Actually **k** is defined to simply and conveniently eliminate $m_i$ from Eq. 12-6 and replace it with **k**. Note that

$$\mathbf{k} \bullet \mathbf{t}_n = p_1 n_1 \mathbf{a}_1 \bullet \mathbf{b}_1 + \ldots = 2\pi p_1 n_1 + \ldots \tag{12-11}$$

Define $p_i = m_i/N_i$, and the restriction that $m_i$ can have $N_i$ integer values between 0 and $N_i-1$ carried over to $p_i$ is

$$0 \le p_i \le (N_i-1)/N_i \tag{12-12}$$

So $p_i$ varies from zero to almost one. Then **k** can be used to label the irreducible representations of the translation group. The **k**th irreducible representation is given by Eq. 12-13 which replaces (is the same as) Eq. 12-6.

$$\Gamma_{\mathbf{k}}\,\{E\,|\,t_n\} = \exp\,(-i\mathbf{k} \bullet \mathbf{t}_n) \tag{12-13}$$

The representation is unitary since the complex conjugate of Eq. 12-13 times itself is clearly 1.

Figure 12-2 shows the allowed values of **k** in a convenient primitive cell in reciprocal space for a 2-dimensional crystal composed of ten by ten unit cells when $\mathbf{a}_1$ is perpendicular to $\mathbf{a}_2$, and for convenience $2a_1 \approx a_2$ so $b_1 \approx 2b_2$. As can be seen there are 100 allowed **k** values. In

1-dimension the **k** values start at 0 and the largest is almost equal to the reciprocal lattice vector **b**. That is, the maximum **k** is **b**(N−1)/N. Normally crystals are not composed of ten unit cells but millions. Thus, the allowed **k** values are discrete but the density is very large and there is a quasi-continuous number of allowed **k** values in reciprocal space.

Repeating, we see that **k is a label of an irreducible representation of the translation operator**. For this cyclic group with $N_1N_2N_3$ symmetry operations, there are the same number of irreducible representations such that the values of **k are restricted as in Eq. 12-12**. We will see in Sections 12-1e and 1f that **k** is also equal to 2π divided by a wavelength, and thus, **k** has a physical as well as a mathematical meaning. Also, we will see that although the allowed values of **k** are restricted, it will sometimes be convenient to discuss the values of **k** filling all of reciprocal space by considering $\mathbf{k}' = \mathbf{k} + \mathbf{K}_n$. For a reciprocal lattice vector, Eq. 12-9, we have

$$\begin{aligned}\exp(-i\mathbf{K}_{n'} \cdot \mathbf{t}_n) &= \exp[i(n'_1\, n_1\mathbf{b}_1 \cdot \mathbf{a}_1 + \ldots)] \\ &= \exp[-i2\pi(n'_1\, n_1 + \ldots)] = 1 \qquad (12\text{-}14)\end{aligned}$$

Thus, we can treat the allowed values of **k** in the reciprocal lattice as periodic since we can treat **k′** and $\mathbf{k} + \mathbf{K}_n$ as identical wave vectors and identical labels of the irreducible representations. Because of this periodicity in reciprocal space we can consider all of the reciprocal space filled with points obtained by $\mathbf{k} + \mathbf{K}_n$ for all integer values of n; this is called the **extended zone** scheme. Of course, it is understood that in any counting of states there are still the correct number of **k** values as discussed; the extended zone is merely a convenience.

The allowed values of **k** in reciprocal space are usually drawn within a more convenient primitive cell than shown in Fig. 12-2. This convenient cell is more symmetrical about **k** = 0. It is formed in the same manner as the Wigner–Seitz cell in direct space, thus it will display the point symmetry more clearly. Such a cell in reciprocal space is called the **Brillouin zone** of the first Brillouin zone. Brillouin zones will be discussed in more detail later. Here we think only in terms of using them to display the allowed values of **k**. Various parts of the rectangular reciprocal lattice in Fig. 12-2 can be moved an amount $\mathbf{K}_n$ to form the Brillouin zone. For example, the allowed values of **k** shown in Fig. 12-2 in the reciprocal lattice cell bounded by $\mathbf{b}_1$ and $\mathbf{b}_2$ can be moved to form a cell about the origin. One quarter of the cell labeled 1 remains where it is. The quarter of the cell labeled 2 is moved an amount $\mathbf{K}_{1,0,0} = \mathbf{b}_1$, i.e., $\mathbf{k}' = \mathbf{k} - \mathbf{K}_{1,0,0}$.

Similarly, region 4 is moved an amount $-\mathbf{b}_2$ and region 3 is moved an amount $-\mathbf{b}_1 - \mathbf{b}_2$. The new cell still has the proper number of allowed **k** values now labeled **k′**. This **k′** labels the irreducible representations in Eq. 12-13 with a **k′** instead of a **k** since

$$\begin{aligned} \exp(i\mathbf{k} \cdot \mathbf{t}_n) &= \exp(i\mathbf{k}' \cdot \mathbf{t}_n) \exp(-i\mathbf{K}_n \cdot \mathbf{t}_n) \\ &= \exp(i\mathbf{k}' \cdot \mathbf{t}_n) \end{aligned} \qquad (12\text{-}15)$$

from Eq. 12-14. The allowed **k′** values are contained in a cell in reciprocal space bounded by $-\mathbf{b}_1/2$ and $+\mathbf{b}_1/2$, etc. From the definition of the allowed **k** values in Eq. 12-11 we have defined $p_i = m_i/N_i$, so $m_i$ is now restricted to have integer values between $-N_i/2$ and $+N_i/2$. Thus

$$-1/2 \le p_i \le +1/2 \qquad (12\text{-}16)$$

The prime in the **k** can be dropped since **k** and **k′** give identical results as can be seen in Eq. 12-15. Henceforth it will always be understood that the allowed **k** values are restricted to the symmetrical cell in reciprocal space. Also note that the cell defined by Eq. 12-16 has slightly more allowed **k** values than the cell defined by Eq. 12-12. These extra values are on the perimeter of the cell (surface for a 3-dimensional cell). This presents no difficulty because the **k** values (for example, $\mathbf{k} = -\mathbf{b}_1/2$ and $+\mathbf{b}_2/2$) differ by a reciprocal lattice vector and are thus considered identical.

The cell in reciprocal space or k-space defined by Eq. 12-10 and 12-16 is called the **first Brillouin zone** or Brillouin zone. The first, second, etc. Brillouin zones are obtained by forming the line (or plane for three dimensions) that is the perpendicular bisector of the line from $\mathbf{k} = 0$ to each reciprocal lattice vector $\mathbf{K}_n$. Such lines are shown in Fig. 12-3 for this 2-dimensional lattice. The equation which describes the lines is

$$\mathbf{k} \cdot \mathbf{K}_n = (1/2)\,|\mathbf{K}_n|^2 \qquad (12\text{-}17)$$

The first Brillouin zone is the smallest area centered about the origin in reciprocal space as can be seen in Fig. 12-3. The area is exactly the correct size to contain the appropriate number of allowed **k** values ($N_1N_2$). Also, as can be seen in Fig. 12-3, the second and higher zones are more complicated in the sense that they are made up of "bits and pieces" of reciprocal space. The requirement on these pieces is that <u>no perpendicular bisector can pass through a piece</u> (there will be energy discontinuities along perpendicular bisectors as will be seen), <u>each zone must have the same area as the first zone, and the various pieces can be transformed to the first zone by a reciprocal lattice vector</u>. Figure 12-3

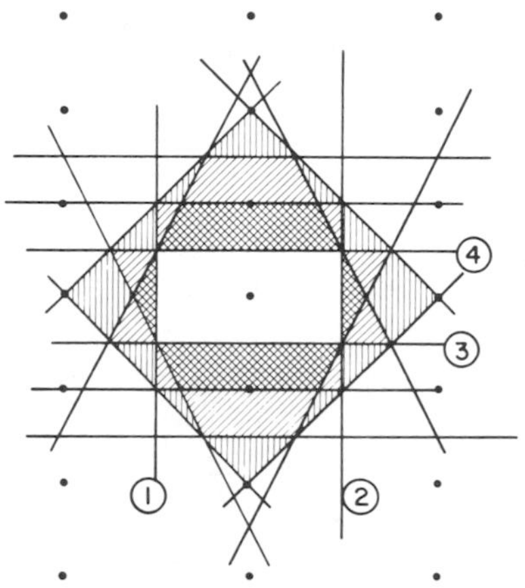


**Fig. 12-3** Construction of Brillouin zones of a 2-dimensional rectangular lattice.

shows the four "bits and pieces" that make up the second zone. For convenience, the second, third etc. Brillouin zones often are not shifted by a reciprocal lattice vector into the first zone. This is called the **extended zone scheme**. If the extended zone scheme is not used, then in the first Brillouin zone, many eigenvalues for each **k** due to the higher lying bands will be found. Thus, in the **reduced zone scheme** (shifting all the points in reciprocal space by a reciprocal lattice vector to the first Brillouin zone), the allowed energy states are a multivalued function of **k**.

### d. Bloch functions

We have the irreducible representations of the translation group and a geometric interpretation of the allowed **k** values which serve as labels for the irreducible representations. Now consider the basis functions.

A basis function of the **k**th irreducible is in general $\phi_{\mathbf{k}}(\mathbf{r})$ and we always mean $\{R \mid t\}\psi(x) = \psi(\{R \mid t\}^{-1}x)$ as in Section 7-2.

$$\{E \mid t_n\}\, \phi_{\mathbf{k}}(\mathbf{r}) = \phi_{\mathbf{k}}(\mathbf{r} - \mathbf{t}_n)$$

$$\{E \mid t_n\}\, \phi_{\mathbf{k}}(\mathbf{r}) = \exp(i\mathbf{k} \cdot \mathbf{t}_n)\, \phi_{\mathbf{k}}(\mathbf{r}) \tag{12-18}$$

where the first equation results from the meaning of the translation operation as in Eq. 7-2 and the second equation is the meaning of a basis function as in Eq. 4-8. clearly the two functions on the right sides are equal. We try a form

$$\phi_{\mathbf{k}}(\mathbf{r}) = \exp(i\,\mathbf{k} \bullet \mathbf{r})\, u_{\mathbf{k}}(\mathbf{r}) \qquad (12\text{-}19)$$

motivated by the fact that if $u_{\mathbf{k}}(r)$ is a constant, Eq. 12-19 is the eigenfunction of the free electron Hamiltonian. Using this form, Eq. 12-18 gives

$$\phi_{\mathbf{k}}(\mathbf{r} - \mathbf{t}_n) = \exp[i\mathbf{k} \bullet (\mathbf{r} - \mathbf{t}_n)]\, u_{\mathbf{k}}(\mathbf{r}\text{-}\mathbf{t}_n)$$

$$\exp(-i\mathbf{k} \bullet \mathbf{t}_n)\, \phi_{\mathbf{k}}(\mathbf{r}) = \exp[i\mathbf{k} \bullet (\mathbf{r} - \mathbf{t}_n)]\, u_{\mathbf{k}}(\mathbf{r}) \qquad (12\text{-}20)$$

Therefore Eq. 12-20 shows

$$u_{\mathbf{k}}(r) = u_{\mathbf{k}}(\mathbf{r} - \mathbf{t}_n) \qquad (12\text{-}21)$$

where $\mathbf{t}_n$ is any symmetry translation operation of the lattice. Thus, the $u_{\mathbf{k}}(r)$ has the periodicity of the direct lattice. A function of the form shown in Eq. 12-19 with the lattice periodicity condition shown in Eq. 12-21 is called a **Bloch function**. These functions are basis functions of the **k**th irreducible representation. Since basis functions have the proper symmetry to be eigenfunctions, we look for the condition on the Bloch functions that will make them eigenfunctions. The wave equation is

$$[\,(-\hbar^2/2m)\,\nabla^2 + V(\mathbf{r})]\,\phi_{\mathbf{k}}(\mathbf{r}) = E(\mathbf{k})\,\phi_{\mathbf{k}}(\mathbf{r}) \qquad (12\text{-}22)$$

where $V(\mathbf{r})$ has the periodicity of the lattice. Substituting Eq. 12-19 into the wave equation, using $\mathbf{p} = -i\hbar\nabla$, and remembering that for an operator $p(e^{ikx}u) = (pe^{ikx})\,u + e^{ikx}(pu)$, one obtains

$$\exp(i\,\mathbf{k} \bullet \mathbf{r})\,[(-\hbar^2/2m)\,\{\nabla^2 + 2i\,\mathbf{k} \bullet \nabla - k^2\} + V(\mathbf{r})]\,u_{\mathbf{k}}(\mathbf{r})$$

$$= \exp(i\,\mathbf{k} \bullet \mathbf{r})\,E(\mathbf{k})\,u_{\mathbf{k}}(r)$$

$$[(-\hbar^2/2m)\,\nabla^2 + (\hbar/m)\,\mathbf{k} \bullet \mathbf{p} + V(\mathbf{r})]\,u_{\mathbf{k}}(\mathbf{r})$$

$$= [E(\mathbf{k}) - \hbar^2k^2/2m]\,u_{\mathbf{k}}(\mathbf{r}) \equiv E(k)'\,u_k(\mathbf{r}) \qquad (12\text{-}23)$$

**Bloch's theorem** states that if the potential energy has the periodicity of the lattice ($\mathbf{t}_n$ is a symmetry operation), then the eigenfunctions are the Bloch functions (Eq. 12-19 with condition Eq. 12-21) where $u_{\mathbf{k}}(r)$ is a solution of Eq. 12-23 over a primitive cell. The solution must join smoothly to the same function in the next cell. $E'$ is the energy of the electron with respect to a free electron at the same k.

The term $\mathbf{k}\bullet\mathbf{p}$ in Eq. 12-23 will lead to the "group of **k**" when point symmetry operations are discussed in the next section. Now we comment that the $\mathbf{k}\bullet\mathbf{p}$ term is qualitatively similar to an orbital angular momentum term $-l(l + 1)/r^2$ in the Hamiltonian of hydrogen atom. The angular momentum term arises in a similar manner to the $\mathbf{k \cdot p}$ term. It can be considered an extra potential energy, and is zero for s-state eigenfunctions for hydrogen, as the $\mathbf{k \cdot p}$ term is, for $\mathbf{k} = 0$. For the important case of free electrons $V(\mathbf{r})$ is constant and $u_{\mathbf{k}}(r)$ is a constant, so the energy and nonnormalized wave function is

$$\phi_{\mathbf{k}}(\mathbf{r}) = \exp(i\mathbf{k} \cdot \mathbf{r}), \qquad E(\mathbf{k}) = \hbar^2 k^2/2m \qquad (12\text{-}24)$$

**e. Energy levels — nearly free electrons**

Figure 12-4 shows a plot of energy versus k for free electrons as in Eq. 12-24 for a 1-dimensional crystal. The energy depends parabolically on k. Also shown are dotted lines representing the Brillouin zone boundaries at $n\pi/a$, where n is an integer other than zero. This would apply to a crystal with a constant potential energy $V(x)$. Once there is a translation operator Eq. 12-1, and cyclic boundary conditions Eq. 12-3, there is a first Brillouin zone $-\pi/a \leq k \leq \pi/a$ which can be used to describe completely the eigenfunctions and eigenvalues resulting from the translational symmetry of the lattice. Thus, various parts of the E versus k curve have to be displaced by a reciprocal lattice vector to bring that part into the first zone. Since $\phi_{\mathbf{k}+\mathbf{K}_n}(r)$ is identical to $\phi_{\mathbf{k}}(r)$ the corresponding energies are the same, $E(\mathbf{k}) = E(\mathbf{k} + \mathbf{K}_n)$. It can be seen that in the first zone E(k) is a multivalenced function. This is similar to the hydrogen atom where all infinite number of s-eigenfunctions transform as the same irreducible representation but have different principal quantum numbers and different energies. In the solid, for one irreducible representation (one value of **k**), there are an infinite number of eigenstates corresponding to different bands. Thus the eigenfunctions for a given **k** are usually given a band index $u_{\mathbf{k}i}(\mathbf{r})$ where i = 1, 2,... corresponding to the lowest energy band, the next lowest, etc. It is instructive to plot the real part of the plane wave wave function for various k-values throughout the zone. Hopefully this will not confuse the reader because the charge density $e^{+ikx}\, e^{-ikx} = 1$ is always constant and independent for k and x for plane waves. Only the real part of $e^{ikx} = \cos kx + i \sin kx$ will be plotted to see how the eigenfunction is modulated. At k = 0 the eigenfunction is 1 and independent of x. (See Fig. 12-5) As k increases to $\pi/a$, the function oscillates more rapidly until at $k = \pi/a$ the function alternnates sign at every nuclear site. This is very similar to the $\pi$-bonding cyclic hydrocarbon system discussed in Chapter 10 where the lowest energy eigenfunction is a bonding function in which all the $p_\pi$-electrons add in phase to

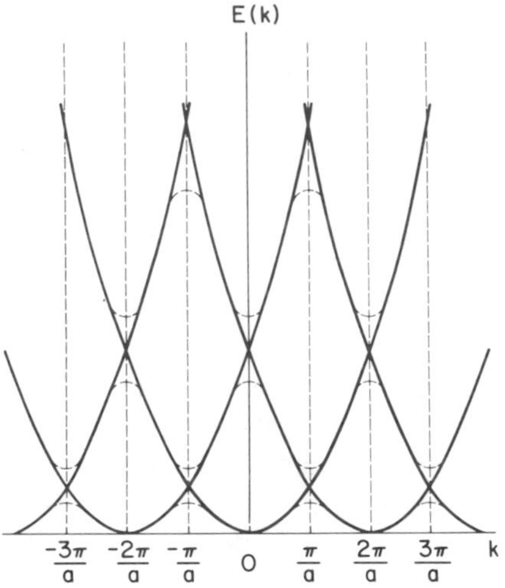


**Fig. 12-4** Energy versus k for free electrons. Also shown in the E versus k for nearby free electrons.

make the MO. The highest energy MO eigenfunction is an antibonding function where all the $p_\pi$-electrons on the nearest neighbor sites are exactly out of phase with each other. Eigenfunctions with intermediate energies have intermediate phase relations among the $p_\pi$-electrons. Figure 12-5 also shows the behavior of the real part of the eigenfunction for $k > \pi/a$. By adding or subtracting a reciprocal lattice vector $\pm$ n $2\pi/a$ to various pieces of the energy function in Fig. 12-4, all the energy levels can be shown in the reduced zone or first Brillouin zone. For example, in order to write the Bloch function for a free electron Eq. 12-24 for the second band in the Brillouin zone, we add or subtract the reciprocal lattice vector $2\pi/a$ to the free electron wave function. This brings the k values within the Brillouin zone. So for the second band,

$$-\pi/a \leq k \leq 0 \quad \phi_k(x) = \exp[ix(k+2\pi/a)] = \exp(ikx)\exp(i2\pi x/a)$$
$$0 \leq k \leq \pi/a \quad \phi_k(x) = \exp[ix(k-2\pi/a)] = \exp(ikx)\exp(-i2\pi x/a) \tag{12-25}$$

These functions are for free electrons and have the Bloch form but $u_k(x)$ is no longer a constant when the free electron function is restricted to a range of $-\pi/a \leq k \leq \pi/a$. Clearly $u_k(x) = u_k(x + na)$. As can be seen in Fig. 12-5 for $k = 2\pi/a$, the second band shows nodes within one primitive cell. Consider the eigenfunction at $k = \pi/a$ where the first and

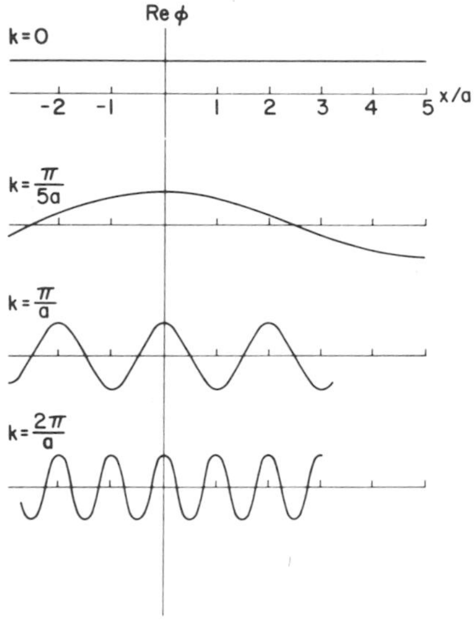


**Fig. 12-5** The real part of the free electron wave function exp(ikx) for several k.

second energy bands are degenerate. The two eigenfunctions transform as the same irreducible representation, and hence there can be off-diagonal matrix elements between the states. So far in this section we are considering V(x) as a constant of zero and there are no matrix elements between these states. However, we can expect this degeneracy to be removed when V(x) is no longer a constant, and the new eigenfunctions will be the sum and difference of the noninteracting eigenfunctions as shown in the treatment of the no-crossing rule in Section 7-5b. Anticipating this result, for $k = \pi/a$ the sum and difference of the eigenfunctions of band 1 and 2 can be formed where $u_{k_2}(x)$ comes from Eq. 12-25, remembering $\exp(ix) = \cos x + i \sin x$,

$$\left.\begin{aligned}\phi_{k_1}(x) &= \exp(i\pi x/a)\\ \phi_{k_2}(x) &= \exp(-i\pi x/a)\end{aligned}\right\} \qquad \begin{aligned}\phi_+ &= \phi_1+\phi_2 \propto \cos(\pi x/a)\\ \phi_- &= \phi_1-\phi_2 \propto \sin(\pi x/a)\end{aligned} \quad (12\text{-}26)$$

Change density is still uniform throughout the primitive cell. However, $\phi_-$ by itself has zero charge density at $x = 0$ where the nucleus is located and maximum charge at the edges of the cell $x = a/2$ between the two nuclei. $\phi_+$ has maximum charge density at the nuclear position and minimum at $x = a/2$. When a nonconstant potential is used, its detailed dependence on x will determine whether $\phi_+$ or $\phi_-$ is lower in energy. An atom that has a highly positive nuclear (or core) potential will cause $\phi_+$ to have a lower energy than $\phi_-$ while atoms with a potential that will cause bonding between them, will cause $\phi_-$ to have the lower energy.

Now we would like to relax the condition that the potential energy is a constant and show formally that the appropriate matrix elements are nonzero and will split the degeneracies discussed. To do this two general and important theorems are proved.

**Theorem 1.** If V(r) has the period of the direct lattice Eq. 12-1, then it can be expanded in a Fourier series in terms of the reciprocal lattice vectors.

$$V(\mathbf{r}) = \Sigma_{\mathbf{K}_n} A_{\mathbf{K}_n} \exp(i\mathbf{K}_n \bullet \mathbf{r})$$

$$V(\mathbf{r}+\mathbf{t}_n) = \Sigma_{\mathbf{K}_n} A_{\mathbf{K}_n} \exp(i\mathbf{K}_n \bullet \mathbf{r}) \exp(i\mathbf{K}_n \bullet \mathbf{t}_n) = V(\mathbf{r}) \tag{12-27}$$

where the second equation follows from Eq. 12-14. Thus, the expansion has correct periodicity and

$$A_{\mathbf{K}_n} = (1/\text{vol cell}) \int_{\text{cell}} V(\mathbf{r}) \exp(-i\mathbf{K}_n \bullet \mathbf{r})\, d\mathbf{r} \tag{12-28}$$

**Theorem 2.** If V(r) has the period of the direct lattice, then

$$\int V(\mathbf{r}) \exp(-i\mathbf{f} \bullet \mathbf{r})\, d\mathbf{r} = 0 \tag{12-29}$$

unless **f** is a reciprocal lattice vector $\mathbf{K}_n$ where upon the value of the integral is given by the coefficients of the Fourier series in the above theorem. The integral is over the sample with volume V. This theorem is proved by substituting for V(**r**) the Fourier expansion in Eq. 12-27 into the integral.

$$\Sigma_{\mathbf{K}_n} \int A_{\mathbf{K}_n} \exp[i(\mathbf{K}_n - \mathbf{f}) \bullet \mathbf{r}]\, d\mathbf{r} = V\, \Sigma_{\mathbf{K}_n} A_{\mathbf{K}_n} \delta_{\mathbf{K}_n,\mathbf{f}} \tag{12-30}$$

The second theorem applied to the periodic crystal, shows that off-diagonal terms in the secular determinant will be nonzero only if the **k**'s of the two Bloch states differ by a reciprocal lattice vector.

Consider the secular determinant for $k = \pi/a$ for the two states in Eq. 12-26 corresponding to the first and second band. These eigenfunctions are degenerate if the periodic potential is constant. However, when the potential has the period of the lattice, there is an off-diagonal matrix in the secular determinant. Writing the secular determinant for the interaction of the first and second band for $0 \le k \le \pi/a$, where $E_1$ and $E_2$

are the energies given by Eq. 12-24,

$$\begin{vmatrix} E_1 - E & D \\ D & E_2 - E \end{vmatrix} = 0$$

$$D = \langle \exp(ikx)\exp(-i2\pi x/a) \mid V(x) \mid \exp(ikx) \rangle \qquad (12\text{-}31)$$

The solutions for the two values of E are given in Section 7-5b. As can be seen, the Bloch functions in the off-diagonal term differ by a reciprocal lattice vector. Thus, $f = 2\pi/a$ in Theorem 2. For $k = \pi/a$, $E_1 = E_2$, the solutions of the secular determinant are $E_\pm = E_1 \pm |D|$ and the eigenfunctions are the sum or difference form shown in Eq. 12-26. The correspondence of each eigenfunction with its appropriate energy depends on the details of the potential. Note that at $k = \pi/a$ the function in the first band is interacting with the function in the second band at $k = -\pi/a$, i.e., <u>the function must always differ by a reciprocal lattice vector as clearly shown in Theorem 2</u>. The interaction of two waves that have a k-difference equal to a reciprocal lattice vector is intimately connected with Bragg diffraction. Further discussions of this point can be found in the references at the end of this chapter.

Consider the next crossing of two free electron eigenvalues which in the reduced zone in Fig. 12-4 occurs at $k = 0$ and $E = (\hbar^2/2m)(2\pi/a)^2$. The free electron eigenvectors are given in Eq. 12-25. As can be seen, they differ by a reciprocal lattice vector and thus an off-diagonal matrix element removes the degeneracy. All the degeneracies in Fig. 12-4 will be removed in this manner and are shown dotted in the figure.

### f. Energy levels — tight binding approximation

If, instead of thinking in terms of a nearly free electron model, the solid is considered to be formed by well-localized, atomic-like eigenstates the same type bands and splittings will occur. We consider this other extreme, the **tight binding approximation,** in this section.

Consider a 1s-state at each nuclear site in the direct lattice. If b(x) is a 1s-state centered at $x = 0$ in the lattice, the 1-dimensional model is used again for convenience, then the periodicity of the $u_k(x)$ can be written into the eigenfunction as

$$\phi_k(x) = \exp(ikx)\, \Sigma_n\, b(x\text{-}na) \qquad (12\text{-}32)$$

At $k = 0$ the eigenfunction of the crystal is just a series of 1s-states adding in phase as the lowest bonding MO state as discussed in Chapter

10. The eigenfunction of the two bands at $k = \pi/a$ are

$$\left.\begin{aligned} \phi_1 &= \exp(i\pi x/a)\, \Sigma_n\, b(x - na) \\ \phi_2 &= \exp(-\, i\pi x/a)\, \Sigma_n\, b(x - na) \end{aligned}\right\} \quad \begin{aligned} \phi_+ &\propto \cos\,(\pi x/a)\, \Sigma\, b(x - na) \\ \phi_- &\propto \cos\,(\pi x/a)\, \Sigma\, b(x - na) \end{aligned} \tag{12-33}$$

These results are similar to the results shown in Eq. 12-26. Note that $\phi_1$ and $\phi_2$ in Eq. 12-33 differ by a reciprocal lattice vector so the off-diagonal matrix element in the secular determinant will be nonzero and a splitting will occur just as in the nearly free electron case.

Figure 12-6 shows schematically the tight binding results at progressively higher energies in the Brillouin zone. At $k = 0$, a 1s-like wave function has been attached to each atom in this 1-dimensional lattice and the figure shows $\phi_k(x)$ from Eq. 12-32. For $k = \pi/5a$, or one fifth of the distance to the Brillouin zone edge, the real part of the wave function is shown modulated by the $e^{ikx}$ term. The imaginary part of the wave function, not shown, is identical but displaced. At $k = \pi/a$ the real and imaginary parts of the wave function are shown. The real part, corresponding to $\phi_+$ in Eq. 12-33, is similar to a 1s antibonding MO since every other atom wave function has an opposite phase. The imaginary part corresponds to $\phi_-$ in Eq. 12-33 and looks like a 2p wave function that is $\sigma$-bonding to its neighbors. Details of the potential will determine which of these two functions will lie lower in energy. Thus the situation is very similar to the free electron case. At $k = 2\pi/a$ the real part of $\phi_k(x)$ looks like 2s functions that are bonding MO and the imaginary looks like 2p functions that are antibonding. At $k = 3\pi/a$, the real part looks like 2s antibonding functions while the imaginary part looks like 3p bonding functions. Thus, pictorially it can be seen how the higher energy levels in the first Brillouin zone can have more modes. The modes resulting from absorbing into the $u_k(x)$ part of the eigenfunction the modes from the $\exp[(ikx)+(iK_nx)]$ part. We also see, in this very simple model, how the energy levels of the bands have a close similarity to the hydrogen atom problem. In real solids the crystal potential will distort the atomic eigenfunctions and the three dimensionality will complicate the problem, but the simple ideas already discussed remain unchanged.

The figures in this section are for a 1-dimensional lattice. Clearly the 3-dimensional case is extremely difficult to draw. However, the 3-dimensional results follow from these considerations in the same way. Consider a tightly bond 1s-function at each site of a simple cubic lattice. At $k = 0$ the lowest energy eigenfunction consists of all of these functions with positive phase. Thus, a 3-dimensional bonding MO is formed. If the

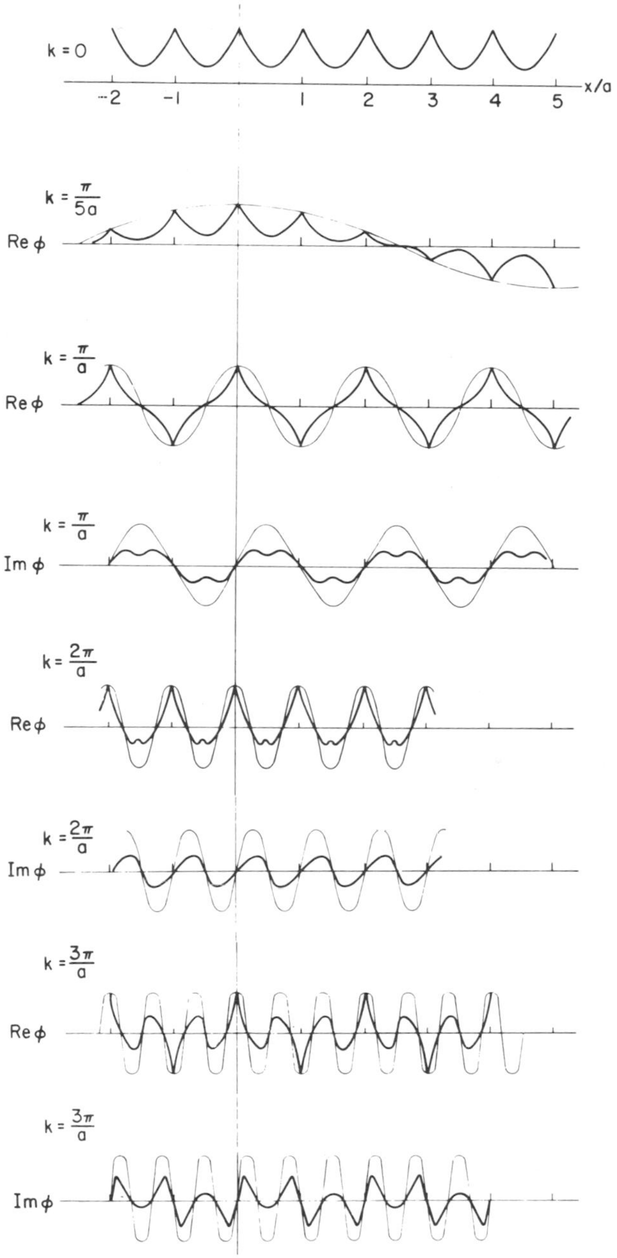


**Fig. 12-6** The real and imaginary parts of a tight binding wave function for various k in one dimension.

functions at $\mathbf{k} = (1, 0, 0)\ (\pi/a)$ are considered then in the x-direction the real part of the wave function looks like antibonding 1s-functions (the phase of each plane of functions alternates), while the imaginary part looks like bonding 2p-functions just as in Fig. 12-6. However in the y- and z-directions the functions still look bonding 1s-functions. (See the Problems.)

In summary it should be reemphasized that the results discussed so far come from the translational symmetry operations. The reciprocal lattice vector $\mathbf{k}$ is a label for the irreducible representations of the translation symmetry operation. One can also give the allowed values of $\mathbf{k}$ a geometric interpretation (discussed more extensively in the next section). The higher bands in the first Brillouin zone are modulated results from the lower bands due to the translation operation.

## 12-2 Symmorphic Space Groups

### a. Introduction

So far, in Section 12-1, a potential energy has been treated that has the translational periodicity of the lattice $V(\mathbf{r} + \mathbf{t}_n) = V(\mathbf{r})$ with cyclic boundary conditions. This has led to Bloch functions as basis functions of this cyclic group, Eqs. 12-19 and 12-21. These can be then taken as eigenfunctions when the $u_{\mathbf{k}}(\mathbf{r})$ satisfy the wave equation Eq. 12-23. Now we would like to consider crystals where there is more symmetry than just translational symmetry. We would like to consider point symmetry operations that bring the unit cell into an identical configuration, i.e., $\{R\,|\,0\}V(\mathbf{r}) = V(\mathbf{r})$. In this section only symmorphic space groups are considered so the space group operations can be separated

$$\{R\,|\,\mathbf{t}_n\} = \{E\,|\,\mathbf{t}_n\}\,\{R\,|\,0\} \tag{12-34}$$

but $\{R\,|\,0\}\{E\,|\,\mathbf{t}_n\} \neq \{E\,|\,\mathbf{t}_n\}\{R\,|\,0\}$ as remarked in Chapter 11.

We have already noted that the point group $P_g$ of a space group G is isomorphic to the factor group G/T which is taken with respect to the invariant subgroup of the translations T. Since irreducible representations of a factor group are also irreducible representations of the full group, we might suspect that all the irreducible representations of G can be found from the knowledge of those of T and $P_g$. This is in fact true. For references to a full proof of this see the Notes. While we do not prove this we will show how one goes about finding all the irreducible representations and the effects on the eigenvalues at all points in the Brillouin zone. In looking for irreducible representations of G we will start so that the elements of T are in block form, i.e., if we have only $\{E \mid t_n\}$ in G then the problem is diagonalized. So the basis functions of G will be Bloch functions or linear combinations of these functions. This procedure simplifies later work and has been often used. For example, if we are dealing with f-atomic functions and the problem relates to a free atom, we would start with basis functions that are spherical harmonics. If the problem relates to a cubic or lower crystal field, we would start with basis functions that transform as partners of the irreducible representations of $O_h$ symmetry. If the problem relates to a particular type of hybrid function, then a different linear combination would be more appropriate.

We proceed by investigating under what conditions Bloch functions are basis functions and eigenfunctions of space group operations. We know that Bloch functions diagonalize the Hamiltonian when just translational symmetry operations are present. Since $\{R \mid 0\}$ commutes with the Hamiltonian, the eigenfunction $\phi_{\mathbf{k}}(R^{-1}\mathbf{r})$ and $\phi(\mathbf{r})$ have the same energy, i.e.,

$$\begin{aligned} H\phi_{\mathbf{k}}(\mathbf{r}) &= E(\mathbf{k})\,\phi_{\mathbf{k}}(\mathbf{r}) \\ \{R \mid 0\}\,H\,\phi_{\mathbf{k}}(\mathbf{r}) &= \{R \mid 0\}\,E(\mathbf{k})\,\phi_{\mathbf{k}}(\mathbf{r}) \\ H\,\phi_{\mathbf{k}}(R^{-1}\mathbf{r}) &= E(\mathbf{k})\,\phi_{\mathbf{k}}(R^{-1}\mathbf{r}) \end{aligned} \qquad (12\text{-}35)$$

Now the form of $\phi_{\mathbf{k}}(R^{-1}\mathbf{r})$ and the conditions on the form must be investigated. When only the translational symmetry is considered, these considerations lead to the wave equation Eq. 12-23. Since the energy $E(\mathbf{k})$ is the same for $\phi_{\mathbf{k}}(\mathbf{r})$ as for $\phi_{\mathbf{k}}(R^{-1}\mathbf{r})$, we might want to write $E(\mathbf{k}) = E(R\mathbf{k})$ where there are as many equations like this as there are symmetry operations of the point group. This particular convention will not be used here but the idea that these different eigenfunctions have the same energy must be borne in mind.

First a few preliminaries: A dot product of two vectors is a scalar, thus if both vectors in the dot product are operated on by the same symmetry operation R, then the scalar does not change its value. This is a special case of the unitary property Eq. 7-7. Thus

$$\mathbf{p} \cdot \mathbf{p} = (R\mathbf{p}) \cdot (R\mathbf{p}), \qquad \mathbf{k} \cdot \mathbf{k} = (R\,\mathbf{k}) \cdot (R\,\mathbf{k}) \quad \text{etc.}$$

$$\mathbf{k} \cdot (R^{-1}\mathbf{r}) = R(\mathbf{k}) \cdot R(R^{-1}\mathbf{r}) = (R\mathbf{k}) \cdot (RR^{-1}\mathbf{r}) = R\mathbf{k} \cdot \mathbf{r} \qquad (12\text{-}36)$$

The last result $\mathbf{Rk \cdot r}$ means $(\mathbf{Rk}) \cdot \mathbf{r}$ where R is not operating on both $\mathbf{k}$ and $\mathbf{r}$, but only on $\mathbf{k}$.

Now the form of $\phi_{\mathbf{k}}(R^{-1}\mathbf{r})$ will be investigated. Again a Bloch form is be taken and the conditions on the form determined.

$$\phi_{\mathbf{k}}(R^{-1}\mathbf{r}) = \exp(i\, \mathbf{k} \cdot R^{-1}\mathbf{r})\, u_{\mathbf{k}}(R^{-1}\mathbf{r}) = \exp(i\, R\,\mathbf{k} \cdot \mathbf{r})\, u_{\mathbf{k}}(R^{-1}\mathbf{r})$$

$$\equiv \exp(i\, R\,\mathbf{k} \cdot \mathbf{r})\, u_{R\mathbf{k}}(\mathbf{r}) \equiv \phi_{R\mathbf{k}}(\mathbf{r}) \qquad (12\text{-}37)$$

where Eq. 12-36 shows $\mathbf{k} \cdot R^{-1}\mathbf{r} = R\mathbf{k} \cdot \mathbf{r}$. Now we ask if this function in Eq. 12-37 is a basis function of the translation symmetry operations of Bloch form with the periodicity of the direct lattice. To investigate this, the function must be operated on by a general translation operation.

$$\{E \mid \mathbf{t}_n\}\, \phi_{R\mathbf{k}}(\mathbf{r}) = \{E \mid \mathbf{t}_n\}\, \phi_{\mathbf{k}}\, (R^{-1}\mathbf{r})$$

$$= \exp[i\, R\,\mathbf{k} \cdot (\mathbf{r} - \mathbf{t}_n)]\, u_{\mathbf{k}}(R^{-1}\mathbf{r} - R^{-1}\mathbf{t}_n)$$

$$= \exp(-i\, R\,\mathbf{k} \cdot \mathbf{t}_n)\, \exp(i\, R\,\mathbf{k} \cdot \mathbf{r})\, u_{\mathbf{k}}(R^{-1}\mathbf{r})$$

$$= \exp(-i\, R\,\mathbf{k} \cdot \mathbf{t}_n)\, [\exp(i\, R\,\mathbf{k} \cdot \mathbf{r})\, u_{R\mathbf{k}}(\mathbf{r})] \qquad (12\text{-}38)$$

where the fact that $R^{-1}\mathbf{t}_n = \mathbf{t}_n{}'$and $u(\mathbf{r}' + \mathbf{t}_n{}') = u(\mathbf{r})'$ is used. Thus, for the translation group, the function $\phi_{R\mathbf{k}}(\mathbf{r})$ can be taken as a basis function of $R\mathbf{k}$th irreducible representation with the Bloch form. We now ask what form the wave equation takes for $\phi_{R\mathbf{k}}(\mathbf{r})$. $\{R \mid 0\}$ operating on both sides of the wave equation as in Eq. 12-35 will yield equations just like Eq. 12-22 and 12-23 with the result

$$\{(-\hbar^2/2m)\, \nabla^2 + (\hbar/m) R\mathbf{k} \cdot \mathbf{p} + V(\mathbf{r})\} u_{R\mathbf{k}}(\mathbf{r}) = [E(\mathbf{k}) - \hbar^2 k^2/2m] u_{R\mathbf{k}}(\mathbf{r}) \qquad (12\text{-}39)$$

This results from Eq. 12-36 and $V(R^{-1}\mathbf{r}) = V(r)$. When determining eigenfunctions, the Hamiltonian must commute with all the symmetry operations of the group. The term $R\mathbf{k} \cdot \mathbf{p}$ has important consequences. When analyzing for eigenfunctions from Eq. 12-39 for a nonzero $\mathbf{k}$, only the symmetry operations that leave $\mathbf{k}$ invariant are the appropriate ones that must be considered. It should always be remembered that $\mathbf{k}$ and $\mathbf{k} + \mathbf{K}_n$ are equivalent.

### b. The group of k

The **group of k** is defined as the subgroup of the point group $P_g$ that leaves $\mathbf{k}$ invariant, i.e.,

$$R\,\mathbf{k} = \mathbf{k} + \mathbf{K}_n \qquad (12\text{-}40)$$

This group is sometimes called the point group of the wave vector **k** and we will write it as P(k). Thus, P(k) contains the symmetry operations of $P_g$ which leave **k** invariant, Eq. 12-40. Note the elements P(k) depend on **k** in magnitude and direction. The irreducible representations of the group of **k** are sometimes called the **small representations**. If P(**k**) is the trivial group with only the element $\{E \mid 0\}$, then **k** is said to be a **general point in the Brillouin zone**. If P(k) contains more than one element and is a larger group then for all neighboring points, **k** is said to be a **special point**. If the points on a line have the same group of **k** with more than one symmetry operation, then one has a **special line** (or symmetry line).

As a simple example of the group of **k**, consider a square lattice. Figure 12-7a shows the Brillouin Zone for a square lattice which results from a square primitive cell. The point symmetry operations are E, $2C_4$, $C_2$, $2\sigma_v$, $2\sigma_d$. If **k** is picked so that it does not lie on the Brillouin zone edges or on any of the symmetry lines, then it a **general point**. In Fig. 12-7 the point symmetry operations on the general point $\mathbf{k}_1$ generate eight **k** values, none of which are equivalent. The group of **k** for $\mathbf{k}_1$ is just $\{E \mid 0\}$ and the irreducible representation is 1. Thus, for a general point, the point symmetry operations add nothing new to the translational symmetry operations. The irreducible representations, basis functions, and eigenfunctions of the space group are completely determined by the translational symmetry, Section 12-1. This will be discussed more fully later. Note that the Bloch type eigenfunctions for the eight **k** values shown in Fig. 12-7a are different from each other but the energies are the same as shown in Eq. 12-35. Generalizing, for an h-order point group and **k** at a general point in the Brillouin zone, the h-Bloch functions $\phi_{R\mathbf{k}}(\mathbf{r})$ form a basis of an h-order irreducible representation of the space group. As discussed in Section 12-1, these Bloch functions can be eigenstates if they satisfy Eq. 12-23 or 12-39. Thus, at general points the eigenfunctions are simple Bloch functions and cannot be linear combinations of such functions. Naturally these eigenfunctions diagonalize the secular equation. Another way of appreciating this fact is to realize that in the secular equation for a general point **k** the off-diagonal matrix elements, Eq. 12-31, must be zero. This is because there is no other **k** that differs from **k** by a reciprocal lattice vector, and Theorem 2 shows that there are no off-diagonal matrix elements.

Before we discuss the irreducible representations of the space groups, it is useful to define the star of **k**. The **star of k** is defined as the set of inequivalent **k** vectors obtained by applying all $\{R \mid 0\}$ of $P_g$ to **k**. Thus in Fig. 12-7a the star of general point **k** consists of all the eight

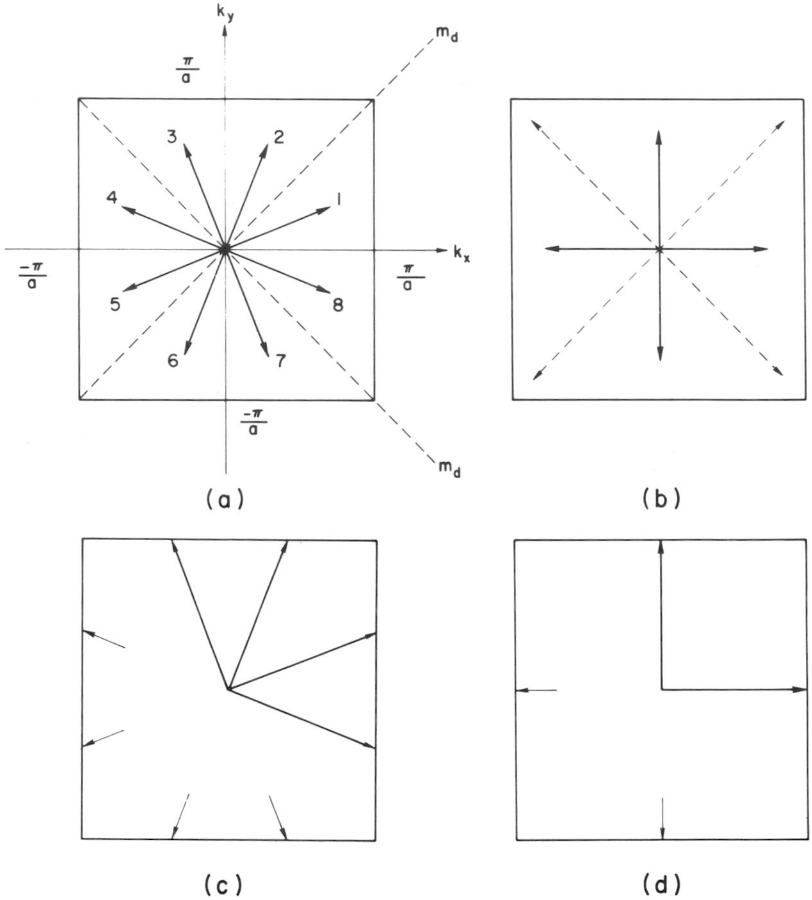


**Fig. 12-7** The first Brillouin zones of a 2-dimensional square lattice. The vectors in the star of **k** are shown for a general k in (a) and for various special **k** values in (b)–(d).

values shown. If $\mathbf{k}_1$ is rotated so that it is on the $k_x$ axis, then $\mathbf{k}_1$ and $\mathbf{k}_8$ are equivalent, Eq. 12-42, as are 2 and 3, 4 and 5, 6 and 7. Thus, if **k** is along $k_x$, the star of **k** consists of four inequivalent **k** vectors. This is shown in Fig. 12-7b where the star of $\mathbf{k} = \alpha(\mathbf{b}_1 + \mathbf{b}_2)$ is also shown, i.e., **k** along a diagonal plane, to consist of four inequivalent values of **k**. Referring back to Fig. 12-7a, if $\mathbf{k}_1$ is not rotated but lengthened until it touches the Brillouin zone boundary at $k_x = \pi/a$, then $\mathbf{k}_1$ and $\mathbf{k}_4$ are equivalent since they differ from each other by a reciprocal lattice vector $\mathbf{b}_1$. Similarly, 2 and 7, 3 and 6, 5 and 8 are equivalent. Then the star of **k** is shown by the four vectors in Fig. 12-7c. Similarly if $\mathbf{k}_1$ were along $k_x$, then the star of k consists of the two inequivalent vectors as shown in Fig. 12-7d. For $\mathbf{k} = (\pi/a)(\mathbf{b}_1 + \mathbf{b}_2)$, the corner of the Brillouin zone, all the other corners of the zone are equivalent by a reciprocal lattice vector translation. Thus the star of **k** consists of only one vector. The same result is found if $\mathbf{k} = 0$, the zone center. It should be noted, that which of the equivalent **k** is considered in the star of **k** is irrelevant, since all **k** in the star are, indeed, equivalent.

### c. Irreducible representations

If $h(P_g)$ is the order of the point group $P_g$ of the space group, and $h(P(k))$ is the order of the group of **k** P(k), and s(k) is the number of wave vectors in the star of **k**, then

$$h(P_g) = [s(k)] \cdot [h(P(k))] \tag{12-41}$$

Hence, the point symmetry operations of $P_g$ are split between those that are in the P(k) and s(k) symmetry operations of $P_g$ resulting in the star of **k**. Usually the particular choice of the s(k) symmetry operations is not unique. For example, Fig. 12-7c shows the four vectors 1, 2, 3, and 8 making up the star of **k**, but 2, 3, 4, and 5 would be an equivalent choice. What is important is to adhere to the choice once it is made.

We now state a fundamental theorem covering the irreducible representations of a symmorphic space group G. As we will see this theorem eliminates the need for explicitly writing down a character table for G, which would be too large at any rate. By the use of this theorem we can obtain the basis functions and degeneracies for all the allowed **k** values in terms of the character tables of the 32 point groups. Suppose we have a Bloch function for an allowed **k**. Associated with this **k** is the group of **k**, P(k), which has irreducible representations, characters, etc. Also associated with this **k** are the s(k) members of the star of **k**and s(k) symmetry operations, $\equiv \{\overline{R}_s \mid 0\}$,from the point group $P_g$ of the space group. Then, $\phi(\mathbf{r})_{\mathbf{k}m}{}^i$ is a Bloch function that transforms as mth row of the ith irreducible unitary representation $\Gamma^i$ of P(k).

**Theorem.** If $\Gamma^i$ is an $l$-dimension irreducible representation, then the set of $[l] \cdot [s(k)]$ functions

$$\{R_s \mid 0\}\phi_{\mathbf{k}m}{}^i(\mathbf{r}) \qquad s = 1, \ldots, s(k) \ \text{and} \ m = 1, \ldots, l \tag{12-42}$$

form a basis of an $[l] \bullet [s(\mathbf{k})]$ dimension irreducible unitary representation of the space group G. Furthermore, all the irreducible representations of G are obtained this way. So we can see that for a given **k**, one need only consider Bloch functions from the translation group and the much smaller group P(k) to get irreducible representations, basis functions, and resulting degeneracies caused by the symmetry operations of the full space group.

The proof of this theorem was first given in 1936 by Seitz for all space groups including nonsymmorphic groups. Simpler proofs can be given for symmorphic space groups. We will not prove this theorem but will show how we obtain all the irreducible representations for general and special points and give detailed examples.

First consider a general point in the Brillouin zone. P(k) contains only $\{E\,|\,0\}$ as a symmetry operations so the only irreducible representation is 1. Obviously $h(P(k)) = 1$ and $s(k) = h(P_g)$, the order of the point group. Thus, for a general point there is only one irreducible representation of G, as in Eq. 12-42. It has dimension $h(P_g)$, and the $h(P_g)$ partners of this irreducible representation have the same eigenvalue. Using the notation in Eq. 12-37 and Eq. 12-42 for the symmetry operations that form the star of **k**, we have $E(\mathbf{k}) = E(\overline{R}_s\mathbf{k})$ for all s-values, which for a general point is the same as the order of the point group of the crystal, $h(P_g)$. The basis functions are $\exp(i\mathbf{k}\bullet\mathbf{r})\, u_{\mathbf{k}}(\mathbf{r})$, $\exp(iR_1\mathbf{k}\bullet\mathbf{r})\, u_{\mathbf{k}}(R_1^{-1}\mathbf{r})$, etc.

Next consider the point $\mathbf{k} = 0$, or any other **k** where the group of **k** is the same as $P_g$. (For the square lattice discussed above this will be the Γ and M points as in Fig. 12-8). For this situation we have only one vector in the star of **k** so $s(k) = 1$ and $h(P_g) = h(P(k))$. The degeneracies and splittings of the linear combinations of Bloch functions that are basis functions are obtained from the dimensions of the various irreducible representations of the group of **k** which for this case is the point group of the crystal. Clearly, there will be some irreducible representations of $P_g$ of dimension greater then one for symmetries higher than orthorhombic. Eigenfunctions that have a degeneracy due to the group of **k**, are said to have an **essential degeneracy**. The essential degeneracy occurs at a single value of **k**. The degeneracy associated with a general point where $E(\mathbf{k}) = E(\overline{R}_s\mathbf{k})$, so h different **k**'s have the same energy, is not called an essential degeneracy. We see that when $P(k) = P_g$, the various splitting, etc. are investigated as has been done many times earlier in this book. Examples will be given shortly.

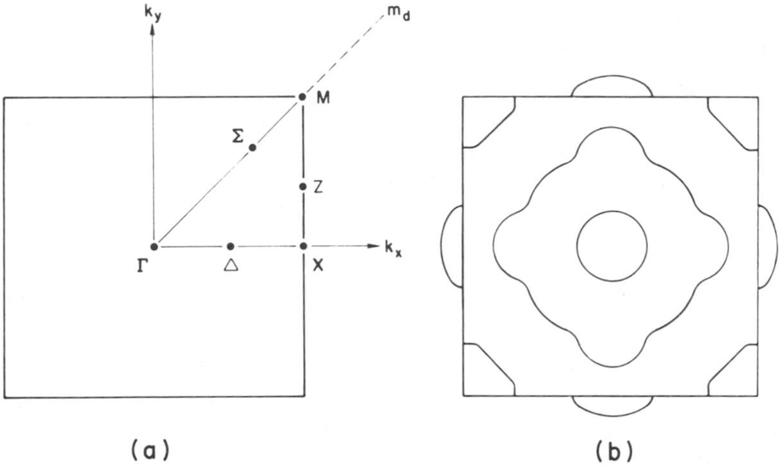


**Fig. 12-8** (a) The first Brillouin zone of a 2-dimensional square lattice showing the labels of the special points and lines. (b) Constant energy surfaces in this Brillouin zone. Three different values of energy are shown.

The last case to consider in general is the case where $1 < h(P(k)) < h(P_g)$. Clearly the types of essential degeneracies will depend on the group of **k** which varies for the different special points and lines. For an $l$-dimensional irreducible representation of P(k), there will be a $[l]\cdot[s(k)]$ degeneracy in the Brillouin zone as the theorem, Eq. 12-42. In the next section a simple example will illustrate these remarks.

Before discussing the example, a word about notation is in order. In the field of solid state physics, the irreducible representations of the group of **k** (the small representations) are often labeled with neither the Mulliken nor the Bethe notation. The BSW (Bouchaert, Smoluskowski, and Wigner) notation is often used. The irreducible representations are labeled with a letter corresponding to the labels of the special point or special line and then number subscripts are given to the letter for the different representations in a manner similar to the Bethe notation. Unfortunately, primes are added for certain of the special points. Care should therefore be taken when reading papers that use the BSW notation.

**d. A simple example**

Figure 12-8a shows a square lattice. We will use this as a simple example to illustrate some of the ideas discussed and to follow. The special points are labeled Γ, M, X and the special lines are labeled Σ, Z, Δ as discussed with respect to Fig. 12-7. For **k** values corresponding to each of these special points and special lines, the group of **k** consists of more than just the identity operation. For the special line Δ, the group of **k** consists of $\{E, \sigma_v{}^y\}$, for Σ, the group of **k** consists of $\{E, \sigma_d\}$, for Z, $\{E, \sigma_v{}^x\}$, for X, $\{E, C_2, \sigma_v{}^x, \sigma_v{}^y\}$. For Γ and M the group of **k** is the same as the point group of the crystal $P_g$. Table 12-2 shows the symmetry operations and character tables which show the small representations of the group of **k** for each of the special points and lines for this square lattice. There is no fundamental need to have these tables here since they are the same tables as in Appendix 3 for the appropriate three point groups of the 32 point groups. They are included in Table 12-2 for convenience and tutorial purposes only. The labels of the small (≡ irreducible) representations are in the BSW notation which is often used in solid state physics.

For special points or lines, the symmetry operations of the group of **k** impose new conditions on the irreducible representation, basis functions, and eigenfunctions. The eigenfunctions are no longer always simple Bloch functions but often linear combinations of such functions. However, these functions will still have the periodicity fo the lattice since the Bloch functions in the linear combinations have the same or equivalent **k** as discussed in the last section.

For example, the group of Δ consists of the two symmetry operations and two irreducible representations shown in Table 12-2. Basis functions of the group of Δ must be even or odd under the symmetry operations. Thus, the eigenfunction of the crystal with **k** along the Δ line must be even or odd and this restricts the type of basis functions.

Along the special line Z, the star of Z consists of four inequivalent values of **k**. Two equivalent values of **k** and the simple Bloch functions associated with these values are

$$\mathbf{k}_a = 1/2\,\mathbf{b}_1 + \alpha\mathbf{b}_2 \qquad \phi_a = \exp(i\,\alpha\,2\pi\,y)\exp(i\pi\,x/a)$$

$$\mathbf{k}_b = -1/2\,\mathbf{b}_1 + \alpha\mathbf{b}_2 \qquad \phi_b = \exp(i\,\alpha\,2\pi\,y)\exp(-i\pi\,x/a) \qquad (12\text{-}43)$$

where $\alpha$ is a positive number between 0 and 1/2. Under the symmetry operations of the group of Z, these two functions transform between themselves forming a reducible representation. Reducing the representation yields and the irreducible representations $Z_1 + Z_2$. Basis functions can immediately be written down by the use of projection operator tech-

**Table 12-2** Character tables for the various special points and lines of the square lattice as defined in Fig. 12-8

IRREDUCIBLE REPRESENTATIONS AT SPECIAL POINTS

| $\Gamma$,M | E | $C_2$ | $2C_4$ | $2\sigma_v$ | $2\sigma_d$ |
|---|---|---|---|---|---|
| $\Gamma_1,M_1$ | 1 | 1 | 1 | 1 | 1 |
| $\Gamma_2,M_2$ | 1 | 1 | 1 | -1 | -1 |
| $\Gamma_3,M_3$ | 1 | 1 | -1 | 1 | -1 |
| $\Gamma_4,M_4$ | 1 | 1 | -1 | -1 | 1 |
| $\Gamma_5,M_5$ | 2 | -2 | 0 | 0 | 0 |

| X | E | $C_2$ | $\sigma_v$ | $\sigma_d$ |
|---|---|---|---|---|
| $X_1$ | 1 | 1 | 1 | 1 |
| $X_2$ | 1 | 1 | -1 | -1 |
| $X_3$ | 1 | -1 | 1 | -1 |
| $X_4$ | 1 | -1 | -1 | 1 |

| $\Delta$<br>$\Sigma$<br>Z | E<br>E<br>E | $\sigma_v^y$<br>$\sigma_d$<br>$\sigma_v^x$ |
|---|---|---|
| $\Delta_1,\Sigma_1,Z_1$ | 1 | 1 |
| $\Delta_2,\Sigma_2,Z_2$ | 1 | -1 |

niques. The result obviously is

$$\phi(Z_1) = \phi_a + \phi_b = 2 \exp(i\,\alpha\, 2\pi\, y) \cos \pi\, x/a$$

$$\phi(Z_2) = \phi_a - \phi_b = 2\, i \exp(i\,\alpha\, 2\pi\, y) \sin \pi\, x/a \qquad (12\text{-}44)$$

which is very similar to the 1-dimensional case in Eq. 12-26 where the periodic potential of the crystal will cause an energy difference between the two functions in Eq. 12-44. The behavior of the eigenfunction at the X point is the same as along the Z line. Thus the eigenfunctions can be taken as in Eq. 12-44 but with y = 0.

The behavior of the energy near a Brillouin zone boundary can be seen. Whenever a mirror plane occurs in the point group of the crystal, the normal component of the energy at a Brillouin zone boundary is zero. In Fig. 12-7, $E(\mathbf{k}_1) = E(\mathbf{k}_4)$ because $\mathbf{k}_1$ and $\mathbf{k}_4$ are in the same star of a general point. Also, $E(\mathbf{k}_4) = E(\mathbf{k}_4 + \mathbf{b}_1)$ which is a point to the right of $\mathbf{k}_1$ on the other side of the Brillouin zone. As $\mathbf{k}_1$ increases in magnitude, the special point Z will be reached where only 1-dimensional irreducible representations are found. So $E(\mathbf{k}_1) = E(\mathbf{k}_4 + \mathbf{b}_1)$ immediately implies that E(k) has a maximum or minimum as it crosses the zone boundary. Thus, the normal component of the energy is zero at a zone boundary that has a mirror plan parallel to it. See the Notes for further reference to this point.

For the M point in Fig. 12-8a the star of M consists of only one wave vector since the four corners of the Brillouin zone are equivalent. The four simple Bloch functions that would be written down for the four **k** values are

$$\mathbf{k}_a = 1/2\,\mathbf{b}_1 + 1/2\,\mathrm{b}_2 \qquad \phi_a = \exp[i\,(\pi/a)\,(x+y)]$$

$$\mathbf{k}_b = -1/2\,\mathbf{b}_1 + 1/2\,\mathrm{b}_2 \qquad \phi_b = \exp[i\,(\pi/a)\,(-x+y)]$$

$$\mathbf{k}_c = -1/2\,\mathbf{b}_1 - 1/2\,\mathrm{b}_2 \qquad \phi_c = \exp[i\,(\pi/a)\,(-x-y)]$$

$$\mathbf{k}_d = -1/2\,\mathbf{b}_1 - 1/2\,\mathrm{b}_2 \qquad \phi_d = \exp[i\,(\pi/a)\,(x-y)] \qquad (12\text{-}45)$$

Under all the symmetry operations of the group of M, these four functions transform among themselves forming a reducible representation. When reduced, one obtains $M_1 + M_4 + M_5$. Again, by the use of projection operators or direct inspection, the wave functions that transform according to these three irreducible representations can be obtained.

$$\phi(M_1) = \phi_a+\phi_b+\phi_c+\phi_d = 4\cos\,(\pi/a)x\ \cos(\pi/a)\,y$$

$$\phi(M_4) = \phi_a-\phi_b+\phi_c-\phi_d = -\sin\,(\pi/a)x\ \sin(\pi/a)\,y$$

$$\phi_\alpha(M_5) = \phi_a+\phi_d-\phi_b-\phi_c = 4i\sin\,(\pi/a)x\ \cos(\pi/a)\,y$$

$$\phi_\beta(M_5) = \phi_a+\phi_b-\phi_c-\phi_d = 4i\cos(\pi/a)x\ \sin(\pi/a)\,y \qquad (12\text{-}46)$$

A nonconstant potential energy will split the degeneracy of eigenfunctions at the M point. However, the two functions that transform as partners of the $M_5$ irreducible representation will not split, giving an essential degeneracy.

Notice that it is not necessary to consider states in the entire Brillouin zone. For a general point, the energy for all h members of the star of **k** is the same. So it is really only necessary to consider 1/h of the conventional Brillouin zone. For example, Fig. 12-8a shows 1/8 of the zone enclosed by the special points and lines that must be considered. The other seven pieces of the first Brillouin zone can be obtained by "symmetry." This possible reduction of the zone describing the allowed **k** vectors is normally ignored and the full Brillouin zone, determined by only the translational symmetry, is used in solid state physics. In fact, the term degeneracy is usually not used to describe the fact that energy for all the **k** values in the star of **k** is the same. It is usually reserved for essential and accidental degeneracies. The motivation for this is clear. It is very convenient and helpful to preserve the shape of the Brillouin zone since it displays the point symmetry in a clear and simple way. This will be very apparent when the energy versus **k** contours are considered but it is

already clear from Fig. 12-8a. Namely 1/8 of the Brillouin zone does not show the point symmetry very clearly.
As a last point for this simple example we consider putting electrons in the lattice and find the energy contours. The result is very simple to visualize since it is similar to rotating the 1-dimensional result of Fig. 12-4 about the energy axis. However, the "square" symmetry does show up. For each **k** two electrons can be put in the lattice, one with spin up and one with spin down. If very few electrons are in the band, the lowest energy states will be filled. The result in Fig. 12-4 would be to fill up the band to **k** values very much smaller than $\pi/a$, which for the 2-dimensional square lattice would be a spherical energy contour shown for small **k** values in Fig. 12-8b. Since the **k** values are so far from the Brillouin zone boundaries, off-diagonal terms in the secular equation Eq. 12-31 are very small and deviations from spherical contours are not observed. As more electrons are put in the solid, large **k** values are occupied. As $\mathbf{k} = \pi/a$ is approached from below, deviations from the parabolic band is observed in the 1-dimensional case as in Fig. 12-4. For the 2-dimensional case, deviations from spherical contours occur for **k** values close to the Brillouin zone boundary which is in the directions of $\mathbf{b}_1$ and $\mathbf{b}_2$. Figure 12-8b shows bulges in the constant energy contours for such **K** values. As more electrons are added, larger **k** values are occupied. The last energy contour shows that some of the energies in certain **k** directions in the second Brillouin zone are lower than energies associated with other **k** values in the first Brillouin. The details of the division between first and second Brillouin zone clearly depend on the off-diagonal terms in the secular equation Eq. 12-21, i.e., the potential energy. However, it is already clear that complicated energy contours (**Fermi surfaces**) can arise in simple lattices due to the discontinuities at the Brillouin zone boundaries. Notice that the normal component of the energy ($\nabla_{\mathbf{k}}E$) is zero at the boundaries. Also notice that the energy of any general point is the same as that for each of the seven other **k** values in the star of **k** has been discussed.

### e. Compatibility relations

Consider the behavior of the eigenfunctions as **k** varies from the special point Γ along Σ to M along Z to X along Δ and back to Γ. The eigenfunctions at Γ transform as the irreducible representations of the group of Γ. When **k** is no longer at the Γ point but along Σ, the number of symmetry operations is much reduced. The group of Σ is a subgroup of the group of Γ so the correlation of the irreducible representations be-

COMPATIBILITY RELATIONS

| REPRESENTATION | COMPATIBLE WITH |
|---|---|
| $\Delta_1$ | $\Gamma_1, \Gamma_3, \Gamma_5; X_1, X_4$ |
| $\Delta_2$ | $\Gamma_2, \Gamma_4, \Gamma_5; X_2, X_3$ |
| $\Sigma_1$ | $\Gamma_1, \Gamma_4, \Gamma_5; M_1, M_4, M_5$ |
| $\Sigma_2$ | $\Gamma_2, \Gamma_3, \Gamma_5; M_2, M_3, M_5$ |
| $Z_1$ | $X_1, X_3; M_1, M_3, M_5$ |
| $Z_2$ | $X_2, X_4; M_2, M_4, M_5$ |

$\Gamma_5$ REDUCES INTO $\Delta_1 + \Delta_2$ OR $\Sigma_1 + \Sigma_2$
$M_5$ REDUCES INTO $\Sigma_1 + \Sigma_2$ OR $Z_1 + Z_2$

(a)

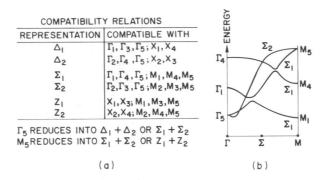

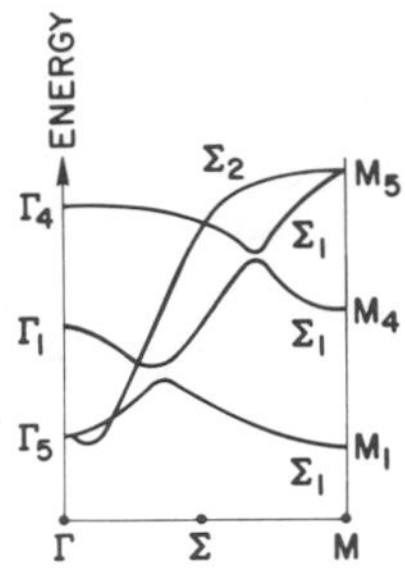


(b)

**Fig. 12-9** (a) Compatibility tables for the irreducible representations shown in Table 12-2 for the square lattice. (b) A schematic of the energy levels along the $\Sigma$-direction of the square lattice.

tween the two groups can be easily determined by standard considerations, as in Section 7-5c. The correlation tables in Appendix 7 show the results. Figure 12-8 shows the character tables for the group of **k** for the points and lines that we discuss. These are the same tables as can be found in Appendix 3 for the point groups $C_{4v}$, $C_{2v}$, and $C_{1h}$. Figure 12-9a shows the results for the irreducible representations of the point groups concerned. These results are usually referred to as **compatibility relations** when referred to the group of **k**.

Figure 12-9b shows an example of what might happen in a crystal. At the $\Gamma$ point the lowest band is assumed to be the doubly degenerate $\Gamma_5$. The eigenfunction making up this band could be $p_x$- and $p_y$-like. As **k** moves from the $\Gamma$ point to the $\Sigma$ line, the 2-fold essential degeneracy must split and the eigenfunctions that were partners of the $\Gamma_5$ irreducible representation must now separately transform as $\Sigma_1$ and $\Sigma_2$ (as shown). In a similar manner the eigenfunctions that transform as $\Gamma_1$ correlate with those that transform as $\Sigma_1$ and $\Gamma_4$ also correlates with $\Sigma_1$. As **k** increases along the $\Sigma$ line Fig. 12-9b shows a case where the eigenvalue of eigenfunction $\Gamma_5\Sigma_2M_5$ is degenerate with the eigenfunction of $\Gamma_5\Sigma_1M_1$, and then at two other **k** values along $\Sigma$ it is again degenerate with two other bands. These crossings are allowed since the eigenfunctions transform as different irreducible representations. This is called an **accidental degeneracy**. The particular value of **k** along $\Sigma$ where these accidental degeneracies occur, depends sensitivity on the potential energy and will vary with temperature, pressure, etc. Also shown in Fig. 12-9b are two no-crossings (Section 7-5b) or repulsions of levels that both transform as $\Sigma_1$. As **k** increases along $\Sigma$ to the M point, eigenfunctions that transform as $\Sigma_1$ and $\Sigma_2$ with different energies suddenly coalesce into the $M_5$ irreducible representation resulting in an essential degeneracy. The lowest energy band in Fig. 12-9b is $\Gamma_5\Sigma_2\Sigma_1M_1$. Note that the accidental degeneracy due to a crossing of the $\Sigma_1$ and $\Sigma_2$ eigenfunctions along the $\Sigma$ line is removed as soon as **k** is moved off the $\Sigma$ line to a general point.

### f. Some 3-dimensional Brillouin zones

Figure 12-10a shows the Brillouin zone for a simple cubic primitive cell. The Brillouin zone is also a simple cubic cell in reciprocal space extending from $-\pi/a$ to $+\pi/a$ in the three directions. The figure shows the special points and lines. There are four special points Γ, R, X, M. The group of Γ and R contains all 48 symmetry operations of the full point group $O_h$. This is indicated in Fig. 12-10b. The group of X and M contain the same symmetry operations as the point group $D_{4h}$. However, it must be understood that the z-axis in the point group $D_{4h}$ refers to the direction from the origin to the point X or to the point M, for the group of X and M. These directions fix the horizontal and vertical planes etc. The six special lines are listed in Fig. 12-10b and shown in Fig. 12-10a. The group of **k** for the various lines is listed, again changes of directions of the z-axis of the conventionally listed point groups must be taken into account. Figure 12-10c shows the compatibility relations between the special point Γ and the three special lines Δ, Λ, and Σ that connect to it. Similarly, the three special lines that connect to the special points M and X. These compatibility tables are useful when the energy levels are discussed in the next subsection.

Note the symmetry elements that contain the special points and lines enclose a volume that is 1/48 of the volume of the Brillouin zone. Only the general points in this volume need be considered since they have the same energy as the 47 other members of the star of **k** for a general point. In the free electron approximation, the lowest energy states will have spheres for constant energy contours in reciprocal space. As the energy increases the contours will get closer to the Brillouin zone face. For the free electron case, where the potential energy is a constant, the constant energy contours will remain spheres. However, for a nonconstant V(**r**) there is an interaction between Bloch functions that differ by a reciprocal lattice vector causing an energy splitting at the zone boundaries just as in the 1-dimensional case, Eq. 12-31. Also the constant energy contours must be normal to the Brillouin zone faces because the mirror planes cause the energy on both sides to be equal. Figure 12-11 shows an

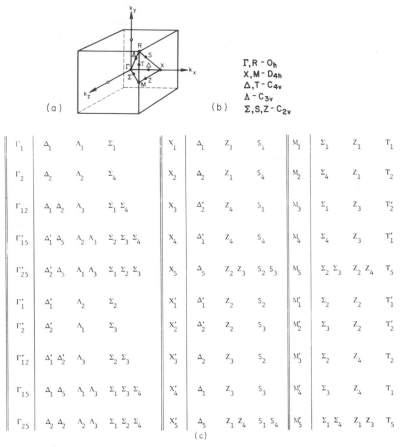

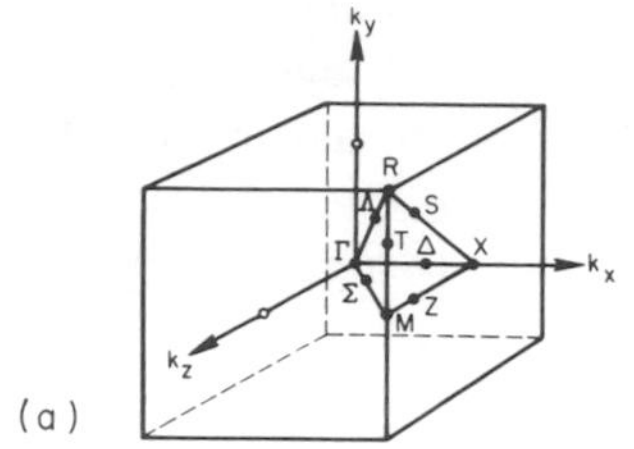


(b) $\Gamma, R - O_h$
$X, M - D_{4h}$
$\Delta, T - C_{4v}$
$\Lambda - C_{3v}$
$\Sigma, S, Z - C_{2v}$

| | | | | | | | | | | | |
|---|---|---|---|---|---|---|---|---|---|---|---|
| $\Gamma_1$ | $\Delta_1$ | $\Lambda_1$ | $\Sigma_1$ | $X_1$ | $\Delta_1$ | $Z_1$ | $S_1$ | $M_1$ | $\Sigma_1$ | $Z_1$ | $T_1$ |
| $\Gamma_2$ | $\Delta_2$ | $\Lambda_2$ | $\Sigma_4$ | $X_2$ | $\Delta_2$ | $Z_1$ | $S_4$ | $M_2$ | $\Sigma_4$ | $Z_1$ | $T_2$ |
| $\Gamma_{12}$ | $\Delta_1\ \Delta_2$ | $\Lambda_3$ | $\Sigma_1\ \Sigma_4$ | $X_3$ | $\Delta'_2$ | $Z_4$ | $S_1$ | $M_3$ | $\Sigma_1$ | $Z_3$ | $T'_2$ |
| $\Gamma'_{15}$ | $\Delta'_1\ \Delta_5$ | $\Lambda_2\ \Lambda_3$ | $\Sigma_2\ \Sigma_3\ \Sigma_4$ | $X_4$ | $\Delta'_1$ | $Z_4$ | $S_4$ | $M_4$ | $\Sigma_4$ | $Z_3$ | $T'_1$ |
| $\Gamma'_{25}$ | $\Delta'_2\ \Delta_5$ | $\Lambda_1\ \Lambda_3$ | $\Sigma_1\ \Sigma_2\ \Sigma_3$ | $X_5$ | $\Delta_5$ | $Z_2\ Z_3$ | $S_2\ S_3$ | $M_5$ | $\Sigma_2\ \Sigma_3$ | $Z_2\ Z_4$ | $T_5$ |
| $\Gamma'_1$ | $\Delta'_1$ | $\Lambda_2$ | $\Sigma_2$ | $X'_1$ | $\Delta'_1$ | $Z_2$ | $S_2$ | $M'_1$ | $\Sigma_2$ | $Z_2$ | $T'_1$ |
| $\Gamma'_2$ | $\Delta'_2$ | $\Lambda_1$ | $\Sigma_3$ | $X'_2$ | $\Delta'_2$ | $Z_2$ | $S_3$ | $M'_2$ | $\Sigma_3$ | $Z_2$ | $T'_2$ |
| $\Gamma'_{12}$ | $\Delta'_1\ \Delta'_2$ | $\Lambda_3$ | $\Sigma_2\ \Sigma_3$ | $X'_3$ | $\Delta_2$ | $Z_3$ | $S_2$ | $M'_3$ | $\Sigma_2$ | $Z_4$ | $T_2$ |
| $\Gamma_{15}$ | $\Delta_1\ \Delta_5$ | $\Lambda_1\ \Lambda_3$ | $\Sigma_1\ \Sigma_3\ \Sigma_4$ | $X'_4$ | $\Delta_1$ | $Z_3$ | $S_3$ | $M'_4$ | $\Sigma_3$ | $Z_4$ | $T_1$ |
| $\Gamma_{25}$ | $\Delta_2\ \Delta_2$ | $\Lambda_2\ \Lambda_3$ | $\Sigma_1\ \Sigma_2\ \Sigma_4$ | $X'_5$ | $\Delta_5$ | $Z_1\ Z_4$ | $S_1\ S_4$ | $M'_5$ | $\Sigma_1\ \Sigma_4$ | $Z_1\ Z_3$ | $T_5$ |

(c)

**Fig. 12-10** (a) The Brillouin zone, the special points and lines for a simple cubic crystal. (b) The point symmetry of the special points and lines. (c) Compatibility tables for the irreducible representations for this example.

example of the effect for a simple orthorhombic Brillouin zone. The generalization to a cubic zone is clear.

Figure 12-12 shows the Brillouin zone for a face-centered cubic lattice, a body-centered cubic lattice, and a simple orthorhombic lattice. Then special points and lines are labeled in an arbitrary but conventional way. The often used conventions and Brillouin zones for all the 14 possibilities can be found in the article by Koster. Note that the Brillouin zone of a body-centered cubic lattice is the same shape cell as the Wigner-Seitz cell, which is in direct, not reciprocal, space of a face-centered cubic

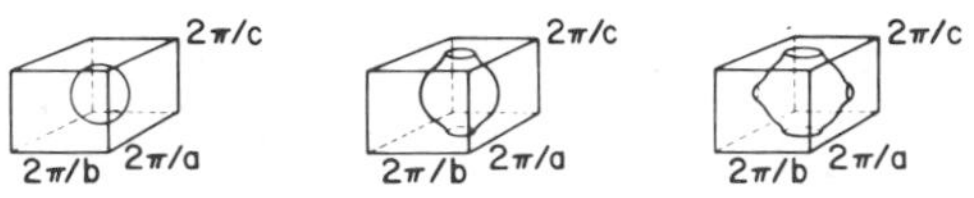


**Fig. 12-11** The Brillouin zone of a simple orthorhombic crystal. Constant energy contours are shown at successively higher energy.

lattice. Also the Brillouin zone of a cubic F-lattice is the same as the Wigner–Seitz cell of an I-lattice.

### g. Energy levels in a simple cubic lattice (plane wave approximation)

In this section we work out the energy versus **k** for some of the simple directions in a simple cubic lattice (see Fig. 12-10). The ideas are really no different from those discussed in Section 12-1e where a 1-dimensional case was discussed. However, here the three dimensions bring in the group of **k** in a simple, but nontrivial way. Just as in Section 12-1e, moving the **k** values back into the first Brillouin zone by a multiple of the reciprocal lattice vector will give several Bloch functions, at a given **k**, that must transform appropriately as the group of **k**. We will carry out the analysis using free electron wave functions. Thus, basis functions will transform as different irreducible representations of the various groups of **k** but will not formally have different energies, since $V(\mathbf{r}) = 0$. However, when the potential is nonzero, splittings will occur at the special points and lines and only the essential degeneracies will remain.

For the simple cubic lattice, a reciprocal lattice vector can be given in component form as $\mathbf{K}_n = (n_1, n_2, n_3)\ 2\pi/a$, where $n_i$ is any interger; **k** can be given as $\mathbf{k} = (k_1, k_2, k_3)\ 2\pi/a$, where $-1/2 \leq k_i \leq 1/2$. Then the energy and wave function for the free electron case is, from Eq. 12-24,

$$E(\mathbf{k}) = (\hbar^2/2m)\,(2\pi/a)^2[(k_1+n_1^2+(k_2+n_2)^2+(k_3+n_3)^2]$$

$$\varepsilon(\mathbf{k}) \equiv 2ma^2E(\mathbf{k})/\hbar^2=(k_1+n_1)^2+(k_2+n_2)^2+(k_3+n_3)^2$$

$$\phi_{\mathbf{k}}(\mathbf{r}) = \exp(2\pi i/a)[(k_1+n_1)x+(k_2+n_2)y+(k_3+n_3)z] \qquad (12\text{-}47)$$

In Eq. 12-47, we have defined a reduced energy $\varepsilon(\mathbf{k})$ that we will use.

Consider the energies and Bloch functions along the $\Delta$-line from $\Gamma$ to X. The energies at the special points and along the special line are

$$\varepsilon(\Gamma) = n_1^2+n_2^2+n_3^2$$

$$\varepsilon(\Delta) = (k_1+n_1)^2+n_2^2+n_3^2 \qquad 0<k_1<1/2$$

$$\varepsilon(X) = (1/2+n_1)^2+n_2^2+n_3^2 \qquad (12\text{-}48)$$

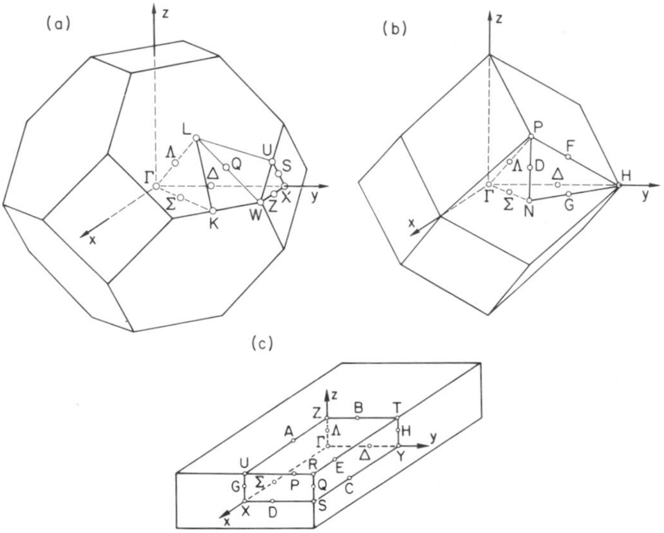


**Fig. 12-12** (a) The Brillouin zone for a face-centered cubic lattice with the special points and lines labeled. (b) The same for a body-centered cubic lattice. (c) The same for a simple orthorhombic lattice.

Start from lowest energy at the Γ point for n = (0, 0, 0), which is zero. The wave function is a constant, so it transforms as the $\Gamma_1$ irreducible representation of the group of Γ which is the point group $O_h$. The group of Δ is $C_{4v}$ and the group of X is $D_{4h}$ as in Fig. 12-10b. As **k** increases from the Γ point along the Δ-line,

$$\varepsilon(\Delta) = k_1^{\,2} \qquad \phi_\Delta(\mathbf{r}) = \exp[(2\pi i/a)k_1 x] \tag{12-49}$$

The result is called A in Fig. 12-13 where (1) indicates a onefold degeneracy. Clearly the Bloch function in Eq. 12-49 transforms into itself under all the symmetry operations of the group of Δ, so it transforms as the $\Delta_1$ irreducible representation. [We are taking the unique axis of the group of Δ ($C_{4v}$) as the x-axis to conform with Fig. 12-10 and Eq. 12-48.] For the X-point, the result from Eq. 12-49, ε(X) = 1/4, is clear. However, we must remember that for the X-point, another part of the extended zone, when moved into the first Brillouin zone by a reciprocal lattice vector, will also have ε(X) = 1/4. This other part is from n = (−1, 0, 0). Thus the energy and two Bloch functions are

$$\varepsilon(X) = 1/4 \qquad \phi_X(\mathbf{r}) = \exp(\pi i x/a) \qquad \phi_X(\mathbf{r}) = \exp(-\pi i x/a) \tag{12-50}$$

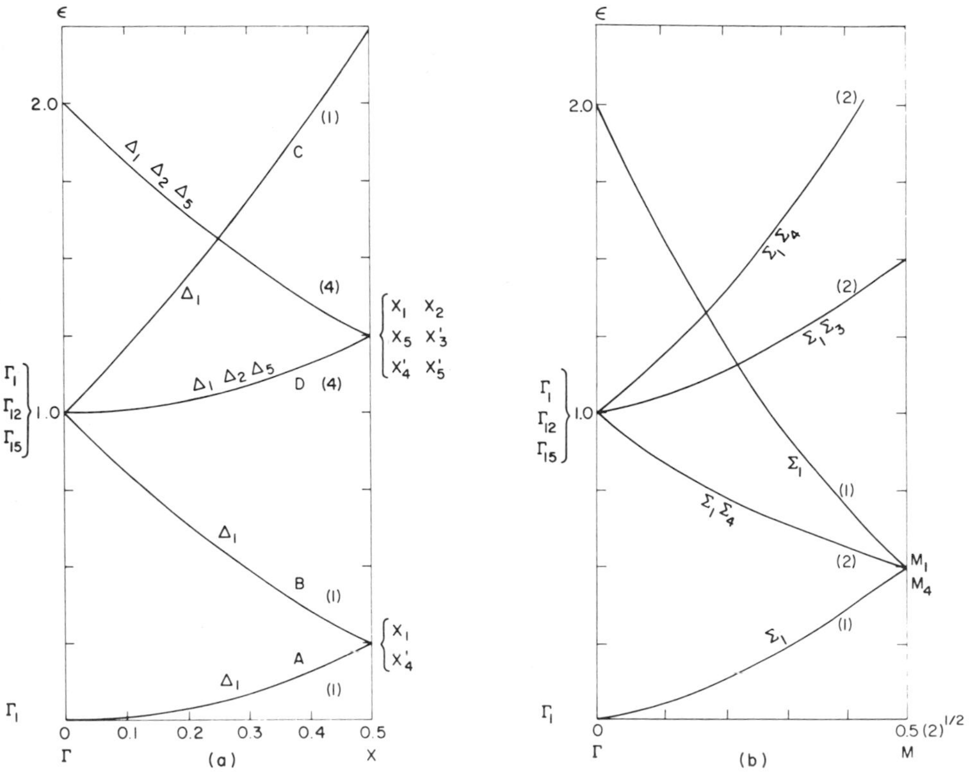


**Fig. 12-13** Reduced energy versus reduced **k** for a simple cubic Bravais lattice in the Δ-direction, [100], and the Λ-direction [111], with zero crystal potential. The numbers in parentheses indicate the degeneracy along the special lines and the letters A, B, etc. are only used in conjunction with the text to help explain the various lines.

Clearly these two functions transform into each other under some of the symmetry operations of the group of $X(D_{4h})$, such as the inversion operation. The resulting representation can be reduced to $X_1 + X_4'$ $(A_{1g} + A_{2u})$. Basis functions can be obtained by projection operator techniques or guess, i.e., the sum and difference of the two functions. The result, with an obvious simplification of notation, is

$$\phi(X_1)=\cos(\pi x/a) \sim 1+O(x^2)+\ldots$$

$$\phi(X_4')=\sin(\pi x/a) \sim x+O(x^3)+\ldots \qquad (12\text{-}51)$$

Note that by expanding the basis functions to the lowest power of x, we can see at a glance under what irreducible representation the various functions transform. In Eq. 12-51, it is clear that any number transforms as the totally symmetric irreducible representation $X_1$ $(A_{1g})$, and that x in our standard approach in Appendix 3 is called z and transforms as the $A_{2u}$ irreducible representation, which now is called $X_4'$.

The curve called B in Fig. 12-13, which starts at $\varepsilon(X) = 1/4$ has

$$\varepsilon(\Delta) = (k_1 - 1)^2 \qquad \phi_\Delta(\mathbf{r}) = \exp\,[(2\pi i/a)\,(k_1 - 1)x] \qquad (12\text{-}52)$$

This function, as that in Eq. 12-49, transforms into itself under all the operations of the group of $\Delta(C_{4v})$, hence is labeled $\Delta_1$ in Fig. 12-13.

The $\Gamma$ point, at $\varepsilon(\Gamma) = 1$, is the first time we get something different from the 1-dimensional case. These are six parts of the extended zone which when brought into the first Brillouin zone by a reciprocal lattice vector will give this value of energy. These are $n = (\pm 1, 0, 0), (0, \pm 1, 0), (0, 0, \pm 1)$. The Bloch functions are trivial to write down, and with an obvious simplification of notation are

$$\phi_1 = \exp(2\pi i/a)x \qquad \phi_3 = \exp(2\pi i/a)y \qquad \phi_5 = \exp(2\pi i/a)z$$

$$\phi_2 = \exp(-2\pi i/a)x \qquad \phi_4 = \exp(-2\pi i/a)y \qquad \phi_6 = \exp(-2\pi i/a)z \quad (12\text{-}53)$$

Since the group of $\Gamma(O_h)$ has only 3-dimensional irreducible representations at most, it is clear that some linear combinations of the functions in Eq. 12-53 are required. Reducing the representation formed by the six functions, we obtain $\Gamma_1 + \Gamma_{12} + \Gamma_{15}$ ($A_{1g} + E_g + T_{1u}$). The appropriate linear combinations yield

$$\phi(\Gamma_1) = \cos(2\pi x/a) + \cos(2\pi y/a) + \cos(2\pi z/a)$$

$$\phi(\Gamma_{12}) = \begin{cases} \cos(2\pi y/a) - \cos(2\pi z/a) \\ \cos(2\pi x/a) - (1/2)[\cos(2\pi y/a) + \cos(2\pi z/a)] \end{cases}$$

$$\phi(\Gamma_{15}) = [\sin(2\pi x/a)\;,\; \sin(2\pi y/a)\;,\; \sin(2\pi z/a)] \qquad (12\text{-}54)$$

Expanding the functions in Eq. 12-54 to lowest order in x, y, and z we obtain what we might have expected all along.

$$\begin{aligned} \phi(\Gamma_1) &\sim 1 = \phi_1 + \phi_2 + \phi_3 + \phi_4 + \phi_5 + \phi_6 \\ \phi(\Gamma_{12}) &\sim y^2 - z^2 = \phi_3 + \phi_4 - \phi_5 - \phi_6 \text{ and} \\ &\sim x^2 - 1/2(y^2 + z^2) = \phi_1 + \phi_2 - 1/2(\phi_3 + \phi_4 + \phi_5 + \phi_6) \\ \phi(\Gamma_{15}) &\sim x = \phi_1 - \phi_2 \text{ and } \sim y = \phi_3 - \phi \text{ and } \sim z = \phi_5 - \phi_6 \end{aligned}$$

Thus, the 6 fold degeneracy at the $\Gamma$ point can be made of $s + p^3 + d^2$ orbitals as in the molecular orbital case for six nearest neighbors in $O_h$ symmetry. To continue the energy to higher values, the curve in Fig. 12-13 marked C comes from the reciprocal lattice point $n = (1, 0, 0)$.

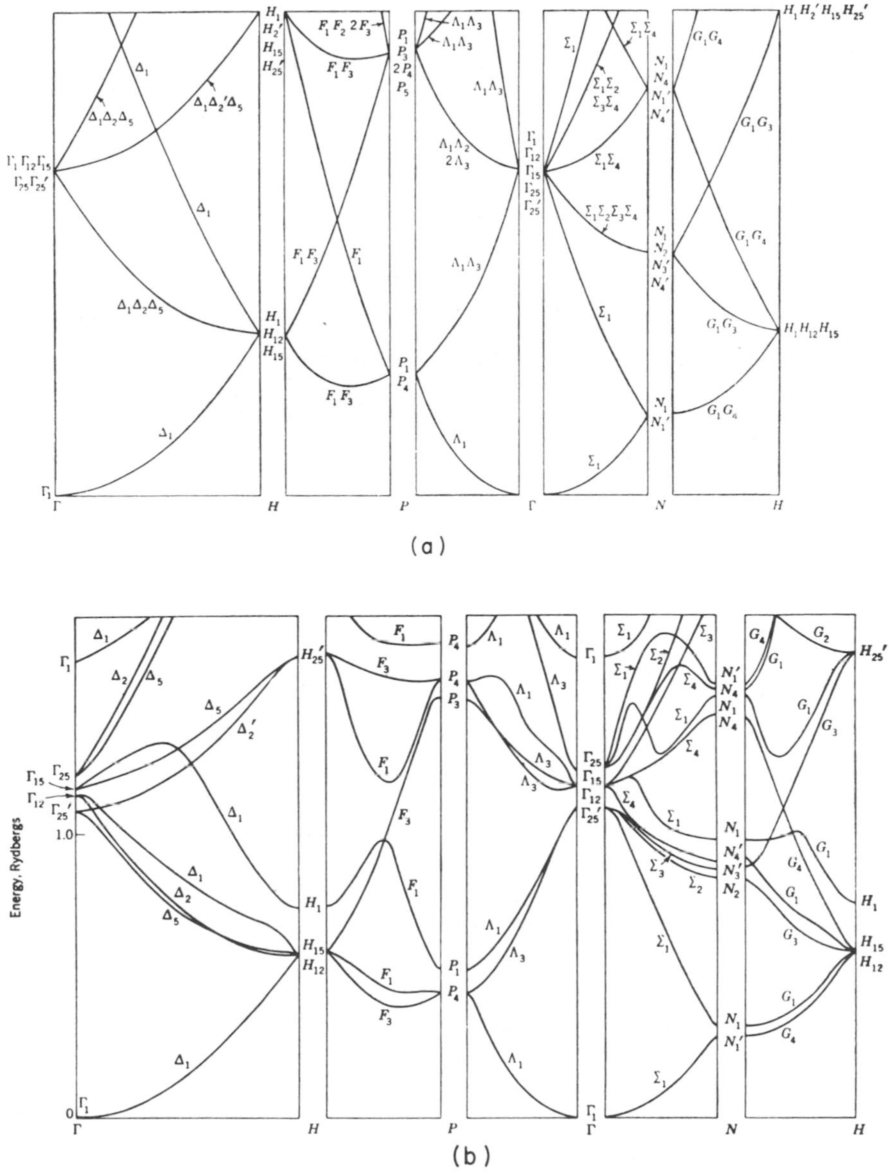


**Fig. 12-14** (a) Energy bands along special directions for a free electron metal with a body-centered cubic lattice. The Brillouin zone is shown in Fig. 12-12b. (b) The same energy bands but for sodium. From "Quantum Theory of Molecules and Solids" by J. C. Slater. Copyright 1963. Used with permission of McGraw-Hill Book Company.

From Eq. 12-47, we have

$$\varepsilon(\Delta) = (1+k_1)^2 \qquad \phi_\Delta(\mathbf{r} = \exp[(2\pi i/a)\,(1+k_1)x] \qquad (12\text{-}55)$$

which transform as $\Delta_1$. The other curve marked D comes from the other four reciprocal lattice translations to the $\Gamma$ point with $\varepsilon(\Gamma) = 1$, namely n $= (0, \pm 1, 0)$ and n $= (0, 0, \pm 1)$. The energy for all four cases is $\varepsilon(\Delta) = 1 + k_1^2$. The four Bloch functions can be written down from Eq. 12-47. Reducing the representations formed by these functions and obtaining the appropriate linear combinations that transform as the irreducible representations we obtain

$$\phi(\Delta_1)=\exp(2\pi ix/a)\,[\cos(2\pi y/a)+\cos(2\pi z/a)]$$

$$\phi(\Delta_2)=\exp(2\pi ix/a)\,[\cos(2\pi y/a)+\cos(2\pi z/a)]$$

$$\phi(\Delta_5)=[\exp(2\pi ix/a)\sin(2\pi y/a),\ \exp(2\pi ix/a)\,\sin(2\pi z/a)] \qquad (12\text{-}56)$$

Again expanding for small x, y, z, and remembering that our principal axis is x in these equations, we have $\phi(\Delta_2) \sim y^2 - z^2$ and $\phi(\Delta_5) \sim (y,z)$. Note that this is the first case where, when $V(\mathbf{r}) \neq 0$, we will get a splitting of the free electron degeneracy along an entire special line into three energy levels. However, the essential degeneracy associated with $\Delta_5$ irreducible representation will remain.

The energy level diagram can be obtained for higher energies as in Fig. 12-13 but an appreciation of the problems and results, hopefully, has been obtained by now. Remember, so far we have only talked about eigenvalues and eigenfunctions along the special line $\Delta$ and its end points. Figure 12-13 also shows $\varepsilon$ versus $\mathbf{k}$ for the special line $\Sigma$ including the end point M. For the M point $k_1 = k_2 = \sqrt{2}/2$. We have $\varepsilon(R) = 1/2$ or two times that of the lowest energy X point. Thus we see how the electrons often will spill into the second band near the X point before the lowest energy band is full. This is schematically illustrated in Fig. 12-8b for the 2-dimensional case. Of course, which states fill first depends on the matrix elements of the potential energy between the various functions.

Similar calculations can be carried out for other lattices. Several examples are suggested in the problems.

### h. Other bands

It is clear, from Fig. 12-13, that when $V(\mathbf{r}) \neq 0$ the picture will become much more complicated due to the splittings of the degeneracies. When a real potential is put into the calculations, a great deal of bending will occur at different points in the Brillouin zone. Sodium was one of the first metals to be worked on extensively and the result was fairly free-electron-like. Figure 12-14 shows the free electron result for a body-centered cubic lattice. Energy is plotted as **k** varies along the special lines of the Brillouin zone. Also shown in the figure is a calculated result for sodium using the augmented plane wave method. The splitting required by symmetry can be seen as well as some strong level repulsions. Subsequent work showed that the other alkali metals were considerably less free-electron-like than sodium. (See the Notes.)

Figure 12-15a shows a calculated band structure for silicon, in two different directions in k-space. Silicon is a face-centered cubic material with the diamond structure. It has just enough electrons to fill the bands up to the $\Gamma_{25}'$ band (valence band). Then, there is a gap between the highest filled valence band ($\Gamma_{25}'$) and the lowest empty (conduction) band; so silicon is a semiconductor. As can be seen the lowest unfilled band is along the $\Delta$ line in the $k = <1, 0, 0>$ directions and transforms as the $\Delta_1$ irreducible representation of the group of $\Delta$. Thus, silicon has six conduction band minima along the six $k = <1, 0, 0>$ type directions. If and electron were excited from the top of the valence bnad ($\Gamma_{25}'$) to the lowest conduction band ($\Delta_1$) by light, conservation of **k** would require a phonon with the difference of **k** to be emitted or absorbed. (The k of light $\approx 0$.)

Germanium has the same crystal structure as silicon. The valence band is quite similar but the conduction band differs. For germanium, the lowest empty band at the $\Gamma$ point has $\Gamma_2'$ character instead of $\Gamma_{15}$ as does silicon (see Fig. 12-15b). The compatability relations require that the various $\Delta$ and $\Lambda$ bands shift along with the $\Gamma$ bands, so the conduction bands between these two semiconductors differ considerably. For germanium the lowest empty conduction band occurs at the L point and has $L_1$ character. Thus, for germanium there are eight conduction band minima along the eight $k = <1, 1, 1>$ type directions at the zone edge. So the lowest conduction bands differ considerably between germanium and silicon.

## 12-3 Nonsymmorphic Space Groups

The next several sections (covering the new degeneracies in nonsymmorphic space groups, the effect of spin-orbit coupling, and the effect of time reversal) will be brief and only attempt to outline the effects on

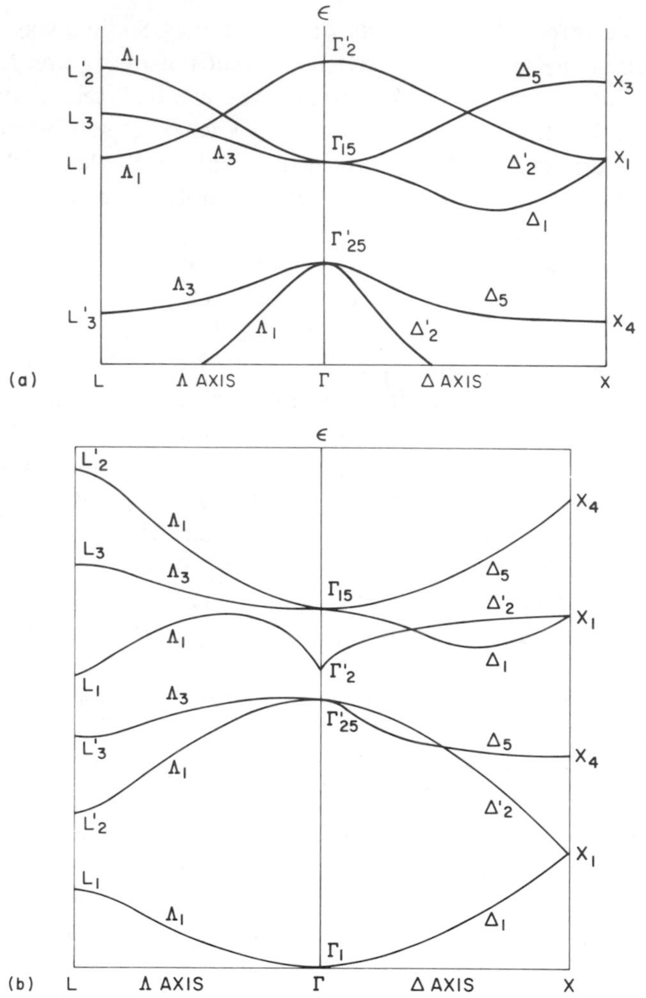


**Fig. 12-15** (a) A calculated band structure for silicon in two directions in k-space. Silicon has a face-centered Bravais lattice so the Brillouin zone is shown in Fig. 12-12a. (b) The same calculations but for germanium. Spin–orbit interaction is not considered in this figure. [F. Herman, Phys. Rev. **95**, 847 (1954).]

the bands caused by these new considerations. Full consideration of any of these areas is rather involved and somewhat specialized.

In considering the bands for space groups that have glide and screw operations, there are no new considerations that arise for **k** within the Brillouin zone. However, an important new consideration arises for **k** values on the surface of the Brillouin zone. We find that, at some special points along some special lines, or on some entire faces, there are only 2-dimensional irreducible representations. This is called **sticking together of bands** due to glide planes and screw axes. When spin–orbit coupling is included, 4-dimensional irreducible representations are obtained for some **k** points on the surface where the group of **k** is lower symmetry than cubic. So there certainly are new considerations that arise. However, we will continue to ignore spin–orbit coupling in this section and discuss it in the next section.

We use as an example a simple 2-dimensional model of a lattice that has a glide plane and show that 1-dimensional irreducible representations cannot be obtained for certain points on the surface of the Brillouin zone. Rather, a 2-dimensional irreducible representaiton is obtained which means there is an essential degeneracy. Figure 12-16a shows the 2-dimensional primitive unit cell and the symmetry operations of the 2-dimensional lattice that we will consider. The "short" international symbol for this lattice is pmg. The small p is used for 2-dimensional primitive cells. The m is for a mirror plane perpendicular to the x-axis ($\sigma_v{}^x$) and the g is a glide plane-reflection across a plane perpendicular to the y-axis ($\sigma_v{}^y$) and a translation b/2 in the y-direction. (For 2-dimensional lattices, there is only one kind of glide plane. There are no diagonal or diamond glides.) As can be seen in the figure the mg symmetry operations imply a 2-fold axis. So the "full" symbol for this lattice is p2mg. The symmetry operations can be written in the usual operator form and are

$$\{E\,|\,0\} \equiv e \qquad \{C_2\,|\,0\} \equiv c$$
$$\{\sigma_v{}^y\,|\,\mathbf{a}_1/2\} \equiv v \qquad \{\sigma_v{}^x\,|\,\mathbf{a}_1/2\} \equiv u \tag{12-57}$$

The origin is taken as the 2-fold axis as is usually the case in the international tables. With respect to this origin, the u-operation may appear as a glide operation but it is not. The translation is parallel to the normal of the reflection plane so the operation is just a reflection across a plane displaced from the origin, as is also apparent in Fig. 12-16a. However, the v-operation is a true glide plane. We should mathematically determine if the four operations in Eq. 12-57 can be expressed, about a new origin, without any translation. See Problem 11. Since this cannot be done, we have a true nonsymorphic space group.

The point group of the space group is mm or $C_{2v}$ thus there are only 1-dimensional irreducible representations. Figure 12-16b shows the Brillouin zone with the special points labeled. Consider the X point which

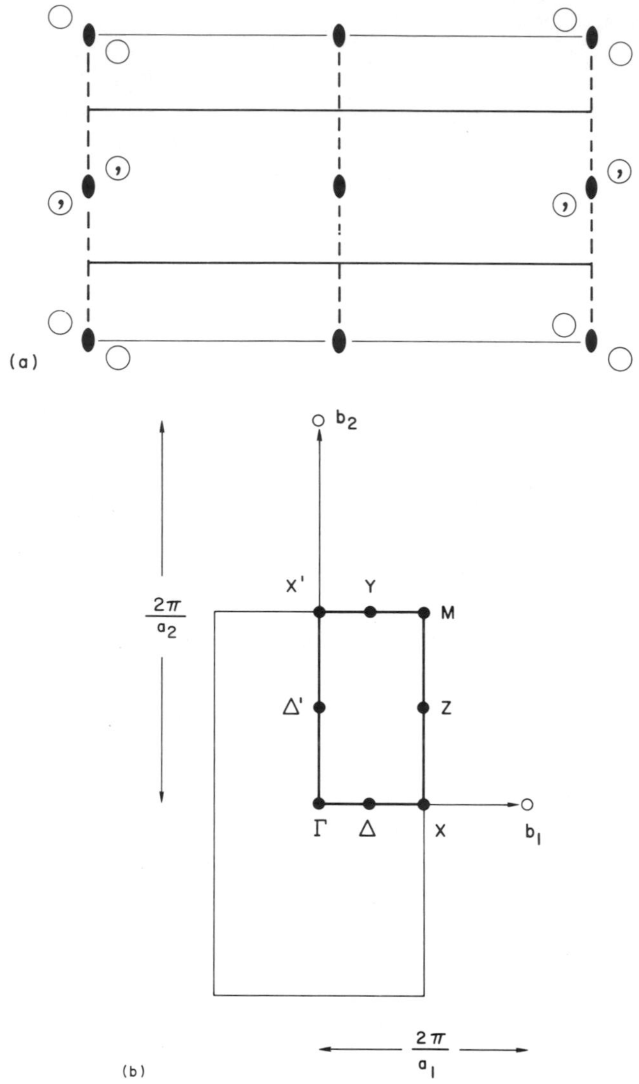


**Fig. 12-16** (a) The unit cell of a 2-dimensional lattice with space group p2mg. (b) The Brillouin zone of the 2-dimensional lattice with space group p2mg.

is a point on the surface of the zone where we anticipate new types of irreducible representations. Let X(x, y) be an eigenfunction at the X-point, i.e., $\mathbf{k} = (\pi/a_1, 0)$. We show that the space group cannot have a 1-dimensional irreducible representation at the X-point. Operating on

X(x, y) with the symmetry operations in Eq. 12-57, we have

$$\begin{aligned} cX(x, y) &= X(-x, -y) \\ vX(x, y) &= X(x+a_1/2, -y) \\ uX(x, y) &= X(-x+a_1/2, y) \end{aligned} \qquad (12\text{-}58)$$

So we see that $uX(x, y) = vcX(x, y)$. Also note that $v^2X(x, y) = X(x, y)$, with the same being true for u and c. If we assume that the representation in 1-dimensional, then $vX(x, y) = \pm X(x, y)$ and $cX(x, y) = \pm X(x, y)$. Hence

$$u^2X(x, y)=vcvcX(x, y)=(\pm 1)^2(\pm 1)^2X(x, y)=X(x, y) \qquad (12\text{-}59)$$

However we also have for an irreducible representation

$$\begin{aligned} u^2X(x,y) &= uX(-x+a_1/2,y)=X(x+a_1,y) \\ &= \exp[i\,(\pi/a_1)a_1]\,X(x,y)=-X(x,y) \end{aligned} \qquad (12\text{-}60)$$

Equations 12-60 and12-59 contradict each other so the representation cannot be 1-dimensional.

For the nonsymorphic groups, the problem arises because some of the symmetry operations are of the form $\{R\,|\,\boldsymbol{\tau}\}$, where $\boldsymbol{\tau}$ is a fraction of the primitive unit cell. When such symmetry operations operate on a Bloch function $\exp(i\,\mathbf{k}\cdot\mathbf{x})\,u_{\mathbf{k}}\,(\mathbf{x})$, a phase factor $\exp(i\,\mathbf{k}\cdot\boldsymbol{\tau})$ is obtained which cannot be handled by our theory of representations in its present form. (A similar problem was encountered when half-integral spin was considered in Chapter 8 where a phase factor $-1$ arises. To handle that situation the group was enlarged.) The situation can be handled in two ways. The first way involves ray (or projective) representation theory. Our present matrix representations of group elements require that

$$\Gamma(A)\,\Gamma(B) = \Gamma(C) \text{ if } AB = C \qquad (12\text{-}61)$$

where A, B, and C are group elements and the Γ's are square matrices. This definition of representations, some times called vector representations, can be enlarged to

$$\Gamma(A)\,\Gamma(B) = \varepsilon\Gamma(C) \text{ if } AB = C \qquad (12\text{-}62)$$

where $\varepsilon$ is a phase factor that depends on A and B. Representations of the type in Eq. 12-62 are called ray or projective representations. See the Notes for references to these representations. The second way to handle the problem associated with nonsymmorphic groups is to artificially enlarge the factor group so that these extra phase factors do not appear. This method is fairly straightforward (see the Notes) but yields more irreducible representations than appropriate for the true space groups. Thus, only those representations which are compatible with the Bloch function of the given **k**, on the surface of the Brillouin zone must be retained.

The result for the X-point is one 2-dimensional irreducible representations. The same situation applies to nonsymmorphic 3-dimensional space groups. For example, Fig. 12-12a shows the Brillouin zone of a face-centered cubic lattice. For a symmorphic space group, the irreducible representations at all the special points and lines can be found from the appropriate one of the 32 point groups as outlined in Section 12-2. Now consider the diamond lattice. The space group is Fd3m ($O_h^7$), which clearly is nonsymmorphic. The group of W contains two 2-dimensional irreducible representations, the group of Z has one 2-dimensional irreducible representation, the group of X has four 2-dimensional irreducible representations. So along the entire line W-Z-X, all the bands are 2-fold degenerate. The character tables for these points as well as other special points and lines on the surface of the Brillouin zone for the hexagonal close packed $P6_3/mmc$ ($D_{6h}^4$) space groups can be found in Koster's article along with reference to the original work.

As will be discussed in the next section, the inclusion of spin–orbit coupling along with the considerations discussed in this section further complicates the problem.

## 12-4 Spin–Orbit Effects on Bands

In this section we discuss the effects on bands of including a spin–orbit coupling term in the Hamiltonian. The effects are usually small in magnitude, thus sometimes are ignored. However, they can be very interesting because new degeneracies and/or splitting occur which can be treated by group theory. We will assume throughout this section that $E(\mathbf{k}) = E(-\mathbf{k})$ which is a result from time reversal for spinless particles. However, in this section we want to consider the electrons as having a spin and we take $\alpha$ as a spin up and $\beta$ as a spin down function. Time reversal then reverses the spin direction and since spin can be treated as a circulating electric current, $E(\mathbf{k})\alpha = E(-\mathbf{k})\beta$ is obtained (from time reversal) and this will be used in this section.

In Section 12-2 when electrons were put into bands two electrons went into each eigenstate. The eigenfunctions were $u_{\mathbf{k}}(\mathbf{r})\exp(i\,\mathbf{k}\cdot\mathbf{r})\,\alpha$ and $u_{\mathbf{k}}(\mathbf{r})\exp(i\,\mathbf{k}\cdot\mathbf{r})\,\beta$, i.e., the electrons spacially have a Bloch function with a

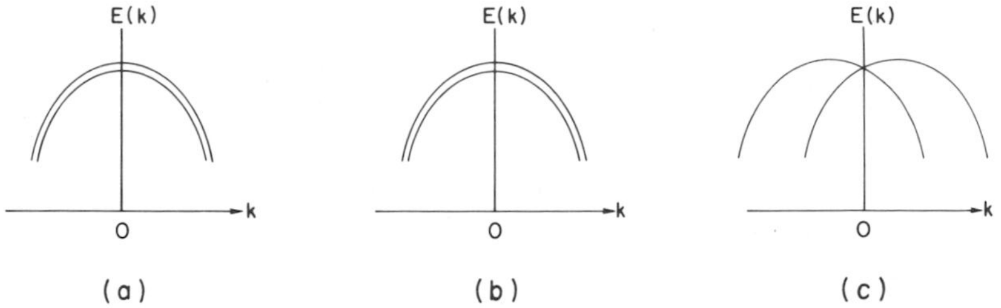


**Fig. 12-17** A schematic diagram showing the degeneracies due to spin–orbit coupling and/or inversion symmetry. (a) No spin–orbit coupling, i or no i; (b) spin–orbit coupling, inversion symmetry; (c) spin–orbit coupling, no inversion symmetry.

spin up or spin down. There are several terms of the same order that result if the four component Dirac equation is put into a two-component form. (See the Notes for references.) However, all these terms do the same thing from a symmetry point of view. The Hamiltonian no longer transforms into itself if the symmetry operations are applied to the orbital terms alone, rather the symmetry operations must be applied to the spin and orbital parts simulatneously just as discussed for the atomic case in Chapter 8. The previous bases functions then transform among themselves under the symmetry operations applied to the spin as well as the orbit part, and the resulting representation is often reducible. We will discuss only the reductions and not basis functions.

First, consider the eigenvalues in general and close to $\mathbf{k} = 0$. We have

$$E(\mathbf{k})\alpha = E(-\mathbf{k})\beta \qquad \text{time reversal (t)} \qquad (12\text{-}63a)$$

$$E(\mathbf{k})\alpha = E(-\mathbf{k})\alpha \qquad \text{inversion (i)} \qquad (12\text{-}63b)$$

therefore

$$E(\mathbf{k})\alpha = E(\mathbf{k})\beta \qquad \text{t + i} \qquad (12\text{-}63c)$$

Equation 12-63a is a general result from time inversion symmetry. If the space group has a center of symmetry, Eq. 12-63b follows since the inversion operation operates on **k** but does not affect the spin. Therefore Eq. 12-63c follows. So for a crystal that has inversion symmetry all bands are at least 2-fold degenerate. However, we also see that if the crystal does not have inversion symmetry there is no degeneracy at an arbitrary **k**. Equation 12-63 only shows that this possibility exists and in the next section such a result is proved. Figure 12-17 shows this result. Note that for **k** = 0, Eq. 12-63a shows that spin up and spin down electrons have the same energy, regardless of the inversion operation. The result pictured in Fig. 12-17c in unexpected. Each band has only a one fold degeneracy at a general **k** which results in band maximum at $\mathbf{k} \neq 0$. This situation apparently occurs in the semiconductor InSb. In the conduction band, the maximum in energy is about $10^{-4}$ eV higher than the **k** = 0 value and occurs at **k** values along the <111> at less then 1% of the distance to the Brillouin zone boundary.

The rest of this section discusses three topics: symmorphic crystals with inversion; nonsymmorphic crystals inversion; crystals without inversion. The addition of spin–orbit coupling and the use of the double groups gives different types of result in each case.

### a. Symmorphic crystals with inversion

The effects of spin–orbit coupling for this class of crystals is straightforward as in Chapter 8 and nothing new can be added to the discussion. If we assume the electron with spin transforms as the $\mathbf{D}^{1/2}$ irreducible representation, to determine the double group irreducible representations, we need only determine the reduction of the direct product $\mathbf{D}^{1/2} \times \Gamma_i$ where $\Gamma_i$ is the single group irreducible representation of the band that is determined without spin–orbit coupling. For example in a simple cubic crystal or in the P, I, and F-cubic lattices, the group of **k** for the Γ point in the Brillouin zone is $O_h$. The group of **k** for the R-point in the P-lattice is also $O_h$. Labeling the single group irreducible representations as $\Gamma_1{}^+$ to $\Gamma_5{}^+$ and $\Gamma_1{}^-$ to $\Gamma_1{}^-$ for five g-representations and the five u-representations, respectively, in the group $O_h$, then $\mathbf{D}^{1/2}$ is conventionally labeled as $\Gamma_6{}^+$. From the multiplication tables in Appendix 4 we have

$$\Gamma_6{}^+ \times \Gamma_1{}^\pm = \Gamma_6{}^\pm \qquad \Gamma_6{}^+ \times \Gamma_4{}^\pm = \Gamma_6{}^\pm + \Gamma_8{}^\pm$$

$$\Gamma_6{}^+ \times \Gamma_2{}^\pm = \Gamma_7{}^\pm \qquad \Gamma_6{}^+ \times \Gamma_5{}^\pm = \Gamma_7{}^\pm + \Gamma_8{}^\pm$$

$$\Gamma_6{}^+ \times \Gamma_3{}^\pm = \Gamma_8{}^\pm \tag{12-64}$$

In a similar manner the group of **k** for any **k** can be treated. For Ge and Si, Fig. 12-15 shows that the top of the valence band occurs at **k** = 0 and transforms as the $\Gamma_{25}{}'$ (BSW notation) whch is $\Gamma_5{}^+$ ($T_{2g}$ in Schoenflies notation) of the point group $O_h$. So the addition of spin–orbit coupling splits this level into 4-fold $\Gamma_8{}^+$ and 2-fold $\Gamma_7{}^+$ level. Figure 12-18a shows the results for the Γ point and along several directions in **k** space for a diamond space group. (However, be aware that the discussion so far only applies to **k** within the Brillouin zone since diamond is not a symmorphic space group. Part b of this section will cover **k** on boundary.)

In summary the addition of spin–orbit coupling to symmorphic space groups that have a center of inversion can be treated by the straightforward methods of Chapter 8 at all points in the Brillouin zone. There is at least a double degeneracy for all values of **k**. Also, splitting of some bands at special points along special lines can occur.

**b. Nonsymmorphic crystals with inversion**

For values of **k** not on the Brillouin zone boundary, these crystals can be treated in the same manner as in Section 12-4a for symmorphic crystals with a center of inversion. When spin–orbit coupling is not included, all of the irreducible representations for the various groups of **k** are the same as those found for the 32 point groups. So the direction product of any of these irreducible representations with $\mathbf{D}^{1/2}$ can be obtained as in Section 12-4a and there will be at least a double degeneracy for all of the **k** values.

The new aspect of the problem comes from **k** values on the Brillouin zone surface. As was seen in Section 12-3, irreducible representations are required that are not found among the irreducible representations of the 32 point groups. Naturally the double groups for these new irreducible representations are also not found among the double groups that have been discussed for the 32 point groups. These new double group representations will not be discussed here but can be found in an article by Koster (see the Notes). Figure 12-18a shows the bands for diamond in certain **k** directions. The space group of diamond (Ge and Si, too) is Fd3m ($O_h{}^7$) which includes diamond glide planes in this face-centered cubic lattice. For **k** at the X point in the [100], there is only one double group irreducible representation $X_5$ which is 4-fold degenerate. So there is more "sticking together" of bands than would have been guessed, even after Section 12-3.

A comment on notation is appropriate. consider the X-point in Fig. 12-12a. For this point in diamond ($O_h{}^7$-Fd3m) the single group irreducible representations are labeled $X_1$, $X_2$, $X_3$, $X_4$ and are new irreducible representations which are not the same as any of those found among the 32 point groups. This is also true, for the double group irreducible representation $X_5$ for this X point. These should not be confused with other irreducible representations for the same X point for face-centered

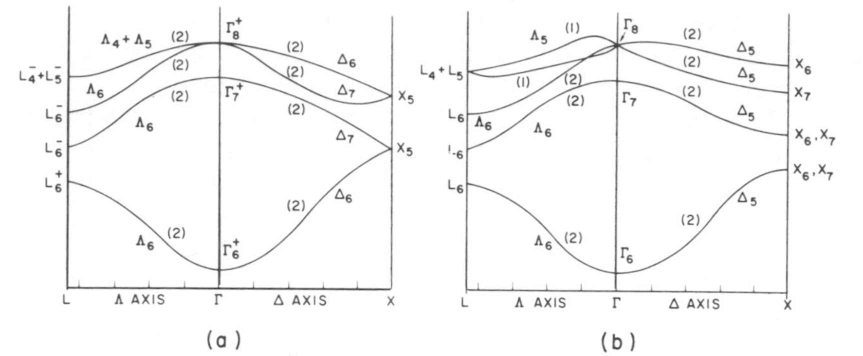


**Fig. 12-18** (a) The band structure for diamond (same structure as Si and Ge) in certain **k**-directions. The space group is Fd3m ($O_h^7$). Spin-orbit coupling is included so the irreducible representations are from the double group. (b) The band structure for a zinc blende type of crystal which has the same atomic positions as diamond except there are two different kinds of atoms, e.g. ZnTe, so the center of inversion is no longer a symmetry operation. The space group is F43m ($T_d^2$). (After Long.)

cubic crystals ($O_h^5$-Fm3m) that are symmorphic. For a symmorphic crystal with point group $O_h$, the group of **k** for the X point is $D_{4h}$ and the irreducible representations usually are labeled $X_i^{\pm}$ with i = 1 to 5, the + and − going with g and u-representations. However, for the zinc blende structure ($T_d^2$-F$\bar{4}$3m), the group of **k** at the same X point is $D_{2d}$ and the irreducible representations usually are labeled $X_i$ with i = 1 to 5. This similarity of notation applied to quite different irreducible representations is a problem in band theory that can be confusing. Care should therefore be exercised.

### c. Crystals without inversion

As already discussed, it is possible for crystals without the symmetry operation of inversion to have singly degenerate bands. Of course not all bands in all **k** directions are required to be singly degenerate. Tests can be made on each band which will depend on the symmetry operations in the group of **k**. This is discussed in the next section; here we merely show some results and how the various irreducible representations are affected in a general way.

We show the results (Fig. 12-19) for the bands close to **k** = 0 in three different directions in the Brillouin zone for a crystal with space group F$\bar{4}$3m ($T_d^2$). The zinc blende structure (GaAs, InSb, etc.) has this

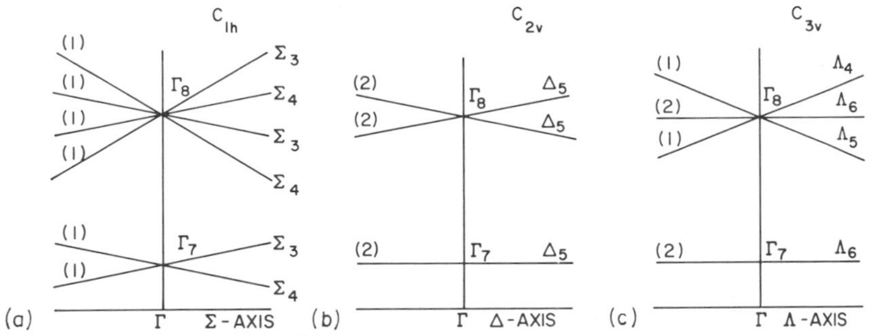


**Fig. 12-19** A schematic energy versus **k**-direction for a face-centered cubic Bravais lattice and a crystal structure without inversion.

space group. As can be seen, this space group is a symmorphic, face-centered, multiply primitive cell. The double group irreducible representations, that can be used as labels of the various bands, can be obtained in the same way as described in Section 12-4a. One takes $\mathbf{D}^{1/2} \times \Gamma_i$ where $\Gamma_i$ is the single group irreducible representation. The resultant direct product can be decomposed into double group irreducible representations for thc particular group of **k** which is one of the 32 point groups. Figure 12-19 shows three directions in **k** space. The group of $\Sigma$, $\Delta$, and $\Lambda$ is $C_{1h}$, $C_{2v}$, and $C_{3v}$, respectively, for this space group. As can be seen, all the irreducible representations are double group representations. However for some of these, the two representations that are the complex conjugate of one another, e.g., $\Sigma_4$ and $\Sigma_4$ as well as $\Lambda_4$ and $\Lambda_5$, are not degenerate. To determine whether these sorts of irreducible representations are degenerate or not requires the involved tests we will discuss in the next section.

Figure 12-18a shows some of the valance band of a diamond type of crystal, space group Fd3m ($O_h^7$), in special directions out to the zone boundary. Figure 12-18b shows a similar figure for a zinc blende type crystal, which has a space group $F\bar{4}3m$ ($T_d^2$). The atom positions in the two crystals are the same except that in zinc blende there are two different kinds of atoms. Thus, in zinc blend there is no center of inversion. Of course this changes the symmetry operations and, as can be seen, effects on the bands. The bands at the $\Gamma$ and L-point of the diamond lattice are labeled according to odd and even irreducible representations, while the irreducible representations for the zinc blende lattice cannot be classified as odd or even. For the zinc blende lattice, Fig. 12-18b shows some 1-dimensional (dougle group) irreducible representations as well as a maximum in valence band in the L-direction. Also see in Figs. 12-17 and 12-19c.

## 12-5 Time Reversal Symmetry

### a. Introduction

The symmetry operation of time reversal was introduced into quantum mechanics by Wigner (see the Notes) and here we follow much of his work in abbreviated form.

We assume that if we reverse the direction of motion (including the "spinning" of the electrons) of all staes, then the same physical laws apply to this time-reversed system. If we have a state function $\psi$, then we denote $\boldsymbol{\theta}\psi$ as the time-reversed state. If $\psi_n$ is an eigenstate of a Hamiltonian $\mathbf{H}\psi_n = E_n\psi_n$, then the time reversal operator $\boldsymbol{\theta}$ can be applied to both sides of this equation. Since we are assuming $\boldsymbol{\theta}$ is a symmetry operation for all closed isolated systems, $\boldsymbol{\theta}$ commutes with $\mathbf{H}$ and $E_n$ is a real number so $\mathbf{H}(\boldsymbol{\theta}\psi_n) = E_n(\boldsymbol{\theta}\psi_n)$. Thus, $\psi_n$ and $\boldsymbol{\theta}\psi_n$ are both eigenfunctions with the same energy. A very important consideration will be discussed later; namely, are $\psi_n$ and $\boldsymbol{\theta}\psi_n$ linearly independent. If they are, a degeneracy arises due to time reversal symmetry. If the two functions are linearly dependent ($\boldsymbol{\theta}\psi_n = \alpha\psi_n$ where $\alpha$ is a constant), then no new degeneracy arises. We would like to determine the form of $\boldsymbol{\theta}$ and its properties and then determine if any new degeneracies arise.

Before we determine the form of $\boldsymbol{\theta}$, it is important to distinguish the time reversal symmetry operation $\boldsymbol{\theta}$ from the more ordinary time displacement by an amount of t. If we have stationary states and eigenvalues denoted by $\psi_n$ and $E_n$, then in the course of a time interval t the state at $t = 0$, $\phi_0$ will evolve to a state $\phi_t$ as

$$\phi_0 = \Sigma a_n \Psi_n \rightarrow \phi_t = \Sigma a_n \Psi_n \exp(-E_n t/\hbar) \qquad (12\text{-}65)$$

This is a displacement of time and is clearly a unitary operation, $(e^{-i})(e^{-i})^\dagger = 1$. As we will see, time reversal is rather different.

### b. Properties of $\boldsymbol{\theta}$

In order to determine some of the properties of the time reversal or time inversion operator, consider the following two different sequences of operations that should be equivalent in a closed physical system: Displacement by time t followed by a time inversion operation should be the same as a time inversion followed by a displacement by and amount of time $-t$. Writing this mathematically in operator form and following the consequences, we have

$$
\begin{aligned}
\exp(-iHt/\hbar)\,\boldsymbol{\theta} &= \boldsymbol{\theta}\exp(iHt/\hbar)\\
\boldsymbol{\theta}^{-1}\exp(-iHt/\hbar)\,\boldsymbol{\theta} &= \exp(+iHt/\hbar)\\
\exp(-\boldsymbol{\theta}^{-1}iHt\boldsymbol{\theta}/\hbar) &= \exp(+iHt/\hbar)\\
-\boldsymbol{\theta}^{-1}iHt\boldsymbol{\theta} &= iHt\\
\boldsymbol{\theta}^{-1}i\boldsymbol{\theta} &= -i \quad \text{or} \quad \boldsymbol{\theta} i = -i\boldsymbol{\theta}
\end{aligned}
\tag{12-66}
$$

The third equality follows because $\boldsymbol{\theta}^{-1}$ and $\boldsymbol{\theta}$ operate on each term in the series expansion of the exponential. Also we have used $\boldsymbol{\theta}t = t\boldsymbol{\theta}$ since we want our time reversal operator to be independent of time, and $\mathbf{H}\boldsymbol{\theta} = \boldsymbol{\theta}\mathbf{H}$ since it is a symmetry operation. Thus, the result in Eq. 12-66 shows that $\boldsymbol{\theta}$ is not a linear operator because a linear operator $\mathbf{L}$ has the property that $\mathbf{L}i = i\mathbf{L}$ and $\mathbf{L}(a\psi + b\phi) = a\mathbf{L}\psi + b\mathbf{L}\phi$, while for $\boldsymbol{\theta}$ we have

$$\boldsymbol{\theta}(a\Psi + b\phi) = a^*\boldsymbol{\theta}\Psi + b^*\boldsymbol{\theta}\phi \tag{12-67}$$

An operator with the property in Eq. 12-66 and 12-67 is called an **anti-linear operator**. The operation of complex conjunction $\mathbf{K}$ is an antilinear operator because $\mathbf{K}i = -i\mathbf{K}$.

We can also show that the time reversal operator is an **antiunitary operator**. That is,

$$<\boldsymbol{\theta}\Psi \,|\, \boldsymbol{\theta}\phi> = <\Psi \,|\, \phi>^* = <\phi \,|\, \Psi> \tag{12-68}$$

We show this by expanding the eigenfunctions in terms of an orthonormal set of functions. The assumptions that we make are that $\boldsymbol{\theta}$ does not affect the orthogonality or the normalization of the functions.

$$
\begin{aligned}
<\boldsymbol{\theta}\Psi \,|\, \boldsymbol{\theta}\phi> &= <\boldsymbol{\theta}\Sigma a_n\Psi_n \,|\, \boldsymbol{\theta}\Sigma b_m\phi_m>\\
&= \Sigma_{n,m} < a_n^*\boldsymbol{\theta}\Psi_n \,|\, b_m^*\boldsymbol{\theta}\phi_m>\\
&= \Sigma a_n b_m^* <\boldsymbol{\theta}\Psi_n \,|\, \boldsymbol{\theta}\phi_m>\\
&= \Sigma\, a_n\, b_m^*\, \delta_{nm} = <\Psi \,|\, \phi>^*
\end{aligned}
\tag{12-69}
$$

We also note that the operation of complex conjunction $\mathbf{K}$, i.e., $\mathbf{K}\psi = \psi^*$, is also antiunitary; $<\mathbf{K}\psi \,|\, \mathbf{K}\phi> = <\psi^* \,|\, \phi^*> = <\psi \,|\, \phi>^*$. $\mathbf{K}$ also has the important property

$$\mathbf{K}^2 = 1 \qquad (12\text{-}70)$$

Now we show that the product of any two antilinear operators is linear. In particular we do this for $\boldsymbol{\theta}\,\mathbf{K}$:

$$\boldsymbol{\theta}\mathbf{K}(a\Psi+b\phi)=\boldsymbol{\theta}(a^*\mathbf{K}\Psi+b^*\mathbf{K}\phi)=a\boldsymbol{\theta}\mathbf{K}\Psi+b\boldsymbol{\theta}\mathbf{K}\phi \qquad (12\text{-}71)$$

The product is also unitary:

$$<\boldsymbol{\theta}\mathbf{K}\Psi \mid \boldsymbol{\theta}\mathbf{K}\phi>=<\boldsymbol{\theta}\Psi \mid \boldsymbol{\theta}\phi>^* = <\Psi \mid \phi> \qquad (12\text{-}72)$$

so the product $\boldsymbol{\theta}\,\mathbf{K} = \mathbf{U}$, where $\mathbf{U}$ is unitary operator or using Eq. 12-70, $\boldsymbol{\theta} = \mathbf{U}\,\mathbf{K}$. That is, any antiunitary operator, in particular the time reversal operator, can be factored into a unitary operator and complex conjugation.

Consider one last general property of $\boldsymbol{\theta}$, namely, the possible values of $\boldsymbol{\theta}^2$. Even though $\boldsymbol{\theta}\psi$ need not be the same state as $\psi$, we want $\boldsymbol{\theta}\,\boldsymbol{\theta}\,\psi$ $(= \boldsymbol{\theta}^2\psi)$ to be the same state as $\psi$. So $\boldsymbol{\theta}^2$ can differ from $\psi$ by a change of phase $\boldsymbol{\theta}^2\psi = c\psi$, where $|c| = 1$. Thus $\boldsymbol{\theta}^2 = c\,\mathbf{1} = \mathbf{U\,K\,U\,K} = \mathbf{U\,U}^*$ relates the unitray operator to c. However since $\mathbf{U}$ is unitary, $\mathbf{U}^* = c\mathbf{U}^\dagger$, operating on the left with $\mathbf{K}$ yields $\mathbf{U} = c\tilde{\mathbf{U}}$ since $\mathbf{U}^\dagger = \tilde{\mathbf{U}}^*$; also taking the transpose $\tilde{\mathbf{U}} = c\mathbf{U}$ and substituting into $\mathbf{U} = c\,\tilde{\mathbf{U}}$ we get; $c = \pm 1$. This yields the general result

$$\boldsymbol{\theta}^2 = \pm 1 \qquad (12\text{-}73)$$

The upper sign in Eq. 12-73 applies to the spin-independent Schodinger equation or the spin-dependent equation for an even number of electrons. If the number of electrons (or the number of particles with half-integral spin) is odd, then the lower sign applies. This is proved in Wigner and elsewhere. See the Notes. Using these facts Kramer's theorem can be proved. The theorem is: In the absence of an external magnetic field (which would remove the time inversion symmetry) all the energy levels of a system containing an odd number of electrons must be at least doubly degenerate. To prove this theorem we note that $\psi$ and $\boldsymbol{\theta}\psi$ have the same energy as discussed earlier in this section. If they are linearly independent a degeneracy occurs. To prove this case, assume that they are linearly dependent, then $\boldsymbol{\theta}\psi = a\psi$. Operating in both sides by $\boldsymbol{\theta}$, using the antilinear property

$$\boldsymbol{\theta}(\boldsymbol{\theta}\Psi) = \boldsymbol{\theta}(a\Psi) = a^*(\boldsymbol{\theta}\Psi) = a^*\,a\,\Psi \qquad (12\text{-}74)$$

However for an odd number of electrons $\theta^2\psi = -\psi$, which, combined with Eq. 12-74, gives $-\psi = a^*a\psi$ which is impossible since $a^*a$ is a positive number. So $\theta\psi$ and $\psi$ must be linearly independent and the theorem is proven.

At this point it would be appropriate to determine the explicit form of $\theta$ with respect to its effect on the quantum mechanical operators, x, p, and spin. We do not do that here (see Wigner) but state and discuss the result. The effects of the time reversal operator can be divided into two categories. The first contains observables that classically depend on even powers of time such as position and energy. The second category includes momentum and angular momentum which depend on an odd power of the time. As one might guess $\theta$ commutes with observables in the first category and yields a minus sign for the second class.

$$\theta \mathbf{x} \theta^{-1} = \mathbf{x} \qquad \theta \mathbf{x} = \mathbf{x}\theta$$

$$\theta \mathbf{p} \theta^{-1} = -\mathbf{p} \qquad \theta \mathbf{p} = -\mathbf{p}\theta$$

$$\theta \sigma \theta^{-1} = -\sigma \qquad \theta \sigma = -\sigma\theta \tag{12-75}$$

It is from these equations that the proper sign in Eq. 12-73 can be obtained for an odd or even number of electrons.

### c. Time reversal and degeneracy

We now discuss the possibility of obtaining irreducible representations of a group with the inclusion of $\theta$ as a symmetry operation. From such irreducible representations we want to determine any new degeneracies. However, because $\theta$ is antilinear and the ordinary spatial symmetry operations R are linear, it is not possible to obtain representations of the group containing both types of operations in the usual sense. We will see why presently. Nevertheless, when antiunitary operations $\theta$R operate on a basis function the result is a linear combination of its partners so a matrix can be defined in the usual way

$$\theta R \Psi_i^m = \Sigma_j \Gamma_m(\theta R)_{ji}\, \Psi_j^m \tag{12-76}$$

as in Eq. 4-8. The matrices in Eq. 12-76 are unitary if the partners are orthogonal. The problem arises when the product of two symmetry operations T and R is considered, with one operation antilinear and the other not. If T is linear Eq. 12-77a is obtained;

$$TR\Psi_i^m = T\,\Sigma\,\Gamma_m(R)_{ji}\,\Psi_j^m = \Sigma\,\Gamma_m(T)_{kj}\Gamma_m(R)_{ji}\Psi_k^m \qquad (12\text{-}77a)$$

or

$$= \Sigma\Gamma_m(T)_{kj}\,\Gamma_m(R)_{ji}{}^*\Psi_k^m \qquad (12\text{-}77b)$$

if T is antilinear Eq. 12-77b is obtained. So $\Gamma(TR) = \Gamma(T)\,\Gamma(R)$ only when T is also linear, since if T is antilinear, we obtain $\Gamma(T)\,\Gamma(R)^*$. To handle this problem, Wigner developed a system of representations called corepresentations and these are treated in his book. Here we discuss the resolution of the problem in simpler terms.

In order to determine if extra degeneracies occur, we want to know when the sets of functions $\psi^m$ and $\theta\psi^m$ are linearly independent. If they are, degeneracy is doubled. If they are linearly dependent, the time reversal operation introduces no new degeneracy. It turns out that this question can be resolved knowing only the unitray representation matrices corresponding to the ordinary symmetry operations R and their complex conjugate, and we need not know anything about matrices representing the antiunitary operations. Three cases arise which can lead to doubling of the degeneracy depending on whether the spin is integral or half integral:

(a) Γ and Γ* are equivalent and can be chosen to be real and identical.

(b) Γ and Γ* are inequivalent.

(c) Γ and Γ* are equivalent but cannot be chosen to be real and identical.

Table 12-3 summarizes the degeneracies that arise due to time reversal symmetry. Note that is case (b), where Γ and Γ* are inequivalent, the degeneracy is doubled independently of the spin. See the Problems for the proofs of some of these results.

As is usually the case in group theory, consideration of the characters of the irreducible representations leads to simplified equations that are easy to use. The Frobenius–Schur test can be used to separate the three cases and we need only know the characters of the square of each ordinary symmetry element R in the group for the particular irreducible representation under consideration. The result, which is fairly easy to

**Table 12-3** Results as to whether there will be additional degeneracy due to the time reversal symmetry operation. (Results for integral spin are for an even number of electrons or for the case where spin is neglected. The half-integral spin case applies to an odd number of electrons.)

| Case | a | b | c |
|---|---|---|---|
| Relation of $\Gamma$ to $\Gamma^*$ | Can be made real and identical | Inequi-valent | Equivalent but cannot be made real |
| Frobenius-Schur Test | h | 0 | -h |
| Herring Test | n | 0 | -n |
| Degeneracies for Integral Spin | none | doubled | doubled |
| Degeneracies for Half-Integral Spin | doubled | doubled | none |

obtain and is left as a problem, is

$$\Sigma_R \chi_i (R^2) = \begin{cases} h & \text{case (a)} \\ 0 & \text{case (b)} \\ -h & \text{case (c)} \end{cases} \tag{12-78}$$

where h is the order of the group. This sum, over all the elements of the simple (unitary) group is usually easy to perform. We only need to know $R^2$ and $\chi(R^2)$ from each class as is easily proven, but for double groups it must be remembered that terms like $(C_2)^2 = \bar{E}$ and not E. Table 12-4 shows the character table of the point group $C_3$ and the result of Eq. 12-78 for each irreducible representation. The presence of time reversal symmetry will double the degeneracy of $\Gamma_2$ and $\Gamma_3$ and this is why these two irreducible representations are usually bracketed and called an E-irreducible representation. For the $\Gamma_1$ irreducible representation, case (a) is obtained and since we are dealing with the single group irreducible representation, the spin is integral so there is no extra degeneracy.

Time reversal symmetry is usually important in causing additional degeneracy in systems with low spatial symmetry, such as the preceding

**Table 12-4** Results of Frobenius–Schur test for point group $C_3$

| $C_3$ | $E$ | $C_3$ | $C_3^2$ | $\epsilon = e^{\frac{2\pi i}{3}}$ |
|---|---|---|---|---|
| $\Gamma_1$ | 1 | 1 | 1 | |
| $\Gamma_2$ | 1 | $\epsilon$ | $\epsilon^*$ | |
| $\Gamma_3$ | 1 | $\epsilon^*$ | $\epsilon$ | |
| $\chi_1(R^2)$ | 1 | 1 | 1 | $\Sigma\chi_1(R^2) = 3$ |
| $\chi_2(R^2)$ | 1 | $\epsilon^*$ | $\epsilon$ | $= 0$ |
| $\chi_3(R^2)$ | 1 | $\epsilon$ | $\epsilon^*$ | $= 0$ |

example of point group $C_3$. For the full rotation group time reversal gives no additional degeneracy.

As discussed in the last section, time reversal effects can be important in the electronic band theory. For this single particle theory the integral spin case applied to band theory if spin is neglected, and the half integral spin case applies if spin effects are to be included in the Hamiltonian. However, in this field the use of the Frobenius–Schur test is impractical due to the essentially infinite number of symmetry operations in the space group. Herring (see the Notes) has shown that the test can be simplified so that the sum need only be performed over some of the symmetry element of the group of **k**. The test is

$$\Sigma_{Q_0} \chi(Q_0^2) = \begin{cases} n & \text{case (a)} \\ 0 & \text{case (b)} \\ -n & \text{case (c)} \end{cases} \qquad (12\text{-}79)$$

where there are n elements $Q_0$ of the space group that take **k** into $-$**k**. So $Q_0^2$ will take **k** into itself and will be a symmetry element of the groups of **k**. Two rules are helpful in evaluating Eq. 12-79:

(1) If the group of **k** contains the inversion i, then $Q_0$ are the elements of this group of **k**;

(2) If the group of **k** does not contain i then $Q_0$ are the elements of the group of **k** times the operation i.

Chapter 13

# THE FULL ROTATION GROUP

*MAGIC THEATER*
*ENTRANCE NOT FOR EVERYBODY*

*H. H. Hesse, "Steppenwolf"*

## 13-1 The Homomorphism between the Special Unitary Group in Two Dimensions, SU(2), and the 3-Dimensional Rotation Group

In Chapter 8 some of the irreducible representations of the rotation group were determined. As noted there, only the irreducible representations corresponding to integral angular momentum were found. In order to obtain the irreducible representations of half-integral angular momentum we will use the method of Weyl. We will show that there is a homomorphism between the 2-dimensional special unitary group and the pure rotation group in three dimensions. Then the irreducible representations of the 2-dimensional unitary group will be obtained. Although this method does not appear to be straightforward, it does enable one to determine the integral as well as half-integral angular momentum irreducible representations.

Let $\mathbf{u}$ be a unitary unimodular matrix of two dimensions. Its general form is

$$\mathbf{u} = \begin{bmatrix} a & b \\ c & d \end{bmatrix} \tag{13-1}$$

Since we want $\mathbf{u}$ to be unitary

$$\mathbf{u}\,\mathbf{u}^\dagger = \begin{bmatrix} a & b \\ c & d \end{bmatrix} \begin{bmatrix} a^* & c^* \\ b^* & d^* \end{bmatrix} = \begin{bmatrix} 1 & 0 \\ 0 & 1 \end{bmatrix} \tag{13-2}$$

From the unitary condition in Eq. 13-2 and the unimodular ("special") condition which means the determinant of **u** is +1, ad−bc = 1, it follows that the most general form of **u** is

$$\mathbf{u} = \begin{bmatrix} a & b \\ -b^* & a^* \end{bmatrix} \tag{13-3a}$$

The unimodular condition on Eq. 3a gives

$$aa^* + bb^* = 1 = |a|^2 + |b|^2 \tag{13-3b}$$

The unimodular unitary matrix Eq. 13-3 is usually called the **special unitary** matrix. So we have a set of special unitary matrices in two dimensions that form a group called SU(2). It is clear that these matrices form a group since the product of any two matrices, as in Eq. 13-3, results in another matrix of this form so the set is closed. A unit matrix exists, each element has an inverse, and matrix multiplication is associative. Note that a and b in Eq. 13-3 are complex hence, there are three independent parameters which will turn out to be related to the three Euler angles.

Consider the Pauli matrices

$$\sigma_x = \begin{bmatrix} 0 & 1 \\ 1 & 0 \end{bmatrix} \qquad \sigma_y = \begin{bmatrix} 0 & i \\ -i & 0 \end{bmatrix} \qquad \sigma_z = \begin{bmatrix} -1 & 0 \\ 0 & 1 \end{bmatrix} \tag{13-4}$$

Every 2-dimensional matrix **h** with zero trace can be expressed as a linear combination of the Pauli matrices, i.e.,

$$\mathbf{h} = x\sigma_x + y\sigma_y + z\sigma_z = \begin{bmatrix} -z & x+iy \\ x-iy & z \end{bmatrix} \tag{13-5}$$

**h** is Hemitian for x, y, z real. If **h** is transformed by an arbitrary special unitary matrix Eq. 13-3, then **h′** is obtained which still has zero trace, i.e.,

$$\mathbf{h}' = \mathbf{uhu}^\dagger = \mathbf{u}\,[\mathbf{r} \bullet \boldsymbol{\sigma}]\,\mathbf{u}^\dagger = \mathbf{r}' \bullet \boldsymbol{\sigma} \tag{13-6a}$$

$$\begin{bmatrix} -z' & x'+iy' \\ x'-iy' & z' \end{bmatrix} = \begin{bmatrix} a & b \\ -b^* & a^* \end{bmatrix} \begin{bmatrix} -z & x+iy \\ x-iy & z \end{bmatrix} \begin{bmatrix} a^* & -b \\ b^* & a \end{bmatrix} \tag{13-6b}$$

Solving Eq. 13-6b explicitly for **x′** in terms of x, y, and z, and similarly for **y′** and **z′**, the result can be written in a form to make it look like a simple rotation of coordinates,

$$\begin{bmatrix} x' \\ y' \\ z' \end{bmatrix} = \begin{bmatrix} \frac{1}{2}(a^2 + a^{*2} - b^2 - b^{*2}) & \frac{1}{2}(a^2 - a^{*2} + b^2 - b^{*2}) & a^*b^* + ab \\ \frac{1}{2}i\,(a^{*2} - a^2 + b^2 - b^{*2}) & \frac{1}{2}(a^2 + a^{*2} + b^2 + b^{*2}) & i\,(a^*b^* - ab) \\ -(a^*b + ab^*) & i\,(a^*b - ab^*) & aa^* - bb^* \end{bmatrix} \begin{bmatrix} x \\ y \\ z \end{bmatrix}$$

$$r' = \mathbf{R}(\mathbf{u})\, r \qquad \text{i.e.} \qquad r_i' = \sum_{j=1}^{3} [R(\mathbf{u})]_{ij} r_j \tag{13-7}$$

We now show that $\mathbf{R(u)}$ as defined in Eq. 13-7 represents a simple rotation of coordinates. (a) $\mathbf{R(u)}$ is real since each term in it is real. This is immediately clear if a and b are written in terms of real and imaginary parts and each of the terms in Eq. 13-7a is computed. (b) Lengths are preserved. This is clear because the determinant of a product equals the product of the determinants so $|\mathbf{h'}| = |\mathbf{u}| \bullet |\mathbf{h}| \bullet |\mathbf{u}^\dagger| = |\mathbf{h}|$. The determinant of u is one, hence, $|\mathbf{h'}| = -(x'^2 + y'^2 + z'^2) = |\mathbf{h}| = -(x^2 + y^2 + z^2)$. Thus, $\mathbf{R(u)}$ in Eq. 13-7 represents a real orthogonal transformation of coordinates. We can also see that it represents a pure rotation (proper rotation as opposed to improper rotation, by noting that the determinant of $\mathbf{R(u)}$ is +1 because as $\mathbf{u}$ goes continuously to the unit matrix (b → 0 and a → 1), $\mathbf{R(u)}$ goes continuously into the 3-dimensional unit matrix.

Thus, we have shown that for a matrix $\mathbf{u}$, that is a member of the group SU(2), there corresponds a 3-dimensional rotation matrix $\mathbf{R(u)}$. Presently we will show that as $\mathbf{u}$ covers all members of SU(2), all possible rotations in three dimensions will be obtained. So there is homomorphism between the group SU(2) and the pure rotation group in three dimensions $O(3)^+$ which was discussed in Chapter 8. We will see that the homomorphism is not an isomorphism since more then one matrix from the SU(2) group corresponds to one member of the $O(3)^+$ group.

To see the homomorphism between unitary unimodular matrices of SU(2) and the $O(3)^+$ group, we investigate $\mathbf{u}$ and the resultant $\mathbf{R(u)}$ matrice. First insist that $\mathbf{u}$ be diagonal so b = 0, and aa* = 1 so a = $\exp(-i\alpha/2)$. The resulting $\mathbf{u}$ and $\mathbf{R(u)}$ from Eq. 13-7 is

$$\mathbf{u}_1(\alpha) = \begin{bmatrix} \exp(-i\alpha/2) & 0 \\ 0 & \exp(i\alpha/2) \end{bmatrix} = \exp[(i\alpha/2)\,\sigma_z]$$

$$\mathbf{R}(\mathbf{u}_1) = \begin{bmatrix} \cos\alpha & \sin\alpha & 0 \\ -\sin\alpha & \cos\alpha & 0 \\ 0 & 0 & 1 \end{bmatrix} \tag{13-8a}$$

and this form of $\mathbf{u}$ leads to a rotation about the z-axis. If we take $\mathbf{u}$ to be real, then the form of $\mathbf{u}$ and the corresponding $\mathbf{R(u)}$ are

$$\mathbf{u}_2(\beta) = \begin{bmatrix} \cos(\beta/2) & -\sin(\beta/2) \\ \sin(\beta/2) & \cos(\beta/2) \end{bmatrix} = \exp[(i\beta/2)\,\sigma_y]$$

$$\mathbf{R}(\mathbf{u}_2) = \begin{bmatrix} \cos\beta & 0 & -\sin\beta \\ 0 & 1 & 0 \\ \sin\beta & 0 & \cos\beta \end{bmatrix} \tag{13-8b}$$

This is a rotation about the y-axis. The last possibility is that b can be imaginary. The resulting **u** and **R(u)** are

$$u_3(\Delta) = \begin{bmatrix} \cos(\Delta/2) & i\sin(\Delta/2) \\ i\sin(\Delta/2) & \cos(\Delta/2) \end{bmatrix} = \exp[(i\Delta/2)\,\sigma_x]$$

$$\mathbf{R}(\mathbf{u}_3) = \begin{bmatrix} 1 & 0 & 0 \\ 0 & \cos\Delta & \sin\Delta \\ 0 & -\sin\Delta & \cos\Delta \end{bmatrix} \qquad \textbf{(13-8c)}$$

which are rotations about the x-axis.

To describe the rotations in terms of Euler angles as in Eq. 8-2, we have $R(\alpha, \beta, \gamma) = R_z(\alpha)R_y(\beta)R_z(\gamma)$ which is the product of three unitary unimodular matrices

$$\mathbf{u}_1(\alpha)\,\mathbf{u}_2(\beta)\,\mathbf{u}_1(\gamma) = \begin{bmatrix} \exp(-i\alpha/2) & 0 \\ 0 & \exp(i\alpha/2) \end{bmatrix} \begin{bmatrix} \cos(\beta/2) & -\sin(\beta/2) \\ \sin(\beta/2) & \cos(\beta/2) \end{bmatrix} \times \begin{bmatrix} \exp(-i\gamma/2) & 0 \\ 0 & \exp(i\gamma/2) \end{bmatrix}$$

$$\equiv \mathbf{u}(\alpha,\beta,\gamma) = \begin{bmatrix} \exp[(-i/2)\,\alpha+\gamma]\cos(\beta/2) & -\exp[(i/2)\,\gamma-\alpha]\sin(\beta/2) \\ \exp[(i/2)\,\alpha-\gamma]\ \sin(\beta/2) & \exp[(i/2)\,\alpha+\gamma]\cos(\beta/2) \end{bmatrix}$$

**(13-9)**

Since all rotations can be described by the Euler angles and $R(\alpha, \beta, \lambda)$, we see that for every rotation in 3-dimensional space, at least one matrix of the SU(2) group is obtained. We say the three unit vectors $\mathbf{e}_x$, $\mathbf{e}_y$, $\mathbf{e}_z$, where $\mathbf{e}_x$ is a column vector with one as the first entry and the next two entries are zero, etc, for $\mathbf{e}_y$ and $\mathbf{e}_z$, span the space of $O(3)^+$. Similarly, we say that the unit vectors that span the space in which the SU(2) matrices are defined are called spinors $\varepsilon$ and $\sigma$

$$\epsilon = \begin{bmatrix} 1 \\ 0 \end{bmatrix} \qquad \delta = \begin{bmatrix} 0 \\ 1 \end{bmatrix} \qquad \textbf{(13-10)}$$

If the components of a spinor are (u, v), then the rotation in terms of the Euler angles $\alpha$, $\beta$, $\gamma$ is just given by the matrix $\mathbf{u}(\alpha, \beta, \alpha)$ in Eq. 13-9 or

$$\begin{bmatrix} u' \\ v' \end{bmatrix} = \mathbf{u}(\alpha, \beta, \gamma) \begin{bmatrix} u \\ v \end{bmatrix} \qquad \textbf{(13-11)}$$

It will turn out that $\mathbf{u}(\alpha, \beta, \gamma)$ is just the irreducible representation for spin 1/2 so it is sometimes written as $\mathbf{D}^{1/2}$.

As a last point concerning the homomorphism between the group SU(2) and $O(3)^+$ we would like to know how many elements of SU(2) correspond to one element of $O(3)^+$. Consider the elements

$$\begin{bmatrix} -1 & 0 \\ 0 & -1 \end{bmatrix}, \qquad \begin{bmatrix} 1 & 0 \\ 0 & 1 \end{bmatrix} \tag{13-12}$$

These two elements form an invariant subgroup of SU(2) and the products of either of these elements with any other element in SU(2) will yield the same element in the $O(3)^+$ group. We can also see the two to one homomorphism by noting that $\mathbf{u} = (+1)$ and $\mathbf{u} = (-1)$ matrices in Eq. 13-12 both correspond to the same rotation in $O(3)^+$ space. This is most easily seen from Eq. 13-6. Thus $\mathbf{R(u)} = \mathbf{R(-u)}$. This comes about because the Euler angles are determined by a rotation up to a multiple of $2\pi$, but the angles in the SU(2) matrices are determined up to a multiple of $\pi$, or by half angles as in Eq. 13-9. Thus, the elements of the double group of $O(3)^+$ can be represented by the matrices of SU(2). The two SU(2) matrices $\mathbf{u}$ and $-\mathbf{u}$ are associated with the same rotation $\mathbf{R(u)}$ of the $O(3)^+$ group. Given the SU(2) matrix $\mathbf{u}$, $\mathbf{R(u)}$ can be determined by Eq. 13-7a. Or given the Euler angles, Eq. 13-9 can be used to determine the SU(2) matrix. Given $\mathbf{R}$ as in Eq. 13-7, $\mathbf{u}$ can be determined up to a sign as shown in the problems.

## 13-2 Irreducible Representations of SU(2)

We would like to obtain the irreducible representations of the SU(2) group. To do this, consider the n+1 polynomials in u and v which are $u^n$, $u^{n-1}v$, $u^{n-2}v^2, \ldots, uv^{n-1}$, $v^n$. Naturally, n is an integer. We note that under a unitary transformation as in Eq. 13-3a,

$$\begin{bmatrix} u' \\ v' \end{bmatrix} = \begin{bmatrix} a & b \\ -b^* & a^* \end{bmatrix} \begin{bmatrix} u \\ v \end{bmatrix} \quad \text{or} \quad \begin{matrix} u' = au + bv \\ v' = -b^*u + a^*v \end{matrix} \tag{13-13}$$

So the n+1 polynomials transform among themselves under an operation of SU(2) and they form an n+1 dimensional representation of the group. Setting n = 2j, the (2j + 1) polynomials are

$$u^{2j},\ u^{2j-1}v^1,\ u^{2j-2}v^2,\ \ldots,\ u^1 v^{2j-1},\ v^{2j} \tag{13-14}$$

where j can be an integer or half-integer and the representation is of dimension 2j + 1. For j = 1/2 the polynomials are u and v; for j = 1, $u^2$, uv, $v^2$; for j = 3/2, $u^3$, $u^2v$, $uv^2$, $v^3$; etc. Thus, the basis functions of the 2j + 1 representation in a short hand, normalized, symmetrical form are

$$f_m^j(u, v) = \frac{u^{j+m}\, v^{j-m}}{[(j+m)!\,(j-m)!]^{1/2}} \qquad m = +j, j-1, \ldots, -j \tag{13-15}$$

As will be seen later, the denominator will be required for the representation to be unitary. Recall from Section 7-2, on the transformation of functions, that if R is a symmetry transformation which describes the transformation of coordinates as $x' = Rx$ when R operates on a function of the coordinates $\psi(x)$, one has $R\psi(R^{-1}x)$. The symmetry operation **u**, Eq. 3a, will operate on the function $f_m{}^j(u,v)$. Since

$$\mathbf{u}^{-1}\begin{bmatrix} u \\ v \end{bmatrix} = \begin{bmatrix} a^* & -b \\ b^* & a \end{bmatrix}\begin{bmatrix} u \\ v \end{bmatrix} = \begin{bmatrix} a^*u - bv \\ b^*u + av \end{bmatrix} \tag{13-16}$$

we have

$$\mathbf{u}f_m^j(u, v) = f_m^j(a^*u - bv,\ b^*u + av) = \frac{(a^*u - bv)^{j+m}\,(b^*u + av)^{j-m}}{[(j+m)!\,(j-m)!]^{1/2}} \tag{13-17}$$

By expanding the right side of Eq. 17 using the binomial theorem and expressing all the functions of u and v in terms of $f_m{}^j$ one obtains

$$\mathbf{u}f_m^j = \Sigma\, U(\mathbf{u})_{m'm}^j\, f_{m'}^j \tag{13-18a}$$

$$U(\mathbf{u})_{m'm}^j = \sum_p (-1)^p \frac{[(j+m)!\,(j-m)!\,(j+m')!\,(j-m')!]^{1/2}}{(j-m'-p)!\,(j+m-p)!\,p!\,(p+m'-m)!}$$

$$\times\ \{a^{j-m'-p}\, a^{*\,j+m-p}\, b^p\, b^{*\,p+m'-m}\} \tag{13-18b}$$

where the sum is over all values of p for which the denominator if finite ($0! = 1$, but fractorial of a negative number is infinite). $U(\mathbf{u})^j$ is the representation we desire. Several important points concerning the representation $U(\mathbf{u})^j$ must be considered. First it must be shown that this representation is indeed unitary. To prove that Eq. 13-18b is unitary, the term $[(j+m)!(j-m)!]^{-1/2}$ of Eq. 13-15 is required. Second, it must be shown that the $U(\mathbf{u})^j$ matrices form an irreducible representation of the SU(2) group. This is done using Schur's lemma, Section 3-3. Finally, it must be shown that no other irreducible representations beside $U(\mathbf{u})^j$ exist. When this is done it is clear that $j = 0, 1/2, 1, 3/2 \ldots$ and no other values are permitted. These three statements are proved clearly in Wigner (see the Notes and the Problems), so they will not be reproduced here.

We would like to determine the character of $U(\mathbf{u})^j$ to prove that it is as anticipated in Chapter 8. Recall that all **u** can be diagonalized by a similarity transformation using a unitary matrix. Thus all **u** can be transformed into the form $\mathbf{u}_1(\alpha)$ as in Eq. 13-8a, and **u** and the corresponding

$\mathbf{u}_1(\alpha)$ are in the same class since all rotations by the same angle are in the same class regardless of axis, as shown in Section 8-2. for b = 0 and a = exp(−i(α/2), only the term with p = 0 occurs in the sum in Eq. 13-18 and this is nonzero only if m′ = m, so

$$U\{\mathbf{u}_1(\alpha)\}_{m'm} = \delta_{m'm}\,[\exp(-i\alpha/2)]^{j-m}\,[\exp(i\alpha/2)]^{j+m} = \delta_{m'm}\exp(im\alpha)$$

$$\chi\,[U\{\mathbf{u}_1(\alpha)\}] = \sum_{m=-j}^{j} \exp(im\alpha) \qquad (13\text{-}19)$$

where the sum is over the 2j + 1 values that differ by an integer. This is just the formula that is obtained in Chapter 8 for integer values of j, but now it is proved for half-integer values as well. (These are half-odd integers but the term half-integer values is often used.)

## 13-3 Wigner Coefficients

In Section 8-4 we showed how eigenfunctions $u_{m_1}^{j_1}$ with angular momentum $j_1$ coupled with those of $j_2$, $v_{m_2}^{j_2}$, to form the correct linear combinations with $J = j_1 + j_2, j_1 + j_2 - 1, \ldots, |j_1 - j_2|$. In this section we would like to calculate explicitly the coefficients of the linear combinations of the products $u_{m_1}v_{m_2}$ that give the eigenfunction $\psi_M^J$. (Note: We will drop the superscripts, for convenience, when possible.) These coefficients are called Wigner, Clebsch–Gordan, or vector coupling coefficients. In Section 8-4b they were obtained by a simple method for a particular example. Here we determine them in general.

In Section 8-4b it was found that $\mathbf{D}^{j_1} \times \mathbf{D}^{j_2} = \Sigma\mathbf{D}^J$, where the sum is $J = j_1 + j_2, j_1 + j_2 - 1, \ldots, |j_1 - j_2|$. This means that a similarity transformation S exists that transforms the direct product into diagonal form, i.e.

$$\mathbf{D}(R)^{j_1} \times \mathbf{D}(R)^{j_2} = S^{-1}\,M(R)\,S \qquad (13\text{-}20)$$

where M(R) is the matrix

$$\begin{bmatrix} \mathbf{D}^{|j_1-j_2|} & 0 & 0 & \cdots & 0 \\ 0 & \mathbf{D}^{|j_1-j_2|+1} & 0 & \cdots & 0 \\ 0 & 0 & \mathbf{D}^{|j_1-j_2|+2} & \cdots & 0 \\ \vdots & \vdots & \vdots & & \vdots \\ 0 & 0 & 0 & \cdots & \mathbf{D}^{j_1+j_2} \end{bmatrix} = M(R) \qquad (13\text{-}21)$$

Since the left side of Eq. 13-20 and M(R) is unitary, one can prove that S is unitary, $S^{-1} = S^\dagger$. (See the Problems.) In fact, we shall see that there is a certain arbitrariness in S which allows the phase to be picked. This freedom can be used to make the coefficients of S real and positive. This is the Condon and Shortley phase choice.

The left side of Eq. 13-20 as well as S and M(R) are matrices of dimension $(2j_1 + 1)(2j_2 + 1) = \Sigma(2J + 1)$. The elements of M(R), in direct product notation, are

$$M(R)_{J'M';JM} = \delta_{JJ'}\, D(R)^{J}_{M'M} \tag{13-22}$$

The rows and columns of $\mathbf{D}^{j_1} \times \mathbf{D}^{j_2}$ are labeled by $m_1$ and $m_2$ so the columns of S must also be labeled similarly. However, because of the way M(R) is labeled, Eq. 13-22, the rows of S must be labeled by J and M, where J goes from $|j_1 - j_2|$, to $j_1 + j_2$ and M from $-J$ to J as usual. (This can be seen in Table 13-1.) A typical term in Eq. 13-20 can be seen to be

$$D(R)^{j_1}_{m_1'm_1}\, D(R)^{j_2}_{m_2'm_2} = \sum_{MM'} \sum_{J} S_{JM;m_1'm_2'}\, D(R)^{J}_{M'M}\, S_{JM;m_1m_2} \tag{13-23}$$

The importance of the matrix S in the similarity transformation, Eq. 13-20, is that it gives the proper linear combinations of the products $u_{m_1}{}^{j_1} v_{m_2}{}^{j_2}$ which transform as the $\mathbf{D}^J$ irreducible representations, i.e.,

$$\Psi^{J}_{M} = \sum_{m_1m_2} S^{*}_{JM;m_1m_2}\, u_{m_1} v_{m_2} \tag{13-24a}$$

$$u_{m_1} v_{m_2} = \sum_{L'M'} S_{L'M';m_1m_2}\, \Psi^{L'}_{M'} \tag{13-24b}$$

We prove that $\psi_M{}^J$ obtained from Eq. 13-24a does indeed transform into itself as the $\mathbf{D}^J$ irreducible representation of the full rotation group. [Note that in Eq. 13-24 we have continued to suppress the superscripts on u, v, and S. The Condon and Shortly notation for S is $(j_1j_2m_1m_2/JM)$.] Equation 13-24b follows from Eq. 13-24a. To prove that Eq. 13-24a is correct, operate on both sides by a symmetry operation $R\,\overline{R}$ where R operates on the coordinates of u, and $\overline{R}$ on those of v and they must be an operation of the same angle about the same axis as discussed in Section 8-4b. The second equality follows from the meaning of Ru, etc; the third from Eq. 13-24b; the fourth from the direct product; the fifth from Eq. 13-20; the sixth from Eq. 13-22; the last shows what we set out to prove. So the S matrix gives the correct linear combinations of $u^{j_1}v^{j_2}$ that transform properly when the coordinate of u and v are simultaneously rotated by the same amount about the same axis. If the Hamiltonian is invariant to such rotations, then $\psi_M{}^J$ are the proper linear combinations of the

**Table 13-1** Formula for the Wigner coefficients $s_{Jm_1m_2}^{j_1j_2}$ for $J = j_1 + 1$

| J | $m_2 = -1$ | 0 | +1 |
|---|---|---|---|
| $j_1 - 1$ | $\sqrt{\frac{(j_1+m_1)(j_1+m_1-1)}{2j_1(2j_1+1)}}$ | $-\sqrt{\frac{(j_1-m_1)(j_1+m_1)}{j_1(2j_1+1)}}$ | $\sqrt{\frac{(j_1-m_1-1)(j_1-m_1)}{2j_1(2j_1+1)}}$ |
| $j_1$ | $\sqrt{\frac{(j_1-m_1+1)(j_1+m_1)}{2j_1(j_1+1)}}$ | $\frac{m_1}{\sqrt{j_1(j_1+1)}}$ | $-\sqrt{\frac{(j_1+m_1+1)(j_1-m_1)}{2j_1(j_1+1)}}$ |
| $j_1 + 1$ | $\sqrt{\frac{(j_1-m_1+1)(j_1-m_1+2)}{(2j_1+1)(2j_1+2)}}$ | $\sqrt{\frac{(j_1-m_1+1)(j_1+m_1+1)}{(2j_1+1)(j_1+1)}}$ | $\sqrt{\frac{(j_1+m_1+1)(j_1+m_1+2)}{(2j_1+1)(2j_1+2)}}$ |

$$\begin{aligned}
R\,\bar{R}\,\Psi_M^J &= \sum_{m_1m_2} S^*_{JM;m_1m_2}\, R\, u_{m_1}\, \bar{R}\, v_{m_2} \\
&= \sum_{m_1m_2}\sum_{m_1'm_2'} S^*_{JM;m_1m_2}\, \mathbf{D}(R)^{j_1}_{m_1'm_1}\, \mathbf{D}(R)^{j_2}_{m_2'm_2}\, u_{m_1'}\, v_{m_2'} \\
&= \sum_{m_1m_2}\sum_{m_1'm_2'}\sum_{J'M'} S^*_{JM;m_1m_2}\, \mathbf{D}(R)^{j_1}_{m_1'm_1}\, \mathbf{D}(R)^{j_2}_{m_2'm_2}\, S_{J'M';m_1'm_2'}\, \Psi^{J'}_{M'} \\
&= \sum_{J'M'} \{S\,\mathbf{D}(R)^{j_1} \times \mathbf{D}(R)^{j_2}\, S^{-1}\}_{J'M';JM}\, \Psi^{J'}_{M'} \\
&= \sum_{J'M'} M(R)_{J'M';JM}\, \Psi^{J'}_{M'} = \sum_{J'M'} \delta_{JJ'}\, \mathbf{D}(R)^J_{M'M}\, \Psi^{J'}_{M'} \\
&= \sum_{M'} \mathbf{D}(R)^J_{M'M}\, \Psi^J_{M'}
\end{aligned} \tag{13-25}$$

composite system. Further symmetries, such as the indistinguishability of electrons, must be considered depending on the problem that is being investigated.

The program now is to determine values for the $S_{JM;m_1m_2}$ coefficients. First apply a rotation to Eq. 13-24a through the z-axis by an amount $\alpha$. The representation is well known, namely, $e^{im\alpha}$.

$$\begin{aligned}
R_\alpha\,\bar{R}_\alpha\,\Psi_M^J = \exp(iM\alpha)\,\Psi_M^J &= \sum_{m_1m_2} S^*_{JM;m_1m_2}\, R_\alpha\, u_{m_1}\, \bar{R}_\alpha\, v_{m_2} \\
&= \sum S^*_{JM;m_1m_2} \exp[i(m_1+m_2)\alpha]\, u_{m_1}\, v_{m_2}
\end{aligned} \tag{13-26}$$

Since $u_{m_1} v_{m_2}$ forms a complete set, we have

$$S_{JM;m_1m_2} = 0 \qquad \text{for } M \neq m_1 + m_2 \tag{13-27a}$$

$$S_{Jm_1+m_2;m_1m_2} = S_{JM;m_1m_2} = S_{Jm_1m_2} \tag{13-27b}$$

Equation 13-27a expresses the selection rule and Eq. 13-27b is a notational change in order to simplify the subscripts. However, it should still be remembered that the superscripts are being suppressed and $s_{Jm_1m_2}^{j_1j_2}$ might be a better way to write Eq. 13-27b. Equation 13-23 could be written in terms of $s_{Jm_1m_2}$ where M′ and M in $\mathbf{D}^J(R)$ would be replaced by $m_1' + m_2'$ and $m_1 + m_2$, respectively.

The S matrix is not uniquely determined, so the phase can be adjusted. Consider a diagonal matrix V, such that $V_{J'M';JM} = \omega_J \delta_{J'J} \delta_{M'M}$

$$V = \begin{bmatrix} \omega_{|j_1-j_2|} & 0 & \cdots & 0 \\ 0 & \omega_{|j_1-j_2|+1} & \cdots & 0 \\ \vdots & \vdots & & \vdots \\ 0 & 0 & \cdots & \omega_{j_1+j_2} \end{bmatrix} \tag{13-28}$$

M(R), Eq. 13-21, commutes with a diagonal matrix. Note that Eq. 13-20 will not be changed if VS replaces S provided $|\omega| = 1$ in Eq. 13-28 which will also keep VS unitary. The elements of $(VS)_{JM;m_1m_2} = \omega_J S_{JM;m_1m_2}$ replace those of S. So Eq. 13-20 becomes $(VS)^{-1}M(VS) = S^1V^{-1}MVS = S^{-1}MV^{-1}VS = S^{-1}MS$. This extra degree of freedom allows a suitable choice of $\omega$ to be picked to make each term in S or s real and positive. We arrange

$$S_{(J)(j_1-j_2);(j_1)(-j_2)} = s_{(J)(j_1)(-j_2)} = |\, s_{(J)(j_1)(-j_2)} \,| \tag{13-29}$$

This choice makes all $s_{Jm_1m_2}$ real and positive which can be checked in the final equation for this quantity. This choice of phase is the Condon and Shortly choice and is usually, but not always, used in most references. See the Notes.

To obtain an algebraic equation for $s_{Jm_1m_2}$ a considerable amount of manipulation and algebra is involved. The details are clearly worked out in Wigner p. 190. Since the notation used here is similar to Wigner's his equations should be followed easily. From the general equation one can check that $s_{Jm_1m_2}$ is indeed real. Table 13-1 shows the formulas for the Wigner coefficients for $s_{Jm_1m_2}$. By inserting the appropriate values, the results in Eq. 8-22 can be obtained. Extensive tables of Wigner coefficients exist and references to them can be found in the notes. Table 13-2 shows the S matrix for several cases. The labeling of the rows and columns, as discussed just after Eq. 13-22 is shown clearly.

**Table 13-2** The S Matrix written in full matrix form for several cases

$j_1 = \frac{1}{2}$, $j_2 = \frac{1}{2}$:

| J | M | $m_1 = \frac{1}{2}$, $m_2 = \frac{1}{2}$ | $m_1 = \frac{1}{2}$, $m_2 = -\frac{1}{2}$ | $m_1 = -\frac{1}{2}$, $m_2 = \frac{1}{2}$ | $m_1 = -\frac{1}{2}$, $m_2 = -\frac{1}{2}$ |
|---|---|---|---|---|---|
| 1 | 1 | 1 | | | |
| 1 | 0 | | $\sqrt{\frac{1}{2}}$ | $\sqrt{\frac{1}{2}}$ | |
| 0 | 0 | | $\sqrt{\frac{1}{2}}$ | $-\sqrt{\frac{1}{2}}$ | |
| 1 | -1 | | | | 1 |

$j_1 = 1$, $j_2 = \frac{1}{2}$:

| J | M | $m_1 = 1$, $m_2 = \frac{1}{2}$ | $m_1 = 1$, $m_2 = -\frac{1}{2}$ | $m_1 = 0$, $m_2 = \frac{1}{2}$ | $m_1 = 0$, $m_2 = -\frac{1}{2}$ | $m_1 = -1$, $m_2 = \frac{1}{2}$ | $m_1 = -1$, $m_2 = -\frac{1}{2}$ |
|---|---|---|---|---|---|---|---|
| $\frac{3}{2}$ | $\frac{3}{2}$ | 1 | | | | | |
| $\frac{3}{2}$ | $\frac{1}{2}$ | | $\sqrt{\frac{1}{3}}$ | $\sqrt{\frac{2}{3}}$ | | | |
| $\frac{1}{2}$ | $\frac{1}{2}$ | | $\sqrt{\frac{2}{3}}$ | $-\sqrt{\frac{1}{3}}$ | | | |
| $\frac{3}{2}$ | $-\frac{1}{2}$ | | | | $\sqrt{\frac{2}{3}}$ | $\sqrt{\frac{1}{3}}$ | |
| $\frac{1}{2}$ | $-\frac{1}{2}$ | | | | $\sqrt{\frac{1}{3}}$ | $-\sqrt{\frac{2}{3}}$ | |
| $\frac{3}{2}$ | $-\frac{3}{2}$ | | | | | | 1 |

## 13-4 Irreducible Tensor Operators

In Chapter 4 the use and transformation properties of **ordinary**, or **Cartesian, tensors** were discussed. A Cartesian tensor, $C_{ijh...}$ was defined in Eq. 4-16 by its transformation properties. A Cartesian tensor of rank two has $3^2 = 9$ components and under an orthogonal coordinate transformation, Eq. 4-29, transforms as (Eq. 4-17)

$$C'_{ij} = \sum_{m,n=1}^{3} R_{im} R_{jn} C_{mm} \qquad (13\text{-}30)$$

In this chapter we use $C_{ij}$ for a Cartesian tensor and reserve the capital T notation for an irreducible tensor. Under a rotation these nine components of $C_{ij}$ transform among themselves and yield a 9-dimensional representation of the rotation group. However, from Chapter 4, we know that this is a reducible representation since the components of $C_{ij}$ transform as the product of Cartesian coordinate two at a time, or as $x_i x_j$. Under a rotation, the coordinates transform as the $\mathbf{D}^1$ irreducible representation of the full rotation group $R(\alpha, \beta, \gamma)$. Thus the nine terms $x_i x_j$, for i and j = 1, 2, 3, transform as $\mathbf{D}^2 \times \mathbf{D}^2 = \mathbf{D}^1 + \mathbf{D}^1 + \mathbf{D}^0$. In fact, the breaking up of a second rank Cartesian tensor into parts that transform irreducibly under rotations is well known. The separate parts of cartesian tensors of ranks 0, 1, and 2 are given by

$$C^0 = \sum_{i=1}^{3} C_{ii}$$

$$(C^1)_k = \tfrac{1}{2}\,[C_{ij} - C_{ji}] \qquad \text{i, j, k cyclic}$$

$$(C^2)_{ij} = \tfrac{1}{2}\,[C_{ij} + C_{ji} - \tfrac{2}{3}\,C^0\,\delta_{ij}] \tag{13-31}$$

Hence, a zero rank tensor with one component, an antisymmetric tensor (axial vector) with three components, and a symmetric second rank tensor with zero trace and five components are obtained. The components of any one of these $C^i$ transform among themselves under all the symmetry operations of $R(\alpha, \beta, \gamma)$. However, these tensors, Eq. 13-31, are not very useful in quantum mechanics. While it is true that their $2l+1$ components transform irreducible among themselves as the $\mathbf{D}^l$ irreducible representation of the full rotation group, nevertheless, the components do not correspond to definite projection quantum numbers. This can be seen by recalling that a first rank tensor has components x, y, z which clearly transform as the $\mathbf{D}^1$ irreducible representation of the full rotation group. However, the linear combination $(x - iy)/\sqrt{2}$, z, $-(x+iy)/\sqrt{2}$ are proportional to the spherical harmonic $Y_m{}^l$ with m = − 1, 0, 1, transform as $\mathbf{D}^1$, and also give projection quantum numbers which are very important when dealing with states that have sharp angular momentum.

We would now like to define an irreducible tensor. Recall, from Section 7-3, that if an operator O when operating on $\psi$ gives $\phi$, $\phi = O\psi$, and if a rotation transforms $\psi$ to $R\psi$ then it transforms the operator to $ROR^{-1}$, Eq. 7-12. Then we define an **irreducible tensor** $T_M{}^L$, where L is an integer, as a set of 2L + 1 functions, which when operated on by the rotation R transform among themselves as the $\mathbf{D}^L$ representation of the full rotation group, i.e.,

$$R\,T^L_M\,R^{-1} = \sum_{M'=-L}^{L} \mathbf{D}(\alpha, \beta, \gamma)^L_{M'M}\,T^L_{M'} \tag{13-32}$$

Note that any operator that is proportional to the spherical harmonics $Y_m{}^l(\theta, \phi)$ is an irreducible tensor operator, see Eq. 8-11. The rotation operator R, as shown in Eq. 7-22, can be written in terms of angular momentum operators as $R_{\mathbf{n}}(\varepsilon) = \exp(i\,\varepsilon\,\mathbf{n} \bullet \mathbf{J}/\hbar)$.

There is another definition of an irreducible tensor which will be shown to be equivalent to that in Eq. 13-32. The set of 2L + 1 functions constitutes an irreducible tensor if the following commutation relations with respect to angular momentum are obeyed:

$$[J_\pm , T^L_M] = [J\,(J+1) - M\,(M \pm 1)]^{1/2}\, T^L_{M\pm1} \tag{13-33a}$$

$$[J_Z , T^L_M] = M\, T^L_M \tag{13-33b}$$

To prove Eq. 13-33b, consider a rotation about the z-axis by an amount $\alpha$, then Eq. 13-32 becomes

$$\exp\,(i\alpha\, J_Z/\hbar)\, T^L_M \exp\,(-i\alpha\, J_Z/\hbar) = \sum_{M'} \mathbf{D}(\alpha, 0, 0)^L_{M'M}\, T^L_{M'}$$

$$= \exp\,(iM\;\alpha/\hbar)\, T^L_M \tag{13-34}$$

since $D^L\,(\alpha, 0, 0)_{M'M} = \exp(in\alpha/\hbar)\delta_{M'M}$. By expanding Eq. 13-34 to first order in $\alpha$ and collecting terms,

$$[1 + (i\alpha\, J_Z/\hbar)]\; T^L_M\; [1 - (i\alpha\, J_Z/\hbar)] = [1 + (iM\alpha/\hbar)]\; T^L_M$$

$$T^L_M + (i\alpha/\hbar)\, J_Z\, T^L_M - (i\alpha/\hbar)\, T^L_M\, J_Z = T^L_M + (iM\alpha/\hbar)\, T^L_M$$

$$J_Z\, T^L_M - T^L_M\, J_Z = M T^L_M \tag{13-35}$$

This last relation is just Eq. 13-33b. The commutation relation Eq. 13-33a is obtained by considering rotations about the x and y axis in Eq. 13-34.

The definition of irreducible tensor, Eq. 13-33, looks very similar to the commutation relations for angular momentum which are

$$[J_Z , J_\pm] = \pm J_\pm \qquad [J_+ , J_-] = 2\, J_Z$$

$$[J^2 , J_M] = 0 \tag{13-36}$$

Thus we can define the three components of an irreducible tensor with L = 1 in terms of angular momentum operators or in terms of the coordinates proportional to the spherical harmonics as in Eq. 13-37.

$$T^1_1 = -\frac{1}{\sqrt{2}}\, J_+ \qquad \text{or} \qquad -\frac{1}{\sqrt{2}}(x + iy)$$

$$T^1_0 = J_Z \qquad \text{or} \qquad z$$

$$T^1_{-1} = \frac{1}{\sqrt{2}}\, J_- \qquad \text{or} \qquad \frac{1}{\sqrt{2}}\,(x - iy) \tag{13-37}$$

Equations 13-33 are satisfied, or writing out the nine equations we have,

$$[J_+, T_1^1] = 0 \qquad [J_-, T_{1,0}^1] = \sqrt{2}\, T_{0,-1}^1$$

$$[J_+, T_{0,-1}^1] = \sqrt{2}\, T_{1,0}^1 \qquad [J_-, T_{-1}^1] = 0$$

$$[J_Z, T_\pm] = \pm T_\pm \qquad [J_Z, T_0^1] = 0 \tag{13-38}$$

An irreducible tensor of rank two can be defined in a similar way as Eq. 13-37, that is, in terms of products of angular momentum or coordinates proportional to the spherical harmonics of order 2.

$$T_2^2 = J_+^2 \qquad \text{or} \qquad (x + iy)^2$$

$$T_1^2 = -(J_Z J_+ + J_+ J_Z) \qquad \text{or} \qquad -2z(x + iy)$$

$$T_0^2 = \sqrt{\tfrac{2}{3}}\,(3J_Z^2 - J^2) \qquad \text{or} \qquad \sqrt{\tfrac{2}{3}}\,(3z^2 - r^2)$$

$$T_{-1}^2 = J_Z J_- + J_- J_Z \qquad \text{or} \qquad 2z(x - iy)$$

$$T_{-2}^2 = J_-^2 \qquad \text{or} \qquad (x - iy)^2 \tag{13-39}$$

As can be seen from Eq. 13-39, care must be exercised when replacing z, x + iy, etc. by $J_Z$, $J_\pm$, etc. since the angular momentum operators do not commute. Symmetrized products must be used.

Before discussing the matrix elements of irreducible tensors in the next section, the multiplication of two irreducible tensors will be discussed. $T(A)_{M_1}^{L_1}$ and $T(B)_{M_2}^{L_2}$ are irreducible tensors that depend on coordinates A and B, respectively. A and B might be angular coordinates of different electrons. The $(2L_1+1)(2L_2+1)$ products $T_{M_1}^{L_1} T_{M_2}^{L_2}$ are transform as the $\mathbf{D}^{L_1} \times \mathbf{D}^{L_2}$ representation of the rotation group. The reduction gives the irreducible tensor $T_M^L$, with L running from $L_1+L_2$ to $|L_1-L_2|$,

$$T_M^L(A, B) = \sum_{M_1} S_{LM;M_1,M-M_1}\, T_{M_1}^{L_1}(A)\, T_{M_2-M_1}^{L_2}(B) \tag{13-40}$$

where the coefficient S is just the Wigner or Clebsch–Gordan coefficient of Section 13-3. Thus, when irreducible tensors are multiplied, the ranks are combined as in the vector model rather than algebraically. Naturally it must be proved that the expression in Eq. 13-40 is indeed an irreducible tensor. This is done by rotating the coordinate system and using the result in Eq. 13-32. The details are left as an exercise. One important result of Eq. 13-40 is the formation of an invariant which can be obtained if $L_1 = L_2$. Then, an invariant $T_0^0 \propto H$, can be obtained in very simple form since the Wigner coefficients $S_{00;M_1-M_1}$ has one term that is dependent on $M_1$

$$H(A, B) = \sum_{M_1} (-1)^{M_1} T_{M_1}^{L_1}(A)\, T_{-M_1}^{L_1}(B) \qquad (13\text{-}41)$$

This is a zero rank tensor, which is a combination of the operators A and B in a form such that if the coordinates are rotated, H(A,B) = H(A,′B′). This is different with what is usually called a scalar. An excellent sample of the use of Eq. 13-41 to obtain the quadrupole Hamiltonian is given in Section 13-5b. As another example of Eq. 13-40, consider both $u_m^{\ 1}$ and $v_m^{\ 1}$ as irreducible tensors with L = 1 (m, n = 0, ±1). Then Eq. 13-40 shows that certain products of these two irreducible tensors form irreducible tensors $T_M^{\ L}$, with L = 2, 1, 0. According to Eq. 13-40 and Eq. 13-41 for $T_0^{\ 0}$, these are

$$T_0^0 = -u_1\, v_{-1} + u_0\, v_0 - u_{-1}\, v_1$$

$$T_0^1 \propto u_1\, v_{-1} - u_{-1}\, v_1 \qquad T_{\pm 1}^1 \propto u_{\pm 1}\, v_0 - u_0\, v_{\pm 1}$$

$$T_0^2 \propto 2\, u_2\, v_{-2} + u_1\, v_{-1} + u_{-1}\, v_1 \qquad T_{\pm 2}^2 \propto u_{\pm 1}\, v_{\pm 2}$$

$$T_{\pm 1}^2 \propto u_{\pm 1}\, v_0 - u_0\, v_{\pm 1}$$

These results are, of course, just the same as those in Eq. 8-22 where two p-electrons were combined to form D, P, and S configurations. The problems are the same.

## 13-5 The Wigner–Eckart Theorem

The Wigner–Eckart theorem is very useful in many branches of physics and chemistry. It gives us the matrix elements of irreducible tensor operators between states of sharp angular momentum. Before we derive the relationship, let us state and discuss some aspects of the theorem. The actual proof of the theorem is straightforward but algebraically long.

The Wigner–Eckart theorem, Eq. 13-42, states that the matrix element of an irreducible tensor operator $T_M^{\ L}$ between two states of angular momentum can be written as

$$\langle\, j'\ m'\ \eta' \mid T_M^L \mid j\ m\ \eta\, \rangle = S_{j'm';Mm}^{Lj} \langle\, j'\ \eta' \parallel T^L \parallel j\ \eta\, \rangle \qquad (13\text{-}42)$$

The Wigner coefficient S, Eq. 13-24, is written out with all the sub and superscripts although we will suppress the superscripts as often as possible. Equation 13-42 could be written in terms of the small s-coefficients, Eq. 13-27, as well. The double barrelled or **reduced matrix element**, $\langle \| \rangle$, is independent of all the projection quantum numbers M, m′, and m as is implied by notation and $\eta$ and $\eta'$ are all the other quantum numbers that label the state but do not involve angular momentum. We will now drop the $\eta$'s. The theorem Eq. 13-42 shows that the matrix element of an irreducible tensor operator can be separated into two parts. The first part is the Wigner coefficient that contains all the projection quantum number dependence and the angular momentum selection rules, in other words, the symmetry properties of the matrix element. The second term contains the physical aspects of the problem. The selection rules from the Wigner coefficients, as discussed in Section 13-3, are the so-called triangle rule on the label of the irreducible representation, Eq. 13-43a, and the condition on the projection quantum numbers Eq. 13-43b,

$$j' = L + j, \quad L + j - 1, \cdots, \quad |L - j| \tag{13-43a}$$

$$m' = M + m \tag{13-43b}$$

The restrictions on the m's, $m \leq j$, naturally holds. We also see from the theorem that the matrix elements between the same jm states of all irreducible tensor operators of the same order (same L) are proportional to each other for the same M. The only difference between such matrix elements will be the ratio of the reduced matrix elements.

The theorem is proved by obtaining recursion relationships for the matrix elements of $T_M{}^L$ and for the Wigner coefficients, and noting that the dependence on the projection quantum in the relations is the same.

### a. Recursion relations

Matrix elements of the commutator $[J_z, T_M{}^L]$ from Eq. 13-33b are

$$\langle j' m' | J_Z T_M^L | j m \rangle - \langle j' m' | T_M^L J_Z | j m \rangle = M \langle j' m' | T_M^L | j m \rangle \tag{13-44}$$

Operating with $J_z$ on the left in the first term but remembering that $J_z$ is a Hermitian operator, thus $\langle \psi | J_z \phi \rangle = \langle J_z \psi | \phi \rangle$ in the second term, and using the usual relations for $J_z$ in Eq. 7-26, and then collecting terms, one has

$$(m' - m - M) \langle j' m' | T_M^L | j m \rangle = 0 \tag{13-45}$$

Hence, the matrix element is zero unless $m' = m + M$, which is the same condition on the projection quantum numbers for the Wigner coefficients to be nonzero.

The same approach is taken for the other type of commutor $[J_{\pm}, T_M{}^L]$ in Eq. 13-33a. The matrix elements are

$$\langle j'\ m' \mid J_{\pm}\ T^L_M \mid j\ m \rangle - \langle j'\ m' \mid T^L_M\ J_{\pm} \mid j\ m \rangle$$

$$= [\ L\ (L+1) - M\ (M \pm 1)\ ]^{1/2}\ \langle j'\ m' \mid T^L_{M\pm1} \mid j\ m \rangle \quad (13\text{-}46)$$

Again operating on the appropriate side with the angular momentum raising and lowering operators and using the results from Eq. 7-26, we obtain Eq. 13-47. However, in obtaining this equation remember that though $J_x$ and $J_y$ are Hermitian, the operators $J_{\pm}$ are not; $(J_{\pm})^{\dagger} = J_{\mp}$, so $<\psi \mid J_{\pm}\phi> = <J_{\pm}{}^{\dagger}\psi \mid \phi> = <J_{\mp}\psi \mid \phi>$.

$$[\ j'\ (j'+1) - m'\ (m' \mp 1)\ ]^{1/2}\ \langle j'\ m' \mp 1 \mid T^L_M \mid j\ m \rangle$$

$$- [\ j\ (j+1) - m\ (m \pm 1)\ ]^{1/2}\ \langle j'\ m' \mid T^L_M \mid j\ m \pm 1 \rangle$$

$$= [L\ (L+1) - M\ (M \pm 1)\ ]^{1/2}\ \langle j'\ m' \mid T^L_{M\pm1} \mid j\ m \rangle$$

(13-47)

Note that according to Eq. 13-45, each term in Eq. 13-47 will be zero unless $m' = m \pm 1$. The recursion relation, given by Eq. 13-47, enables all of the matrix elements of $T_M{}^L$ to be calculated for a given j′ and j in terms of any one matrix element.

Now the same type of approach is applied to the Wigner coefficients. Rewriting Eq. 13-24a in the notation of this section,

$$\psi^{j'}_{m'} = \Sigma\ S^{jL}_{j'm';mM}\ u^j_m v^L_M \quad (13\text{-}48)$$

where we have used our phase convention. The equation naturally has the usual selection rules of the Wigner coefficients, Eq. 13-43, which determine terms that are zero. Operating on both sides of Eq. 13-48 with $j'_{\mp} = j_{\mp} + L_{\mp}$, we obtain

$$[\ j'\ (j'+1) - m'\ (m' \mp 1)\ ]^{1/2}\ \psi^{j'}_{m'\mp1}$$

$$= \sum_{m,M} [\ j\ (j+1) - m\ (m \mp 1)\ ]^{1/2}\ S^{jL}_{j'm';mM}\ u^j_{m\mp1}\ v^L_M$$

$$+ \sum_{m,M} [\ L\ (L+1) - M\ (M \mp 1)\ ]^{1/2}\ S^{jL}_{j'm';mM}\ u^j_m\ v^L_{M\mp1} \quad (13\text{-}49)$$

On the left side of Eq. 13-49, $\psi_{m'\mp1}{}^{j'}$ is written in terms uv by using Eq.

13-48. In the first term on the right side of Eq. 13-49 set $m \mp 1 = \mu$ and $M = \lambda$. In the second term on the right side of Eq. 13-49 set $m = \mu$ and $M \mp 1 = \lambda$

$$\sum_{\mu,\lambda} [\, j'(j'+1) - m'(m' \mp 1)\,]^{1/2}\, S^{jL}_{j'm' \mp 1;\mu\lambda}\, u^{j}_{\mu}\, v^{L}_{\lambda}$$

$$= \sum_{\mu,\lambda} [\, j(j+1) - \mu(\mu \pm 1)\,]^{1/2}\, S^{jL}_{j'm';\mu \pm 1 \lambda}\, u^{j}_{\mu}\, v^{L}_{\lambda}$$

$$+ \sum_{\mu,\lambda} [\, L(L+1) - \lambda(\lambda \pm 1)\,]^{1/2}\, S^{jL}_{j'm';\mu\lambda \pm 1}\, u^{j}_{\mu}\, v^{L}_{\lambda} \qquad (13\text{-}50)$$

Since the product function $u_{\mu}{}^{j}\, v_{\lambda}{}^{L}$ is in all the terms, we have a condition on the coefficients. Taking the $\mu = m$ and $\lambda = M$ terms for the coefficients on both sides and transposing the first term on the right to the left side of the equation, we have

$$[\, j'(j'+1) - m'(m' \mp 1)\,]^{1/2}\, S^{jL}_{j'm' \mp 1;mM}$$

$$- [\, j(j+1) - m(m \pm 1)\,]^{1/2}\, S^{jL}_{j'm';m \pm 1 \lambda}$$

$$= [\, L(L+1) - M(M \pm 1)\,]^{1/2}\, S^{jL}_{j'm';mM \pm 1} \qquad (13\text{-}51)$$

This is a recursion relation for the Wigner coefficients. For a given j′, j, L the equation relates the coefficients for changes in the projection quantum numbers. Note that in each term the Wigner coefficient is zero unless $m' = m + M \pm 1$ which is the same condition as noted for the terms in Eq. 13-47. Also note, in comparing Eqs. 13-51 and 13-47, that the coefficients of the same terms are identical. So the matrix elements of irreducible tensor operators and the Wigner coefficients have the same independence on the projection quantum numbers and the constant of proportionality between the two can involve only j′, j, and L but not the projection quantum numbers. Hence, Eq. 13-42 is shown to be true.

### b. Examples of the Wigner–Eckart theorem

Here we give only a few examples of this theorem which is extensively used.

**Example 1. Operator Equivalents.** In Section 4-6 the potential energy experienced by an impurity ion in a crystal was discussed. The nonspherical terms in the Hamiltonian $V_M{}^L$ can be written in terms of spherical harmonics (discussed in Section 8-2) and typically are of the form $V_M{}^L = f(r)\, Y_m{}^l\, (\theta\phi)$. For f-electrons, only potential energy terms with L $\leq$ 6 need be considered. Some of the typical terms encountered are:

$$V_0^2 = \Sigma\ (3\ z^2 - r^2)$$

$$V_0^4 = \Sigma\ (35\ z^4 - 30\ r^2\ z^2 + 3\ r^4)$$

$$V_0^6 = \Sigma\ (231\ z^6 - 315\ r^2\ z^4 + 105\ r^4\ z^2 - 5\ r^6)$$

$$V_6^6 = \Sigma\ (x^6 - 15\ x^4\ y^2 + 15\ x^2\ y^4 - y^6) \tag{13-52}$$

where the sum is over the coordinates of all the electrons, and the z-axis is the principal axis as usual. The problem is to evaluate the matrix elements of these crystal field potential energy terms between sharp angular momentum states that are described in terms of $|JM\rangle$. The electronic states can be described in terms of sharp angular momentum because for the 4f-electrons, the crystal field terms (Eq. 13-52) are much weaker then the spin-orbit coupling. Thus, the 4f-electrons can be treated to a first approximation as in the free atom. Then for a given quantum state with a J, there is a 2J + 1 degeneracy that will be lifted by terms of the type in Eq. 13-52; so matrix elements of the type $\langle JM' | V_M{}^L | JM\rangle$ must be evaluated. Since the potential energy terms transform under rotations, as spherical harmonics they transform as irreducible tensors. Thus, the matrix elements can be evaluated relatively easily using the method of operator equivalents. The idea is that the matrix elements of all irreducible tensors of a given value of L between the same states are proportional to each other. The constant of proportionality is just the ratio of the reduced matrix elements. So the crystal field matrix elements can be evaluated up to a constant by using an irreducible tensor composed of angular momentum operators. The relationships in Eq. 13-37 and 13-39 show an irreducible tensor, with L = 1 and 2, respectively, written in terms of Cartesian coordinates and angular momentum operators. The only caution in writing an irreducible tensor in terms of angular momentum operators, given the tensor in terms of Cartesian coordinates, is that the angular momentum operators do not commute among themselves, so the symmetrized product must be used as in Eq. 13-39 for $T_{\pm 1}{}^2$. For higher order terms, a certain amount of algebra is required, this has been done and is tabulated in the references in the Notes.

Consider for example the $r^2z^2$ term in Eq. 13-52 which is $x^2z^2 + y^2z^2 + z^4$. For the $x^2z^2$ term, the symmetrized angular momentum is

$$x^2 z^2 \rightarrow \tfrac{1}{6}\,[J_x^2 J_z^2 + J_z^2 J_x^2 + J_x J_z J_x J_z + J_z J_x J_x J_z$$

$$+ J_x J_z J_z J_x + J_z J_x J_z J_x]$$

$$= \tfrac{1}{6}\,[3 J_x^2 J_z^2 + 3 J_z^2 J_x^2 - 2 J_x^2 + 3 J_y^2 - 2 J_z^2] \qquad (13\text{-}53)$$

The equality was obtained by using the commutation relations. Using this procedure the operator equivalents for the first two relations in Eq. 13-52 between states of a given J are

$$V_0^2 = \Sigma\,(3 z^2 - r^2) \rightarrow \alpha\,[3 J_z^2 - J(J+1)]$$

$$V_0^4 = \Sigma\,(35 z^4 - 30 r^2 z^2 + 3 r^4) \rightarrow \beta\,[35 J_z^4 - 30 J(J+1) J_z^2$$

$$- 6 J(J+1) + 3 J^2 (J+1)^2] \qquad (13\text{-}54)$$

where the $\alpha$ and $\beta$ are constants that must be calculated. They are usually evaluated for the simplest state, usually M = J. For $Ce^{3+}$ $4f^1$, the ground state if $^2F_{5/2}$ and $\alpha = [-2/35]<r^2>$ and $\beta = [2/(7)(45)]<r^4>$, where $<r^n>$ is the exception value of $r^n$ over the radial part of the 4f electron. The sixth order terms in Eq. 13-52 have zero matrix elements with the J = 5/2 states. The matrix element is

$$\langle JM' \mid \Sigma\, 3 z^2 - r^2 \mid JM \rangle = \alpha\,\langle JM' \mid 3 J_z^2 - J(J+1) \mid JM \rangle$$

$$= \alpha\,[3 M^2 - J(J+1)]\,\delta_{M'M} \qquad (13\text{-}55)$$

This is fairly simple result. See the Notes for references to the literature.

**Example 2. Nuclear Quadrupole Interaction.** The calculation of the Hamiltonian that involves the interaction of the nuclear quadrupole moment with the electric field gradient determined by the electrons outside the nucleus will be presented here. The calculation is a good example of the use of irreducible tensor operators and of the Wigner–Eckart theorem.

Figure 13-1 defines the angles and distances that will be used to describe the Coulomb interaction between a small volume of nuclear charge with a corresponding volume of electron charge. The subscripts e and n refer to electron and nuclear coordinates, respectively, in an obvious way as in Fig. 13-1. Writing the interaction energy and expanding 1/r in terms of Legendre polynomials$P(\theta_{en})^l$ where,for example, $P^2 = (3\cos^2\theta_{en} - 1)/2$ as Eq. 8-9 and Appendix 8, we have the following:

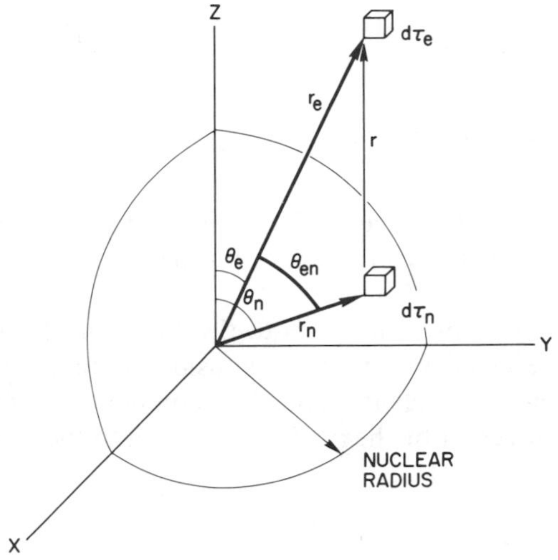


**Fig. 13-1** Coordinates used to describe the interaction of a volume of nuclear charge with a volume of electronic charge.

$$\mathbf{H}_{en} = \int_{\tau_n} \int_{\tau_e} (\rho_e \, \rho_n \, d\, \tau_e \, d\, \tau_n / r) \tag{13-56a}$$

$$\begin{aligned} 1/r &= (r_n^2 + r_e^2 + 2\, r_n \, r_e \cos\, \theta_{en})^{-1/2} \\ &= (1/r_e) + (r_n/r_e^2)\, P^1 + (r_n^2/r_e^3)\, P^2 + \ldots \end{aligned} \tag{13-56b}$$

$$\mathbf{H}_{en} = \sum_l \mathbf{H}^l_{en} = \sum_l \int_{\tau_n} \int_{\tau_e} (\rho_e \, \rho_n \, r_n^l / r_e^{\,l+1})\, P^l(\theta_{en}) \, d\, \tau_e \, d\, \tau_n \tag{13-56c}$$

Equation 13-56c shows how the interaction energy is written in terms of multipole moments. We will be concerned with the $l = 2$ or electric quadrupole moment term. The terms in the Hamiltonian Eq. 13-56c can be simplified so that the electron and nuclear coordinates appear separately, by the use of the spherical harmonic addition theorem

$$P^l(\theta_{en}) = (4\pi/2l{+}1) \sum_{m=-l}^{l} (-1)^m \, Y^l_{-m}(\theta_e, \phi_e)\, Y^l_{+m}(\theta_n, \phi_n) \tag{13-57}$$

where the $Y_m{}^l$ are spherical harmonics. Thus the Hamiltonian is written in terms of the scalar product of two irreducible tensors defined as

$$\mathbf{H}^l_{en} = \mathbf{Q}^l \bullet \mathbf{F}^l = \sum_{m=-l}^{l} (-1)^m \, Q^l_m \, F^l_{-m} \tag{13-58a}$$

$$Q^l_m = (4\pi/2l+1)^{1/2} \int_{\tau_n} \rho_n\, r^l_n\, Y^l_m(\theta_n, \phi_n)\, d\tau_n \tag{13-58b}$$

$$F^l_m = (4\pi/2l+1)^{1/2} \int_{\tau_e} (\rho_e / r_e^{l+1})\, Y^l_m(\theta_e, \phi_e)\, d\tau_e \tag{13-58c}$$

These equations refer to the general $2^l$-pole interaction and are an example of the formation of an invariant from irreducible tensors as in Eq. 13-41

We specialize these equations to the $l = 2$ or electron quadrupole moment. All the electron charge is considered to be outside a small sphere surrounding the nucleus. Thus, the components of the gradient of the electric field at center of the nucleus, $\nabla E$, due to the electrons external to the nucleus appears in $\mathbf{H}_{en}{}^2$

$$\mathbf{H}^2_{en} = Q_0(\nabla E)_0 - Q_{+1}(\nabla E)_{-1} - Q_{-1}(\nabla E)_{+1} + Q_{+2}(\nabla E)_{-2} + Q_{-2}(\nabla E)_{+2} \tag{13-59}$$

where the superscripts have been dropped. These terms in general are

$$Q_0 = \frac{1}{2}\int_{\tau_n} \rho_n\, r^2_n (3\cos^2\theta_n - 1)\, d\tau_n = \frac{1}{2}\int_{\tau_n} \rho_n (3z^2_n - r^2_n)\, d\tau_n$$

$$Q_{\pm1} = \mp(3/2)^{1/2}\int \rho_n\, z_n\, x_{n\pm}\, d\tau_n \qquad Q_{\pm2} = (3/8)^{1/2}\int \rho_n\, x^2_{n\pm}\, d\tau_n \tag{13-60}$$

where $x_\pm = x \pm y$. Similarly, the components of the electric field gradient can be written, using $E_\pm = E_x \pm E_y$ and $\partial_\pm = \partial/\partial x \pm i\partial/\partial y$,

$$(\nabla E)_0 = \frac{1}{2}\int (\rho_e/r_e^3)(3\cos^2\theta_e - 1)\, d\tau_e = -\frac{1}{2}(\partial E_z / \partial z)$$

$$(\nabla E)_{\pm1} = \mp(3/2)^{1/2}\int (\rho_e / r_e^3)\cos\theta_e \sin\theta_e (\cos\phi_e \pm i\sin\phi_e)\, d\tau_e = \pm(6)^{-1/2}\,\partial_\pm E_z$$

$$(\nabla E)_{\pm2} = (3/8)^{1/2}\int (\rho_e / r_e^3)\sin^2\theta_e (\cos\phi_e \pm i\sin\phi_e)^2\, d\tau_e = [-1/2\,(6)^{1/2}]\,\partial_\pm E_\pm \tag{13-61}$$

The Hamiltonian in Eq. 13-56c can be expressed in other ways. For example, the $\pm$ terms could be combined together or the terms can be written using Cartesian coordinates. For the latter case one obtains

$$\mathbf{H}^2_{en} = -\frac{1}{6}\sum_{i,j=1}^{3} Q_{ij}(\nabla E)_{ij} \tag{13-62a}$$

$$Q_{ij} = \int \rho_n (3x_{ni}\, x_{nj} - \delta_{ij}\, r^2_n)\, d\tau_n \tag{13-62b}$$

$$(\nabla E)_{ij} = -\int \frac{\rho_e}{r_e^5}(3x_{ei}\,x_{ej} - \delta_{ij}\,r_e^2)\,d\tau_e \tag{13-62c}$$

The expressions in Eq. 13-62 are more familiar than those in Eqs. 13-59 to 13-61 but have the disadvantages that Cartesian tensors have compared to irreducible tensors.

We would like to have more useable expressions for the Hamiltonian than those in Eqs. 13-60 and 13-61. To evaluate the terms in Eq. 13-60 the nuclear wave function must be known. That is a difficult problem and should be separated from the problem discussed here, which is the interaction energy of a nuclear quadrupole moment with an external field gradient. To separate the problems, the operator equivalent technique is used. The proportionality between the Cartesian coordinates in Eq. 13-60 and the nuclear angular momentum operators is used where the constant of proportionality is a reduced matrix element, Eq. 13-42, that contains all the nuclear wave function information. Thus, Eq. 13-60 becomes

$$Q_0 = C\,(3\,I_z^2 - I^2)/2 \qquad Q_{\pm 2} = C\sqrt{\frac{3}{8}}\,I_\pm^2$$

$$Q_{\pm 1} = \mp C\sqrt{\frac{3}{2}}\,(I_z\,I_\pm + I_\pm\,I_z)/2 \tag{13-63}$$

By convention the nuclear quadrupole moment Q is defined by the expectation value of $3z^2 - r^2$ over the nuclear wave function with the projection quantium number taking its maxium value (m = I).

$$e\,Q \equiv \int_{\tau_n} (\rho_n)_{I,m=I}\,(3z_n^2 - r_n^2)\,d\,\tau_n = C\,\langle\, I, m = I \mid 3I_z^2 - I^2 \mid I, m = I\,\rangle$$

$$e\,Q = C\,[3\,I^2 - I\,(I+1)] \qquad C = e\,Q/I\,(2\,I - 1) \tag{13-64}$$

This constant contains the physical aspects of the nuclear charge density. In an analogous manner the elements of ∇E in Eq. 13-61 can be evaluated. If the nucleus is in a molecule where the angular momentum **J** specifies the orientation of the molecule, then the term $q_J$ specifies the electric field gradient for the projection quantum number taking its maximum value m = J. So

$$q_J \equiv \int_{\tau_e} (\rho_e)_{J,m=J}\,[(3\cos^2\theta_e - 1)/r_e^3]\,d\,\tau_e = C'\,\langle\, J, m=J \mid 3\,J_z^2 - J^2 \mid J, m=J\,\rangle \tag{13-65}$$

Thus, Eq. 13-61 can be written in a more useful manner as follows:

$$(\nabla E)_0 = [e\, q_J/2\, J\, (2\, J - 1)]\, [3\, J_z^2 - J\, (J + 1)]$$

$$(\nabla E)_{\pm 1} = \mp (3/2)^{1/2}\, [e\, q_J/2J\, (2\, J - 1)]\, [J_z\, J_\pm + J_\pm\, J_z]$$

$$(\nabla E)_{\pm 2} = (3/2)^{1/2}\, [e\, q_J/2\, J\, (2\, J - 1)]\, J_\pm^{\,2} \tag{13-66}$$

Matrix elements of the Hamiltonian between states of different nuclear or rotational projection ($I_z$ or $J_z$) quantum numbers can now be calculated easily.

The quadrupole Hamiltonian in the form shown in Eq. 13-62 can also be expressed in terms of the same reduced matrix elements and angular momentum operators as

$$Q_{ij} = [e\, Q/I\, (2\, I - 1)]\, [(3/2)\, (I_i\, I_j + I_j\, I_i) - \delta_{ij}\, I\, (I + 1)]$$

$$(\nabla E)_{ij} = [-e\, q_J/J\, (2\, J - 1)]\, [(3/2)\ \ J_i\, J_j + J_j\, J_i) - \delta_{ij}\, J\, (J + 1)] \tag{13-67}$$

The interaction Hamiltonian can also be written in the form

$$\mathbf{H}_{en}^2 = [e^2\, q_J\, Q/2\, I\, (2\, I - 1)\, J\, (2\, J - 1)]\, [3\, (\mathbf{I} \cdot \mathbf{J})^2 + (3/2)\, (\mathbf{I} \cdot \mathbf{J}) - I\, (I + 1)\, J\, (J + 1)] \tag{13-68}$$

This equation is obtained by expanding **I** and **J** in terms of $I_z$, $I_\pm$, etc. and using Eq. 13-59 with Eqs. 13-63 and 13-66.

## 13-6 Survey of 3j and Racah Coefficients

The coupling of electrons with different orbitals in many electron atoms is very important in atomic physics. For the two-electron case the Wigner coefficients are all that are required. However, for three or more electrons, the problems become very difficult. Racah coefficients are used when coupling three angular momenta. (Racah coefficients are proportional to the 6j coefficients.) The 9j symbols are used when four angular momenta are coupled; they are also useful when transforming from an *ls* to a jj-coupling scheme. The 12j symbols are also defined and used when five angular momenta are coupled, etc. We will only, very briefly, mention 3j and 6j symbols. The Notes should be consulted for references to this interesting and important field.

The expression for the Wigner coefficient in Eq. 13-24 are somewhat symmetric in $j_1$, $j_2$, and J. One can define the 3j symbol as follows:

$$\begin{pmatrix} j_1 & j_2 & j_3 \\ m_1 & m_2 & m_3 \end{pmatrix} = [(-1)^{j_1-j_2-m_3} (2j_3+1)^{-1/2}] S^{j_1 j_2}_{j_3 m_1 m_2} \delta_{m_1+m_2+m_3,0} \tag{13-69}$$

The Wigner coefficient is zero unless $j_3 = j_1+j_2, j_1+j_2-1, \ldots, |j_1-j_2|$. This is the so-called vector triangle rule. Thus, we take the 3j coefficients as zero if this condition is not fulfilled and it is also zero if $|m_i| > j_i$. The angular momenta that enter into the 3j symbol are not entirely symmetric. However, if two j's are interchanged together with their corresponding m's, the values of the 3j symbol is unchanged if $j_1+j_2+j_3$ is even and it changes sign if $j_1+j_2+j_3$ is odd. That is,

$$(-1)^{j_1+j_2+j_3} \begin{pmatrix} j_1 & j_2 & j_3 \\ m_1 & m_2 & m_3 \end{pmatrix} = \begin{pmatrix} j_2 & j_1 & j_3 \\ m_2 & m_1 & m_3 \end{pmatrix} = \begin{pmatrix} j_1 & j_3 & j_2 \\ m_1 & m_3 & m_2 \end{pmatrix} = \begin{pmatrix} j_3 & j_2 & j_1 \\ m_3 & m_2 & m_1 \end{pmatrix} \tag{13-70}$$

The 3j symbols have other symmetries. See the Problems. Extensive tabulations of 3j symbols have been published and references are given in the Notes.

The 6j symbols arise when three angular momenta are coupled together. Suppose there are three particles labled u, v, w. They are to be coupled together to give a function that transforms as the J irreducible representation. We can couple them in three different ways. First the angular momentum of u and v can be coupled to give an angular momentum j, and the third particle can be coupled to j to form a total angular momentum J. This will be written as X(uv, j | w, JM). The second way to couple the particles is to couple v and w first and the couple u. The resultant function is Y(vw, j | u, JM). The third way is Z(uw, j | v, JM). These three functions X, Y, and Z are not the same even though J and M are the same for the three functions. A priori, none of them is closer to the true eigenfunction than the others. Of course, the actual physical problem may cause one of these functions to be a much better approximation. For example if the atom has 2s2p6d electrons, 2s coupled first to 2p and then to 6d would give a reasonable approximation to the eigenfunction. However, each function can be expressed as a linear combination of all of the other states, for example Z(uw, j′ | v, J′M′). Thus

$$X(j \mid JM) = \sum_{j'} \sum_{J'M'} c(j\,JM; j'\,J'\,M')\, Z(j' \mid J'\,M') \tag{13-71}$$

where we have dropped the u, v, and w since the angular momentum description is independent of these states. In this sum the coefficients for $J \neq J'$ or $M \neq M'$ are zero since these inequalities mean the functions on both sides of the equation will transform as different irreducible representations or different rows of the same irreducible representation, respectively. So the sum on J′ and M′ can be dropped. Also the coefficient in Eq. 13-71 is independent of M because $\langle X(j\,|\,JM) \mid Z(j'\,|\,JM)\rangle$ does not depend on M since they are partners of a unitary irreducible representation, Eq. 4-14. Thus, the coefficient can be rewritten as $c(jjj')^j$. The 6j symbol can be defined in terms of these coefficients as

$$\begin{Bmatrix} J & j_2 & j' \\ j_1 & j_3 & j \end{Bmatrix} = (-1)^{2j_1}\, c\,(j, j')^J\,(2\,j+1)^{-1/2}\,(2\,j'+1)^{-1/2} \tag{13-72}$$

where $j_i$ with i = 1, 2, 3 is the angular momentum of the three particles that are coupled. As implied in the expression, the 6j symbol is independent of $m_{j_1}$ and M. It is sometimes called the regrouping coefficient, which is an instructive name since the 6j symbol describes the overlap of a function (obtained by grouping $j_1$, $j_2$, and $j_3$ in a certain manner to obtain J) with another function obtained by grouping them in a different manner to obtain the same J. As originally defined by Racah, the Racah coefficient W is

$$W\,(j_1\,j_2\,l_2\,l_1\,;\,j_3\,l_3) = (-1)^{j_1+j_2+l_1+l_2} \begin{Bmatrix} j_1 & j_2 & j_3 \\ l_1 & l_2 & l_3 \end{Bmatrix} \tag{13-73}$$

The 6j symbol can be written explicitly in terms of 3j symbols. Rotenberg et al. (see the Notes) should be consulted for this relationship as well as similar ones for the 9j and 12j symbols. The 9j symbols are used for coupling four angular momentum together. For two equivalent particles, they are used to express the eigenfunctions in LS-coupling in terms of jj-coupling eigenfunctions. They are also used in computing matrix elements of products of two irreducible tensor operators. General references in the Notes should be consulted for further study of the use of group theory in theory of angular momentum.

## Notes

Wigner, Chapter 15, is an excellent reference for the information in Sections 13-1 and 13-2. Chapter 17 of the same book is the best reference for Section 13-3 on Wigner coefficients. Material similar to Sections 13-1 through 13-3 also can be found in Falicov, Chapter 8; Hamermesh, Chapter 9; Heine, Chapter 2 and Section 20; and Tinkham, Chapters 5 and 6. Parts of Chapters 3–5 of Di Bartolo should also be consulted.

A number of books dealing primarily with the theory of angular momentum are listed in the references and they should be consulted. For more extensive discussions of the Wigner coefficients, irreducible tensor operators, the Wigner–Eckart theorem with many applications, and the n-j symbols see in particular Brink and Satchler, Edmonds, and Rose. Discussions of the various phase conventions of the Wigner coefficients can also be found in these books.

There are many tabulations of the n-j symbols. In particular, Rotenberg, et al. have an extensive tabulation of the 3-j and 6-j symbols as well as a interesting discussion of n-j symbols.

The nuclear quadrupole problem in Section 13-5b can be found in N. F. Ramsey, "Molecular Beams" [Oxford Univ. Press (Clarendon), London and New York, 1956] as well as other sources.

Operator equivalents were originally discussed by Stevens, Proc. Phys. Soc. **A65**, 209 (1952). for a review of these techniques see Hutchings, Solid State Phys. **16**, 227 (1964), and Buckmaster, Chatterjee, and Shing, Phys. Stat. Sol. **13**, 9 (1072).

There is a very extensive literature on the use of the SU(3) group in particle physics. Probably the best way to find recent articles and references is to look in the Review of Modern Physics for articles. Also consult: B. Ram, Amer. J. Phys. **35**, 16 (1957): T. J. Nelson, Amer. J. Phys. **36** 791 (1968); R. D. Young, Amer. J. Phys. **41**, 472 (1973); M. J. Englefield, "Group Theory and the Coulomb Problem" (Wiley, New York, 1972).

## Problems

**1.** Show explicitly that the SU(2) group is indeed a group.

**2.** Prove, explicitly, that $R(\mathbf{u})$ in Eq. 13-7 represents an orthogonal rotation of coordinates.

**3.** Given a $\mathbf{u}$, $\mathbf{R}(\mathbf{u})$ can be found as in Eq. 13-7. Find the inverse of this relation, i.e., $\mathbf{u}(\mathbf{R})$. (Hint: see Cornwell, Chapter 8.) Note how the sign of $\mathbf{u}$ is undetermined when obtained from $\mathbf{R}$.

**4.** (a) Prove that $U(\mathbf{u})^j$ in Eq. 13-18 is unitary. (b) Show that this representation is irreducible. (c) Show that there are no other irreducible representations of SU(2) besides $U(\mathbf{u})^j$.

**5.** So far, in discussing continuous groups, we have completely ignored the problem that is encountered in converting sums over a finite number of symmetry operations to integrals for continuous groups. (a) Discuss the need for the Hurwitz invariant integral. (b) Find the invariant density for an axial rotation group. (c) Find the invariant density function for a rotation about an arbitrary axis. (See Wigner, Chapter 10, and Tinkham, Section 5-2.)

**6.** If $A = S^{-1}BS$ and A and B are unitary, show that S must be unitary. (Hint: see Wigner, p. 78.)

**7.** Show that the expresion in Eq. 13-40 is indeed an irreducible tensor of rank L.

**8.** (a) Show that an irreducible tensor is Hermitain if $(T_m{}^L)^\dagger = (-1)^m T_{-m}{}^L$. (b) Show that $T_L{}^L = (T_1{}^1)^L$. This relation provides a systematic procedure for generating irreducible tensor operators from $T_1{}^1$. By using the lowering operator, the $T_M{}^L$ can be obtained.

**9.** For an irreducible tensor operator given in terms of angular momentum operators $T(J)_M{}^L$ as in Eqs. 13-37 and 13-39, show that

$$\langle J' \| T(J)^L \| J \rangle = [L!L!(2J+L+1)!/2^L(2L)!(2J-1)!]^{1/2}\delta(J', J).$$

**10.** Show that the equality in Eq. 13-53 is true. Then show that $r^2z^2 \rightarrow (1/6)[6J(J+1)J_z{}^2+J(J+1) -5 J_z{}^2]$ using the fact that $J_x{}^2+J_y{}^2+J_z{}^2=J^2$ which, operating on states in a given J manifold, gives a constant $J(J+1)$.

**11.** Show that the form of the nuclear quadrupole Hamiltonian shown in Eq. 13-68 is equivalent to that shown in Eqs. 13-59, 13-63, and 13-66.

**12.** Show that for a nucleus in a solid at a site of axial symmetry, the quadrupole Hamiltonian is $\mathbf{H}_Q=[eQq/4I(2I-1)][3I_z{}^2-I(I+1)]$ where $q = \Sigma e_i(3 \cos^2\theta_i-1)/r_i{}^3$. The symbols $e_i$, $\theta_i$, $r_i$ are the charge, azimuthal angle, and distance to ith charge.

**13.** Consider the same nucleus as in the previous problem but also in a strong external magnetic field **H**. Then the total Hamiltonian is $\boldsymbol{H} = g\beta\mathbf{H} \cdot \mathbf{I} + H_Q$, where $H_Q$ is given in the previous problem. Using first order perturbation theory, $H_Q \ll g\beta\mathbf{H} \cdot \mathbf{I}$, show how $H_Q$ changes the energy levels. Show that the absorption spectra has a term $\overline{A}(3\cos^2\theta-1)$ where $\theta$ is the angle between the principal axis of the crystal and the external magnetic field. Show that $\overline{A}$ can be measured even if the sample consists ofa power of these crystals. (The appreciation that powders can be used often in place of single crystals is important. See M. H. Cohen and F. Reif in Solid State Physics **5**, 321 (1957) and G. Burns and B. A. Scott, Phys. Rev. Lett. **25**, 1191 (1970) and the references quoted there.

**14.** (a) Prove the relation in Eq. 13-70 for the 3j symbols. (b) for the 3j symbols prove that

$$\begin{pmatrix} j_1 & j_2 & j_3 \\ m_1 & m_2 & m_3 \end{pmatrix} = \begin{pmatrix} j_2 & j_3 & j_1 \\ m_2 & m_3 & m_1 \end{pmatrix} = \begin{pmatrix} j_3 & j_1 & j_2 \\ m_3 & m_1 & m_2 \end{pmatrix}$$

(c) prove that

$$(-1)^{j_1+j_2+j_3} \begin{pmatrix} j_1 & j_2 & j_3 \\ m_1 & m_2 & m_3 \end{pmatrix} = \begin{pmatrix} j_1 & j_2 & j_3 \\ -m_1 & -m_2 & -m_3 \end{pmatrix}$$

(d) Show that

$$\langle Y^{l'}_{m'}(\theta, \phi) \mid T^{L}_{M} \mid Y^{l}_{m}(\theta, \phi) \rangle = (-1)^{-m'} [(2l'+1)(2L+1)(2l+1)/4\pi]^{1/2}$$

$$\times \begin{pmatrix} l' & L & l \\ -m' & M & m \end{pmatrix} \begin{pmatrix} l' & L & l \\ 0 & 0 & 0 \end{pmatrix}$$

and write this equation in terms of a reduced matrix element.

Appendix 1

# CRYSTAL SYSTEMS

## The Seven Crystal Systems

The seven crystal systems are listed with their properties and the types of Bravais lattices. The symbols for the Bravais lattice are: P, primitive; I, body centered; F, face centered; C (or A or B) for side centered, C if the xy face is centered, A if the yz face is centered, and B if the xz face is centered; R for a rhombohedral cell (it is primitive because there is only one lattice point per R-cell). There is a major confusion about the terms rhombohedral and trigonal which we try to clarify later and a minor confusion in the monoclinic system which we clarify now assumng some knowledge of the 14 Bravais lattices as well as the seven crystal systems.

For the monoclinic Bravais lattice one can choose the 2-fold axis, or unique axis, as the c-axis. This is the principal axis which is the convention in this book and is the usual solid state physics convention. In the stereograms and the international tables which is called the "first setting." In which case the B- or A-face can be centered, and conventionally one picks the B-face which is listed in the table. However, crystallographers tend to call the 2-fold axis the b-axis, therefore the C-face is centered, which is called the "second setting" in stereograms and the international tables.

The "properties" of each crystal system are derived from the basic symmetry condition of each crystal system. We list the basic symmetry condition for each crystal system, using the international notation because this notation displays the idea clearly.

| | | |
|---|---|---|
| trigonal | : | 1-fold or $\bar{1}$-axis |
| monoclinic | : | 2-fold or $\bar{2}$-axis |
| orthorhombic | : | three 2-fold or $\bar{2}$-axes |
| tetragonal | : | 4-fold or $\bar{4}$-axis |
| cubic | : | four 3-fold axes |
| trigonal | : | 3-fold or $\bar{3}$-axis |
| hexagonal | : | 6-fold or $\bar{6}$-axis |

Occasionally some confusion occurs because a crystal structure is reported, for example, with a = b = c and all the angles = $\pi/2$ but the space group is not cubic. This is because a = b = c only within the error of the measurement but from certain conditions on the x-ray reflections one can tell that the crystal does not have four 3-fold axes, rather has some other symmetry.

Now that the basic symmetry conditions are listed for each crystal system we can clarify the trigonal–rhombohedral–hexagonal confusion. The symmetry conditions listed above for the trigonal (3-fold symmetry) and the hexagonal (6-fold symmetry) crystal systems give the exact same conditions on the axes of the cell, namely a = b ≠ c and $\alpha = \beta = \pi/2$ and $\gamma = 2\pi/3$. Thus one might argue that it does not make sense to define two different crystal systems. However, it does make sense because as defined the hexagonal crystal system can not be centered in any way, keeping the 6-fold symmetry, so there is only one hexagonal Bravais lattice which is primitive and labeled with a P. On the other hand the trigonal crystal system can be centered with lattice points at the positions (2/3, 1/3, 2/3) and (1/3, 2/3, 1/3) still keeping the 3-fold symmetry. This cell has three lattice points in it, but a rhombohedral cell with one lattice point can be constructed. Thus, the rhombohedral cell is a primitive cell but is given the label R to remind one that it is rhombohedral. There are other crystal structures that also have a 3-fold or $\bar{3}$-axis and thus are trigonal but are not centered so are not rhombohedral. Thus, all rhombohedral structures are trigonal but not all trigonal structures are rhombohedral. However, some extra confusion is added by calling the axes a = b ≠ c and $\alpha = \beta = \pi/2$ and $\gamma = 2\pi/3$ hexagonal axes no matter what crystal system they are applied to. One more confusion arises from the fact that rhombohedral axes are sometimes not easy to work with. Thus, rhombohedral crystals sometimes are not referred to their primitive cells but referred to their original trigonal unit cells with three lattice points per cell. However, as mentioned above, the axes of these cells are called hexagonal axes. In the international symbol for the space group trigonal systems will always have a 3 or $\bar{3}$ and hexagonal systems will always have a 6 or $\bar{6}$. Chapter 11 should be consulted for more details.

The conventions for directions are: [u,v,w] is one particular direction in space given by $u\mathbf{a} + v\mathbf{b} + w\mathbf{c}$; <u,v,w> is a set of directions in space related by symmetry. Thus, in a cubic system [1,1,1] is the direction $\mathbf{a} + \mathbf{b} + \mathbf{c}$ while <1,1,1> is the eight directions [1,1,1], $[\bar{1},1,1]$, $[1\bar{1}1]$, $[11\bar{1}]$, etc., where $\bar{1}$ is a shorthand way of writing $-1$. A plane is described by the direction perpendicular to the plane. Thus, (u,v,w) is a plane perpendicular to the $u\mathbf{a} + v\mathbf{b} + w\mathbf{c}$ direction. For example, the plane containing the yz-axes is called (1,0,0).

| Crystal system | Properties | Types of Bravais lattices |
|---|---|---|
| Triclinic | $a \neq b \neq c$<br>$\alpha \neq \beta \neq \gamma \neq \pi/2$ | P |
| Monoclinic | $a \neq b \neq c$<br>$\alpha = \beta = \pi/2 \neq \gamma$ | P, B |
| Orthorhombic | $a \neq b \neq c$<br>$\alpha = \beta = \gamma = \pi/2$ | P, I, F, C |
| Tetragonal | $a = b \neq c$<br>$\alpha = \beta = \gamma = \pi/2$ | P, I |
| Trigonal | | |
| Rhombohedral axes | $a = b = c$<br>$\alpha = \beta = \gamma < 2\pi/3 \neq \pi/2$ or $\pi/3$ | R |
| Hexagonal axes | $a = b \neq c$<br>$\alpha = \beta = \pi/2; \gamma = 2\pi/3$ | P |
| Hexagonal | $a = b \neq c$<br>$\alpha = \beta = \pi/2; \gamma = 2\pi/3$ | P |
| Cubic | $a = b = c$<br>$\alpha = \beta = \gamma = \pi/2$ | P, I, F |

## The 14 Bravais Lattices

The 14 Bravais lattices are listed, classified among the seven crystal systems. The Γ notation is used as well as the notation P, I, F, C (A or B), R for the Bravais lattice that we have been using throughout this book. The Γ notation is used by some solid state people. $\mathbf{a}_1$, $\mathbf{a}_2$, and $\mathbf{a}_3$

are the fundamental periods of the lattice (primitive lattice vectors) and for each conventionally multiprimitive Bravais lattice, the primitive lattice vectors are given in terms of a coordinate system that is the coordinate system of the simple primitive lattice of that crystal system. See Chapter 11 for further details.

Triclinic

1. $\Gamma_t$(P). $\mathbf{a}_1, \mathbf{a}_2, \mathbf{a}_3$ are arbitrary.

Monoclinic

2. $\Gamma_m$ (P). $\mathbf{a}_3$ is perpendicular to $\mathbf{a}_1$ and $\mathbf{a}_2$.
3. $\Gamma_m^b$ (B). $\mathbf{a}_1 = (a, 0, 0,)$ $\mathbf{a}_2 = (0, b, 0), \mathbf{a}_3 = (a/2, 0, c/2)$.

Orthorhombic

4. $\Gamma_0$ (P). $\mathbf{a}_1 = (a, 0, 0)$, $\mathbf{a}_2 = (0, b, 0), \mathbf{a}_3 = (0, 0, c)$.
5. $\Gamma_0^b$ (C). $\mathbf{a}_1 = (a/2, b/2, 0)$, $\mathbf{a}_2 = (a/2, -b/2, 0), \mathbf{a}_3 = (0, 0, c)$.
6. $\Gamma_0^v$ (I). $\mathbf{a}_1 = (a/2, b/2, c/2)$, $\mathbf{a}_2 = (a/2, -b/2, c/2), \mathbf{a}_3 = (a/2, b/2, -c/2)$.
7. $\Gamma_0^f$ (F). $\mathbf{a}_1 = (a/2, b/2, 0)$, $\mathbf{a}_2 = (0, b/2, c/2), \mathbf{a}_3 = (a/2, 0, c/2)$.

Tetragonal

8. $\Gamma_q$ (P). $\mathbf{a}_1 = (a, 0, 0)$, $\mathbf{a}_2 = (0, a, 0), \mathbf{a}_3 = (0, 0, b)$.
9. $\Gamma_q^v$ (I). $\mathbf{a}_1 = (a/2, a/2, b/2)$, $\mathbf{a}_2 = (a/2, -a/2, b/2), \mathbf{a}_3 = (a/2, a/2, -b/2)$.

Cubic

10. $\Gamma_c$ (P). $\mathbf{a}_1 = (a, 0, 0)$, $\mathbf{a}_2 = (0, a, 0), \mathbf{a}_3 = (0, 0, a)$.
11. $\Gamma_c^v$ (I). $\mathbf{a}_1 = \frac{1}{2} a (1, 1, 1)$, $\mathbf{a}_2 = \frac{1}{2} a (1, -1, 1), \mathbf{a}_3 = \frac{1}{2} a (1, 1, -1)$.
12. $\Gamma_c^f$ (F). $\mathbf{a}_1 = \frac{1}{2} a (1, 1, 0)$, $\mathbf{a}_2 = \frac{1}{2} a (0, 1, 1), \mathbf{a}_3 = \frac{1}{2} a (1, 0, 1)$.

Rhombohedral (Trigonal)

13. $\Gamma_{rh}$ (R). $\mathbf{a}_1 = (2a/3, a/3, c/3)$, $\mathbf{a}_2 = (-a/3, a/3, c/3), \mathbf{a}_3 = (-a/3, -2a/3, c/3)$. $\mathbf{a}_1, \mathbf{a}_2, \mathbf{a}_3$ are of equal length.

Hexagonal

14. $\Gamma_h$ (P). $\mathbf{a}_1 = (a, 0, 0)$, $\mathbf{a}_2 = (0, a, 0), \mathbf{a}_3 = (0, 0, c)$. The angle between $\mathbf{a}_2$ and $\mathbf{a}_3$ is 120°.

## Stereograms of the 32 Point Groups

Each point group is described with the Schoenflies as well as international (Hermann–Mauguin) notation. The point groups are grouped according to the seven crystal systems. The monoclinic point groups are listed with the c-axis as the principal axis (first setting) and also with the b-axis as the principal axis (second setting). See Chapter 1 for further details. The following diagrams are from "The International Tables for X-Ray Crystallography."

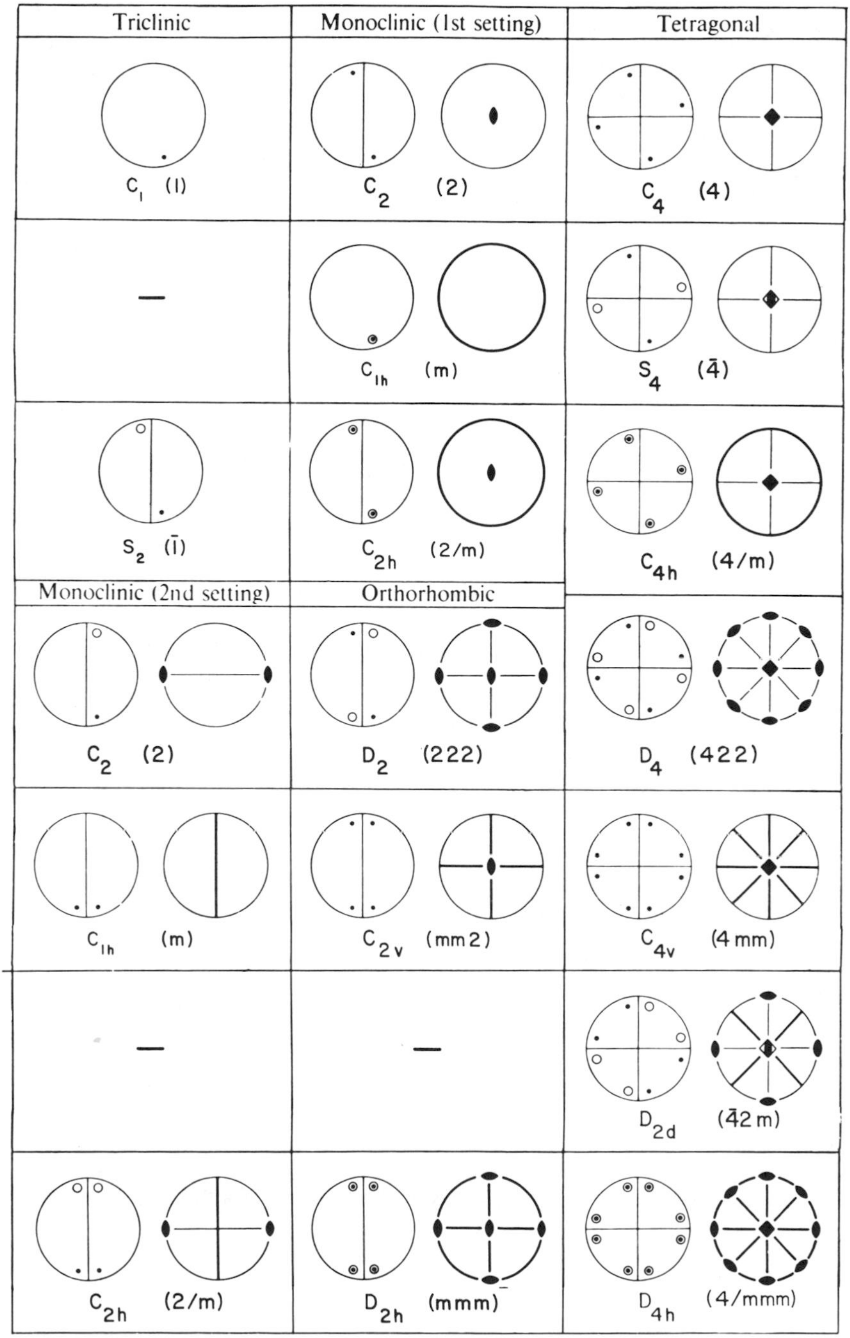

| Triclinic | Monoclinic (1st setting) | Tetragonal |
|---|---|---|
| $C_1$  (1) | $C_2$   (2) | $C_4$   (4) |
| — | $C_{1h}$   (m) | $S_4$   ($\bar{4}$) |
| $S_2$  ($\bar{1}$) | $C_{2h}$   (2/m) | $C_{4h}$   (4/m) |

| Monoclinic (2nd setting) | Orthorhombic | |
|---|---|---|
| $C_2$   (2) | $D_2$   (222) | $D_4$   (422) |
| $C_{1h}$   (m) | $C_{2v}$   (mm2) | $C_{4v}$   (4mm) |
| — | — | $D_{2d}$   ($\bar{4}$2m) |
| $C_{2h}$   (2/m) | $D_{2h}$   (mmm) | $D_{4h}$   (4/mmm) |

| Trigonal | Hexagonal | Cubic |
|---|---|---|
| 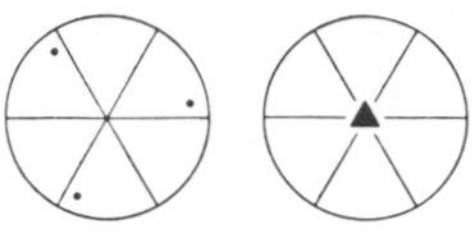 $C_3$ (3) | 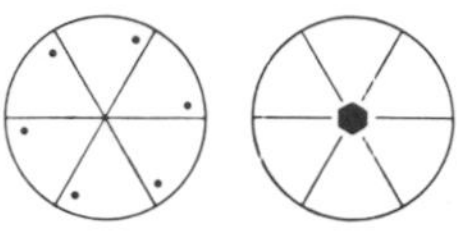 $C_6$ (6) | 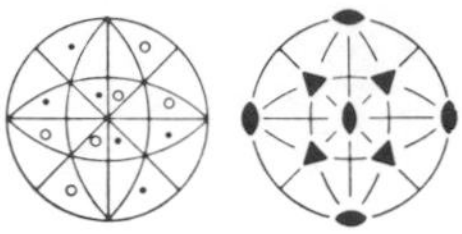 T (23) |
| — | 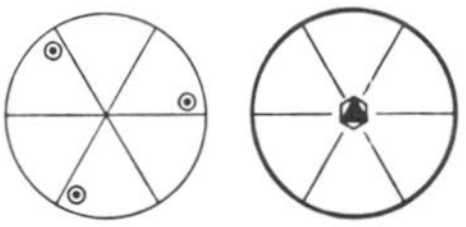 $C_{3h}$ ($\bar{6}$) | — |
| 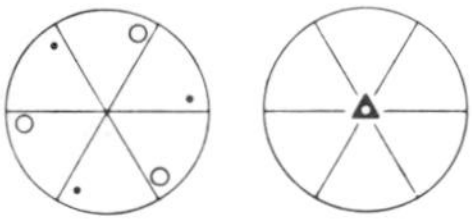 $S_6$ ($\bar{3}$) | 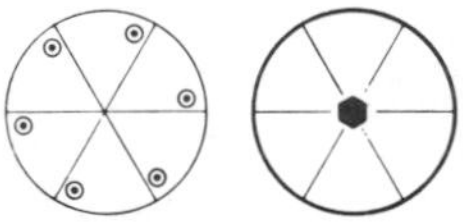 $C_{6h}$ (6/m) | 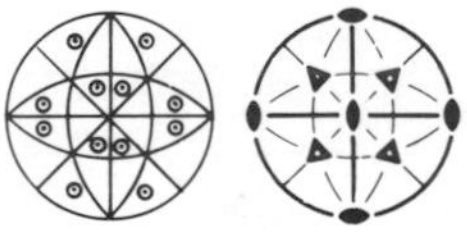 $T_h$ (m3) |
| 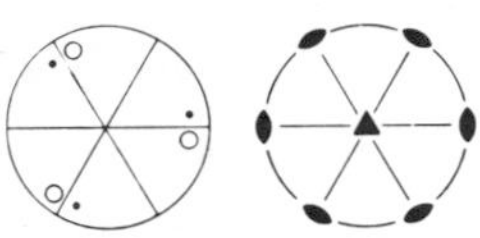 $D_3$ (32) | 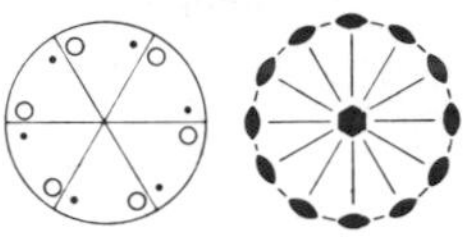 $D_6$ (622) | 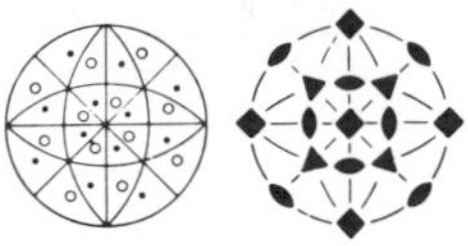 O (432) |
| 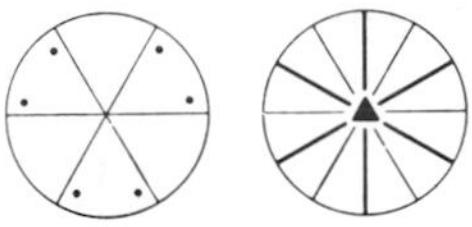 $C_{3v}$ (3m) | 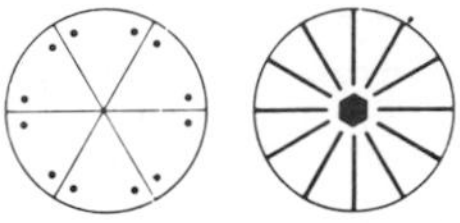 $C_{6v}$ (6mm) | — |
| — | 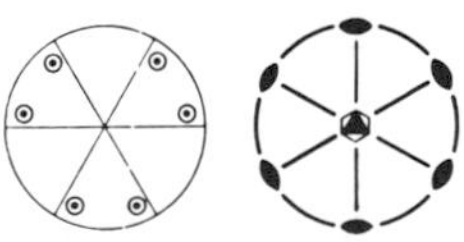 $D_{3h}$ ($\bar{6}$m2) | 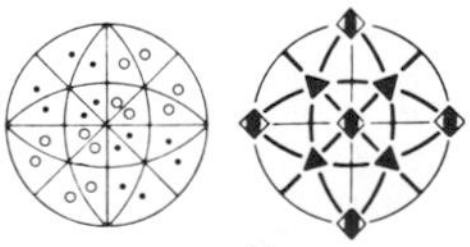 $T_d$ ($\bar{4}$3m) |
| 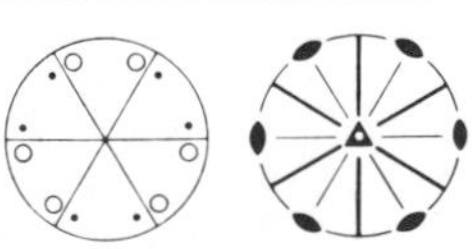 $D_{3d}$ ($\bar{3}$m) | 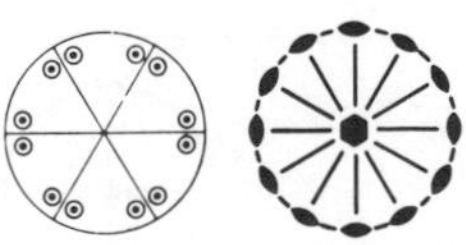 $D_{6h}$ (6/mmm) | 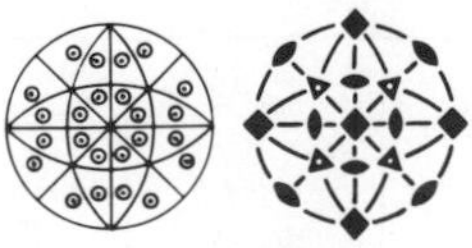 $O_h$ (m3m) |

Appendix 2

# THE 32 POINT GROUPS

This is a list of the 32 crystallographic point groups. The first three columns show a symbol for each point group in three notations: Schoenflies notation which is used throughout this book; the international (or Hermann–Mauguin) notation; the full international notation (the symmetry operations in the full symbol can be derived from those in the short symbol). For a discussion of the Schoenflies notation see Chapter 1; for the international notation see Chapter 11. The next column lists all of the symmetry operations, grouped by classes, for each point group. The generating elements for each point group are listed next (all of the symmetry operations of the point group can be obtained by various products of these generating elements where E is understood to be included). See Chapters 1 and 4 for further discussion. The last column shows the numbers of the space groups that have each of the point groups shown.

| Schoenflies | International | Full Int. | Symmetry Elements | Generating Elements | Space Groups |
|---|---|---|---|---|---|
| Triclinic | | | | | |
| $C_1$ | 1 | 1 | E | E | 1 |
| $S_2$ $(C_i)$ | $\bar{1}$ | $\bar{1}$ | E i | i | 2 |
| Monoclinic | | | | | |
| $C_2$ | 2 | 2 | E $C_2$ | $C_2$ | 3-5 |
| $C_{1h}$ $(C_s)$ | m | m | E $\sigma_h$ | $\sigma_h$ | 6-9 |
| $C_{2h}$ | 2/m | $\frac{2}{m}$ | E $C_2$ i $\sigma_h$ | i $C_2$ | 10-15 |
| Orthorhombic | | | | | |
| $D_2$ (V) | 222 | 222 | E $C_2$ $C_2'$ $C_2'$ | $C_2$ $C_2^y$ | 16-24 |
| $C_{2v}$ | mm2 | mm2 | E $C_2$ $\sigma_v$ $\sigma_v$ | $C_2$ $\sigma_v^y$ | 25-46 |
| $D_{2h}$ $(V_h)$ | mmm | $\frac{222}{mmm}$ | E $C_2$ $C_2'$ $C_2'$ i $\sigma_h$ $\sigma_v$ $\sigma_v$ | i $C_v^y$ $C_2$ | 47-74 |
| Tetragonal | | | | | |
| $C_4$ | 4 | 4 | E $2C_4$ $C_2$ | $C_4$ | 75-80 |
| $S_4$ | $\bar{4}$ | $\bar{4}$ | E $2S_4$ $C_2$ | $S_4^3$ | 81-82 |
| $C_{4h}$ | 4/m | $\frac{4}{m}$ | E $2C_4$ $C_2$ i $2S_4$ $\sigma_h$ | i $C_4$ | 83-88 |
| $D_4$ | 422 | 422 | E $2C_4$ $C_2$ $2C_2'$ $2C_2''$ | $C_2^y$ $C_4$ | 89-98 |
| $C_{4v}$ | 4mm | 4mm | E $2C_4$ $C_2$ $2\sigma_v$ $2\sigma_d$ | $\sigma_v^y$ $C_4$ | 99-110 |
| $D_{2d}(V_d)$ | $\bar{4}2m$ | $\bar{4}2m$ | E $C_2$ $2C_2'$ $2\sigma_d$ $2S_4$ | $C_2^y$ $S_4^3$ | 111-122 |
| $D_{4h}$ | 4/mmm | $\frac{422}{mmm}$ | E $2C_4$ $C_2$ $2C_2'$ $2C_2''$ i $2S_4$ $\sigma_h$ $2\sigma_v$ $2\sigma_d$ | i $C_2^y$ $C_4$ | 123-142 |
| Trigonal (Rhombohedral) | | | | | |
| $C_3$ | 3 | 3 | E $2C_3$ | $C_3$ | 143-146 |
| $S_6(C_{3i})$ | $\bar{3}$ | $\bar{3}$ | E $2C_3$ i $2S_6$ | i $C_3$ | 147-148 |
| $D_3$ | 32 | 32 | E $2C_3$ $3C_2$ | $C_2^y$ $C_3$ | 149-155 |
| $C_{3v}$ | 3m | 3m | E $2C_3$ $3\sigma_v$ | $\sigma_v^y$ $C_3$ | 156-161 |
| $D_{3d}$ | $\bar{3}m$ | $\bar{3}\frac{2}{m}$ | E $2C_3$ $3C_2$ i $2S_6$ $3\sigma_v$ | i $C_2^y$ $C_3$ | 162-167 |
| Hexagonal | | | | | |
| $C_6$ | 6 | 6 | E $2C_6$ $2C_3$ $C_2$ | $C_2$ $C_3$ | 168-173 |
| $C_{3h}$ | $\bar{6}$ | $\bar{6}$ | E $2C_3$ $\sigma_h$ $2S_3$ | $\sigma_h$ $C_3$ | 174 |
| $C_{6h}$ | 6/m | $\frac{6}{m}$ | E $2C_6$ $2C_3$ $C_2$ i $2S_3$ $2S_6$ $\sigma_h$ | i $C_2$ $C_3$ | 175-176 |
| $D_6$ | 622 | 622 | E $2C_6$ $2C_3$ $C_2$ $3C_2'$ $3C_2''$ | $C_2$ $C_2^y$ $C_3$ | 177-182 |
| $C_{6v}$ | 6mm | 6mm | E $2C_6$ $2C_3$ $C_2$ $3\sigma_v$ $3\sigma_d$ | $C_2$ $\sigma_v^y$ $C_3$ | 183-186 |
| $D_{3h}$ | $\bar{6}m2$ | $\bar{6}m2$ | E $2C_3$ $3C_2$ $\sigma_h$ $2S_3$ $3\sigma_v$ | $C_2^y$ $\sigma_h$ $C_3$ | 187-190 |
| $D_{6h}$ | 6/mmm | $\frac{622}{mmm}$ | E $2C_6$ $2C_3$ $C_2$ $3C_2'$ $3C_2''$ i $2S_3$ $2S_6$ $\sigma_h$ $3\sigma_v$ $3\sigma_d$ | i $C_2^y$ $C_2$ $C_3$ | 191-194 |
| Cubic | | | | | |
| T | 23 | 23 | E $8C_3$ $3C_2$ | $C_2$ $C_3$[111] | 195-199 |
| $T_h$ | m3 | $\frac{2}{m}\bar{3}$ | E $8C_3$ $3C_2$ i $8S_6$ $3\sigma_h$ | i $C_2$ $C_3$[111] | 200-206 |
| O | 432 | 432 | E $8C_3$ $3C_2$ $6C_2$ $6C_4$ | $C_4$ $C_3$[111] | 207-214 |
| $T_d$ | $\bar{4}3m$ | $\bar{4}3m$ | E $8C_3$ $3C_2$ $6\sigma_d$ $6S_4$ | $S_4^3$ $C_3$[111] | 215-220 |
| $O_h$ | m3m | $\frac{4}{m}\bar{3}\frac{2}{m}$ | E $8C_3$ $3C_2$ $6C_2$ $6C_4$ i $8S_6$ $3\sigma_h$ $6\sigma_d$ $6S_4$ | i $C_4$ $C_3$[111] | 221-230 |

Appendix 3

# CHARACTER TABLES

## Character Tables for a Number of Point Groups

Character tables are given for a number of point groups. Included among these tables are those for the 32 crystallographic point groups. These can be found by the symbol for the point group, in the international notation, in square brackets on the right. In general, the Schoenflies notation is used for the point group and the symmetry operation (see Chapter 1) and the Mulliken or chemical notation is used for the irreducible representations. The 1-dimensional irreducible representations are labeled with the letters A or B, the 2-dimensional ones with E, the 3-dimensional ones with T (F is used by some authors). The subscripts g and u occur for point groups with a center of inversion as a symmetry operation. If the irreducible representation is symmetric (transforms into +1 times itself) under the inversion operations, then it is g (gerade); if it is antisymmetric, then it is u (ungerade). See Section 4-2. See Wilson, Decius, and Cross, Appendix 10, for an explanation of the number subscripts.

## Irreducible Representation in Other Notations

We list the symbols used for irreducible representations for the point groups $O_h$ and $D_{4h}$ in the Mulliken, Bethe, and Bouckaert–Smoluchowski–Wigner (BSW) notation. Only the first two notations are used in this book, see Chapter 11 for further discussion. However, the BSW notation is often used in solid state physics works. The extension of the BSW notation for various special points and lines in the Brillouin zone can be found in Koster's article given in the bibliography. (The BSW notation is slightly more complicated than implied in this table because a special point or line can have $D_{4h}$ symmetry but if the point or line is labeled by a different letter, this different letter will be used in place of M.

For example, in the simple cubic Brillouin zone the special points M and X both have $D_{4h}$ symmetry. Thus, the irreducible representations at X will be given by $X_1$, $X_2$,..., $X_5'$ in the same manner as $M_1$,..., $M_5'$ in the table.)

## Non–Axial Groups

| $C_1$ | E | [1] |
|---|---|---|
| A | 1 | |

| $S_2$ | E | i | | [$\bar{1}$] |
|---|---|---|---|---|
| $A_g$ | 1 | 1 | $R_x$, $R_y$, $R_z$ | $\{x^2, y^2, z^2$; xy, xz, yz$\}$ |
| $A_u$ | 1 | -1 | x, y, z | |

| $C_{1h}$ | E | $\sigma_h$ | | [m] |
|---|---|---|---|---|
| A' | 1 | 1 | x, y, $R_z$ | $\{x^2, y^2$; $z^2$, xy$\}$ |
| A" | 1 | -1 | z, $R_x$, $R_y$ | yz, xz |

## $C_n$ Groups

| $C_2$ | E | $C_2$ | | [2] |
|---|---|---|---|---|
| A | 1 | 1 | z, $R_z$ | $x^2$, $y^2$, $z^2$, xy |
| B | 1 | -1 | x, y, $R_x$, $R_y$ | yz, xz |

| $C_3$ | E | $C_3$ | $C_3^2$ | $\epsilon = \exp(2\pi i/3)$ | [3] |
|---|---|---|---|---|---|
| A | 1 | 1 | 1 | z, $R_z$ | $x^2+y^2$, $z^2$ |
| E $\{$ | 1 | $\epsilon$ | $\epsilon^*$ | x+iy, $R_x$+i$R_y$ $\}$ | $(x^2-y^2$, xy) |
| | 1 | $\epsilon^*$ | $\epsilon$ | x-iy, $R_x$-i$R_y$ | (yz, xz) |

| $C_4$ | E | $C_4$ | $C_2$ | $C_4^3$ | | [4] |
|---|---|---|---|---|---|---|
| A | 1 | 1 | 1 | 1 | z, $R_z$ | $x^2+y^2$, $z^2$ |
| B | 1 | -1 | 1 | -1 | | $x^2-y^2$, xy |
| E $\{$ | 1 | i | -1 | -i | x+iy, $R_x$+i$R_y$ $\}$ | (yz, xz) |
| | 1 | -i | -1 | i | x-iy, $R_x$-i$R_y$ | |

| $C_5$ | E | $C_5$ | $C_5^2$ | $C_5^3$ | $C_5^4$ | $\varepsilon = \exp(2\pi i/5)$ | |
|---|---|---|---|---|---|---|---|
| A | 1 | 1 | 1 | 1 | 1 | z, $R_z$ | $x^2+y^2$, $z^2$ |
| $E_1$ | 1 | $\varepsilon$ | $\varepsilon^2$ | $\varepsilon^{2*}$ | $\varepsilon^*$ | x+iy, $R_x+iR_y$ | (yz, xz) |
| | 1 | $\varepsilon^*$ | $\varepsilon^{2*}$ | $\varepsilon^*$ | $\varepsilon$ | x-iy, $R_x-iR_y$ | |
| $E_2$ | 1 | $\varepsilon^2$ | $\varepsilon^*$ | $\varepsilon$ | $\varepsilon^{2*}$ | | $(x^2-y^2$, xy) |
| | 1 | $\varepsilon^{2*}$ | $\varepsilon$ | $\varepsilon^*$ | $\varepsilon^2$ | | |

| $C_6$ | E | $C_6$ | $C_3$ | $C_2$ | $C_3^2$ | $C_6^5$ | $\varepsilon = \exp(2\pi i/6)$ | [6] |
|---|---|---|---|---|---|---|---|---|
| A | 1 | 1 | 1 | 1 | 1 | 1 | z, $R_z$ | $x^2+y^2$, $z^2$ |
| B | 1 | -1 | 1 | -1 | 1 | -1 | | |
| $E_1$ | 1 | $\varepsilon$ | $-\varepsilon^*$ | -1 | $-\varepsilon$ | $\varepsilon^*$ | x+iy, $R_z+iR_y$ | (xz, yz) |
| | 1 | $\varepsilon^*$ | $-\varepsilon$ | -1 | $-\varepsilon^*$ | $\varepsilon$ | x-iy, $R_x-iR_y$ | |
| $E_2$ | 1 | $-\varepsilon^*$ | $-\varepsilon$ | 1 | $-\varepsilon^*$ | $-\varepsilon$ | | $(x^2-y^2$, xy) |
| | 1 | $-\varepsilon$ | $-\varepsilon^*$ | 1 | $-\varepsilon$ | $-\varepsilon^*$ | | |

| $C_7$ | E | $C_7$ | $C_7^2$ | $C_7^3$ | $C_7^4$ | $C_7^5$ | $C_7^6$ | $\varepsilon = \exp(2\pi i/7)$ | |
|---|---|---|---|---|---|---|---|---|---|
| A | 1 | 1 | 1 | 1 | 1 | 1 | 1 | z, $R_z$ | $x^2+y^2$, $z^2$ |
| $E_1$ | 1 | $\varepsilon$ | $\varepsilon^2$ | $\varepsilon^3$ | $\varepsilon^{3*}$ | $\varepsilon^{2*}$ | $\varepsilon^*$ | x+iy, $R_x+iR_y$ | (xz, yz) |
| | 1 | $\varepsilon^*$ | $\varepsilon^{2*}$ | $\varepsilon^{3*}$ | $\varepsilon^3$ | $\varepsilon^2$ | $\varepsilon$ | x-iy, $R_x-iR_y$ | |
| $E_2$ | 1 | $\varepsilon^2$ | $\varepsilon^{3*}$ | $\varepsilon^*$ | $\varepsilon$ | $\varepsilon^3$ | $\varepsilon^{2*}$ | | $(x^2-y^2$, xy) |
| | 1 | $\varepsilon^{2*}$ | $\varepsilon^3$ | $\varepsilon$ | $\varepsilon^*$ | $\varepsilon^{3*}$ | $\varepsilon^2$ | | |
| $E_3$ | 1 | $\varepsilon^3$ | $\varepsilon^*$ | $\varepsilon^2$ | $\varepsilon^{2*}$ | $\varepsilon$ | $\varepsilon^{3*}$ | | |
| | 1 | $\varepsilon^{3*}$ | $\varepsilon$ | $\varepsilon^{2*}$ | $\varepsilon^2$ | $\varepsilon^*$ | $\varepsilon^3$ | | |

| $C_8$ | E | $C_8$ | $C_4$ | $C_8^3$ | $C_2$ | $C_8^5$ | $C_4^3$ | $C_8^7$ | $\varepsilon = \exp(2\pi i/8)$ | |
|---|---|---|---|---|---|---|---|---|---|---|
| A | 1 | 1 | 1 | 1 | 1 | 1 | 1 | 1 | z, $R_z$ | $x^2+y^2$, $z^2$ |
| B | 1 | -1 | 1 | -1 | 1 | -1 | 1 | -1 | | |
| $E_1$ | 1 | $\varepsilon$ | i | $-\varepsilon^*$ | -1 | $-\varepsilon$ | -i | $\varepsilon^*$ | x+iy, $R_x+iR_y$ | (xz, yz) |
| | 1 | $\varepsilon^*$ | -i | $-\varepsilon$ | -1 | $-\varepsilon^*$ | i | $\varepsilon$ | x-iy, $R_x-iR_y$ | |
| $E_2$ | 1 | i | -1 | -i | 1 | i | -1 | -i | | $(x^2-y^2$, xy) |
| | 1 | -i | -1 | i | 1 | -i | -1 | i | | |
| $E_3$ | 1 | $-\varepsilon$ | i | $\varepsilon^*$ | -1 | $\varepsilon$ | -i | $-\varepsilon^*$ | | |
| | 1 | $-\varepsilon^*$ | -i | $\varepsilon$ | -1 | $\varepsilon^*$ | i | $-\varepsilon$ | | |

## $C_{nv}$ Groups

| $C_{2v}$ | E | $C_2$ | $\sigma_v(xz)$ | $\sigma'_v(yz)$ | | [mm2] |
|---|---|---|---|---|---|---|
| $A_1$ | 1 | 1 | 1 | 1 | z | $x^2, y^2, z^2$ |
| $A_2$ | 1 | 1 | -1 | -1 | $R_z$ | xy |
| $B_1$ | 1 | -1 | 1 | -1 | x, $R_y$ | xz |
| $B_2$ | 1 | -1 | -1 | 1 | y, $R_x$ | yz |

| $C_{3v}$ | E | $2C_3$ | $3\sigma_v$ | | [3m] |
|---|---|---|---|---|---|
| $A_1$ | 1 | 1 | 1 | z | $x^2+y^2, z^2$ |
| $A_2$ | 1 | 1 | -1 | $R_z$ | |
| E | 2 | -1 | 0 | (x, y) ($R_x$, $R_y$) | ($x^2-y^2$, xy) (xz, yz) |

| $C_{4v}$ | E | $2C_4$ | $C_2$ | $2\sigma_v$ | $2\sigma_d$ | | [4mm] |
|---|---|---|---|---|---|---|---|
| $A_1$ | 1 | 1 | 1 | 1 | 1 | z | $x^2+y^2, z^2$ |
| $A_2$ | 1 | 1 | 1 | -1 | -1 | $R_z$ | |
| $B_1$ | 1 | -1 | 1 | 1 | -1 | | $x^2-y^2$ |
| $B_2$ | 1 | -1 | 1 | -1 | 1 | | xy |
| E | 2 | 0 | -2 | 0 | 0 | (x, y) ($R_x$, $R_y$) | (xz, yz) |

| $C_{5v}$ | E | $2C_5$ | $2C_5^2$ | $5\sigma_v$ | $\alpha=2\cos 72°$ | $\beta=2\cos 144°$ |
|---|---|---|---|---|---|---|
| $A_1$ | 1 | 1 | 1 | 1 | z | $x^2+y^2, z^2$ |
| $A_2$ | 1 | 1 | 1 | -1 | $R_z$ | |
| $E_1$ | 2 | $\alpha$ | $\beta$ | 0 | (x, y) ($R_x$, $R_y$) | (xz, yz) |
| $E_2$ | 2 | $\beta$ | $\alpha$ | 0 | | ($x^2-y^2$, xy) |

| $C_{6v}$ | E | $2C_6$ | $2C_3$ | $C_2$ | $3\sigma_v$ | $3\sigma_d$ | | [6mm] |
|---|---|---|---|---|---|---|---|---|
| $A_1$ | 1 | 1 | 1 | 1 | 1 | 1 | z | $x^2+y^2, z^2$ |
| $A_2$ | 1 | 1 | 1 | 1 | -1 | -1 | $R_z$ | |
| $B_1$ | 1 | -1 | 1 | -1 | 1 | -1 | | |
| $B_2$ | 1 | -1 | 1 | -1 | -1 | 1 | | |
| $E_1$ | 2 | 1 | -1 | -2 | 0 | 0 | (x, y) ($R_x$, $R_y$) | (xz, yz) |
| $E_2$ | 2 | -1 | -1 | 2 | 0 | 0 | | ($x^2-y^2$, xy) |

## $C_{nh}$ Groups

| $C_{2h}$ | E | $C_2$ | i | $\sigma_h$ | | [2/m] |
|---|---|---|---|---|---|---|
| $A_g$ | 1 | 1 | 1 | 1 | $R_z$ | $x^2, y^2, z^2$, xy |
| $B_g$ | 1 | -1 | 1 | -1 | $R_x$, $R_y$ | xz, yz |
| $A_u$ | 1 | 1 | -1 | -1 | z | |
| $B_u$ | 1 | -1 | -1 | 1 | x, y | |

| $C_{3h}$ | E | $C_3$ | $C_3^2$ | $\sigma_h$ | $S_3$ | $S_3^5$ | $\varepsilon = \exp(2\pi i/3)$ | [$\bar{6}$] |
|---|---|---|---|---|---|---|---|---|
| A' | 1 | 1 | 1 | 1 | 1 | 1 | $R_z$ | $x^2+y^2$, $z^2$ |
| E' | 1 | $\varepsilon$ | $\varepsilon^*$ | 1 | $\varepsilon$ | $\varepsilon^*$ | x+iy | $(x^2-y^2, xy)$ |
| | 1 | $\varepsilon^*$ | $\varepsilon$ | 1 | $\varepsilon^*$ | $\varepsilon$ | x-iy | |
| A" | 1 | 1 | 1 | -1 | -1 | -1 | z | |
| E" | 1 | $\varepsilon$ | $\varepsilon^*$ | -1 | $-\varepsilon$ | $-\varepsilon^*$ | $R_x+iR_y$ | (xz, yz) |
| | 1 | $\varepsilon^*$ | $\varepsilon$ | -1 | $-\varepsilon^*$ | $-\varepsilon$ | $R_x-iR_y$ | |

| $C_{4h}$ | E | $C_4$ | $C_2$ | $C_4^3$ | i | $S_4^3$ | $\sigma_h$ | $S_4$ | | [4/m] |
|---|---|---|---|---|---|---|---|---|---|---|
| $A_g$ | 1 | 1 | 1 | 1 | 1 | 1 | 1 | 1 | $R_z$ | $x^2+y^2$, $z^2$ |
| $B_g$ | 1 | -1 | 1 | -1 | 1 | -1 | 1 | -1 | | $x^2-y^2$, xy |
| $E_g$ | 1 | i | -1 | -i | 1 | i | -1 | -i | $R_x+iR_y$ | (xz, yz) |
| | 1 | -i | -1 | i | 1 | -i | -1 | i | $R_x-iR_y$ | |
| $A_u$ | 1 | 1 | 1 | 1 | -1 | -1 | -1 | -1 | z | |
| $B_u$ | 1 | -1 | 1 | -1 | -1 | 1 | -1 | 1 | | |
| $E_u$ | 1 | i | -1 | -i | -1 | -i | 1 | i | x+iy | |
| | 1 | -i | -1 | i | -1 | i | 1 | -i | x-iy | |

# $D_n$ Groups

| $D_2$ | E | $C_2$ | $C_2(y)$ | $C_2(x)$ | | [222] |
|---|---|---|---|---|---|---|
| A | 1 | 1 | 1 | 1 | | $x^2$, $y^2$, $z^2$ |
| $B_1$ | 1 | 1 | -1 | -1 | z, $R_z$ | xy |
| $B_2$ | 1 | -1 | 1 | -1 | y, $R_y$ | xz |
| $B_3$ | 1 | -1 | -1 | 1 | x, $R_x$ | yz |

| $D_3$ | E | $2C_3$ | $3C_2'$ | | [32] |
|---|---|---|---|---|---|
| $A_1$ | 1 | 1 | 1 | | $x^2+y^2$, $z^2$ |
| $A_2$ | 1 | 1 | -1 | z, $R_z$ | |
| E | 2 | -1 | 0 | (x, y) $(R_x, R_y)$ | $(x^2-y^2, xy)$ (xz, yz) |

| $D_4$ | E | $2C_4$ | $C_2(\equiv C_4^2)$ | $2C_2'$ | $2C_2''$ | | [422] |
|---|---|---|---|---|---|---|---|
| $A_1$ | 1 | 1 | 1 | 1 | 1 | | $x^2+y^2$, $z^2$ |
| $A_2$ | 1 | 1 | 1 | -1 | -1 | z, $R_z$ | |
| $B_1$ | 1 | -1 | 1 | 1 | -1 | | $x^2-y^2$ |
| $B_2$ | 1 | -1 | 1 | -1 | 1 | | xy |
| E | 2 | 0 | -2 | 0 | 0 | (x, y) $(R_x, R_y)$ | (xz, yz) |

| $D_5$ | $E$ | $2C_5$ | $2C_5^2$ | $5C_2'$ | $\alpha = 2\cos 72°$ | $\beta = 2\cos 144°$ |
|---|---|---|---|---|---|---|
| $A_1$ | 1 | 1 | 1 | 1 | | $x^2+y^2, z^2$ |
| $A_2$ | 1 | 1 | 1 | -1 | $z, R_z$ | |
| $E_1$ | 2 | $\alpha$ | $\beta$ | 0 | $(x, y)$ $(R_x, R_y)$ | $(xz, yz)$ |
| $E_2$ | 2 | $\beta$ | $\alpha$ | 0 | | $(x^2-y^2, xy)$ |

| $D_6$ | $E$ | $2C_6$ | $2C_3$ | $C_2$ | $3C_2'$ | $3C_2''$ | | [622] |
|---|---|---|---|---|---|---|---|---|
| $A_1$ | 1 | 1 | 1 | 1 | 1 | 1 | | $x^2+y^2, z^2$ |
| $A_2$ | 1 | 1 | 1 | 1 | -1 | -1 | $z, R_z$ | |
| $B_1$ | 1 | -1 | 1 | -1 | 1 | -1 | | |
| $B_2$ | 1 | -1 | 1 | -1 | -1 | 1 | | |
| $E_1$ | 2 | 1 | -1 | -2 | 0 | 0 | $(x, y)$ $(R_x, R_y)$ | $(xz, yz)$ |
| $E_2$ | 2 | -1 | -1 | 2 | 0 | 0 | | $(x^2-y^2, xy)$ |

# $D_{nd}$ Groups

| $D_{2d}$ | $E$ | $2S_4$ | $C_2$ | $2C_2'$ | $2\sigma_d$ | | $[\bar{4}2m]$ |
|---|---|---|---|---|---|---|---|
| $A_1$ | 1 | 1 | 1 | 1 | 1 | | $x^2+y^2, z^2$ |
| $A_2$ | 1 | 1 | 1 | -1 | -1 | $R_z$ | |
| $B_1$ | 1 | -1 | 1 | 1 | -1 | | $x^2-y^2$ |
| $B_2$ | 1 | -1 | 1 | -1 | 1 | $z$ | $xy$ |
| $E$ | 2 | 0 | -2 | 0 | 0 | $(x, y)$ $(R_x, R_y)$ | $(xz, yz)$ |

| $D_{3d}$ | $E$ | $2C_3$ | $3C_2'$ | $i$ | $2S_6$ | $3\sigma_d$ | | $[\bar{3}m]$ |
|---|---|---|---|---|---|---|---|---|
| $A_{1g}$ | 1 | 1 | 1 | 1 | 1 | 1 | | $x^2+y^2, z^2$ |
| $A_{2g}$ | 1 | 1 | -1 | 1 | 1 | -1 | $R_z$ | |
| $E_g$ | 2 | -1 | 0 | 2 | -1 | 0 | $(R_x, R_y)$ | $(x^2-y^2, xy)$ $(xz, yz)$ |
| $A_{1u}$ | 1 | 1 | 1 | -1 | -1 | -1 | | |
| $A_{2u}$ | 1 | 1 | -1 | -1 | -1 | 1 | $z$ | |
| $E_u$ | 2 | -1 | 0 | -2 | 1 | 0 | $(x, y)$ | |

| $D_{4d}$ | $E$ | $2S_8$ | $2C_4$ | $2S_8^3$ | $C_2$ | $4C_2'$ | $4\sigma_d$ | | |
|---|---|---|---|---|---|---|---|---|---|
| $A_1$ | 1 | 1 | 1 | 1 | 1 | 1 | 1 | | $x^2+y^2, z^2$ |
| $A_2$ | 1 | 1 | 1 | 1 | 1 | -1 | -1 | $R_z$ | |
| $B_1$ | 1 | -1 | 1 | -1 | 1 | 1 | -1 | | |
| $B_2$ | 1 | -1 | 1 | -1 | 1 | -1 | 1 | $z$ | |
| $E_1$ | 2 | $\sqrt{2}$ | 0 | $-\sqrt{2}$ | -2 | 0 | 0 | $(x, y)$ | |
| $E_2$ | 2 | 0 | -2 | 0 | 2 | 0 | 0 | | $(x^2-y^2, xy)$ |
| $E_3$ | 2 | $-\sqrt{2}$ | 0 | $\sqrt{2}$ | -2 | 0 | 0 | $(R_x, R_y)$ | $(xz, yz)$ |

$\alpha = 2\cos 72°$ $\beta = 2\cos 144°$

| $D_{5d}$ | $E$ | $2C_5$ | $2C_5^2$ | $5C_2'$ | $i$ | $2S_{10}^3$ | $2S_{10}$ | $5\sigma_d$ | | |
|---|---|---|---|---|---|---|---|---|---|---|
| $A_{1g}$ | 1 | 1 | 1 | 1 | 1 | 1 | 1 | 1 | | $x^2+y^2, z^2$ |
| $A_{2g}$ | 1 | 1 | 1 | -1 | 1 | 1 | 1 | -1 | $R_z$ | |
| $E_{1g}$ | 2 | $\alpha$ | $\beta$ | 0 | 2 | $\alpha$ | $\beta$ | 0 | $(R_x, R_y)$ | $(xz, yz)$ |
| $E_{2g}$ | 2 | $\beta$ | $\alpha$ | 0 | 2 | $\beta$ | $\alpha$ | 0 | | $(x^2-y^2, xy)$ |
| $A_{1u}$ | 1 | 1 | 1 | 1 | -1 | -1 | -1 | -1 | | |
| $A_{2u}$ | 1 | 1 | 1 | -1 | -1 | -1 | -1 | 1 | $z$ | |
| $E_{1u}$ | 2 | $\alpha$ | $\beta$ | 0 | -2 | $-\alpha$ | $-\beta$ | 0 | $(x, y)$ | |
| $E_{2u}$ | 2 | $\beta$ | $\alpha$ | 0 | -2 | $-\beta$ | $-\alpha$ | 0 | | |

| $D_{6d}$ | $E$ | $2S_{12}$ | $2C_6$ | $2S_4$ | $2C_3$ | $2S_{12}^5$ | $C_2$ | $6C_2'$ | $6\sigma_d$ | | |
|---|---|---|---|---|---|---|---|---|---|---|---|
| $A_1$ | 1 | 1 | 1 | 1 | 1 | 1 | 1 | 1 | 1 | | $x^2+y^2, z^2$ |
| $A_2$ | 1 | 1 | 1 | 1 | 1 | 1 | 1 | -1 | -1 | $R_z$ | |
| $B_1$ | 1 | -1 | 1 | -1 | 1 | -1 | 1 | 1 | -1 | | |
| $B_2$ | 1 | -1 | 1 | -1 | 1 | -1 | 1 | -1 | 1 | $z$ | |
| $E_1$ | 2 | $\sqrt{3}$ | 1 | 0 | -1 | $-\sqrt{3}$ | -2 | 0 | 0 | $(x, y)$ | |
| $E_2$ | 2 | 1 | -1 | -2 | -1 | 1 | 2 | 0 | 0 | | $(x^2-y^2, xy)$ |
| $E_3$ | 2 | 0 | -2 | 0 | 2 | 0 | -2 | 0 | 0 | | |
| $E_4$ | 2 | -1 | -1 | 2 | -1 | -1 | 2 | 0 | 0 | | |
| $E_5$ | 2 | $-\sqrt{3}$ | 1 | 0 | -1 | $\sqrt{3}$ | -2 | 0 | 0 | $(R_x, R_y)$ | $(xz, yz)$ |

# $D_{nh}$ Groups

[mmm]

| $D_{2h}$ | $E$ | $C_2$ | $C_2(y)$ | $C_2(x)$ | $i$ | $\sigma(xy)$ | $\sigma(xz)$ | $\sigma(yz)$ | | |
|---|---|---|---|---|---|---|---|---|---|---|
| $A_g$ | 1 | 1 | 1 | 1 | 1 | 1 | 1 | 1 | | $x^2, y^2, z^2$ |
| $B_{1g}$ | 1 | 1 | -1 | -1 | 1 | 1 | -1 | -1 | $R_z$ | $xy$ |
| $B_{2g}$ | 1 | -1 | 1 | -1 | 1 | -1 | 1 | -1 | $R_y$ | $xz$ |
| $B_{3g}$ | 1 | -1 | -1 | 1 | 1 | -1 | -1 | 1 | $R_x$ | $yz$ |
| $A_u$ | 1 | 1 | 1 | 1 | -1 | -1 | -1 | -1 | | |
| $B_{1u}$ | 1 | 1 | -1 | -1 | -1 | -1 | 1 | 1 | $z$ | |
| $B_{2u}$ | 1 | -1 | 1 | -1 | -1 | 1 | -1 | 1 | $y$ | |
| $B_{3u}$ | 1 | -1 | -1 | 1 | -1 | 1 | 1 | -1 | $x$ | |

[$\bar{6}$m2]

| $D_{3h}$ | $E$ | $2C_3$ | $3C_2$ | $\sigma_h$ | $2S_3$ | $3\sigma_v$ | | |
|---|---|---|---|---|---|---|---|---|
| $A_1'$ | 1 | 1 | 1 | 1 | 1 | 1 | | $x^2+y^2, z^2$ |
| $A_2'$ | 1 | 1 | -1 | 1 | 1 | -1 | $R_z$ | |
| $E'$ | 2 | -1 | 0 | 2 | -1 | 0 | $(x, y)$ | $(x^2-y^2, xy)$ |
| $A_1''$ | 1 | 1 | 1 | -1 | -1 | -1 | | |
| $A_2''$ | 1 | 1 | -1 | -1 | -1 | 1 | $z$ | |
| $E''$ | 2 | -1 | 0 | -2 | 1 | 0 | $(R_x, R_y)$ | $(xz, yz)$ |

| $D_{4h}$ | E | $2C_4$ | $C_2$ | $2C'_2$ | $2C''_2$ | i | $2S_4$ | $\sigma_h$ | $2\sigma_v$ | $2\sigma_d$ | | [4/mmm] |
|---|---|---|---|---|---|---|---|---|---|---|---|---|
| $A_{1g}$ | 1 | 1 | 1 | 1 | 1 | 1 | 1 | 1 | 1 | 1 | | $x^2+y^2$, $z^2$ |
| $A_{2g}$ | 1 | 1 | 1 | -1 | -1 | 1 | 1 | 1 | -1 | -1 | $R_z$ | |
| $B_{1g}$ | 1 | -1 | 1 | 1 | -1 | 1 | -1 | 1 | 1 | -1 | | $x^2-y^2$ |
| $B_{2g}$ | 1 | -1 | 1 | -1 | 1 | 1 | -1 | 1 | -1 | 1 | | xy |
| $E_g$ | 2 | 0 | -2 | 0 | 0 | 2 | 0 | -2 | 0 | 0 | $(R_x, R_y)$ | (xz, yz) |
| $A_{1u}$ | 1 | 1 | 1 | 1 | 1 | -1 | -1 | -1 | -1 | -1 | | |
| $A_{2u}$ | 1 | 1 | 1 | -1 | -1 | -1 | -1 | -1 | 1 | 1 | z | |
| $B_{1u}$ | 1 | -1 | 1 | 1 | -1 | -1 | 1 | -1 | -1 | 1 | | |
| $B_{2u}$ | 1 | -1 | 1 | -1 | 1 | -1 | 1 | -1 | 1 | -1 | | |
| $E_u$ | 2 | 0 | -2 | 0 | 0 | -2 | 0 | 2 | 0 | 0 | (x, y) | |

| $D_{5h}$ | E | $2C_5$ | $2C_5^2$ | $5C'_2$ | $\sigma_h$ | $2S_5$ | $2S_5^3$ | $5\sigma_v$ | $\alpha = 2\cos 72°$ | $\beta = 2\cos 144°$ |
|---|---|---|---|---|---|---|---|---|---|---|
| $A'_1$ | 1 | 1 | 1 | 1 | 1 | 1 | 1 | 1 | | $x^2+y^2$, $z^2$ |
| $A'_2$ | 1 | 1 | 1 | -1 | 1 | 1 | 1 | -1 | $R_z$ | |
| $E'_1$ | 2 | $\alpha$ | $\beta$ | 0 | 2 | $\alpha$ | $\beta$ | 0 | (x, y) | |
| $E'_2$ | 2 | $\beta$ | $\alpha$ | 0 | 2 | $\beta$ | $\alpha$ | 0 | | $(x^2-y^2, xy)$ |
| $A''_1$ | 1 | 1 | 1 | 1 | -1 | -1 | -1 | -1 | | |
| $A''_2$ | 1 | 1 | 1 | -1 | -1 | -1 | -1 | 1 | z | |
| $E''_1$ | 2 | $\alpha$ | $\beta$ | 0 | -2 | $-\alpha$ | $-\beta$ | 0 | $(R_x, R_y)$ | (xz, yz) |
| $E''_2$ | 2 | $\beta$ | $\alpha$ | 0 | -2 | $-\beta$ | $-\alpha$ | 0 | | |

| $D_{6h}$ | E | $2C_6$ | $2C_3$ | $C_2$ | $3C'_2$ | $3C''_2$ | i | $2S_3$ | $2S_6$ | $\sigma_h(xy)$ | $3\sigma_d$ | $3\sigma_v$ | | [6/mmm] |
|---|---|---|---|---|---|---|---|---|---|---|---|---|---|---|
| $A_{1g}$ | 1 | 1 | 1 | 1 | 1 | 1 | 1 | 1 | 1 | 1 | 1 | 1 | | $x^2+y^2$, $z^2$ |
| $A_{2g}$ | 1 | 1 | 1 | 1 | -1 | -1 | 1 | 1 | 1 | 1 | -1 | -1 | $R_z$ | |
| $B_{1g}$ | 1 | -1 | 1 | -1 | 1 | -1 | 1 | -1 | 1 | -1 | 1 | -1 | | |
| $B_{2g}$ | 1 | -1 | 1 | -1 | -1 | 1 | 1 | -1 | 1 | -1 | -1 | 1 | | |
| $E_{1g}$ | 2 | 1 | -1 | -2 | 0 | 0 | 2 | 1 | -1 | -2 | 0 | 0 | $(R_x, R_y)$ | (xz, yz) |
| $E_{2g}$ | 2 | -1 | -1 | 2 | 0 | 0 | 2 | -1 | -1 | 2 | 0 | 0 | | $(x^2-y^2, xy)$ |
| $A_{1u}$ | 1 | 1 | 1 | 1 | 1 | 1 | -1 | -1 | -1 | -1 | -1 | -1 | | |
| $A_{2u}$ | 1 | 1 | 1 | 1 | -1 | -1 | -1 | -1 | -1 | -1 | 1 | 1 | z | |
| $B_{1u}$ | 1 | -1 | 1 | -1 | 1 | -1 | -1 | 1 | -1 | 1 | -1 | 1 | | |
| $B_{2u}$ | 1 | -1 | 1 | -1 | -1 | 1 | -1 | 1 | -1 | 1 | 1 | -1 | | |
| $E_{1u}$ | 2 | 1 | -1 | -2 | 0 | 0 | -2 | -1 | 1 | 2 | 0 | 0 | (x, y) | |
| $E_{2u}$ | 2 | -1 | -1 | 2 | 0 | 0 | -2 | 1 | 1 | -2 | 0 | 0 | | |

## $S_n$ Groups

| $S_4$ | E | $S_4$ | $C_2$ | $S_4^3$ | | $[\bar{4}]$ |
|---|---|---|---|---|---|---|
| A | 1 | 1 | 1 | 1 | $R_z$ | $x^2+y^2$, $z^2$ |
| B | 1 | -1 | 1 | -1 | z | $x^2-y^2$, xy |
| E | 1 | i | -1 | -i | $x+iy$, $R_x+iR_y$ | (xz, yz) |
| | 1 | -i | -1 | i | $x-iy$, $R_x-iR_y$ | |

| $S_6$ | E | $C_3$ | $C_3^2$ | i | $S_6^5$ | $S_6$ | $\varepsilon=\exp(2\pi i/3)$ | $[\bar{3}]$ |
|---|---|---|---|---|---|---|---|---|
| $A_g$ | 1 | 1 | 1 | 1 | 1 | 1 | $R_z$ | $x^2+y^2$, $z^2$ |
| $E_g$ | 1 | $\varepsilon$ | $\varepsilon^*$ | 1 | $\varepsilon$ | $\varepsilon^*$ | $R_x+iR_y$ | $(x^2-y^2$, xy) (xz, yz) |
| | 1 | $\varepsilon^*$ | $\varepsilon$ | 1 | $\varepsilon^*$ | $\varepsilon$ | $R_x-iR_y$ | |
| $A_u$ | 1 | 1 | 1 | -1 | -1 | -1 | z | |
| $E_u$ | 1 | $\varepsilon$ | $\varepsilon^*$ | -1 | $-\varepsilon$ | $-\varepsilon^*$ | $x+iy$ | |
| | 1 | $\varepsilon^*$ | $\varepsilon$ | -1 | $-\varepsilon^*$ | $-\varepsilon$ | $x-iy$ | |

| $S_8$ | E | $S_8$ | $C_4$ | $S_8^3$ | $C_2$ | $S_8^5$ | $C_4^3$ | $S_8^7$ | $\varepsilon=\exp(2\pi i/8)$ | |
|---|---|---|---|---|---|---|---|---|---|---|
| A | 1 | 1 | 1 | 1 | 1 | 1 | 1 | 1 | $R_z$ | $x^2+y^2$, $z^2$ |
| B | 1 | -1 | 1 | -1 | 1 | -1 | 1 | -1 | z | |
| $E_1$ | 1 | $\varepsilon$ | i | $-\varepsilon^*$ | -1 | $-\varepsilon$ | -i | $\varepsilon^*$ | $x+iy$ | |
| | 1 | $\varepsilon^*$ | -i | $-\varepsilon$ | -1 | $-\varepsilon^*$ | i | $\varepsilon$ | $x-iy$ | |
| $E_2$ | 1 | i | -1 | -i | 1 | i | -1 | -i | | $(x^2-y^2$, xy) |
| | 1 | -i | -1 | i | 1 | -i | -1 | i | | |
| $E_3$ | 1 | $-\varepsilon$ | i | $\varepsilon^*$ | -1 | $\varepsilon$ | -i | $-\varepsilon^*$ | $R_x+iR_y$ | (xz, yz) |
| | 1 | $-\varepsilon^*$ | -i | $\varepsilon$ | -1 | $\varepsilon^*$ | i | $-\varepsilon$ | $R_x-iR_y$ | |

## Cubic Groups

| T | E | $4C_3$ | $4C_3^2$ | $3C_2$ | $\varepsilon=\exp(2\pi i/3)$ | [23] |
|---|---|---|---|---|---|---|
| A | 1 | 1 | 1 | 1 | | $x^2+y^2+z^2$ |
| E | 1 | $\varepsilon$ | $\varepsilon^*$ | 1 | | $(2z^2-x^2-y^2$, $x^2-y^2)$ |
| | 1 | $\varepsilon^*$ | $\varepsilon$ | 1 | | |
| T | 3 | 0 | 0 | -1 | (x, y, z), ($R_x$, $R_y$, $R_z$) | (xy, xz, yz) |

| $T_h$ | E | $4C_3$ | $4C_3^2$ | $3C_2$ | i | $4S_6^5$ | $4S_6$ | $3\sigma_h$ | $\varepsilon=\exp(2\pi i/3)$ | [m3] |
|---|---|---|---|---|---|---|---|---|---|---|
| $A_g$ | 1 | 1 | 1 | 1 | 1 | 1 | 1 | 1 | | $x^2+y^2+z^2$ |
| $E_g$ | 1 | $\varepsilon$ | $\varepsilon^*$ | 1 | 1 | $\varepsilon$ | $\varepsilon^*$ | 1 | | $(2z^2-x^2-y^2$, $x^2-y^2)$ |
| | 1 | $\varepsilon^*$ | $\varepsilon$ | 1 | 1 | $\varepsilon^*$ | $\varepsilon$ | 1 | | |
| $T_g$ | 3 | 0 | 0 | -1 | 3 | 0 | 0 | -1 | ($R_x$, $R_y$, $R_z$) | (xy, xz, yz) |
| $A_u$ | 1 | 1 | 1 | 1 | -1 | -1 | -1 | -1 | | |
| $E_u$ | 1 | $\varepsilon$ | $\varepsilon^*$ | 1 | -1 | $-\varepsilon$ | $-\varepsilon^*$ | -1 | | |
| | 1 | $\varepsilon^*$ | $\varepsilon$ | 1 | -1 | $-\varepsilon^*$ | $-\varepsilon$ | -1 | | |
| $T_u$ | 3 | 0 | 0 | -1 | -3 | 0 | 0 | 1 | (x, y, z) | |

| $T_d$ | $E$ | $8C_3$ | $6\sigma_d$ | $6S_4$ | $3C_2$ | | [$\bar{4}3m$] |
|---|---|---|---|---|---|---|---|
| $O$ | $E$ | $8C_3$ | $6C_2'$ | $6C_4$ | $3C_2(\equiv C_4^2)$ | | [432] |
| $A_1$ | 1 | 1 | 1 | 1 | 1 | | $x^2+y^2+z^2$ |
| $A_2$ | 1 | 1 | -1 | -1 | 1 | | |
| $E$ | 2 | -1 | 0 | 0 | 2 | | $(2z^2-x^2-y^2, x^2-y^2)$ |
| $T_1$ | 3 | 0 | -1 | 1 | -1 | $(x, y, z)$ $(R_x, R_y, R_z)$ | |
| $T_2$ | 3 | 0 | 1 | -1 | -1 | $(x, y, z)$ in Td | $(xy, xz, yz)$ |

| $O_h$ | $E$ | $8C_3$ | $6C_2$ | $6C_4$ | $3C_2(\equiv C_4^2)$ | $i$ | $6S_4$ | $8S_6$ | $3\sigma_h$ | $6\sigma_d$ | | [$m3m$] |
|---|---|---|---|---|---|---|---|---|---|---|---|---|
| $A_{1g}$ | 1 | 1 | 1 | 1 | 1 | 1 | 1 | 1 | 1 | 1 | | $x^2+y^2+z^2$ |
| $A_{2g}$ | 1 | 1 | -1 | -1 | 1 | 1 | -1 | 1 | 1 | -1 | | |
| $E_g$ | 2 | -1 | 0 | 0 | 2 | 2 | 0 | -1 | 2 | 0 | | $(2z^2-x^2-y^2, x^2-y^2)$ |
| $T_{1g}$ | 3 | 0 | -1 | 1 | -1 | 3 | 1 | 0 | -1 | -1 | $(R_x, R_y, R_z)$ | |
| $T_{2g}$ | 3 | 0 | 1 | -1 | -1 | 3 | -1 | 0 | -1 | 1 | | $(xz, yz, xy)$ |
| $A_{1u}$ | 1 | 1 | 1 | 1 | 1 | -1 | -1 | -1 | -1 | -1 | | |
| $A_{2u}$ | 1 | 1 | -1 | -1 | 1 | -1 | 1 | -1 | -1 | 1 | | |
| $E_u$ | 2 | -1 | 0 | 0 | 2 | -2 | 0 | 1 | -2 | 0 | | |
| $T_{1u}$ | 3 | 0 | -1 | 1 | -1 | -3 | -1 | 0 | 1 | 1 | $(x, y, z)$ | |
| $T_{2u}$ | 3 | 0 | 1 | -1 | -1 | -3 | 1 | 0 | 1 | -1 | | |

# Linear Groups

| $C_{\infty v}$ | $E$ | $2C_\infty(\phi)$ | $2C_\infty(2\phi)$ | $\cdots$ | $\infty\sigma_v$ | | |
|---|---|---|---|---|---|---|---|
| $A_1$ $(\Sigma^+)$ | 1 | 1 | 1 | $\cdots$ | 1 | $z$ | $x^2+y^2$, $z^2$ |
| $A_2$ $(\Sigma^-)$ | 1 | 1 | 1 | $\cdots$ | -1 | $R_z$ | |
| $E_1$ $(\Pi)$ | 2 | $2\cos\phi$ | $2\cos 2\phi$ $\cdots$ | | 0 | $(x, y)$ $(R_x, R_y)$ | $(xz, yz)$ |
| $E_2$ $(\Delta)$ | 2 | $2\cos 2\phi$ | $2\cos 4\phi$ $\cdots$ | | 0 | | $(x^2-y^2, xy)$ |
| $\cdots$ | $\cdots$ | $\cdots$ | $\cdots$ | $\cdots$ | $\cdots$ | | |
| $\cdots$ | $\cdots$ | $\cdots$ | $\cdots$ | $\cdots$ | $\cdots$ | | |
| $E_n$ | 2 | $2\cos n\phi$ | $2\cos 2n\phi$ $\cdots$ | | 0 | | |

| $D_{\infty h}$ | $E$ | $2C_\infty(\phi)$ | $\cdots$ | $\infty\sigma_v$ | $i$ | $2S_\infty(\phi)$ | $\cdots$ | $\infty C_2'$ | | |
|---|---|---|---|---|---|---|---|---|---|---|
| $A_{1g}$ $(\Sigma_g^+)$ | 1 | 1 | $\cdots$ | 1 | 1 | 1 | $\cdots$ | 1 | | $x^2+y^2$, $z^2$ |
| $A_{2g}$ $(\Sigma_g^-)$ | 1 | 1 | $\cdots$ | -1 | 1 | 1 | $\cdots$ | -1 | $R_z$ | |
| $E_{1g}$ $(\Pi_g)$ | 2 | $2\cos\phi$ | $\cdots$ | 0 | 2 | $-2\cos\phi$ | $\cdots$ | 0 | $(R_x, R_y)$ | $(xz, yz)$ |
| $E_{2g}$ $(\Delta_g)$ | 2 | $2\cos 2\phi$ | $\cdots$ | 0 | 2 | $2\cos 2\phi$ | $\cdots$ | 0 | | $(x^2-y^2, xy)$ |
| $\cdots$ | $\cdots$ | $\cdots$ | $\cdots$ | $\cdots$ | $\cdots$ | $\cdots$ | $\cdots$ | $\cdots$ | | |
| $E_{ng}$ | 2 | $2\cos n\phi$ | $\cdots$ | 0 | 2 | $(-1)^n 2\cos n\phi$ | $\cdots$ | 0 | | |
| $\cdots$ | $\cdots$ | $\cdots$ | $\cdots$ | $\cdots$ | $\cdots$ | $\cdots$ | $\cdots$ | $\cdots$ | | |
| $A_{1u}$ $(\Sigma_u^+)$ | 1 | 1 | $\cdots$ | 1 | -1 | -1 | $\cdots$ | -1 | $z$ | |
| $A_{2u}$ $(\Sigma_u^-)$ | 1 | 1 | $\cdots$ | -1 | -1 | -1 | $\cdots$ | 1 | | |
| $E_{1u}$ $(\Pi_u)$ | 2 | $2\cos\phi$ | $\cdots$ | 0 | -2 | $2\cos\phi$ | $\cdots$ | 0 | $(x, y)$ | |
| $E_{2u}$ $(\Delta_u)$ | 2 | $2\cos 2\phi$ | $\cdots$ | 0 | -2 | $-2\cos 2\phi$ | $\cdots$ | 0 | | |
| $\cdots$ | $\cdots$ | $\cdots$ | $\cdots$ | $\cdots$ | $\cdots$ | $\cdots$ | $\cdots$ | $\cdots$ | | |
| $E_{nu}$ | 2 | $2\cos n\phi$ | $\cdots$ | 0 | -2 | $(-1)^{n+1} 2\cos n\phi$ | $\cdots$ | 0 | | |
| $\cdots$ | $\cdots$ | $\cdots$ | $\cdots$ | $\cdots$ | $\cdots$ | $\cdots$ | $\cdots$ | $\cdots$ | | |

## Icosahedral Group

| $I_h$ | E | $12C_5$ | $12C_5^2$ | $20C_3$ | $15C_2$ | i | $12S_{10}^3$ | $12S_{10}$ | $20S_6$ | $15\sigma$ | $\alpha=\frac{1}{2}(1+\sqrt{5})$ | $\beta=\frac{1}{2}(1-\sqrt{5})$ |
|---|---|---|---|---|---|---|---|---|---|---|---|---|
| $A_g$ | 1 | 1 | 1 | 1 | 1 | 1 | 1 | 1 | 1 | 1 | | $x^2+y^2+z^2$ |
| $A_{1g}$ | 3 | $\alpha$ | $\beta$ | 0 | −1 | 3 | $\alpha$ | $\beta$ | 0 | −1 | $(R_x, R_y, R_z)$ | |
| $T_{2g}$ | 3 | $\beta$ | $\alpha$ | 0 | −1 | 3 | $\beta$ | $\alpha$ | 0 | −1 | | |
| $G_g$ | 4 | −1 | −1 | 1 | 0 | 4 | −1 | −1 | 1 | 0 | | |
| $H_g$ | 5 | 0 | 0 | −1 | 1 | 5 | 0 | 0 | −1 | 1 | | $\left\{ \begin{array}{l} 2z^2-x^2-y^2, \\ x^2-y^2, \\ xy, yz, zx \end{array} \right\}$ |
| $A_u$ | 1 | 1 | 1 | 1 | 1 | −1 | −1 | −1 | −1 | −1 | | |
| $T_{1u}$ | 3 | $\alpha$ | $\beta$ | 0 | −1 | −3 | $-\alpha$ | $-\beta$ | 0 | 1 | (x, y, z) | |
| $T_{2u}$ | 3 | $\beta$ | $\alpha$ | 0 | −1 | −3 | $-\beta$ | $-\alpha$ | 0 | 1 | | |
| $G_u$ | 4 | −1 | −1 | 1 | 0 | −4 | 1 | 1 | −1 | 0 | | |
| $H_u$ | 5 | 0 | 0 | −1 | 1 | −5 | 0 | 0 | 1 | −1 | | |

$I_h = I \times C_i$ [In I (x,y,z) transforms as $T_1$]

## Irreducible Representation in Other Notations

### $O_h$

| Mulliken (Chemical) | Bethe | BSW |
|---|---|---|
| $A_{1g}$ | $\Gamma_1^+$ | $\Gamma_1$ |
| $A_{2g}$ | $\Gamma_2^+$ | $\Gamma_2$ |
| $E_g$ | $\Gamma_3^+$ | $\Gamma_{12}$ |
| $T_{1g}$ | $\Gamma_4^+$ | $\Gamma_{15}'$ |
| $T_{2g}$ | $\Gamma_5^+$ | $\Gamma_{25}'$ |
| $A_{1u}$ | $\Gamma_1^-$ | $\Gamma_1'$ |
| $A_{2u}$ | $\Gamma_2^-$ | $\Gamma_2'$ |
| $E_u$ | $\Gamma_3^-$ | $\Gamma_{12}'$ |
| $T_{1u}$ | $\Gamma_4^-$ | $\Gamma_{15}$ |
| $T_{2u}$ | $\Gamma_5^-$ | $\Gamma_{25}$ |
| $E_{1/2}^+$ | $\Gamma_6^+$ | |
| $E_{5/2}^+$ | $\Gamma_7^+$ | |
| $G^+$ | $\Gamma_8^+$ | |
| $E_{1/2}^-$ | $\Gamma_6^-$ | |
| $E_{5/2}^-$ | $\Gamma_7^-$ | |
| $G^-$ | $\Gamma_8^-$ | |

### $D_{4h}$

| Mulliken (Chemical) | Bethe | BSW |
|---|---|---|
| $A_{1g}$ | $\Gamma_1^+$ | $M_1$ |
| $A_{2g}$ | $\Gamma_2^+$ | $M_2$ |
| $B_{1g}$ | $\Gamma_3^+$ | $M_3$ |
| $B_{2g}$ | $\Gamma_4^+$ | $M_4$ |
| $E_g$ | $\Gamma_5^+$ | $M_5$ |
| $A_{1u}$ | $\Gamma_1^-$ | $M_1'$ |
| $A_{2u}$ | $\Gamma_2^-$ | $M_2'$ |
| $B_{1u}$ | $\Gamma_3^-$ | $M_3'$ |
| $B_{2u}$ | $\Gamma_4^-$ | $M_4'$ |
| $E_u$ | $\Gamma_5^-$ | $M_5'$ |

# Appendix 4

## SPACE GROUPS

The Schoenflies and international symbol for each of the 230 space groups is given. The space groups are numbered from 1 to 230 and the symmorphic space groups are underlined. See Chapter 11 for a detailed discussion of the notation.

TRICLINIC SYSTEM

| No. of space group | Schoenflies symbol | Standard short symbol |
|---|---|---|
| 1 | $C_1^1$ | P1 |
| 2 | $C_i^1$ | P$\bar{1}$ |

MONOCLINIC SYSTEM

z-axis unique (1st setting)

| No. of space group | Schoenflies symbol | Standard short symbol |
|---|---|---|
| 3 | $C_2^1$ | P2 |
| 4 | $C_2^2$ | P$2_1$ |
| 5 | $C_2^3$ | B2 |
| 6 | $C_s^1$ | Pm |
| 7 | $C_s^2$ | Pb |
| 8 | $C_s^3$ | Bm |
| 9 | $C_s^4$ | Bb |
| 10 | $C_{2h}^1$ | P2/m |
| 11 | $C_{2h}^2$ | P$2_1$/m |
| 12 | $C_{2h}^3$ | B2/m |
| 13 | $C_{2h}^4$ | P2/b |
| 14 | $C_{2h}^5$ | P$2_1$/b |
| 15 | $C_{2h}^6$ | B2/b |

ORTHORHOMBIC SYSTEM

| No. of space group | Schoenflies symbol | Standard short symbol |
|---|---|---|
| 16 | $D_2^1=V^1$ | P222 |
| 17 | $D_2^2=V^2$ | $P222_1$ |
| 18 | $D_2^3=V^3$ | $P2_12_12$ |
| 19 | $D_2^4=V^4$ | $P2_12_12_1$ |
| 20 | $D_2^5=V^5$ | $C222_1$ |
| 21 | $D_2^6=V^6$ | C222 |
| 22 | $D_2^7=V^7$ | F222 |
| 23 | $D_2^8=V^8$ | I222 |
| 24 | $D_2^9=V^9$ | $I2_12_12_1$ |
| 25 | $C_{2v}^1$ | Pmm2 |
| 26 | $C_{2v}^2$ | $Pmc2_1$ |
| 27 | $C_{2v}^3$ | Pcc2 |
| 28 | $C_{2v}^4$ | Pma2 |
| 29 | $C_{2v}^5$ | $Pca2_1$ |
| 30 | $C_{2v}^6$ | Pnc2 |
| 31 | $C_{2v}^7$ | $Pmn2_1$ |
| 32 | $C_{2v}^8$ | Pba2 |
| 33 | $C_{2v}^9$ | $Pna2_1$ |
| 34 | $C_{2v}^{10}$ | Pnn2 |
| 35 | $C_{2v}^{11}$ | Cmm2 |

ORTHORHOMBIC SYSTEM

| No. of space group | Schoenflies symbol | Standard short symbol |
|---|---|---|
| 36 | $C_{2v}^{12}$ | $Cmc2_1$ |
| 37 | $C_{2v}^{13}$ | Ccc2 |
| 38 | $C_{2v}^{14}$ | Amm2 |
| 39 | $C_{2v}^{15}$ | Abm2 |
| 40 | $C_{2v}^{16}$ | Ama2 |
| 41 | $C_{2v}^{17}$ | Aba2 |
| 42 | $C_{2v}^{18}$ | Fmm2 |
| 43 | $C_{2v}^{19}$ | Fdd2 |
| 44 | $C_{2v}^{20}$ | Imm2 |
| 45 | $C_{2v}^{21}$ | Iba2 |
| 46 | $C_{2v}^{22}$ | Ima2 |
| 47 | $D_{2h}^1=V_h^1$ | Pmmm |
| 48 | $D_{2h}^2=V_h^2$ | Pnnn |
| 49 | $D_{2h}^3=V_h^3$ | Pccm |
| 50 | $D_{2h}^4=V_h^4$ | Pban |
| 51 | $D_{2h}^5=V_h^5$ | Pmma |
| 52 | $D_{2h}^6=V_h^6$ | Pnna |
| 53 | $D_{2h}^7=V_h^7$ | Pmna |
| 54 | $D_{2h}^8=V_h^8$ | Pcca |
| 55 | $D_{2h}^9=V_h^9$ | Pbam |

ORTHORHOMBIC SYSTEM

| No. of space group | Schoenflies symbol | Standard short symbol |
|---|---|---|
| 56 | $D_{2h}^{10}=V_h^{10}$ | Pccn |
| 57 | $D_{2h}^{11}=V_h^{11}$ | Pbcm |
| 58 | $D_{2h}^{12}=V_h^{12}$ | Pnnm |
| 59 | $D_{2h}^{13}=V_h^{13}$ | Pmmn |
| 60 | $D_{2h}^{14}=V_h^{14}$ | Pbcn |
| 61 | $D_{2h}^{15}=V_h^{15}$ | Pbca |
| 62 | $D_{2h}^{16}=V_h^{16}$ | Pnma |
| 63 | $D_{2h}^{17}=V_h^{17}$ | Cmcm |
| 64 | $D_{2h}^{18}=V_h^{18}$ | Cmca |
| 65 | $D_{2h}^{19}=V_h^{19}$ | Cmmm |
| 66 | $D_{2h}^{20}=V_h^{20}$ | Cccm |
| 67 | $D_{2h}^{21}=V_h^{21}$ | Cmma |
| 68 | $D_{2h}^{22}=V_h^{22}$ | Ccca |
| 69 | $D_{2h}^{23}=V_h^{23}$ | Fmmm |
| 70 | $D_{2h}^{24}=V_h^{24}$ | Fddd |
| 71 | $D_{2h}^{25}=V_h^{25}$ | Immm |
| 72 | $D_{2h}^{26}=V_h^{26}$ | Ibam |
| 73 | $D_{2h}^{27}=V_h^{27}$ | Ibca |
| 74 | $D_{2h}^{28}=V_h^{28}$ | Imma |

TETRAGONAL SYSTEM

| No. of space group | Schoenflies symbol | Standard short symbol |
|---|---|---|
| 75 | $C_4^1$ | $P4$ |
| 76 | $C_4^2$ | $P4_1$ |
| 77 | $C_4^3$ | $P4_2$ |
| 78 | $C_4^4$ | $P4_3$ |
| 79 | $C_4^5$ | $I4$ |
| 80 | $C_4^6$ | $I4_1$ |
| 81 | $S_4^1$ | $P\bar{4}$ |
| 82 | $S_4^2$ | $I\bar{4}$ |
| 83 | $C_{4h}^1$ | $P4/m$ |
| 84 | $C_{4h}^2$ | $P4_2/m$ |
| 85 | $C_{4h}^3$ | $P4/n$ |
| 86 | $C_{4h}^4$ | $P4_2/n$ |
| 87 | $C_{4h}^5$ | $I4/m$ |
| 88 | $C_{4h}^6$ | $I4_1/a$ |
| 89 | $D_4^1$ | $P422$ |
| 90 | $D_4^2$ | $P42_12$ |
| 91 | $D_4^3$ | $P4_122$ |
| 92 | $D_4^4$ | $P4_12_12$ |
| 93 | $D_4^5$ | $P4_222$ |
| 94 | $D_4^6$ | $P4_22_12$ |
| 95 | $D_4^7$ | $P4_322$ |
| 96 | $D_4^8$ | $P4_32_12$ |
| 97 | $D_4^9$ | $I422$ |
| 98 | $D_4^{10}$ | $I4_122$ |

TETRAGONAL SYSTEM

| No. of space group | Schoenflies symbol | Standard short symbol |
|---|---|---|
| 99 | $C_{4v}^1$ | $P4mm$ |
| 100 | $C_{4v}^2$ | $P4bm$ |
| 101 | $C_{4v}^3$ | $P4_2cm$ |
| 102 | $C_{4v}^4$ | $P4_2nm$ |
| 103 | $C_{4v}^5$ | $P4cc$ |
| 104 | $C_{4v}^6$ | $P4nc$ |
| 105 | $C_{4v}^7$ | $P4_2mc$ |
| 106 | $C_{4v}^8$ | $P4_2bc$ |
| 107 | $C_{4v}^9$ | $I4mm$ |
| 108 | $C_{4v}^{10}$ | $I4cm$ |
| 109 | $C_{4v}^{11}$ | $I4_1md$ |
| 110 | $C_{4v}^{12}$ | $I4_1cd$ |
| 111 | $D_{2d}^1=V_d^1$ | $P\bar{4}2m$ |
| 112 | $D_{2d}^2=V_d^2$ | $P\bar{4}2C$ |
| 113 | $D_{2d}^3=V_d^3$ | $P\bar{4}2_1m$ |
| 114 | $D_{2d}^4=V_d^4$ | $P\bar{4}2_1c$ |
| 115 | $D_{2d}^5=V_d^5$ | $P\bar{4}m2$ |
| 116 | $D_{2d}^6=V_d^6$ | $P\bar{4}c2$ |
| 117 | $D_{2d}^7=V_d^7$ | $P\bar{4}b2$ |
| 118 | $D_{2d}^8=V_d^8$ | $P\bar{4}n2$ |
| 119 | $D_{2d}^9=V_d^9$ | $I\bar{4}m2$ |
| 120 | $D_{2d}^{10}=V_d^{10}$ | $I\bar{4}c2$ |
| 121 | $D_{2d}^{11}=V_d^{11}$ | $I\bar{4}2m$ |
| 122 | $D_{2d}^{12}=V_d^{12}$ | $I\bar{4}2d$ |

TETRAGONAL SYSTEM

| No. of space group | Schoenflies symbol | Standard short symbol |
|---|---|---|
| 123 | $D_{4h}^1$ | $P4/mmm$ |
| 124 | $D_{4h}^2$ | $P4/mcc$ |
| 125 | $D_{4h}^3$ | $P4/nbm$ |
| 126 | $D_{4h}^4$ | $P4/nnc$ |
| 127 | $D_{4h}^5$ | $P4/mbm$ |
| 128 | $D_{4h}^6$ | $P4/mnc$ |
| 129 | $D_{4h}^7$ | $P4/nmm$ |
| 130 | $D_{4h}^8$ | $P4/ncc$ |
| 131 | $D_{4h}^9$ | $P4_2/mmc$ |
| 132 | $D_{4h}^{10}$ | $P4_2/mcm$ |
| 133 | $D_{4h}^{11}$ | $P4_2/nbc$ |
| 134 | $D_{4h}^{12}$ | $P4_2/nnm$ |
| 135 | $D_{4h}^{13}$ | $P4_2/mbc$ |
| 136 | $D_{4h}^{14}$ | $P4_2/mnm$ |
| 137 | $D_{4h}^{15}$ | $P4_2/nmc$ |
| 138 | $D_{4h}^{16}$ | $P4_2/ncm$ |
| 139 | $D_{4h}^{17}$ | $I4/mmm$ |
| 140 | $D_{4h}^{18}$ | $I4/mcm$ |
| 141 | $D_{4h}^{19}$ | $I4_1/amd$ |
| 142 | $D_{4h}^{20}$ | $I4_1/acd$ |

TRIGONAL SYSTEM

| No. of space group | Schoenflies symbol | Standard short symbol |
|---|---|---|
| 143 | $C_3^1$ | $P3$ |
| 144 | $C_3^2$ | $P3_1$ |
| 145 | $C_3^3$ | $P3_2$ |
| 146 | $C_3^4$ | $R3$ |
| 147 | $C_{3i}^1$ | $P\bar{3}$ |
| 148 | $C_{3i}^2$ | $R\bar{3}$ |
| 149 | $D_3^1$ | $P312$ |
| 150 | $D_3^2$ | $P321$ |
| 151 | $D_3^3$ | $P3_112$ |
| 152 | $D_3^4$ | $P3_121$ |
| 153 | $D_3^5$ | $P3_212$ |
| 154 | $D_3^6$ | $P3_221$ |
| 155 | $D_3^7$ | $R32$ |
| 156 | $C_{3v}^1$ | $P3m1$ |
| 157 | $C_{3v}^2$ | $P31m$ |
| 158 | $C_{3v}^3$ | $P3c1$ |
| 159 | $C_{3v}^4$ | $P31c$ |
| 160 | $C_{3v}^5$ | $R3m$ |
| 161 | $C_{3v}^6$ | $R3c$ |
| 162 | $D_{3d}^1$ | $P\bar{3}1m$ |
| 163 | $D_{3d}^2$ | $P\bar{3}1c$ |
| 164 | $D_{3d}^3$ | $P\bar{3}m1$ |
| 165 | $D_{3d}^4$ | $P\bar{3}c1$ |
| 166 | $D_{3d}^5$ | $R\bar{3}m$ |
| 167 | $D_{3d}^6$ | $R\bar{3}c$ |

HEXAGONAL SYSTEM

| No. of space group | Schoenflies symbol | Standard short symbol |
|---|---|---|
| 168 | $C_6^1$ | $P6$ |
| 169 | $C_6^2$ | $P6_1$ |
| 170 | $C_6^3$ | $P6_5$ |
| 171 | $C_6^4$ | $P6_2$ |
| 172 | $C_6^5$ | $P6_4$ |
| 173 | $C_6^6$ | $P6_3$ |
| 174 | $C_{3h}^1$ | $P\bar{6}$ |
| 175 | $C_{6h}^1$ | $P6/m$ |
| 176 | $C_{6h}^2$ | $P6_3/m$ |
| 177 | $D_6^1$ | $P622$ |
| 178 | $D_6^2$ | $P6_122$ |
| 179 | $D_6^3$ | $P6_522$ |
| 180 | $D_6^4$ | $P6_222$ |
| 181 | $D_6^5$ | $P6_422$ |
| 182 | $D_6^6$ | $P6_322$ |
| 183 | $C_{6v}^1$ | $P6mm$ |
| 184 | $C_{6v}^2$ | $P6cc$ |
| 185 | $C_{6v}^3$ | $P6_3cm$ |
| 186 | $C_{6v}^4$ | $P6_3mc$ |
| 187 | $D_{3h}^1$ | $P\bar{6}m2$ |
| 188 | $D_{3h}^2$ | $P\bar{6}c2$ |
| 189 | $D_{3h}^3$ | $P\bar{6}2m$ |
| 190 | $D_{3h}^4$ | $P\bar{6}2c$ |

HEXAGONAL SYSTEM

| No. of space group | Schoenflies symbol | Standard short symbol |
|---|---|---|
| 191 | $D_{6h}^1$ | $P6/mmm$ |
| 192 | $D_{6h}^2$ | $P6/mcc$ |
| 193 | $D_{6h}^3$ | $P6_3/mcm$ |
| 194 | $D_{6h}^4$ | $P6_3/mmc$ |

CUBIC SYSTEM

| No. of space group | Schoenflies symbol | Standard short symbol |
|---|---|---|
| 195 | $T^1$ | P23 |
| 196 | $T^2$ | F23 |
| 197 | $T^3$ | I23 |
| 198 | $T^4$ | $P2_13$ |
| 199 | $T^5$ | $I2_13$ |
| 200 | $T_h^1$ | Pm3 |
| 201 | $T_h^2$ | Pn3 |
| 202 | $T_h^3$ | Fm3 |
| 203 | $T_h^4$ | Fd3 |
| 204 | $T_h^5$ | Im3 |
| 205 | $T_h^6$ | Pa3 |
| 206 | $T_h^7$ | Ia3 |
| 207 | $O^1$ | P432 |
| 208 | $O^2$ | $P4_232$ |
| 209 | $O^3$ | F432 |
| 210 | $O^4$ | $F4_132$ |
| 211 | $O^5$ | I432 |
| 212 | $O^6$ | $P4_332$ |

CUBIC SYSTEM

| No. of space group | Schoenflies symbol | Standard short symbol |
|---|---|---|
| 213 | $O^7$ | $P4_132$ |
| 214 | $O^8$ | $I4_132$ |
| 215 | $T_d^1$ | $P\bar{4}3m$ |
| 216 | $T_d^2$ | $F\bar{4}3m$ |
| 217 | $T_d^3$ | $I\bar{4}3m$ |
| 218 | $T_d^4$ | $P\bar{4}3n$ |
| 219 | $T_d^5$ | $F\bar{4}3c$ |
| 220 | $T_d^6$ | $I\bar{4}3d$ |
| 221 | $O_h^1$ | Pm3m |
| 222 | $O_h^2$ | Pn3n |
| 223 | $O_h^3$ | Pm3n |
| 224 | $O_h^4$ | Pn3m |
| 225 | $O_h^5$ | Fm3m |
| 226 | $O_h^6$ | Fm3c |
| 227 | $O_h^7$ | Fd3m |
| 228 | $O_h^8$ | Fd3c |
| 229 | $O_h^9$ | Im3m |
| 230 | $O_h^{10}$ | Ia3d |

Appendix 5

# MATRICES, VECTOR SPACES, AND LINEAR OPERATORS

In this appendix we cover, in a review form, the above mentioned topics. For a more detailed and rigorous discussion see: Wigner [Chapters 1–4]; Rose [Chapter 1]; DiBartolo [Chapter 1]; Knox and Gold [Appendix 1]; Chisholm [Chapter 2].

## Matrices

A matrix is an array of numbers

$$\begin{array}{lllllll} & & & & & j^{th}\text{ column} & \\ & M_{11} & M_{12} & M_{13} & \cdots & M_{ig} & \\ M = & M_{21} & M_{22} & & & & \\ & \cdot & & & & \cdot & \\ & \cdot & & & & \cdot & \\ & \cdot & & & & \cdot & \\ & M_{1} & \cdots & & & M_{ij} & i^{th}\text{ row} \end{array} \qquad \text{(A5-1)}$$

where $M_{ij}$ is the element of the ith row and jth column. If M has m rows and n columns it is said to be an m by n matrix or an m × n matrix. If m = n it is a square matrix. The elements may be complex.

Matrix multiplication $\gamma = \alpha\beta$, where

$$\gamma_{ik} = \Sigma_j \, \alpha_{ij} \, \beta_{jk} \qquad \text{(A5-2)}$$

Notice that $\alpha$ and $\beta$ must be conformable, that is $\alpha$ must have as many columns as $\beta$ rows. In general, matrix multiplication is not commutive (i.e., $\alpha\beta \neq \beta\alpha$) but it is associative, i.e., $\alpha(\beta\gamma) = (\alpha\beta)\gamma$. See Wigner [Chapter 1]. Two special matrices are: the null matrix where all elements are zero; the unit matrix **1**, which is square, where all diagonal elements are ones and all the off-diagonal elements are zeros, i.e., $(\mathbf{1})_{ij} = \delta_{ij}$.

For square matrices: The inverse $M^{-1}$ of matrix M, is a matrix such that $M^{-1}M = \mathbf{1} = MM^{-1}$, thus a matrix and its inverse commute. The trace of M is $\Sigma_i M_{ii} \equiv \text{Tr } M$ (also called character). The determinant of M is usually written as det M or $|M|$. The inverse of M exists if $|M| \neq 0$.

For any matrix M: The complex conjugate is written $M^*$ and is $(M^*)_{ij} = M_{ij}^*$. The transpose is written $\widetilde{M}$ and is $(\widetilde{M})_{ij} = M_{ji}$. The adjoint is written $M^\dagger$ and is $(M^\dagger)_{ij} = M_{ji}^*$ or $M^\dagger = \widetilde{M}^*$.

Some special matrices are (note a Hermitian matrix is also called self-adjoint):

| | | | |
|---|---|---|---|
| A matrix is real, | if | $M = M^*$ | |
| A matrix is symmetric, | if | $M = \widetilde{M}$ | |
| A matrix is orthogonal, | if | $\widetilde{M} = M^{-1}$ | (or $\widetilde{M}M = \mathbf{1}$) |
| A matrix is unitary, | if | $M^\dagger = M^{-1}$ | (or $M^\dagger M = \mathbf{1}$) |
| A matrix is hermitian, | if | $M^\dagger = M$ | |

## Linear Spaces

A **linear space** is a set of elements **u**, **v**, **w**, ... which have the following properties: Multiplication of any element of the set by a complex number results in another element of the set. For any **u** and **v**, $\mathbf{u} + \mathbf{v} = \mathbf{z}$ where **z** also is a member of the set. Associativity $(\mathbf{u} + \mathbf{v}) + \mathbf{w} = \mathbf{u} + (\mathbf{v} + \mathbf{w})$ is obeyed as is $\mathbf{u} + \mathbf{v} = \mathbf{v} + \mathbf{u}$. The set of elements can be finite or infinite in number such as all the vectors in a plane from a point.

The n vectors $e_1, e_2, ..., e_n$ are linearly independent if no set of numbers, $k_i$, exists so that $k_1e_1 + k_2e_2 + ... + k_ne_n = 0$. Thus, we can have an n-dimensional linear space where it is possible to choose n vectors that are linearly independent but not n + 1.

For an n-dimensional vector space an arbitrary vector **t** can be expressed in terms of the projection of its components, $a_i$, on a set of n-linearly independent basis vectors $e_i$.

$$\mathbf{t} = a_1e_1 + a_2e_2 + ... + a_ne_n \tag{A5-3}$$

where we choose to represent the **components** $a_i$ as a n × 1 column matrix and the **basis** vectors $e_i$ as a 1 × n row matrix

$$\mathbf{e} = [e_1, e_2, ..., e_n] \qquad \mathbf{a} = \begin{bmatrix} a_1 \\ a_2 \\ \vdots \\ a_n \end{bmatrix}$$

$$\mathbf{t} = \mathbf{e}\,\mathbf{a} \tag{A5-4}$$

Now we can also describe the same **t** in terms of a different basis, a primed basis

$$\mathbf{t} = \Sigma_i\, a_i e_i = \Sigma_j\, a_j{}'\, e_j{}'$$
$$\mathbf{t} = \mathbf{e}\,\mathbf{a} = \mathbf{e}'\,\mathbf{a}' \qquad \text{(A5-5)}$$

where the sum is over the n linearly independent vectors in either basis. One basis set can be expressed linearly in terms of the other set. This defines the second basis

$$e_i{}' = \Sigma_j\, e_j\, A_{ji} \qquad i=1,2,\ldots,n$$
$$\mathbf{e}' = \mathbf{e}\,\mathbf{A} \qquad \text{(A5-6)}$$

where **A** is an n × n matrix. From Eqs. A5-5 and A5-6

$$\mathbf{t} = \Sigma_j\, a_j{}'\, e_j{}' = \Sigma_i\, \Sigma_j\, a_j{}'\, e_i\, A_{ij}$$
$$a_i = \Sigma_j\, A_{ij}\, a_j{}' \qquad \text{or} \qquad \mathbf{a} = \mathbf{A}\,\mathbf{a}' \qquad \text{(A5-7)}$$

Thus, we obtain the following expressions for the components and basis vectors which we summarize here:

$$\mathbf{e}' = \mathbf{e}\,\mathbf{A} \qquad \mathbf{e} = \mathbf{e}'\,\mathbf{A}^{-1}$$
$$\mathbf{a}' = \mathbf{A}^{-1}\,\mathbf{a} \qquad \mathbf{a} = \mathbf{A}\,\mathbf{a}' \qquad \text{(A5-8)}$$

(It should be noted that for orthogonal transformations $\mathbf{A}^{-1}$ = A.)

## Linear Operators

An operator T transforms, or maps, a vector **u** in a vector space into a new vector **v** in that space. Thus, there is a rule which associates with every vector **u** a new vector **v**. We can write

$$\mathbf{v} = \mathrm{T}\,\mathbf{u} \qquad \text{(A5-9)}$$

The operator T is a **linear operator** if

$$\mathrm{T}(\mathbf{w} + \mathbf{z}) = \mathrm{T}\mathbf{w} + \mathrm{T}\mathbf{z}$$
$$\mathrm{T}(k\,\mathbf{w}) = k\,\mathrm{T}\,\mathbf{w} \qquad \text{(A5-10)}$$

where **w** and **z** are any vectors in the space and k is a number.

A linear operator can be expressed in terms of a matrix called the matrix representation of T. The effect of the operator on a basis vector $e_i$ can be considered.

$$\mathbf{v} = \Sigma_i\, v_i\, e_i \qquad \mathbf{u} = \Sigma_j\, u_j\, e_j$$
$$\mathbf{v} = \mathbf{T}\mathbf{u} = \Sigma_i\, u_i\, (\mathrm{T}\, e_i) = \Sigma_i\, u_i\, e_i{}' \qquad \text{(A5-11)}$$

The effect of the operator on the basis vectors forms a new set of basis vectors. So from Eq. A5-6, 7, and 8 we have

$$v_i = \Sigma_j A_{ij} u_j \quad \text{or} \quad \mathbf{v} = \mathbf{A}\,\mathbf{u} \qquad \text{(A5-12)}$$

The n × n matrix **A** is **matrix representation** of the operator T in Eq.A5-9.

If we transform to a new basis so that $\mathbf{e}' = \mathbf{e}\,\mathbf{B}$, $\mathbf{v}' = \mathbf{B}^{-1}\,\mathbf{v}$, $u' = \mathbf{B}^{-1}\,\mathbf{u}$ etc. as in Eq. A5-8. Then Eq. A5-12 becomes

$$\mathbf{v}' = \mathbf{A}'\,\mathbf{u}' \qquad \mathbf{v}' = (\mathbf{B}^{-1}\,\mathbf{A}\,\mathbf{B})\mathbf{u}' \qquad \text{(A5-13)}$$

$A'$ and A are equivalent representations, differing only by a choice of basis.

## Vector Space

A **vector space** is a linear space where a scalar product also is defined. (Sometimes called inner product or Hermitian scalar product.) The scalar product $<\mathbf{u} | \mathbf{v}>$ has the following properties: (a) $<\mathbf{u} | \mathbf{v}> = <\mathbf{v} | \mathbf{u}>^*$; (b) $<\mathbf{u} | k\,\mathbf{v}> = k<\mathbf{u} | \mathbf{v}>$ and $<k\mathbf{u} | \mathbf{v}> = k^*<\mathbf{u} | \mathbf{v}>$ where k is a number; (c) $<\mathbf{u}+\mathbf{v} | \mathbf{w}> = <\mathbf{u} | \mathbf{w}> + <\mathbf{v} | \mathbf{w}>$; (d) $<\mathbf{u} | \mathbf{u}>$ is a real and positive number. Note that this is different from the normal "dot" product where a complex conjugate sign would not appear.

We list some definitions and properties of vector spaces. (1) The length of a vector is $|\mathbf{u}| \equiv \sqrt{<\mathbf{u} | \mathbf{u}>}$; (2) Two vectors are orthogonal if $<\mathbf{u} | \mathbf{v}> = 0$; (3) We will always pick an orthogonal basis normalized to one $<e_i | e_j> = \delta_{ij}$. (4) The scalar product can now be clearly stated.

$$<\mathbf{u} | \mathbf{v}> = <[\Sigma_i u_i e_i] \mid [\Sigma_j v_j e_j]> = \Sigma\, u_i^*\, v_j\, \delta_{ij} = \Sigma\, u_i^*\, v_i \qquad \text{(A5-14)}$$

Again note how this differs from the "dot" product.

## Unitary Operators

If **U** is a unitary matrix, $\mathbf{U}^\dagger\mathbf{U} = 1$, we have

$$\Sigma_j\, (U^\dagger)_{ij}\, U_{jk} = \delta_{ik} = \Sigma_j\, U_{ji}^*\, U_{jk} \qquad \text{(A5-15)}$$

We want to show that for a unitary linear operator

$$<U\mathbf{u} | U\mathbf{v}> = <\mathbf{u} | \mathbf{v}> \qquad \text{(A5-16)}$$

A unitary operator is an operator that generates a unitary matrix in Eq. A5-12 as a transformation matrix. Expressing the arbitary vectors in terms of the basis vectors, writing the scalar product in terms of components, gathering terms, of using Eq. A5-15, we have for an orthonormal basis set.

$$\begin{aligned}\langle U\mathbf{u} \mid U\mathbf{v}\rangle &= \langle [\Sigma\ U u_i e_i] \mid [\Sigma\ U v_j e_j]\rangle = [\Sigma_{ik}\ u_i\ U_{ki} e_k]^* [\Sigma_{jl}\ v_j U_{lj} e_l] \\ &= \Sigma_{ijkl}\ u_i^* v_j\ U_{ki}^* U_{lj} e_k^* e_l = \Sigma_{ijk}\ u_i^*\ v_j\ U_{ki}^*\ U_{kj} = \Sigma_i\ u_i^*\ v_i \\ &= \langle \mathbf{u} \mid \mathbf{v}\rangle \end{aligned} \qquad \text{(A5-17)}$$

Thus, for an orthonormal basis, Eq. A5-16 is proved where U is a unitary operator. The physical importance of a unitary transformation is now clear via Eq. A5-16. A **unitary transformation** on a vector space is an orthogonal transformation in the sense that the lengths of vector and the angles between them remain the same after the transformation.

## Hermitian Operators

For a Hermitian matrix, **H**, we have $\mathbf{H}^\dagger = \mathbf{H}$ so $(H^\dagger)_{ij} = H_{ij} = H_{ji}^*$. A Hermitian operator is an operator that generates a Hermitian matrix in Eq. A5-12 as a transformation matrix. For a Hermitian operator we want to show that

$$\langle H\mathbf{u} \mid \mathbf{v}\rangle = \langle \mathbf{u} \mid H\mathbf{v}\rangle \qquad \text{(A5-18)}$$

This can be done in a manner completely analogous to Eq. A5-17 using $\langle e_i \mid e_j\rangle = \delta_{ij}$. In this proof one shows that for any linear operator T,

$$\langle T\mathbf{u} \mid \mathbf{v}\rangle = \langle \mathbf{u} \mid T^\dagger \mathbf{v}\rangle \qquad \text{(A5-19)}$$

Using this relation one could directly prove the result in Eq. A5-16, i.e. $\langle U\mathbf{u} \mid U\mathbf{v}\rangle = \langle \mathbf{u} \mid U^\dagger U\mathbf{v}\rangle = \langle \mathbf{u} \mid \mathbf{v}\rangle$ if U is unitary.

## Other Points about Operators

Equation A5-13 describes how operators transform in different bases. In quantum mechanics the operators are usually unitary so

$$\mathbf{A}' = \mathbf{B}^{-1}\mathbf{A}\mathbf{B} = \mathbf{B}^\dagger\mathbf{A}\mathbf{B} \qquad \text{(A5-20)}$$

When calculating matrix elements in quantum mechanics the choice of basis is arbitrary so the most convenient one should be used. Equation A5-20 can be used to transform the operators.

Another useful result is if H is Hermitian then U is unitary in

$$U = e^{iH} \tag{A5-21}$$

By an exponential operator one means $e^T = \Sigma\ 1/n!\ T^n$ and remember that $e^T e^V = e^{T+V}$ only if T and V commute. However, T and $-T$ commute, so $(e^T)^{-1} = e^{-T}$. The theorem, Eq. A5-21, is easily proved, remembering that $H^n = H \times H \times H \ldots$,

$$U^\dagger = \sum_{n=0}^{\infty} (1/n!)\ [(iH)^n]^\dagger = \Sigma\ (1/n!)\ (-iH)^n = e^{-iH} \tag{A5-22}$$

so $U^\dagger = U^{-1}$. QED In the same manner the converse can be proved, namely in Eq. A5-21 if U is unitary then H is Hermitian.

## Representations and Transformation of Functions

Matrix representations of the groups can be obtained by considering how the basis vectors or the components transform. The relation between the two results is in Eq. A5-8 and some examples of this are in Chapter 3. The standard form of the irrepresentations in this book are obtained by considering how the orthogonal basis vectors transform, $\mathbf{e}'$, $\mathbf{eA}$ from Eq. A5-8, in Euclidean 3-space. Thus, if the transformation is a symmetry operation R we have

$$\mathbf{Re} = \mathbf{e}' = \mathbf{eR} \equiv \mathbf{e}\,\Gamma(R) \tag{A5-23}$$

where $\mathbf{e}$ is $1 \times 3$ row matrix as in Eq. A5-4. The expression Eq. A5-23 can be usually thought of just as $\mathbf{e}' = \mathbf{e}\Gamma(R)$. For each of the h symmetry operations a $\Gamma(R)$ can be obtained which form a $3 \times 3$ representation of the group. For the noncubic groups this will be a reducible representation. To obtain other representations d, f, g, etc. type functions may be considered.

When a symmetry operation is applied to one basis function a linear combination of its partners are obtained.

$$R\psi_i = \Sigma_j\ \psi_j\ \Gamma(R)_{ji} \tag{A5-24}$$

Writing the functions as a $1 \times n$ one can write

$$R[\psi_1, \psi_2, \ldots \psi_n] = [\psi_1, \psi_2, \ldots \psi_n]\ \Gamma(R) \tag{A5-25}$$

which is the same form as the equation for the basis vectors, Eq. A5-23. (Thus, the functions transform cogrediently to the basis vectors. The components transform contragredently to the basis vectors, Eq. A5-8.) In Section 4-4 we prove that the set of matrices in Eq. A5-24 and 25 indeed is a representation of the group. It should be noted that in this proof if one reverses the order of the subscripts, i.e. $R\psi_i = \Sigma\ \Gamma(R)_{ij}\ \psi_j$, then one obtains $\Gamma(C)_{ik} = \Sigma\ \Gamma(B)_{ij}\ \Gamma(A)_{jk} = \Gamma(BA)_{ik}$. This is opposite to the order of the symmetry operations since we have C = AB in the proof Eq. 4-9.

# Appendix 6

## DIRECT PRODUCT TABLES

The reduction of the direct products of the irreducible representations of the point groups O, $D_6$, $D_4$ is given. The double group irreducible representations are included. These results can immediately be extended to the point groups $O_h$, $D_{6h}$, $D_{4h}$ since only g and u-subscripts are added to the irreducible representations. The direct product rules for these subscripts are $g \times g = g$, $g \times u = u$, $u \times u = g$. (See Wilson, Decius, and Cross, Appendix 10, for the extensive rules for the results for the direct product of two irreducible representations.)

The symmetric and antisymmetric direct products of irreducible representations are also given for these point groups. See Chapter 6 for a detailed discussion of the symmetric and antisymmetric product.

### Reduced Product Tables

| $D_4$ | $A_1$ | $A_2$ | $B_1$ | $B_2$ | $E$ | $E_{1/2}$ | $E_{3/2}$ |
|---|---|---|---|---|---|---|---|
| $A_1$ | $A_1$ | $A_2$ | $B_1$ | $B_2$ | $E$ | $E_{1/2}$ | $E_{3/2}$ |
| $A_2$ | $A_2$ | $A_1$ | $B_2$ | $B_1$ | $E$ | $E_{1/2}$ | $E_{3/2}$ |
| $B_1$ | $B_1$ | $B_2$ | $A_1$ | $A_2$ | $E$ | $E_{3/2}$ | $E_{1/2}$ |
| $B_2$ | $B_2$ | $B_1$ | $A_2$ | $A_1$ | $E$ | $E_{3/2}$ | $E_{1/2}$ |
| $E$ | $E$ | $E$ | $E$ | $E$ | $A_1 + A_2 + B_1 + B_2$ | $E_{1/2} + E_{3/2}$ | $E_{1/2} + E_{3/2}$ |
| $E_{1/2}$ | $E_{1/2}$ | $E_{1/2}$ | $E_{3/2}$ | $E_{3/2}$ | $E_{1/2} + E_{3/2}$ | $A_1 + A_2 + E$ | $B_1 + B_2 + E$ |
| $E_{3/2}$ | $E_{3/2}$ | $E_{3/2}$ | $E_{1/2}$ | $E_{1/2}$ | $E_{1/2} + E_{3/2}$ | $B_1 + B_2 + E$ | $A_1 + A_2 + E$ |

| O | $A_1$ | $A_2$ | $E$ | $T_1$ | $T_2$ | $E_{1/2}$ | $E_{5/2}$ | $G$ |
|---|---|---|---|---|---|---|---|---|
| $A_1$ | $A_1$ | $A_2$ | $E$ | $T_1$ | $T_2$ | $E_{1/2}$ | $E_{5/2}$ | $G$ |
| $A_2$ | $A_2$ | $A_1$ | $E$ | $T_2$ | $T_1$ | $E_{5/2}$ | $E_{1/2}$ | $G$ |
| $E$ | $E$ | $E$ | $A_1 + A_2 + E$ | $T_1 + T_2$ | $T_1 + T_2$ | $G$ | $G$ | $E_{1/2} + E_{5/2} + G$ |
| $T_1$ | $T_1$ | $T_2$ | $T_1 + T_2$ | $A_1 + E + T_1 + T_2$ | $A_2 + E + T_1 + T_2$ | $E_{1/2} + G$ | $E_{5/2} + G$ | $E_{1/2} + E_{5/2} + 2G$ |
| $T_2$ | $T_2$ | $T_1$ | $T_1 + T_2$ | $A_2 + E + T_1 + T_2$ | $A_1 + E + T_1 + T_2$ | $E_{5/2} + G$ | $E_{1/2} + G$ | $E_{1/2} + E_{5/2} + 2G$ |
| $E_{1/2}$ | $E_{1/2}$ | $E_{5/2}$ | $G$ | $E_{1/2} + G$ | $E_{5/2} + G$ | $A_1 + T_1$ | $A_2 + T_2$ | $E + T_1 + T_2$ |
| $E_{5/2}$ | $E_{5/2}$ | $E_{1/2}$ | $G$ | $E_{5/2} + G$ | $E_{1/2} + G$ | $A_2 + T_2$ | $A_1 + T_1$ | $E + T_1 + T_2$ |
| $G$ | $G$ | $G$ | $E_{1/2} + E_{5/2} + G$ | $E_{1/2} + E_{5/2} + 2G$ | $E_{1/2} + E_{5/2} + 2G$ | $E + T_1 + T_2$ | $E + T_1 + T_2$ | $A_1 + A_2 + E + 2T_1 + 2T_2$ |

| $D_6$ | $A_1$ | $A_2$ | $B_1$ | $B_2$ | $E_1$ | $E_2$ | $E_{1/2}$ | $E_{3/2}$ | $E_{5/2}$ |
|---|---|---|---|---|---|---|---|---|---|
| $A_1$ | $A_1$ | $A_2$ | $B_1$ | $B_2$ | $E_1$ | $E_2$ | $E_{1/2}$ | $E_{3/2}$ | $E_{5/2}$ |
| $A_2$ | $A_2$ | $A_1$ | $B_2$ | $B_1$ | $E_1$ | $E_2$ | $E_{1/2}$ | $E_{3/2}$ | $E_{5/2}$ |
| $B_1$ | $B_1$ | $B_2$ | $A_1$ | $A_2$ | $E_2$ | $E_1$ | $E_{5/2}$ | $E_{3/2}$ | $E_{1/2}$ |
| $B_2$ | $B_2$ | $B_1$ | $A_2$ | $A_1$ | $E_2$ | $E_1$ | $E_{5/2}$ | $E_{3/2}$ | $E_{1/2}$ |
| $E_1$ | $E_1$ | $E_1$ | $E_2$ | $E_2$ | $A_1 + A_2 + E_2$ | $B_1 + B_2 + E_1$ | $E_{1/2} + E_{3/2}$ | $E_{1/2} + E_{5/2}$ | $E_{3/2} + E_{5/2}$ |
| $E_2$ | $E_2$ | $E_2$ | $E_1$ | $E_1$ | $B_1 + B_2 + E_1$ | $A_1 + A_2 + E_2$ | $E_{3/2} + E_{5/2}$ | $E_{1/2} + E_{5/2}$ | $E_{1/2} + E_{3/2}$ |
| $E_{1/2}$ | $E_{1/2}$ | $E_{1/2}$ | $E_{5/2}$ | $E_{5/2}$ | $E_{1/2} + E_{3/2}$ | $E_{3/2} + E_{5/2}$ | $A_1 + A_2 + E$ | $B_1 + B_2 + E_2$ | $E_1 + E_2$ |
| $E_{3/2}$ | $E_{3/2}$ | $E_{3/2}$ | $E_{3/2}$ | $E_{3/2}$ | $E_{1/2} + E_{5/2}$ | $E_{1/2} + E_{5/2}$ | $B_1 + B_2 + E_2$ | $A_1 + A_2 + E$ | $E_1 + E_2$ |
| $E_{5/2}$ | $E_{5/2}$ | $E_{5/2}$ | $E_{1/2}$ | $E_{1/2}$ | $E_{3/2} + E_{5/2}$ | $E_{1/2} + E_{3/2}$ | $E_1 + E_2$ | $E_1 + E_2$ | $A_1 + A_2 + E_1 + E_2$ |

## Symmetric and Antisymmetric Direct Product

| $O^*$ and $T_d^*$ | | |
|---|---|---|
| $\Gamma$ | $[\Gamma^2]_S$ | $\{\Gamma^2\}_{AS}$ |
| $A_1$ | $A_1$ | . . . |
| $A_2$ | $A_1$ | . . . |
| $E$ | $A_1 + E$ | $A_2$ |
| $T_1$ | $A_1 + E + T_2$ | $T_1$ |
| $T_2$ | $A_1 + E + T_2$ | $T_1$ |
| $E_{1/2}$ | $T_1$ | $A_1$ |
| $E_{5/2}$ | $T_1$ | $A_1$ |
| $G$ | $A_2 + 2T_1 + T_2$ | $A_1 + E + T_2$ |

| $D_4^*$ | | | $D_6^*$ | | |
|---|---|---|---|---|---|
| $\Gamma$ | $[\Gamma^2]_S$ | $\{\Gamma^2\}_{AS}$ | $\Gamma$ | $[\Gamma^2]_S$ | $\{\Gamma^2\}_{AS}$ |
| $A_1$ | $A_1$ | . . . | $A_1$ | $A_1$ | . . . |
| $A_2$ | $A_1$ | . . . | $A_2$ | $A_1$ | . . . |
| $B_1$ | $A_1$ | . . . | $B_1$ | $A_1$ | . . . |
| $B_2$ | $A_1$ | . . . | $B_2$ | $A_1$ | . . . |
| $E$ | $A_1 + B_1 + B_2$ | $A_2$ | $E_1$ | $A_1 + E_2$ | $A_2$ |
| $E_{1/2}$ | $A_2 + E$ | $A_1$ | $E_2$ | $A_1 + E_2$ | $A_2$ |
| $E_{3/2}$ | $A_2 + E$ | $A_1$ | $E_{1/2}$ | $A_2 + E_1$ | $A_1$ |
| | | | $E_{3/2}$ | $A_2 + E_1$ | $A_1$ |
| | | | $E_{5/2}$ | $A_2 + E_1 + E_2$ | $A_1$ |

Appendix 7

## CORRELATION TABLES

These tables show how the irreducible representations of various point groups correlate with the irreducible representations of some of their subgroups. For some of the correlations care must be taken as to which axes and planes go into each other. (See Section 7-5c.)

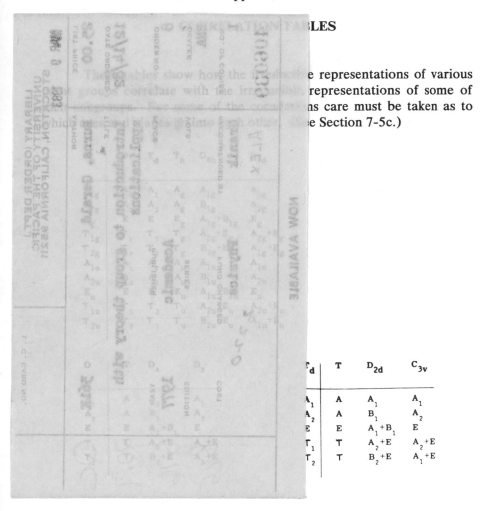

| $O_h$ | $O$ | $T_d$ | $T_h$ | $D_{4h}$ | $D_{3d}$ |
|---|---|---|---|---|---|
| $A_{1g}$ | $A_1$ | $A_1$ | $A_g$ | $A_{1g}$ | $A_{1g}$ |
| $A_{2g}$ | $A_2$ | $A_2$ | $A_g$ | $B_{1g}$ | $A_{2g}$ |
| $E_g$ | $E$ | $E$ | $E_g$ | $A_{1g}+B_{1g}$ | $E_g$ |
| $T_{1g}$ | $T_1$ | $T_1$ | $T_g$ | $A_{2g}+E_g$ | $A_{2g}+E_g$ |
| $T_{2g}$ | $T_2$ | $T_2$ | $T_g$ | $B_{2g}+E_g$ | $A_{1g}+E_g$ |
| $A_{1u}$ | $A_1$ | $A_2$ | $A_u$ | $A_{1u}$ | $A_{1u}$ |
| $A_{2u}$ | $A_2$ | $A_1$ | $A_u$ | $B_{1u}$ | $A_{2u}$ |
| $E_u$ | $E$ | $E$ | $E_u$ | $A_{1u}+B_{1u}$ | $E_u$ |
| $T_{1u}$ | $T_1$ | $T_2$ | $T_u$ | $A_{2u}+E_u$ | $A_{2u}+E_u$ |
| $T_{2u}$ | $T_2$ | $T_1$ | $T_u$ | $B_{2u}+E_u$ | $A_{1u}+E_u$ |

| $O$ | $D_4$ | $D_3$ |
|---|---|---|
| $A_1$ | $A_1$ | $A_1$ |
| $A_2$ | $B_1$ | $A_2$ |
| $E$ | $A_1+B_1$ | $E$ |
| $T_1$ | $A_2+E$ | $A_2+E$ |
| $T_2$ | $B_2+E$ | $A_1+E$ |

| $T_d$ | $T$ | $D_{2d}$ | $C_{3v}$ |
|---|---|---|---|
| $A_1$ | $A$ | $A_1$ | $A_1$ |
| $A_2$ | $A$ | $B_1$ | $A_2$ |
| $E$ | $E$ | $A_1+B_1$ | $E$ |
| $T_1$ | $T$ | $A_2+E$ | $A_2+E$ |
| $T_2$ | $T$ | $B_2+E$ | $A_1+E$ |

| $T_h$ | $T$ | $D_{2h}$ | $S_6$ |
|---|---|---|---|
| $A_g$ | $A$ | $A_g$ | $A_g$ |
| $E_g$ | $E$ | $2A_g$ | $E_g$ |
| $T_g$ | $T$ | $B_{1g}+B_{2g}+B_{3g}$ | $A_g+E_g$ |
| $A_u$ | $A$ | $A_u$ | $A_u$ |
| $E_u$ | $E$ | $2A_u$ | $E_u$ |
| $T_u$ | $T$ | $B_{1u}+B_{2u}+B_{3u}$ | $A_u+E_u$ |

| $T$ | $D_2$ | $C_3$ | $C_2$ |
|---|---|---|---|
| $A$ | $A$ | $A$ | $A$ |
| $E$ | $2A$ | $E$ | $2A$ |
| $T$ | $B_1+B_2+B_3$ | $A+E$ | $A+2B$ |

| $D_{6h}$ | $D_6$ | $C'_2$<br>$D_{3h}$ | $C''_2$<br>$D_{3h}$ | $C'_2$<br>$D_{3d}$ | $C''_2$<br>$D_{3d}$ | $C_{6h}$ | $C_{6v}$ | $\sigma_v \rightarrow \sigma\,(yz)$<br>$\sigma_h \rightarrow \sigma\,(xy)$<br>$D_{2h}$ |
|---|---|---|---|---|---|---|---|---|
| $A_{1g}$ | $A_1$ | $A'_1$ | $A'_1$ | $A_{1g}$ | $A_{1g}$ | $A_g$ | $A_1$ | $A_g$ |
| $A_{2g}$ | $A_2$ | $A'_2$ | $A'_2$ | $A_{2g}$ | $A_{2g}$ | $A_g$ | $A_2$ | $B_{1g}$ |
| $B_{1g}$ | $B_1$ | $A''_1$ | $A''_2$ | $A_{1g}$ | $A_{2g}$ | $B_g$ | $B_2$ | $B_{2g}$ |
| $B_{2g}$ | $B_2$ | $A''_2$ | $A''_1$ | $A_{2g}$ | $A_{1g}$ | $B_g$ | $B_1$ | $B_{3g}$ |
| $E_{1g}$ | $E_1$ | $E''$ | $E''$ | $E_g$ | $E_g$ | $E_{1g}$ | $E_1$ | $B_{2g}+B_{3g}$ |
| $E_{2g}$ | $E_2$ | $E'$ | $E'$ | $E_g$ | $E_g$ | $E_{2g}$ | $E_2$ | $A_g+B_{1g}$ |
| $A_{1u}$ | $A_1$ | $A''_1$ | $A''_1$ | $A_{1u}$ | $A_{1u}$ | $A_u$ | $A_2$ | $A_u$ |
| $A_{2u}$ | $A_2$ | $A''_2$ | $A''_2$ | $A_{2u}$ | $A_{2u}$ | $A_u$ | $A_1$ | $B_{1u}$ |
| $B_{1u}$ | $B_1$ | $A'_1$ | $A'_2$ | $A_{1u}$ | $A_{2u}$ | $B_u$ | $B_1$ | $B_{2u}$ |
| $B_{2u}$ | $B_2$ | $A'_2$ | $A'_1$ | $A_{2u}$ | $A_{1u}$ | $B_u$ | $B_2$ | $B_{3u}$ |
| $E_{1u}$ | $E_1$ | $E'$ | $E'$ | $E_u$ | $E_u$ | $E_{1u}$ | $E_1$ | $B_{2u}+B_{3u}$ |
| $E_{2u}$ | $E_2$ | $E''$ | $E''$ | $E_u$ | $E_u$ | $E_{2u}$ | $E_2$ | $A_u+B_{1u}$ |

| $D_{4h}$ | $D_4$ | $C'_2$<br>$D_{2h}$ | $C''_2$<br>$D_{2h}$ | $C'_2$<br>$D_{2d}$ | $C''_2$<br>$D_{2d}$ | $C_{4h}$ | $C_{4v}$ | $C_4$ | $S_4$ |
|---|---|---|---|---|---|---|---|---|---|
| $A_{1g}$ | $A_1$ | $A_g$ | $A_g$ | $A_1$ | $A_1$ | $A_g$ | $A_1$ | $A$ | $A$ |
| $A_{2g}$ | $A_2$ | $B_{1g}$ | $B_{1g}$ | $A_2$ | $A_2$ | $A_g$ | $A_2$ | $A$ | $A$ |
| $B_{1g}$ | $B_1$ | $A_g$ | $B_{1g}$ | $B_1$ | $B_2$ | $B_g$ | $B_1$ | $B$ | $B$ |
| $B_{2g}$ | $B_2$ | $B_{1g}$ | $A_g$ | $B_2$ | $B_1$ | $B_g$ | $B_2$ | $B$ | $B$ |
| $E_g$ | $E$ | $B_{2g}+B_{3g}$ | $B_{2g}+B_{3g}$ | $E$ | $E$ | $E_g$ | $E$ | $E$ | $E$ |
| $A_{1u}$ | $A_1$ | $A_u$ | $A_u$ | $B_1$ | $B_1$ | $A_u$ | $A_2$ | $A$ | $B$ |
| $A_{2u}$ | $A_2$ | $B_{1u}$ | $B_{1u}$ | $B_2$ | $B_2$ | $A_u$ | $A_1$ | $A$ | $B$ |
| $B_{1u}$ | $B_1$ | $A_u$ | $B_{1u}$ | $A_1$ | $A_2$ | $B_u$ | $B_2$ | $B$ | $A$ |
| $B_{2u}$ | $B_2$ | $B_{1u}$ | $A_u$ | $A_2$ | $A_1$ | $B_u$ | $B_1$ | $B$ | $A$ |
| $E_u$ | $E$ | $B_{2u}+B_{3u}$ | $B_{2u}+B_{3u}$ | $E$ | $E$ | $E_u$ | $E$ | $E$ | $E$ |

| $D_{3h}$ | $D_3$ | $C_{3h}$ | $C_{3v}$ | $C_{2v}$ ($\sigma_h \to \sigma(yz)$) |
|---|---|---|---|---|
| $A'_1$ | $A_1$ | $A'$ | $A_1$ | $A_1$ |
| $A'_2$ | $A_2$ | $A'$ | $A_2$ | $B_2$ |
| $E'$ | $E$ | $E'$ | $E$ | $A_1+B_2$ |
| $A''_1$ | $A_1$ | $A''$ | $A_2$ | $A_2$ |
| $A''_2$ | $A_2$ | $A''$ | $A_1$ | $B_1$ |
| $E''$ | $E$ | $E''$ | $E$ | $A_2+B_1$ |

| $D_{3d}$ | $S_6$ | $D_3$ | $C_{3v}$ | $C_{2h}$ |
|---|---|---|---|---|
| $A_{1g}$ | $A_g$ | $A_1$ | $A_1$ | $A_g$ |
| $A_{2g}$ | $A_g$ | $A_2$ | $A_2$ | $B_g$ |
| $E_g$ | $E_g$ | $E$ | $E$ | $A_g+B_g$ |
| $A_{1u}$ | $A_u$ | $A_1$ | $A_2$ | $A_u$ |
| $A_{2u}$ | $A_u$ | $A_2$ | $A_1$ | $B_u$ |
| $E_u$ | $E_u$ | $E$ | $E$ | $A_u+B_u$ |

| $C_{6h}$ | $C_6$ | $S_6$ | $C_{3h}$ | $C_{2h}$ |
|---|---|---|---|---|
| $A_g$ | $A$ | $A_g$ | $A'$ | $A_g$ |
| $B_g$ | $B$ | $A_g$ | $A''$ | $B_g$ |
| $E_{1g}$ | $E_1$ | $E_g$ | $E''$ | $2B_g$ |
| $E_{2g}$ | $E_2$ | $E_g$ | $E'$ | $2A_g$ |
| $A_u$ | $A$ | $A_u$ | $A''$ | $A_u$ |
| $B_u$ | $B$ | $A_u$ | $A'$ | $B_u$ |
| $E_{1u}$ | $E_1$ | $E_u$ | $E'$ | $2B_u$ |
| $E_{2u}$ | $E_2$ | $E_u$ | $E''$ | $2A_u$ |

| $C_{6v}$ | $C_6$ | $C_{3v}$ ($\sigma_v$) | $C_{3v}$ ($\sigma_d$) | $C_{2v}$ ($\sigma_v \to \sigma(xz)$) |
|---|---|---|---|---|
| $A_1$ | $A$ | $A_1$ | $A_1$ | $A_1$ |
| $A_2$ | $A$ | $A_2$ | $A_2$ | $A_2$ |
| $B_1$ | $B$ | $A_1$ | $A_2$ | $B_1$ |
| $B_2$ | $B$ | $A_2$ | $A_1$ | $B_2$ |
| $E_1$ | $E_1$ | $E$ | $E$ | $B_1+B_2$ |
| $E_2$ | $E_2$ | $E$ | $E$ | $A_1+A_2$ |

| $D_6$ | $C_6$ | $D_3$ ($C'_2$) | $D_3$ ($C''_2$) | $D_2$ | $C_3$ |
|---|---|---|---|---|---|
| $A_1$ | $A$ | $A_1$ | $A_1$ | $A$ | $A$ |
| $A_2$ | $A$ | $A_2$ | $A_2$ | $B_1$ | $A$ |
| $B_1$ | $B$ | $A_1$ | $A_2$ | $B_2$ | $A$ |
| $B_2$ | $B$ | $A_2$ | $A_1$ | $B_3$ | $A$ |
| $E_1$ | $E_1$ | $E$ | $E$ | $B_2+B_3$ | $E$ |
| $E_2$ | $E_2$ | $E$ | $E$ | $A+B_1$ | $E$ |

| $D_4$ | $C_4$ | $D_2$ ($C'_2$) | $D_2$ ($C''_2$) |
|---|---|---|---|
| $A_1$ | $A$ | $A$ | $A$ |
| $A_2$ | $A$ | $B_1$ | $B_1$ |
| $B_1$ | $B$ | $A$ | $B_1$ |
| $B_2$ | $B$ | $B_1$ | $A$ |
| $E$ | $E$ | $B_2+B_3$ | $B_2+B_3$ |

| $D_{2d}$ | $S_4$ | $D_2$ ($C_2 \to C_2$) | $C_{2v}$ |
|---|---|---|---|
| $A_1$ | A | A | $A_1$ |
| $A_2$ | A | $B_1$ | $A_2$ |
| $B_1$ | B | A | $A_2$ |
| $B_2$ | B | $B_1$ | $A_1$ |
| E | E | $B_2+B_3$ | $B_1+B_2$ |

| $D_{2h}$ | $D_2$ | $C_{2h}$ ($C_2$) | $C_{2h}$ ($C_2(x)$) | $C_{2v}$ ($C_2$) | $C_{2v}$ ($C_2(x)$) |
|---|---|---|---|---|---|
| $A_g$ | A | $A_g$ | $A_g$ | $A_1$ | $A_1$ |
| $B_{1g}$ | $B_1$ | $A_g$ | $B_g$ | $A_2$ | $B_1$ |
| $B_{2g}$ | $B_2$ | $B_g$ | $B_g$ | $B_1$ | $B_2$ |
| $B_{3g}$ | $B_3$ | $B_g$ | $A_g$ | $B_2$ | $A_2$ |
| $A_u$ | A | $A_u$ | $A_u$ | $A_2$ | $A_2$ |
| $B_{1u}$ | $B_1$ | $A_u$ | $B_u$ | $A_1$ | $B_2$ |
| $B_{2u}$ | $B_2$ | $B_u$ | $B_u$ | $B_2$ | $B_1$ |
| $B_{3u}$ | $B_3$ | $B_u$ | $A_u$ | $B_1$ | $A_1$ |

| $C_{4h}$ | $C_4$ | $S_4$ | $C_{2h}$ |
|---|---|---|---|
| $A_g$ | A | A | $A_g$ |
| $B_g$ | B | B | $A_g$ |
| $E_g$ | E | E | $2B_g$ |
| $A_u$ | A | B | $A_u$ |
| $B_u$ | B | A | $A_u$ |
| $E_u$ | E | E | $2B_u$ |

| $C_{4v}$ | $C_4$ | $C_{2v}$ ($\sigma_v$) | $C_{2v}$ ($\sigma_d$) |
|---|---|---|---|
| $A_1$ | A | $A_1$ | $A_1$ |
| $A_2$ | A | $A_2$ | $A_2$ |
| $B_1$ | B | $A_1$ | $A_2$ |
| $B_2$ | B | $A_2$ | $A_1$ |
| E | E | $B_1+B_2$ | $B_1+B$ |

| $C_{3h}$ | $C_3$ | $C_s$ |
|---|---|---|
| A' | A | A' |
| E' | E | 2A' |
| A" | A | A" |
| E" | E | 2A" |

| $C_6$ | $C_3$ | $C_2$ |
|---|---|---|
| A | A | A |
| B | A | B |
| $E_1$ | E | 2B |
| $E_2$ | E | 2A |

| $D_3$ | $C_3$ | $C_2$ |
|---|---|---|
| $A_1$ | A | A |
| $A_2$ | A | B |
| E | E | A+B |

| $S_6$ | $C_3$ | $C_i$ |
|---|---|---|
| $A_g$ | A | $A_g$ |
| $E_g$ | E | $2A_g$ |
| $A_u$ | A | $A_u$ |
| $E_u$ | E | $2A_u$ |

| $C_{3v}$ | $C_3$ | $C_s$ |
|---|---|---|
| $A_1$ | A | A |
| $A_2$ | A | A" |
| E | E | A'+A" |

| $C_4$ | $C_2$ |
|---|---|
| A | A |
| B | A |
| E | 2B |

| $S_4$ | $C_2$ |
|---|---|
| A | A |
| B | A |
| E | 2B |

| $D_2$ | $C_2$ | $C_2(x)$ |
|---|---|---|
| A | A | A |
| $B_1$ | A | B |
| $B_2$ | B | B |
| $B_3$ | B | A |

Appendix 8

# SPHERICAL HARMONICS

The spherical harmonics are given for $l$ = 0, 1, 2, and 3. The appropriate linear combinations are also given that transform as the partners of the appropriate irreducible representations of the point group $O_h$.

The angles $\theta$ and $\phi$ are defined in Fig. 8-1b. The Legendre polynomials are functions of $\theta$ only and are defined as,

$$P^l = \frac{1}{2^l l!} \frac{d^l(\cos^2\theta - 1)}{d(\cos\theta)^l} \tag{A8-1a}$$

The spherical harmonics are functions of $\theta$ and $\phi$ and are defined for m = $l, l-1, ..., -l$ as

$$Y_m{}^l = (-1)^m \left[\frac{(l-m)!}{(l+m)!} \frac{2l+1}{4\pi}\right]^{1/2} e^{im\phi} \sin^m \theta \frac{d^m(P^l)}{d(\cos\theta)^m} \tag{A8-1b}$$

Both functions are normalized to one

$$\int_0^{\pi} (P^l)^2 \sin\theta \, d\theta = 1 = \int_{\phi=0}^{2\pi} \int_0^{\pi} (Y_m{}^{l*} Y_m{}^l) \sin\theta \, d\theta \, d\phi \tag{A8-2}$$

Some values of $P^l$ are

$$P^0 = 1 \qquad P^1 = \cos\theta$$
$$P^2 = \tfrac{1}{2}(3\cos^2\theta - 1) \qquad P^3 = \tfrac{1}{2}(5\cos^3\theta - 3\cos\theta)$$

Both functions are also normalized, i.e., the integrals in Eq. A8-2 become $\langle P^j | P^l \rangle = 2\delta_{jl}/(2l+1)$ and $\langle Y_m{}^j | Y_n{}^l \rangle = \delta_{jl}\delta_{mn}$. Some other useful properties of the Legendre polynomials and spherical harmonics are:

$$P^l(\theta = 0) = 1 \qquad P(\theta) = (-1)^l P^l(\pi - \theta)$$
$$Y_m{}^l(\theta = 0, \phi) = \delta_{m,0}(2l+1/4\pi)^{1/2} \qquad Y_0{}^l = (2l+1/4\pi)^{1/2} P^l$$
$$(Y_m{}^l)^* = (-1)^m Y_{-m}{}^l \qquad Y_m{}^l(\pi-\theta, \phi+\pi) = (-1)^l Y(\theta,\phi)$$

The spherical harmonics are given for $l$ = 0, 1, 2, and 3. The appropriate linear combinations that transform as the partners of the appropriate irreducible representations of the point group $O_h$ are also given.

## Spherical Harmonics

$$Y_0^0 = \sqrt{\frac{1}{4\pi}}$$

$$\alpha_1 = \sqrt{\frac{3}{4\pi}}$$

$$Y_{-1}^1 = \alpha_1 \frac{1}{\sqrt{2}} \frac{(x-iy)}{r} = \alpha_1 \frac{1}{\sqrt{2}} \sin\theta\, e^{-i\phi}$$

$$Y_0^1 = \alpha_1 \frac{z}{r} = \alpha_1 \cos\theta$$

$$Y_1^1 = -\alpha_1 \frac{1}{\sqrt{2}} \frac{(x+iy)}{r} = -\alpha_1 \frac{1}{\sqrt{2}} \sin\theta\, e^{+i\phi}$$

$$\alpha_2 = \sqrt{\frac{5}{4\pi}}$$

$$Y_{-2}^2 = \alpha_2 \sqrt{\frac{3}{8}} \frac{(x-iy)^2}{r^2} = \alpha_2 \sqrt{\frac{3}{8}} \sin^2\theta\, e^{-2i\phi}$$

$$Y_{-1}^2 = \alpha_2 \sqrt{\frac{3}{2}} \frac{z(x-iy)}{r^2} = \alpha_2 \sqrt{\frac{3}{2}} \sin\theta \cos\theta\, e^{-i\phi}$$

$$Y_0^2 = \alpha_2 \sqrt{\frac{1}{4}} \frac{3z^2-r^2}{r^2} = \alpha_2 \sqrt{\frac{1}{4}} (3\cos^2\theta-1)$$

$$Y_1^2 = -\alpha_2 \sqrt{\frac{3}{2}} \frac{z(x+iy)}{r^2} = -\alpha_2 \sqrt{\frac{3}{2}} \sin\theta \cos\theta\, e^{+i\phi}$$

$$Y_2^2 = \alpha_2 \sqrt{\frac{3}{8}} \frac{(x+iy)^2}{r^2} = \alpha_2 \sqrt{\frac{3}{8}} \sin^2\theta\, e^{+2i\phi}$$

$$\alpha_3 = \sqrt{\frac{7}{4\pi}}$$

$$Y_{-3}^3 = \alpha_3 \sqrt{\frac{5}{16}} \frac{(x-iy)^3}{r^3} = \alpha_3 \sqrt{\frac{5}{16}} \sin^3\theta\, e^{-3i\phi}$$

$$Y_{-2}^3 = \alpha_3 \sqrt{\frac{15}{8}} \frac{z(x-iy)^2}{r^3} = \alpha_3 \sqrt{\frac{15}{8}} \sin^2\theta \cos\theta\, e^{-2i\phi}$$

$$Y_{-1}^3 = \alpha_3 \sqrt{\frac{3}{16}} \frac{(5z^2-r^2)(x-iy)}{r^3} = \alpha_3 \sqrt{\frac{3}{16}} \sin\theta(5\cos^2\theta-1)e^{-i\phi}$$

$$Y_0^3 = \alpha_3 \sqrt{\frac{1}{4}} \frac{z(5z^2-3r^2)}{r^3} = \alpha_3 \sqrt{\frac{1}{4}} \cos\theta(5\cos^2\theta-3)$$

$$Y_1^3 = -\alpha_3\sqrt{\frac{3}{16}}\,\frac{(5z^2-r^2)(x+iy)}{r^3} = -\alpha_3\sqrt{\frac{3}{16}}\,\sin\theta(5\cos^2\theta-1)e^{+i\phi}$$

$$Y_2^3 = \alpha_3\sqrt{\frac{15}{8}}\,\frac{z(x+iy)^2}{r^3} = \alpha_3\sqrt{\frac{15}{8}}\,\sin^2\theta\,\cos\theta\,e^{+2i\phi}$$

$$Y_3^3 = -\alpha_3\sqrt{\frac{5}{16}}\,\frac{(x+iy)^3}{r^3} = -\alpha_3\sqrt{\frac{5}{16}}\,\sin^3\theta\,e^{+i\phi}$$

| $\ell$ | Irred. Rep. of $O_h$ | Partners |
|---|---|---|
| 0 | $A_{1g}$ $(\Gamma_1^+)$ | $Y_0^0$ |
| 1 | $T_{1u}$ $(\Gamma_4^-)$ | $Y_{-1}^1$ |
| | | $Y_0^1$ |
| | | $-Y_1^1$ |
| 2 | $E_g$ $(\Gamma_3^+)$ | $\frac{1}{\sqrt{2}}(Y_2^2 + Y_{-2}^2)$ |
| | | $Y_0^2$ |
| | $T_{2g}$ $(\Gamma_5^+)$ | $\frac{1}{\sqrt{2}}(Y_2^2 - Y_{-2}^2)$ |
| | | $Y_1^2$ |
| | | $Y_{-1}^2$ |

| | | |
|---|---|---|
| 3 | $A_{2u}$ $(\Gamma_2^-)$ | $\frac{1}{\sqrt{2}}(Y_2^3 - Y_{-2}^3)$ |
| | $T_{1u}$ $(\Gamma_4^-)$ | $\begin{cases} \sqrt{3/8}\; Y_1^3 + \sqrt{5/8}\; Y_{-3}^3 \\ \sqrt{3/8}\; Y_{-1}^3 + \sqrt{5/8}\; Y_3^3 \\ Y_0^3 \end{cases}$ |
| | $T_{2u}$ $(\Gamma_5^-)$ | $\begin{cases} \sqrt{5/8}\; Y_1^3 - \sqrt{3/8}\; Y_{-3}^3 \\ \sqrt{5/8}\; Y_{-1}^3 - \sqrt{3/8}\; Y_3^3 \\ \frac{1}{\sqrt{2}}(Y_2^3 + Y_{-2}^3) \end{cases}$ |

Appendix 9

# TANABE–SUGANO DIAGRAMS

Energy (E) versus cubic crystal field (10Dq) for the $3d^n$ ions are given in units of B (a Racah parameter). Each diagram is for a given ratio of C/B where both C and B are Racah coefficients which are linear combinations of Slater integrals for different two-electron radial moments of the electron distribution. The ratio written is approximately correct for the ion shown. The notation for the energy levels on the left is the weak field scheme, while on the right the strong field notation is given. All the energies are measured from the ground state. Thus, when the ground state changes for some value of 10Dq the figure appears to have a fold in it. For further discussion see Chapter 8 or Tanabe and Sugano, J. Phys. Soc. Japan **9**, 753 (1954).

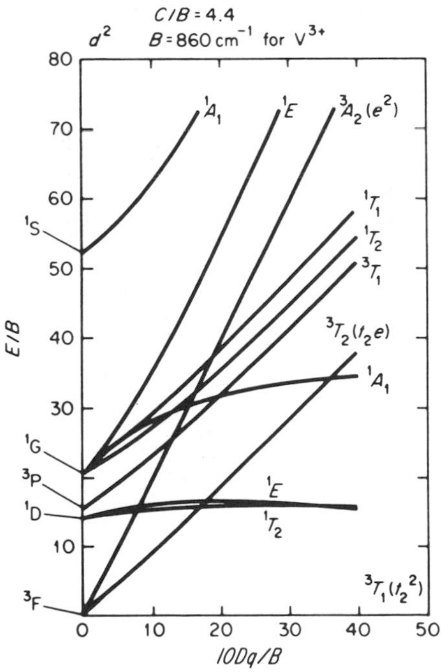

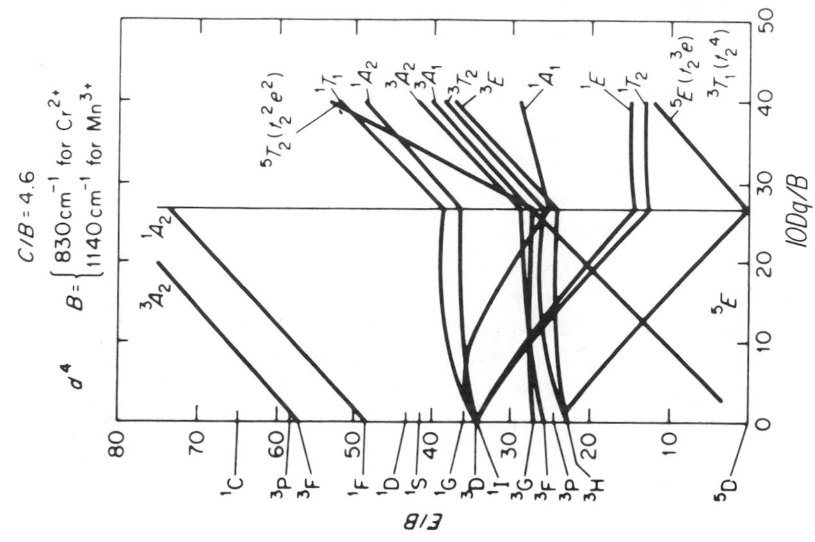
d4
C/B = 4.6
B = 830 cm-1 for Cr2+
1140 cm-1 for Mn3+
E/B
10Dq/B

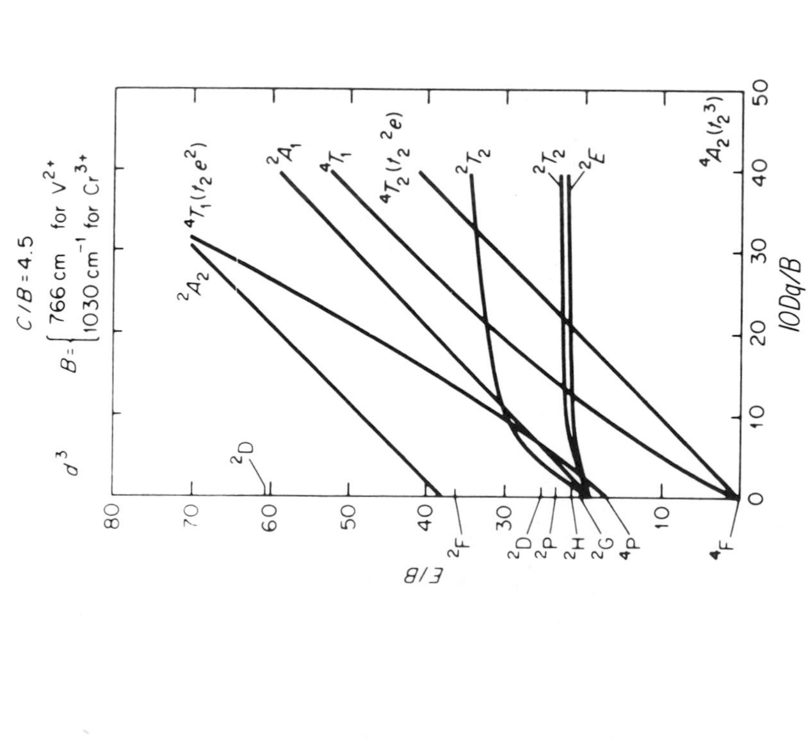
d3
C/B = 4.5
B = 766 cm for V2+
1030 cm-1 for Cr3+
E/B
10Dq/B

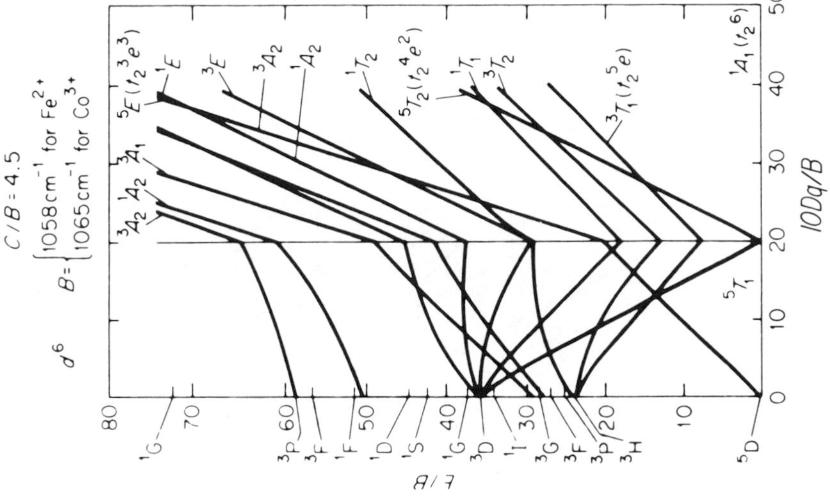
d6
C/B = 4.5
B = 1058 cm-1 for Fe2+
1065 cm-1 for Co3+
E/B
10Dq/B
5D
5T1
1A1(t2 6)
3T1(t2 5e)
5T2(t2 4e2)
5E(t2 3e3)

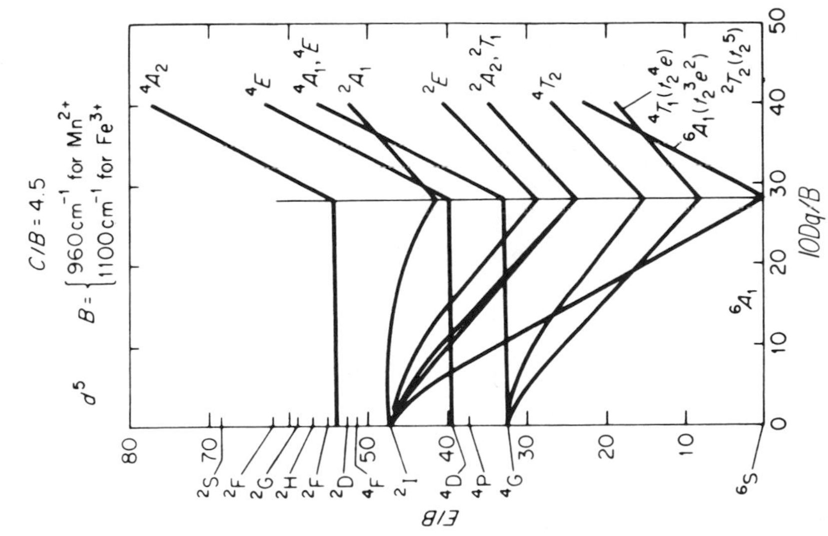
d5
C/B = 4.5
B = 960 cm-1 for Mn2+
1100 cm-1 for Fe3+
E/B
10Dq/B
6S
6A1
6A1(t2 3e2)
4T1(t2 4e)
2T2(t2 5)
4A2
4E
4A1, 4E

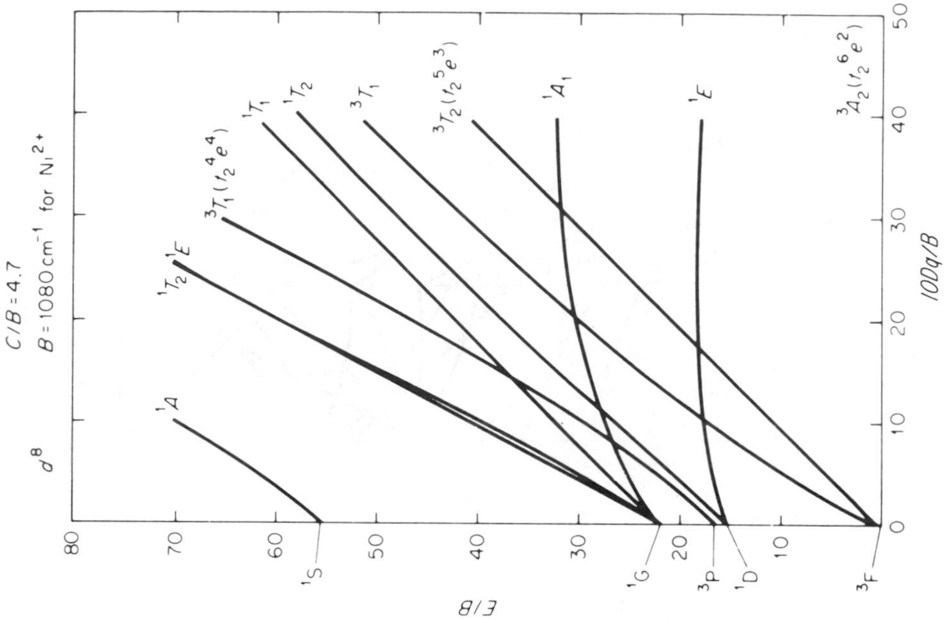

d^8
C/B = 4.7
B = 1080 cm^-1 for Ni^2+
E/B
10Dq/B
^1S
^1G
^3P
^1D
^3F
^1A
^1T_2 ^1E
^3T_1(t_2^4 e^4)
^1T_1
^1T_2
^3T_1
^3T_2(t_2^5 e^3)
^1A_1
^1E
^3A_2(t_2^6 e^2)


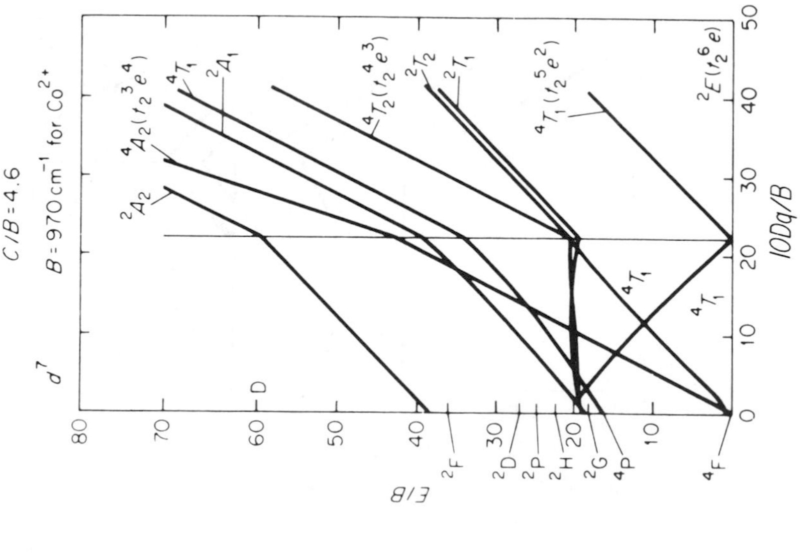

d^7
C/B = 4.6
B = 970 cm^-1 for Co^2+
E/B
10Dq/B
^2F
^2D
^2P
^2H
^2G
^4P
^4F
D
^2A_2
^4A_2(t_2^3 e^4)
^4T_1
^2A_1
^4T_2(t_2^4 e^3)
^2T_2
^2T_1
^4T_1(t_2^5 e^2)
^2E(t_2^6 e)
^4T_1
^4T_1

Appendix 10

# DOUBLE GROUP CHARACTER TABLES

The complete character table for the double group $T_d^*$ and $O^*$ is given. From this table, the character table for the double group $O_h^*$ can be immediately obtained. Shortened version of the double group character tables are given for some other point groups. These shortened versions only show the new irreducible representations, the symmetry operations, and the characters. These shortened tables should be added to the bottom of the appropriate single group character table in Appendix 3. The discussion in Chapter 8 or the observation of the complete double group character table for $O^*$ shown here, demonstrates exactly how this shortened version should be adjoined to the single group character table in Appendix 3. Note that there are many different notations for the irreducible representations of the double groups, so care should be taken when reading other sources.

| | $T_d^*$ | $E$ | $R$ | $4C_3$ | $4C_3^2$ | $3C_2$ | $3S_4$ | $3S_4^3$ | $6\sigma_d$ |
|---|---|---|---|---|---|---|---|---|---|
| | | | | $4C_3R$ | $4C_3^2R$ | $3C_2R$ | $3S_4R$ | $3S_4^3R$ | $6\sigma_dR$ |
| | $O^*$ | $E$ | $R$ | $4C_3$ | $4C_3^2$ | $3C_2$ | $3C_4$ | $3C_4^3$ | $6C_2'$ |
| | | | | $4C_3^2R$ | $4C_3R$ | $3C_2R$ | $3C_4^3R$ | $3C_4R$ | $6C_2'R$ |
| | | $E$ | $R$ | $8C_3$ | $8C_3R$ | $6C_2$ | $6C_4$ | $6C_4R$ | $12C_2'$ |
| $(\Gamma_1)$ | $A_1$ | 1 | 1 | 1 | 1 | 1 | 1 | 1 | 1 |
| $(\Gamma_2)$ | $A_2$ | 1 | 1 | 1 | 1 | 1 | -1 | -1 | -1 |
| $(\Gamma_3)$ | $E_1$ | 2 | 2 | -1 | -1 | 2 | 0 | 0 | 0 |
| $(\Gamma_4)$ | $T_1$ | 3 | 3 | 0 | 0 | -1 | 1 | 1 | -1 |
| $(\Gamma_5)$ | $T_2$ | 3 | 3 | 0 | 0 | -1 | -1 | -1 | 1 |
| $(\Gamma_6)$ | $E_{1/2}$ | 2 | -2 | 1 | -1 | 0 | $\sqrt{2}$ | $-\sqrt{2}$ | 0 |
| $(\Gamma_7)$ | $E_{5/2}$ | 2 | -2 | 1 | -1 | 0 | $-\sqrt{2}$ | $\sqrt{2}$ | 0 |
| $(\Gamma_8)$ | $G$ | 4 | -4 | -1 | 1 | 0 | 0 | 0 | 0 |

| | $T^*$ | $E$ | $R$ | $4C_3$ | $4C_3R$ | $4C_3^2$ | $4C_3^2R$ | $3C_2$ | $\epsilon = \exp(2\pi i/3)$ |
|---|---|---|---|---|---|---|---|---|---|
| | | | | | | | | $3C_2R$ | |
| $(\Gamma_4)$ | $E_{1/2}$ | 2 | -2 | 1 | -1 | -1 | 1 | 0 | |
| $(\Gamma_5)$ | $G$ | { 2 | -2 | $\epsilon$ | $-\epsilon$ | $-\epsilon^*$ | $\epsilon^*$ | 0 } | |
| | | { 2 | -2 | $\epsilon^*$ | $-\epsilon^*$ | $-\epsilon$ | $\epsilon$ | 0 } | |

| | $D_{3h}^*$ | $E$ | $R$ | $S_3$ | $S_3^5$ | $C_3$ | $C_3^2$ | $\sigma_h$ | $3C_2'$ | $3\sigma_v$ |
|---|---|---|---|---|---|---|---|---|---|---|
| | | | | $S_3^5R$ | $S_3R$ | $C_3^2R$ | $C_3R$ | $\sigma_hR$ | $3C_2'R$ | $3\sigma_vR$ |
| | $C_{6v}^*$ | $E$ | $R$ | $C_6$ | $C_6^5$ | $C_3$ | $C_3^2$ | $C_2$ | $3\sigma_v$ | $3\sigma_d$ |
| | | | | $C_6^5R$ | $C_6R$ | $C_3^2R$ | $C_3R$ | $C_2R$ | $3\sigma_vR$ | $3\sigma_dR$ |
| | $D_6^*$ | $E$ | $R$ | $C_6$ | $C_6^5$ | $C_3$ | $C_3^2$ | $C_2$ | $3C_2'$ | $3C_2''$ |
| | | | | $C_6^5R$ | $C_6R$ | $C_3^2R$ | $C_3R$ | $C_2R$ | $3C_2'R$ | $3C_2''R$ |
| $(\Gamma_7)$ | $E_{1/2}$ | 2 | -2 | $\sqrt{3}$ | $-\sqrt{3}$ | 1 | -1 | 0 | 0 | 0 |
| $(\Gamma_8)$ | $E_{3/2}$ | 2 | -2 | 0 | 0 | -2 | 2 | 0 | 0 | 0 |
| $(\Gamma_9)$ | $E_{5/2}$ | 2 | -2 | $-\sqrt{3}$ | $\sqrt{3}$ | 1 | -1 | 0 | 0 | 0 |

| | $D_{2d}^*$ | $E$ | $R$ | $S_4$ | $S_4^3$ | $C_2$ | $2C_2'$ | $2\sigma_d$ |
|---|---|---|---|---|---|---|---|---|
| | | | | $S_4^3R$ | $S_4R$ | $C_2R$ | $2C_2'R$ | $2\sigma_dR$ |
| | $C_{4v}^*$ | $E$ | $R$ | $C_4$ | $C_4^3$ | $C_2$ | $2\sigma_v$ | $2\sigma_d$ |
| | | | | $C_4^3R$ | $C_4R$ | $C_2R$ | $2\sigma_vR$ | $2\sigma_dR$ |
| | $D_4^*$ | $E$ | $R$ | $C_4$ | $C_4^3$ | $C_2$ | $2C_2'$ | $2C_2''$ |
| | | | | $C_4^3R$ | $C_4R$ | $C_2R$ | $2C_2'R$ | $2C_2''R$ |
| $(\Gamma_6)$ | $E_{1/2}$ | 2 | -2 | $\sqrt{2}$ | $-\sqrt{2}$ | 0 | 0 | 0 |
| $(\Gamma_7)$ | $E_{3/2}$ | 2 | -2 | $-\sqrt{2}$ | $\sqrt{2}$ | 0 | 0 | 0 |

| | $C_{3v}^*$ | $E$ | $R$ | $C_3$ | $C_3^2$ | $3\sigma_v$ | $3\sigma_vR$ |
|---|---|---|---|---|---|---|---|
| | | | | $C_3^2R$ | $C_3R$ | | |
| | $D_3^*$ | $E$ | $R$ | $C_3$ | $C_3^2$ | $3C_2$ | $3C_2R$ |
| | | | | $C_3^2R$ | $C_3R$ | | |
| $(\Gamma_4)$ | $E_{1/2}$ | 2 | -2 | 1 | -1 | 0 | 0 |
| $(\Gamma_5)$ | $E_{3/2}$ | 1 | -1 | -1 | 1 | $i$ | $-i$ |
| | | 1 | -1 | -1 | 1 | $-i$ | $i$ |

| | $C_{2v}^*$ | $E$ | $R$ | $C_2$ | $\sigma_v$ | $\sigma_d$ |
|---|---|---|---|---|---|---|
| | | | | $C_2R$ | $\sigma_vR$ | $\sigma_dR$ |
| | $D_2^*$ | $E$ | $R$ | $C_2$ | $C_2(y)$ | $C_2(x)$ |
| | | | | $C_2R$ | $C_2(y)R$ | $C_2(x)R$ |
| $(\Gamma_5)$ | $E_{1/2}$ | 2 | -2 | 0 | 0 | 0 |

| $C_{\infty v}^*$ | $E$ | $2C_\infty(\phi)$ | $2C_\infty(2\phi)$ | . . . | $\infty\sigma_v$ | $R$ | $2C_\infty(\phi)R$ | . . . |
|---|---|---|---|---|---|---|---|---|
| $A_1\Sigma^+$ | 1 | 1 | 1 | . . . | 1 | 1 | 1 | . . . |
| $A_2\Sigma^-$ | 1 | 1 | 1 | . . . | -1 | 1 | 1 | . . . |
| $E_1\Pi$ | 2 | $2\cos\phi$ | $2\cos 2\phi$ | . . . | 0 | 2 | $2\cos\phi$ | . . . |
| $E_2\Delta$ | 2 | $2\cos 2\phi$ | $2\cos 2\cdot2\phi$ | . . . | 0 | 2 | $2\cos 2\phi$ | . . . |
| $E_3\phi$ | 2 | $2\cos 3\phi$ | $2\cos 2\cdot3\phi$ | . . . | 0 | 2 | $2\cos 3\phi$ | . . . |
| . . | . | . . . | . . . . | . . . | . | . | . . . | . . . |
| $E_{1/2}$ | 2 | $2\cos 1\phi/2$ | $2\cos\phi$ | . . . | 0 | -2 | $-2\cos 1\phi/2$ | . . . |
| $E_{3/2}$ | 2 | $2\cos 3\phi/2$ | $2\cos 3\phi$ | . . . | 0 | -2 | $-2\cos 3\phi/2$ | . . . |
| $E_{5/2}$ | 2 | $2\cos 5\phi/2$ | $2\cos 5\phi$ | . . . | 0 | -2 | $-2\cos 5\phi/2$ | . . . |

$[D_{\infty h} = C_{\infty v} \times C_i]$

# BIBLIOGRAPHY

## General Group Theory


Bethe, H. A., "Splitting of Terms in Crystals." Consultants Bureau, New York, 1958. (Originally from Ann. Phys. 3, 133 (1929).)

Bhagavantam, S., and Venkatarayudu, T., "Theory of Groups and Its Application to Physical Problems." Academic Press, New York, 1969.

Bishop, D. M., "Group Theory and Chemistry." Oxford Univ. (Clarendon) Press, London and New York, 1973.

Chisholm, C. D. H., "Group Theoretical Techniques in Quantum Chemistry." Academic Press, New York, 1976.

Cotton, F. A., "Chemical Applications of Group Theory." 2nd ed. Wiley (Interscience), New York, 1971.

Falicov, L. M., "Group Theory and Its Physical Applications." Univ. of Chicago Press, Chicago, Illinois, 1966.

Hall, G. G., "Applied Group Theory." Amer. Elsevier, New York, 1967.

Hall, L. H., "Group Theory and Symmetry in Chemistry." McGraw-Hill, New York, 1969.

Hamermesh, M., "Group Theory and Its Applications to Physical Problems." Addison-Wesley, Massachusetts, 1962.

Heine, V., "Group Theory in Quantum Mechanics." Pergamon, Oxford, 1960.

Hochstrasser, R. M., "Molecular Aspects of Symmetry." Benjamin, New York, 1966.

Leech, J. W., Newman, D. J., "How to Use Groups." Methuen and Co., London, 1969.

Lomont, J. S., "Application of Finite Groups." Academic Press, New York, 1959.

Lyubarskii, G. Ya., "The Application of Group Theory in Physics." Pergamon, Oxford, 1960.

McWeeny, R., "Symmetry: An Introduction to Group Theory and Its Applications." Pergamon, New York, 1963.

Meijer, P. H. E., and Bauer, E., "Group Theory: The Application to Quantum Mechanics." North-Holland Publ., Amsterdam, 1962.

Schonland, D. S., "Molecular Symmetry." Van Nostrand–Reinhold, Princeton, New Jersey, 1971.

Tinkham, M., "Group Theory and Quantum Mechanics." McGraw-Hill, New York, 1964.

Wigner, E. P., "Group Theory." Academic Press, New York, 1959.

Wilson, E. B., Jr., Decius, J. C., and Cross, P. C., "Molecular Vibrations — The Theory of Infrared and Raman Spectra." McGraw-Hill, New York, 1955.

**These sources include reprints:**

Cracknell, A. P., "Applied Group Theory." Pergamon, Oxford, 1968.

Knox, R. S., Gold. A., "Symmetry in the Solid State." Benjamin, New York, 1964.

**Physical Properties**

Bhagavantam, S., "Crystal Symmetry and Physical Properties." Academic Press, New York, 1966.

Birss, R. R., "Symmetry and Magnetism." North-Holland Publ., Amsterdam, 1964.

Fumi, G. F., Acta Cryst. 5, 44 (1952).

Mason, W. P., "Crystal Physics of Interaction Processes." Academic Press, New York, 1966.

Nye, J. F., "Physical Properties of Crystals." Oxford Univ. Press, London and New York, 1957.

Wooster, W. A., "Tensors and Group Theory for the Physical Properties of Crystals." Oxford Univ. Press (Clarendon), London and New York, 1973.

**Solid State Physics and the 230 Space Groups**

Bradley, C. J., and Cracknell, A. P., "The Mathematical Theoryof Symmetry in Solids." Oxford Univ. Press, London and New York, 1972.

Brillouin, L., "Wave Propagation in Periodic Structures." Dover, New York, 1953.

Burns, G., and Glazer, A. M., "Space Groups for Solid State Scientists." Academic Press, New York, 1978.

Cornwell, T. F., "Group Theory and Electronic Energy Bands in Solids." North-Holland Publ., Amsterdam, 1969.

DiBartolo, B., "Optical Interactions in Solids." Wiley (Interscience), New York, 1968.

Hendry, N. F. M., and Lonsdale, D. (eds.), "International Tables for X-Ray Crystallography, Vol. I." The Kynoch Press, Birmingham, 1952.

Jones, H., "The Theory of Brillouin Zones and Electronic States in Crystals." North-Holland Publ., Amsterdam, 1962.

Kittel, C., "Quantum Theory of Solids." Wiley, New York, 1963.

Koster, G. F., "Solid State Physics," Vol. 5 (Seitz and Turnbull, eds.). Academic Press, New York, 1957.

Kovalev, "Irreducible Representations of the Space Groups." Gordon and Breach, New York, 1965. (Originally published by Academy of Sciences USSR Press, 1961.)

Lax, M., "Symmetry Principles in Solid State and Molecular Physics." Wiley (Interscience), New York, 1974.

Long,D., "Energy Bands in Semiconductors." Wiley (Interscience), New York, 1968.

Miller, S. C., and Love, W. F., "Irreducible Representations of Space Groups and Co-Representations of Magnetic Space Groups." Pruett Press, 1967.

Pincherle, L., "Electronic Energy Bands in Solids." Macdonald, 1971.

Slater, J. C., "Quantum Theory of Molecules and Solids, Vol. 2. McGraw-Hill, New York, 1965.

Shubnikov, A. V., Belov, N. V., et al., "Colored Symmetry." Pergamon Press, Oxford, 1964.

Streitwolf, H. W., "Group Theory in Solid State Physics." Macdonald, 1967.

Weinreich, G., "Solids: Elementary Theory for Advanced Students." Wiley, New York, 1965.

Wyckoff, G., "Crystal Structures," Vols. 1-4 (2nd ed). Wiley, New York, 1962-1966.

Zak, J., Casher, A., Gluck, M., and Gur, Y., "The Irreducible Representations of Space Groups." Benjamin, New York, 1969.

**Primarily on the Quantum Theory of Angular Momentum**

Biedenharm, L. C., and Van Dam, H., "Quantum Theory of Angular Momentum." Academic Press, New York, 1965. (Mostly reprints.)

Brink, D. M., and Satchler, G. R., "Angular Momentum." Oxford Univ. Press.

Condon, E. U., and Shortly, G. H., "The Theory of Atomic Spectra." Cambridge Univ. Press, London and New York, 1951.

Edmonds, A. R., "Angular Momentum in Quantum Mechanics." Princeton Univ. Press, Princeton, New Jersey, 1957.

Fang, U., and Racah, G., "Irreducible Tensorial Sets." Academic Press, New York, 1959.

Gelfand, I. M., Minlos,R. A. and Shapiro, Z. Y., "Representations of the Rotation and Lorentz Groups." Macmillan, New York, 1963. (Originally published in Moscow, 1958.)

Judd, B. R., "Operator Techniques in Atomic Spectroscopy." McGraw-Hill, New York, 1963.

Loebl, E. M., (ed.), "Group Theory and Its Applications." Academic Press, New York, 1968, 1971.

Rose, M. E., "Elementary Theory of Angular Momentum." Wiley, New York, 1957.

Rotenberg, M., Bivins, R., Metropolis, N., and Wooten, J. K., Jr., "The 3–j and 6–j Symbols." MIT Press, Cambridge, Massachusetts, 1959.

# INDEX

H



I



J



K



L



M



N